Antioxidants in Health and Disease

Volume 2

Special Issue Editors

Maurizio Battino
Francesca Giampieri

MDPI • Basel • Beijing • Wuhan • Barcelona • Belgrade

MDPI

Special Issue Editors
Maurizio Battino
Università Politecnica delle Marche
Italy

Francesca Giampieri
Università Politecnica delle Marche
Italy

Editorial Office
MDPI
St. Alban-Anlage 66
Basel, Switzerland

This edition is a reprint of the Special Issue published online in the open access journal *Nutrients* (ISSN 2072-6643) from 2016–2018 (available at: http://www.mdpi.com/journal/nutrients/special_issues/antioxidants_health_disease).

For citation purposes, cite each article independently as indicated on the article page online and as indicated below:

Lastname, F.M.; Lastname, F.M. Article title. *Journal Name* **Year**, *Article number*, page range.

First Edition 2018

Volume 2
ISBN 978-3-03842-939-5 (Pbk)
ISBN 978-3-03842-940-1 (PDF)

Volume 1–2
ISBN 978-3-03842-941-8 (Pbk)
ISBN 978-3-03842-942-5 (PDF)

Table of Contents

About the Special Issue Editors

Maurizio Battino, PhD, Associate Professor of Biochemistry in the Department of Clinical Sciences, Faculty of Medicine, UNIVPM (Italy), has been the Director of the Centre for Health and Nutrition, Universidad Europea del Atlantico, Santander (Spain), since December 2014. He obtained a BSc in Bologna (1984) and a PhD in Catania (1990) and completed a post-doctoral training in Granada (1994); he also obtained a MSc in International Communication Technology in Medicine (2011) and was awarded a Doctor Honoris Causa degree by the University of Medicine and Pharmacy of Bucharest (2008). He currently reviews scientific articles for over 30 peer-reviewed journals, serves as the Editor-in-Chief for the Journal of Berry Research, the Mediterranean Journal of Nutrition & Metabolism, and Diseases, and as an Associate Editor for Molecules; he is also an editorial board member of Food Chemistry, Plant Food for Human Nutrition, Nutrition and Aging, and the International Journal of Molecular Sciences. In 2015, 2016, and 2017, he has been recognized as a Thomson Reuters Highly Cited Researcher.

Francesca Giampieri, PhD, works as a Post-Doctoral Research Fellow at the Department of Clinical Science, at the Polytechnic University of Marche (Ancona, Italy). She graduated in Biological Sciences and pursued a Specialization in Food Science at the Polytechnic University of Marche. She currently reviews scientific articles for over 20 peer-reviewed journals, serves as an Associate Editor for the Journal of Berry Research, and is an editorial board member of Molecules, Nutrients, the Annals of Translational Medicine, and the International Journal of Molecular Sciences (Bioactives and Nutraceuticals). She has extensive experience in the field of chemistry, in the assessment of the nutritional and phytochemical composition of different foodstuffs, in the field of biochemistry, in the evaluation of the role of dietary bioactive compounds in human health and, in particular, in the analysis of the biological mechanisms related to oxidative stress.

Preface to "Antioxidants in Health and Disease"

Oxidative stress is defined as an imbalance between the production of free radicals and the necessary antioxidant defenses. Free radicals are chemical species with one or more mismatched electrons that generally damage multiple cellular components, whereas antioxidants are reducing molecules that neutralize free radicals by donating an electron. Oxidative stress can lead to a wide range of biological effects: adaptation, by upregulating the natural defense system, which may protect, completely or in part, against cellular damage; tissue injury, by damaging all molecular targets (DNA, proteins, lipids, cell membranes, and several enzymes); cell death, by activating the processes of necrosis and apoptosis. However, accumulating evidence implicates that free radicals, under physiological and pathological conditions, are able to regulate several signaling pathways, affecting a variety of cellular processes, such as proliferation, metabolism, differentiation, survival, antioxidant and anti-inflammatory response, iron homeostasis, and DNA damage response. In few words, the generation of ROS, within certain boundaries, is essential to maintain cellular homeostasis. This new and more complex view of the role of oxidative stress in biological processes confirms once again the importance of a stable equilibrium between oxidant production and antioxidant defenses to preserve health and longevity.

Because of the cellular damage induced by oxidative stress, there is much interest in the so-called functional foods, encompassing dietary antioxidants, for preventing human disease. The consumption of dietary antioxidants, such as vitamin C, Vitamin E, β-carotene, and polyphenols, has been indeed associated with an improvement of inflammation, a reduction of atherosclerosis progression, a decrease in cellular proliferation and metastatization, and an amelioration of lipid metabolism. In other words, antioxidants modulate several pathways involved in cellular metabolism, survival, and proliferation, maintain well-being, and protect the human body against the development of the most common chronic pathologies, such as metabolic syndrome, diabetes, cancer, and cardiovascular diseases.

The goal of this book is to demonstrate that the consumption of food rich in antioxidants provide health benefits and should be widely recommended as part of a healthy diet.

Maurizio Battino and Francesca Giampieri
Special Issue Editors

nutrients

MDPI

Article

Dietary Vitamin C, E and β-Carotene Intake Does Not Significantly Affect Plasma or Salivary Antioxidant Indices and Salivary C-Reactive Protein in Older Subjects

Anna Gawron-Skarbek [1,*], Agnieszka Guligowska [2], Anna Prymont-Przymińska [3], Małgorzata Godala [4], Agnieszka Kolmaga [4], Dariusz Nowak [5], Franciszek Szatko [1] and Tomasz Kostka [2]

[1] Department of Hygiene and Health Promotion, Medical University of Lodz, Hallera St. 1, Łódź 90-647, Poland; franciszek.szatko@umed.lodz.pl

[2] Department of Geriatrics, Medical University of Lodz, Pieniny St. 30, Łódź 90-993, Poland; agnieszka.guligowska@umed.lodz.pl (A.G.); tomasz.kostka@umed.lodz.pl (T.K.)

[3] Department of General Physiology, Medical University of Lodz, Mazowiecka St. 6/8, Łódź 92-215, Poland; anna.przyminska@umed.lodz.pl

[4] Department of Hygiene of Nutrition and Epidemiology, Medical University of Lodz, Hallera St. 1, Łódź 90-647, Poland; malgorzata.godala@umed.lodz.pl (M.G.); agnieszka.kolmaga@umed.lodz.pl (A.K.)

[5] Department of Clinical Physiology, Medical University of Lodz, Mazowiecka St. 6/8, Łódź 92-215, Poland; dariusz.nowak@umed.lodz.pl

* Correspondence: anna.gawron@umed.lodz.pl; Tel.: +48-42-678-1688

Received: 1 June 2017; Accepted: 6 July 2017; Published: 9 July 2017

Abstract: It is not clear whether habitual dietary intake influences the antioxidant or inflammatory status. The aim of the present study was to assess the impact of antioxidative vitamins C, E, and β-carotene obtained from daily food rations on plasma and salivary Total Antioxidant Capacity (TAC), uric acid and salivary C-reactive protein (CRP). The study involved 80 older subjects (66.9 ± 4.3 years), divided into two groups: group 1 ($n = 43$) with lower and group 2 ($n = 37$) with higher combined vitamins C, E and β-carotene intake. A 24-h dietary recall was obtained from each individual. TAC was assessed simultaneously with two methods in plasma (Ferric Reducing Ability of Plasma—FRAP, 2.2-diphenyl-1-picryl-hydrazyl—DPPH) and in saliva (FRAS and DPPHS test). Lower vitamin C intake corresponded to higher FRAS. There were no other correlations between vitamins C, E or β-carotene intake and antioxidant indices. Salivary CRP was not related to any antioxidant indices. FRAS was decreased in group 2 ($p < 0.01$) but no other group differences for salivary or for plasma antioxidant parameters and salivary CRP were found. Habitual, not extra supplemented dietary intake does not significantly affect plasma or salivary TAC and salivary CRP.

Keywords: plasma total antioxidant capacity; saliva; uric acid; C-reactive protein; diet; vitamin C intake; vitamin E intake; β-carotene; DPPH; FRAP

1. Introduction

There are numerous interventional studies assessing the potential influence of various nutritional compounds added to food [1,2] or beverages [3] in Daily Food Rations (DFR) on antioxidant capacity. Increased antioxidant status has been associated with high consumption of fruit, vegetables and plant oils as main food sources of antioxidative compounds [4,5]. Vitamin C, E and β-carotene are representative dietary antioxidants so their high content in the DFR is expected to enhance antioxidant potential in body fluids, cells and tissues. However, limited information is available on whether different antioxidant capacities found in different body fluids reflect a habitual dietary intake of antioxidants [6,7]. Subjects following a naturally antioxidant-rich diet might experience different biological effects than individuals being supplemented by multivitamins and minerals [8]. There are many external (diet, cigarette smoking) and internal (biochemical disorders) factors that might affect the final result, and preclude an unequivocal conclusion of whether habitual dietary intake without any special regimens is also associated with higher antioxidant status. Oxidative stress and inflammatory conditions are inter-related [9,10]. One of them may appear before or after the other, but the two usually occur together, resulting in both of them taking part in the pathogenesis of many chronic diseases. Complex biochemical interactions between pro- and antioxidants result in a relatively stable homeostasis state. It may be generally assumed that the inflammatory indices and accompanied prooxidants are low when the systemic antioxidant potential is strong enough to counteract these undesirable conditions. Dietary modification may affect inflammatory processes and protect against chronic diseases [11]. It is thought that the protective effects of fruit and vegetable consumption result from the presence of low-molecular antioxidants such as α-tocopherol, ascorbic acid, or β-carotene, as well as non-vitamin antioxidants, such as polyphenols and anthocyanins, or from the synergy of several different antioxidant compounds [5]. Other reports indicate that vitamin C, especially in doses exceeding daily recommended dietary allowance, may exert a prooxidant effect [12].

C-reactive protein (CRP) is an acute-phase protein that increases during inflammatory disorders [13,14]. CRP has been identified as a hallmark of systemic inflammation and is used as a risk bio-marker of different health conditions: cardiovascular disease [15], periodontitis [16,17], metabolic syndrome or diabetes mellitus [18]. Usually it is assessed in plasma but new research attitude appeals to noninvasive CRP or antioxidant parameters determination techniques using saliva samples [19]. Saliva may represent an alternative means for evaluation of the impact of dietary antioxidant intake on the plasma antioxidant defense system.

The variety of methods assessing antioxidant defense system provides a range of results which are at times inconsistent. The assessment of Total Antioxidant Capacity (TAC) may be a better approach than determining the capacities of individual antioxidants. An increased antioxidant capacity in body fluid may not necessarily be a desirable condition if it reflects a response to increased oxidative stress/inflammation. Similarly, a decrease may not necessarily be an undesirable condition if the measurement reflects decreased production of reactive species. These complications suggest that a "battery" of measurements is going to be more sufficient to adequately assess oxidative stress, as well as the antioxidant barrier level, in biological systems than any single measurement of antioxidant status [20]. The content of Uric Acid (UA), the strongest endogenous antioxidant, contributing about 70% of plasma and salivary TAC [21,22], should also be taken into consideration.

The aim of the study was to assess the impact of nutrients, mostly the antioxidative vitamins C, E and β-carotene, obtained from DFR on plasma and salivary TAC, UA and salivary CRP in older adults.

2. Materials and Methods

2.1. Patients

The study was carried out in 80 patients (66.9 ± 4.3 years), 86% of whom were females. The subjects had been treated in Outpatient Geriatric Clinic of the Medical University of Lodz (Łódź, Poland) and selected from a group of subjects participating in the healthy lifestyle workshops organized under the governmental program for the Social Activity of the Elderly (2014–2020) who volunteered to undergo a detailed dietary and laboratory (blood plasma and saliva) assessment. The subjects were consecutively recruited based on inclusion criteria and combined value of vitamin C, E and β-carotene intake (see below) in order to obtain balanced sex composition of the two groups, differing in combined intake value of antioxidant vitamins.

Some patients suffered from hypercholesterolemia (n = 48), arterial hypertension (n = 39), osteoarthritis (n = 33), thyroid insufficiency (n = 26), osteoporosis (n = 19), duodenal and gastric conditions (n = 14), diabetes mellitus (n = 14) and heart failure (n = 11). All diagnosed diseases were in stable phase and pharmacologically controlled. The treatment usually involved angiotensin-converting enzyme inhibitors (n = 25), levothyroxine (n = 26), statins (n = 23), diuretics (n = 22), beta-blockers (n = 18), aspirin (n = 17), calcium channel blockers (n = 9), proton pump inhibitors (n = 7), oral antidiabetic drugs—metformin (n = 9) and sulfonylureas (n = 6).

None of the subjects was diagnosed with tobacco addiction, active inflammatory processes (plasma CRP < 3 mg·L^{-1}), renal dysfunction, disability or dementia. None used any special diet. The study had been approved by the local ethics committee (RNN/73/15/KE) and informed consent was obtained from each subject. The investigations were carried out following the rules of the Declaration of Helsinki of 1975, revised in 2008.

2.2. Study Protocol and Measurements

The examinations took place in the Department of Geriatrics and the laboratory measurements were performed in the Department of Clinical Physiology, in the Central Scientific Laboratory and in the University Hospital and Educational Center, all at the Medical University of Lodz. The subjects reported to the Center between 8.00–10.00 a.m. after overnight fasting and rest for at least 12 h before blood and saliva collection. The time window between teeth cleaning and non-stimulated saliva sample collection was never shorter than 1.5 h. A comprehensive assessment, including age, sex, drug use, smoking and dietary habits was performed with each subject [23]. A 24-h dietary recall from the day before the examination was obtained from each individual.

2.2.1. Anthropometric Data

Height and weight were measured and the Body Mass Index (BMI) was calculated (overweight was for BMI in the range 25–30 kg·m^{-2}, obesity for BMI over 30 kg·m^{-2}). Measurements of waist and hip circumference were taken and Waist-to-Hip Ratio (WHR) was computed as an index of visceral obesity (diagnosed if WHR > 0.8 in females or >1.0 in males).

2.2.2. Plasma UA, CRP and Lipid Profile Determinations

Blood samples (about 9 mL) were drawn from the antecubital vein and collected for further TAC measurements into Vacuette tubes with lithium heparin or into vacutainer tubes with K3 EDTA for other tests (Vacutest, Kima, Italy). Thereafter the samples were incubated for 30 min at 37 °C and then centrifuged (10 min, 4 °C, 2880× g). The resultant plasma samples for TAC measurements (approximately 4 mL) were stored at −80 °C, for no longer than three months [24,25] and the rest was used to assess UA, CRP concentration and lipid profile parameters.

Enzymatic methods were used to determine plasma total cholesterol (TCh), triglycerides (TG) and UA concentration (BioMaxima S.A. diagnostic kit, Lublin, Poland with Dirui CS-400 equipment). High-density lipoprotein cholesterol (HDL-Ch) was measured by the precipitation method (BioMaxima S.A. diagnostic kit). Low-density lipoprotein cholesterol (LDL-Ch) was estimated using the Friedewald formula. Plasma CRP was measured by immunoassay (BioMaxima S.A. diagnostic kit, Lublin, Poland with Dirui CS-400 analyzer, Jilin, China).

2.2.3. Plasma TAC

Plasma TAC measurements were performed using two spectrophotometric methods: Ferric Reducing Ability of Plasma (FRAP) [21] with some modifications [24], and 2.2-diphenyl-1-picryl-hydrazyl test (DPPH) [24,25]. The details of both methods are described elsewhere [24,26].

2.2.4. Salivary TAC

Saliva samples (approximately 5mL) were centrifuged to separate all debris (10 min, 4 °C, 1125× g) [27]. The supernatant was stored at −80 °C max. for 30 days. Salivary TAC also was measured spectrophotometrically using the same equipment (Ultrospec III with Spectro-Kinetics software—LKB Biochrom Pharmacia, Cambridge, UK) and two methods, as for plasma TAC. For Ferric Reducing Ability of Saliva (FRAS) 120 μL of saliva were added to 900 μL of FRAS reagent, but deionized water was not used.

For the 2.2-diphenyl-1-picryl-hydrazyl test of saliva (DPPHS), as for DPPH [24], 200 μL of saliva was required for the deproteinization process; however, for the singular assay, 25 μL of deproteinized saliva were added to 975 μL of DPPH reagent mixture.

To enhance the data reliability, all results were calculated as a mean from three separate experiments. The salivary and plasma TAC assays were performed within the same time frame.

2.2.5. Salivary UA

Salivary UA (SUA) was analyzed using the MaxDiscovery™ Uric Acid Assay Kit (Bioo Scientific, Austin, TX, USA). Hydrogen peroxide, liberated by the action of uricase, reacted with a chromogenic dye using peroxidase to form a visibly colored (red) dye product. The absorbance was measured at 520 nm and the result was proportional to SUA concentration [28].

2.2.6. Salivary CRP

The salivary CRP assays (ELISA Kit—Salimetrics, PA, USA) were based on the colorimetric CRP peroxidase reaction on the substrate tetramethylbenzidine. Optical density was read on a standard VICTOR™ ×4 multifunctional microplate reader (Perkin Elmer, Waltham, MA, USA) at λ = 450 nm. The amount of CRP peroxidase detected was directly proportional to the amount of CRP present in the saliva sample [29].

2.2.7. Nutritional Evaluation

A 24-h recall questionnaire was used to register and then encode the intake of food, beverages, and supplements during the preceding day. The intake of energy, nutrients, vitamins, minerals in the DFR was calculated using the Diet 5.0 software package (developed by the National Food and Nutrition Institute, Warsaw, Poland) and compared with recommendations [30,31]. The degree of insufficient intake of analyzed antioxidative vitamins was estimated according to the following age and sex standards: EAR (the Estimated Average Requirement) for vitamin C (<60 mg/<75 mg, for females/males respectively) and AI (the Adequate Intake) for vitamin E (<8 mg/<10 mg) [30]. No dietary advice was given for the cases before a 24-h recall.

A further extra comparative analysis was run between the two subgroups. Based on a median (Me) value of vitamin C, E and β-carotene intake, a patient received '0' (if the intake was <Me) or '1' point (if the intake was ≥Me). Next the points were added and based on the sum result (min = 0, max = 3) the group was divided into group 1 (n = 43), with a lower vitamin intake ($\sum$ = 0 or 1), and group 2 (n = 37), with a higher vitamin intake ($\sum$ = 2 or 3). The two groups were identical with regard to sex profile (14% of males in each group).

2.3. Statistical Analysis

Data were verified for normality of distribution and equality of variances. Variables that did not meet the assumption of normality were analyzed with non-parametric statistics. Correlations between nutrient intake and age, BMI, WHR, lipid and antioxidant indices in plasma, and antioxidant parameters and CRP in saliva, were analyzed with the Spearman's rank correlation coefficient. The Mann–Whitney test was used to compare the mean values of numerical variables between group 1 and group 2. The results of the quantitative variables were presented as a mean ± standard deviation

(SD) and $p < 0.05$ was considered statistically significant for all analyses. The statistical analysis was performed using Statistica version 10 CSS software (StatSoft Polska Sp. z o.o., Kraków, Poland).

3. Results

3.1. Baseline Groups Characteristics

Detailed demographic, anthropometric and laboratory characteristics of studied groups are shown in Table 1. The two subgroups did not differ with regard to age. Over 1/3 of the group were diagnosed with obesity, and further 0.4 of the group with overweight. Visceral obesity was found in almost 3/4 of the group. Groups 1 and 2 were similar with regard to the anthropometric and lipid profile parameters except for TG: group 2 had a lower TG concentration ($p < 0.01$).

Table 1. Baseline characteristics of the study groups.

Variable	All ($n = 80$)	Group 1 ($n = 43$)	Group 2 ($n = 37$)
Age (years)	66.9 ± 4.3 ($60.0 \div 79.0$)	67.2 ± 4.3 ($60.0 \div 77.0$)	66.7 ± 4.4 ($61.0 \div 79.0$)
Body Mass Index (kg·m^{-2})	29.3 ± 5.2 ($21.4 \div 44.0$)	29.8 ± 5.6 ($21.4 \div 44.0$)	28.7 ± 4.8 ($22.6 \div 39.1$)
Waist circumference (cm)	92.3 ± 12.9 ($71.5 \div 130.0$)	94.2 ± 13.7 ($71.5 \div 130.0$)	90.0 ± 11.5 ($72.0 \div 123.0$)
Waist-to-Hip Ratio	0.87 ± 0.09 ($0.71 \div 1.07$)	0.88 ± 0.09 ($0.71 \div 1.07$)	0.86 ± 0.09 ($0.74 \div 1.07$)
Total Cholesterol (mg·dL^{-1})	182.2 ± 36.6 ($100.5 \div 285.3$)	177.6 ± 37.2 ($100.5 \div 247.2$)	187.6 ± 35.5 ($119.7 \div 285.3$)
LDL-Cholesterol (mg·dL^{-1})	114.6 ± 33.3 ($45.7 \div 196.7$)	108.0 ± 33.9 ($45.7 \div 172.5$)	122.3 ± 31.3 ($59.1 \div 196.7$)
HDL-Cholesterol (mg·dL^{-1})	45.1 ± 13.3 ($17.4 \div 78.3$)	43.0 ± 13.7 ($19.7 \div 76.9$)	47.5 ± 12.6 ($17.4 \div 78.3$)
Triglycerides (mg·dL^{-1})	123.3 ± 54.7 ($30.5 \div 249.3$)	138.0 ± 48.8 ($48.4 \div 244.6$)	106.1 ± 56.9 [†] ($30.5 \div 249.3$)

Data are presented as mean $\pm$ SD (min $\div$ max). [†]—$p < 0.01$ as compared to group 1.

3.2. Antioxidant Indices and Salivary CRP

Table 2 presents the mean values of plasma and salivary antioxidant indices and salivary CRP. FRAS was decreased in group 2 ($r = 2.9$; $p < 0.01$) but no other intergroup differences were found for salivary or for plasma antioxidant parameters. Salivary CRP did not differ between groups.

Table 2. Plasma and salivary antioxidant indices and salivary CRP concentrations.

Plasma	All ($n = 80$)	Group 1 ($n = 43$)	Group 2 ($n = 37$)	Saliva	All ($n = 80$)	Group 1 ($n = 43$)	Group 2 ($n = 37$)
FRAP (mmol FeCl$_2$·L^{-1})	1.21 ± 0.21 ($0.81 \div 1.80$)	1.25 ± 0.23 ($0.81 \div 1.80$)	1.17 ± 0.17 ($0.85 \div 1.63$)	FRAS (mmol FeCl$_2$·L^{-1})	5.99 ± 2.81 ($2.11 \div 19.08$)	6.75 ± 3.18 ($3.01 \div 19.08$)	5.11 ± 2.00 [†] ($2.11 \div 11.49$)
DPPH (% reduction)	23.4 ± 5.8 ($8.6 \div 35.6$)	24.3 ± 6.2 ($8.6 \div 35.6$)	22.5 ± 5.2 ($15.0 \div 34.4$)	DPPHS (% reduction)	27.4 ± 14.5 ($3.5 \div 68.9$)	27.7 ± 14.0 ($9.8 \div 68.1$)	27.2 ± 15.3 ($3.5 \div 68.9$)
UA (mg·dL^{-1})	4.47 ± 1.16 ($1.69 \div 7.38$)	4.54 ± 1.02 ($1.96 \div 6.34$)	4.39 ± 1.32 ($1.69 \div 7.38$)	SUA (mg·dL^{-1})	9.15 ± 4.16 ($0.42 \div 22.33$)	9.96 ± 4.13 ($4.33 \div 22.33$)	8.06 ± 4.01 ($0.42 \div 16.19$)
CRP (mg·dL^{-1})	<0.3	<0.3	<0.3	Salivary CRP (ng mL^{-1})	2.23 ± 1.86 ($0.35 \div 8.82$)	2.22 ± 1.90 ($0.35 \div 7.90$)	2.24 ± 1.83 ($0.47 \div 8.82$)

Data are presented as mean $\pm$ SD (min $\div$ max). FRAP—Ferric Reducing Ability of Plasma; DPPH—2.2-diphenyl-1-picryl-hydrazyl test of plasma; UA—Uric Acid; CRP—C-reactive protein; FRAS—Ferric Reducing Ability of Saliva; DPPHS—2.2-diphenyl-1-picryl-hydrazyl test of saliva; SUA—Salivary Uric Acid. [†]—$p < 0.01$ as compared to group 1.

3.3. Nutritional Characteristics

Generally, the study group was rather well nourished (71% covered the minimum demand for energy, 69% for total protein, 56% for dietary fiber, 41% for magnesium, 75% for zinc; according to the recommendations for the elderly). The percentage of the group with deficient vitamin E consumption of the AI standard was 54% (48% in females and 91% in males), while 23% were vitamin C deficient according to the EAR standard (23% in females and 18% in males). A detailed analysis of kind of fruit and vegetables common chosen by the study group indicated tomatoes, peppers, onion, potatoes, soup greens, cabbage, seasonal fruit (apples, raspberries, strawberries, cherries) as main sources of vitamin

C and β-carotene (the average mass of fruit and vegetables jointly in about 2/3 of the study group was at range 600–900 g per day), with plant oils (mostly oilseed rape, olive oil) as sources of vitamin E.

Several similarities were found between the groups regarding the absolute values of the energy obtained from particular macronutrients intake (19% from proteins, 29% from fat and 51% from carbohydrates), for total fat, saturated and monounsaturated fatty acids, vitamin B_{12}, sodium and manganese. Total energy ($p < 0.001$), total protein ($p < 0.001$), total carbohydrates ($p < 0.01$), sucrose ($p < 0.001$), dietary fiber ($p < 0.001$), polyunsaturated fatty acids ($p < 0.01$), cholesterol ($p < 0.05$) and water ($p < 0.001$) were significantly higher in group 2, as was the intake of some minerals (potassium, calcium, phosphorus, magnesium, iron, zinc, copper, iodine) and vitamins (B vitamins except for B_{12}, vitamin A and D ($p < 0.05$)). As expected, vitamin C (84.5 ± 69.9 mg vs. 186.1 ± 66.9 mg), E (6.4 ± 2.2 mg vs. 10.7 ± 3.2 mg) and β-carotene intake (3627 ± 3773 μg vs. 6494 ± 3887 μg) were also significantly higher in group 2 ($p < 0.001$). After adjustment for nutritional density characteristics (calculation per 1000 kcal), significantly higher intake in group 2 remained for vitamin C, E, β-carotene, sucrose, dietary fiber, potassium, copper, vitamin B_6 and folic acid.

3.4. Correlations for Antioxidant Indices and Salivary CRP in the Study Group (n = 80)

Age positively correlated only with salivary antioxidant indices: FRAS ($r = 0.27$), DPPHS ($r = 0.23$) and SUA ($r = 0.28$) but not with plasma antioxidants. Subjects with higher BMI had increased salivary CRP ($r = 0.27$), and those with higher TG had increased FRAP ($r = 0.35$) and UA ($r = 0.23$). Individuals with visceral obesity were characterized with higher UA ($r = 0.30$).

Lower calcium ($r = -0.26$), magnesium ($r = -0.24$) and vitamin B_{12} ($r = -0.27$) intake were related to higher salivary CRP, without their impact on any antioxidant parameters. Lower dietary fiber ($r = -0.23$), zinc ($r = -0.27$) and vitamin C ($r = -0.26$) intake corresponded only to higher FRAS. There were no other correlations between vitamins C, E or β-carotene intake and antioxidant indices or salivary CRP. Salivary CRP did not relate to any antioxidant indices, neither in saliva nor in plasma ($p > 0.05$). Instead, all plasma antioxidant indices (FRAP, DPPH, UA) correlated positively with their saliva analogues (FRAS, DPPHS, SUA) ($p < 0.05$).

4. Discussion

To the best of our knowledge, this is one of very few studies that assesses TAC by two different established methods in plasma (FRAP and DPPH) and in saliva (FRAS and DPPHS test) in a group of relatively healthy adults. It also performs the first simultaneous assessment of plasma and salivary UA and CRP in the context of dietary antioxidant intake. Our present findings indicate that a higher level of dietary vitamin C intake had an adverse effect on FRAS, but that the intake of other antioxidative vitamins from an habitual dietary intake did not affect the TAC or UA of plasma or saliva. Salivary CRP was not related to the identified level of antioxidant compounds in diet, but higher CRP levels were associated with lower calcium, magnesium and vitamin B_{12} consumption from the DFR. The nutritional status of group 2 was significantly superior to group 1 but generally the antioxidant status of both groups, besides FRAS index, was comparable. Also salivary CRP concentration, regardless of the combined vitamins C, E and β-carotene intake difference, was at a similar level in each group.

The knowledge about positive effect of dietary vitamins intake on good health conditions seems to be indisputable [32,33]. However their impact on the antioxidant potential and inflammatory indices is not so obvious. Recently, diet and CRP, in particular high sensitivity CRP (hs-CRP) are of increasing research interest. There are relatively few studies regarding salivary CRP, especially relating to habitual dietary intake, not to modified daily diet. Usually they concern a certain oral health or cardiac disorders or some dietary interventions. Salivary CRP as well as salivary TAC assessment in view of its noninvasive technique of samples collection seems to be appealing new research direction. In the study by Mazidi et al. [34] the increase in serum hs-CRP was associated with lower level of total dietary fiber and vitamins C, E, A intake (not as in the present study). The hs-CRP concentrations were likely modulated by dietary intake, including dietary sugar, polyunsaturated fatty acids, fiber and antioxidant intake. It is possible that higher PUFA intake may be related to the intensified oxidative stress and to the reduction of inflammation but the available data are full of discrepancies [35]. In our study, PUFA intake was significantly higher in group 2 with higher combined antioxidant vitamins intake but no significant correlation was found between plasma and salivary antioxidant indices nor salivary CRP and polyunsaturated fatty acids. Other reports also indicated that high intake of carotenoids and vitamin C, but not of vitamin E, seems to decrease the level of circulating hs-CRP [36]. In a crossover intervention by Valtueña et al. [37] plasma CRP decreased during the high-TAC diet. Instead, Stringa et al. [38] assessed whether total dietary antioxidant capacity (assessed by dietary FRAP) and serum UA were associated with low-grade chronic inflammation expressed as serum hs-CRP. The results, similarly as in our paper, demonstrated no association between dietary FRAP and hs-CRP levels but contrary to our findings increased levels of UA were observed in subjects with higher levels of hs-CRP. Zhang et al., identified that applying standard diet recommended by guidelines and high fruit and soybean products diet intervention yielded no different effects on serum UA [39].

Data regarding the influence of the dietary antioxidant compounds in DFR on antioxidant parameters, particularly those associated with saliva, is also scarce [40–42]. Stedile et al. [40] reported a positive correlation between dietary TAC, including vitamin C and polyphenols, and plasma TAC in healthy young women. Presumably, the endogenous defenses were fully functional in young subjects. Kamodyová et al. [6] reported that single intake of vitamin C (250 mg) had a positive influence on TAC in healthy participants. A study by Carrión-García et al. [43] in a group of healthy volunteers assessed the relationship between non-enzymatic antioxidant capacity (NEAC) estimated by two different dietary assessment methods (FRAP and trolox equivalent antioxidant capacity) and NEAC plasma levels: statistically significant but relatively weak, positive correlations were found between dietary FRAP (either derived from the food frequency questionnaire, or the 24-h recall) and plasma FRAP, particularly in the fruit and vegetables food groups. As the optimal TAC level for the human body is unknown, our results suggesting a lack of relationship between antioxidative dietary vitamin intake and most of the plasma and salivary antioxidant parameters cannot reduce the significance of habitual dietary intake solely on the basis of its failure to modulate antioxidant potential in vivo. Perhaps considering that it is desirable for human body to have a high TAC level, this area should be investigated further.

An unexpected negative correlation between dietary vitamin C intake and FRAS should also be explored. Sinha et al. [44] reported a positive correlation between dietary vitamin C intake and plasma ascorbic acid (AA) level, as well as some interrelationships between various plasma antioxidants: for instance, a positive association between β-carotene and α-tocopherol, and an inverse one between plasma AA and plasma UA. This observation was similar to another finding in which serum UA decreased in elderly subjects after they were supplemented with high doses of vitamin C [45]. Strawberries added to the usual diet as a source of vitamin C did not increase fasting non-urate plasma antioxidant activity [46]. Wang et al. [7] reported that plasma TAC (determined by VCEAC—vitamin C equivalent antioxidant capacity) was positively associated with dietary intakes of γ-tocopherol and β-carotene, as well as with plasma α-tocopherol and UA, in overweight and apparently healthy postmenopausal women. Our findings do not indicate any relationship between vitamin C consumption and the level of UA or SUA, but a negative relationship is indicated between vitamin C and FRAS (mainly contributed by SUA). At present we are not able to fully explain why this may be, i.e., a lower vitamin C intake is associated with only FRAS and not the other assessed salivary or plasma TAC indexes. Moreover, this correlation disappears in subgroups 1 and 2, but the negative trend remains.

As vitamin C may well contribute to eliminating UA, we may assume that higher vitamin C intake causes a decrease in FRAS, not in FRAP: in saliva, the FRAS test found SUA to be the predominant antioxidant (71.6%) while the FRAP method found the plasma UA to be less predominate (64.0%) [47]. Hence it is reasonable to assume that a link exists between vitamin C intake and salivary and plasma TAC level including SUA/UA that remains unknown for now.

On the other hand, our result might serve as an example of the theory of hormesis, according to which high antioxidant potential is an effect of an undesirable increase in prooxidant concentration, which is possible among the cases with lower vitamin C consumption. However, the question remains why this effect was visible only in saliva, only visible using the FRAS test, but did not appear in plasma. Several explanations are possible: one being the characteristics of methodology used (FRAS based on the ferrous ions reaction), and another the fact that the local prooxidant effect of vitamin C associated with the Haber-Weiss reaction may be stronger in the saliva environment than in plasma, resulting in intensified hydroxyl radical production and the loss of FRAS. Saliva is also more likely to be exposed to bacterial flora, probably generating reactive oxygen species. Wang et al., found that plasma TAC measured by VCEAC gave a better representation of plasma antioxidant levels than ORAC (oxygen radical absorbance capacity) or FRAP assay. However, TAC measured by FRAP correlated only with UA, while more correlations were found by VCEAC [7].

To avoid missing the possible resultant effect of various dietary antioxidative compounds, the different TAC assessment methods should be in future studies accompanied by the particular plasma antioxidant concentration assays.

It should be also noted that while both DPPH and FRAP tests measure the TAC, they reflect somewhat different physiological properties. As neither of the methods for TAC assessment measures all the antioxidants occurring in body fluids, the simultaneous use of both the FRAP and DPPH assays, in spite of their limitations, enhances the completeness and reliability of measurement. For instance Sinha et al., concluded that for people consuming large amounts of vitamin C, plasma AA is not an appropriate biomarker of dietary vitamin C [44].

Despite its strengths, such as its complexity (simultaneously applying two analytical methods in two body fluids, using a number of assessed parameters, the age-, sex- and anthropometric-comparable groups) the study also has some limitations, two being the limited number of subjects and the cross-sectional design of the study. It should be also noticed that our subjects were volunteers, who were probably healthier and fitter than a random sample, as well as more willing to participate in such studies. Nonetheless, bearing in mind the percentage of subjects deficient in vitamin E (54%) and vitamin C intake (23%) it may be assumed that, despite their mean vitamin C intake being more than adequate, the groups were not as well-nourished as could be expected. Moreover, the heterogeneity of the pharmacotherapy could interfere with the results. It was not feasible to find older subjects entirely free from common age-related ailments or using similar drugs and treatment regimens (the average senior suffers from 3–4 coexistent diseases). Nevertheless, the diseases diagnosed in our study group were in a stable phase and pharmacologically controlled.

5. Conclusions

A non-supplemented diet based on habitual dietary intake does not significantly affect plasma or salivary TAC and salivary CRP. The known health benefits of a natural, antioxidant-rich diet may be not related to plasma or salivary antioxidant potential. Further prospective studies are needed to examine these potential relationships.

Acknowledgments: This work was supported by the Medical University of Lodz under Grant No. 502-03/6-024-01/502-64-072 and partially by the Healthy Ageing Research Centre project under Grant REGPOT-2012-2013-1, 7FP. We would like to thank Andrzej Olczyk from Department of Hygiene and Health Promotion for his assistance in material collection and transport, Hanna Jerczyńska from the Central Scientific Laboratory, Agnieszka Sobczak, graduate of Faculty of Medical Laboratory, Alina Prylińska and Anna Piłat from the University Hospital and Educational Center, for their laboratory analysis support, as well as for the assistance of Department of Geriatrics medical staff, all at Medical University of Lodz.

Author Contributions: A.G.-S. and T.K. conceived and designed the experiments; A.G., A.P.-P. performed the experiments; A.G.-S., A.G., A.K. and M.G. analyzed the data; D.N., F.S. and T.K. contributed reagents/materials/analysis tools; A.G.-S. wrote the paper; T.K. revised the manuscript.

Conflicts of Interest: The authors declare no conflict of interest.

References

1. Skulas-Ray, A.C.; Kris-Etherton, P.M.; Teeter, D.L.; Chen, C.Y.; Vanden Heuvel, J.P.; West, S.G. A high antioxidant spice blend attenuates postprandial insulin and triglyceride responses and increases some plasma measures of antioxidant activity in healthy, overweight men. *J. Nutr.* **2011**, *141*, 1451–1457. [CrossRef] [PubMed]

2. Bozonet, S.M.; Carr, A.C.; Pullar, J.M.; Vissers, M.C. Enhanced human neutrophil vitamin C status, chemotaxis and oxidant generation following dietary supplementation with vitamin C-rich sungold kiwifruit. *Nutrients* **2015**, *7*, 2574–2588. [CrossRef] [PubMed]

3. Suraphad, P.; Suklaew, P.O.; Ngamukote, S.; Adisakwattana, S.; Mäkynen, K. The effect of isomaltulose together with green tea on glycemic response and antioxidant capacity: A single-blind, crossover study in healthy subjects. *Nutrients* **2017**, *9*, 464. [CrossRef]

4. Dauchet, L.; Peneau, S.; Bertrais, S.; Vergnaud, A.C.; Estaquio, C.; Kesse-Guyot, E.; Czernichow, S.; Favier, A.; Faure, H.; Galan, P.; et al. Relationships between different types of fruit and vegetable consumption and serum concentrations of antioxidant vitamins. *Br. J. Nutr.* **2008**, *100*, 633–641. [CrossRef] [PubMed]

5. Harasym, J.; Oledzki, R. Effect of fruit and vegetable antioxidants on total antioxidant capacity of blood plasma. *Nutrition* **2014**, *30*, 511–517. [CrossRef] [PubMed]

6. Kamodyová, N.; Tóthová, L.; Celec, P. Salivary markers of oxidative stress and antioxidant status: Influence of external factors. *Dis. Markers* **2013**, *34*, 313–321. [CrossRef] [PubMed]

7. Wang, Y.; Yang, M.; Lee, S.G.; Davis, C.G.; Kenny, A.; Koo, S.I.; Chun, O.K. Plasma total antioxidant capacity is associated with dietary intake and plasma level of antioxidants in postmenopausal women. *J. Nutr. Biochem.* **2012**, *23*, 1725–1731. [CrossRef] [PubMed]

8. Li, M.; Li, Y.; Wu, Z.; Huang, W.; Jiang, Z. Effects of multi-nutrients supplementation on the nutritional status and antioxidant capability of healthy adults. *Wei Sheng Yan Jiu* **2012**, *41*, 60–64. [PubMed]

9. Biswas, S.K. Does the interdependence between oxidative stress and inflammation explain the antioxidant paradox? *Oxid. Med. Cell. Longev.* **2016**, *2016*, 5698931. [CrossRef] [PubMed]

10. Siti, H.N.; Kamisah, Y.; Kamsiah, J. The role of oxidative stress, antioxidants and vascular inflammation in cardiovascular disease (a review). *Vascul. Pharmacol.* **2015**, *71*, 40–56. [CrossRef] [PubMed]

11. Wood, A.D.; Strachan, A.A.; Thies, F.; Aucott, L.S.; Reid, D.M.; Hardcastle, A.C.; Mavroeidi, A.; Simpson, W.G.; Duthie, G.G.; Macdonald, H.M. Patterns of dietary intake and serum carotenoid and tocopherol status are associated with biomarkers of chronic low-grade systemic inflammation and cardiovascular risk. *Br. J. Nutr.* **2014**, *112*, 1341–1352. [CrossRef] [PubMed]

12. Wróblewski, K. Can the administration of large doses of vitamin C have a harmful effect? *Polski Merkur. Lek.* **2005**, *19*, 600–603.

13. Ridker, P.M. Clinical application of c-reactive protein for cardiovascular disease detection and prevention. *Circulation* **2003**, *107*, 363–369. [CrossRef] [PubMed]

14. Tonetti, M.S.; D'Aiuto, F.; Nibali, L.; Donald, A.; Storry, C.; Parkar, M.; Suvan, J.; Hingorani, A.D.; Vallance, P.; Deanfield, J. Treatment of periodontitis and endothelial function. *N. Engl. J. Med.* **2007**, *356*, 911–920. [CrossRef] [PubMed]

15. Miller, C.S.; Foley, J.D.; Floriano, P.N.; Christodoulides, N.; Ebersole, J.L.; Campbell, C.L.; Bailey, A.L.; Rose, B.G.; Kinane, D.F.; Novak, M.J.; et al. Utility of salivary biomarkers for demonstrating acute myocardial infarction. *J. Dent. Res.* **2014**, *93*, 72S–79S. [CrossRef] [PubMed]

16. Ebersole, J.L.; Kryscio, R.J.; Campbell, C.; Kinane, D.F.; McDevitt, J.; Christodoulides, N.; Floriano, P.N.; Miller, C.S. Salivary and serum adiponectin and c-reactive protein levels in acute myocardial infarction related to body mass index and oral health. *J. Periodontal Res.* **2017**, *52*, 419–427. [CrossRef] [PubMed]

17. Nguyen, T.T.; Ngo, L.Q.; Promsudthi, A.; Surarit, R. Salivary oxidative stress biomarkers in chronic periodontitis and acute coronary syndrome. *Clin. Oral. Investig.* **2016**. [CrossRef] [PubMed]

18. Dezayee, Z.M.; Al-Nimer, M.S. Saliva c-reactive protein as a biomarker of metabolic syndrome in diabetic patients. *Indian J. Dent. Res.* **2016**, *27*, 388–391. [CrossRef] [PubMed]

19. Battino, M.; Ferreiro, M.S.; Gallardo, I.; Newman, H.N.; Bullon, P. The antioxidant capacity of saliva. *J. Clin. Periodontol* **2002**, *29*, 189–194. [CrossRef] [PubMed]

20. Prior, R.L.; Cao, G. In vivo total antioxidant capacity: Comparison of different analytical methods. *Free Radic. Biol. Med.* **1999**, *27*, 1173–1181. [CrossRef]

21. Benzie, I.F.; Strain, J.J. The ferric reducing ability of plasma (FRAP) as a measure of "Antioxidant power": The FRAP assay. *Anal. Biochem.* **1996**, *239*, 70–76. [CrossRef] [PubMed]

22. Moore, S.; Calder, K.A.; Miller, N.J.; Rice-Evans, C.A. Antioxidant activity of saliva and periodontal disease. *Free Radic. Res.* **1994**, *21*, 417–425. [CrossRef] [PubMed]

23. Stelmach, W.; Kaczmarczyk-Chalas, K.; Bielecki, W.; Drygas, W. The impact of income, education and health on lifestyle in a large urban population of poland (CINDI programme). *Int. J. Occup. Med. Environ. Health* **2004**, *17*, 393–401. [PubMed]

24. Chrzczanowicz, J.; Gawron, A.; Zwolinska, A.; de Graft-Johnson, J.; Krajewski, W.; Krol, M.; Markowski, J.; Kostka, T.; Nowak, D. Simple method for determining human serum 2,2-diphenyl-1-picryl-hydrazyl (DPPH) radical scavenging activity—Possible application in clinical studies on dietary antioxidants. *Clin. Chem. Lab. Med.* **2008**, *46*, 342–349. [CrossRef] [PubMed]

25. Schlesier, K.; Harwat, M.; Bohm, V.; Bitsch, R. Assessment of antioxidant activity by using different in vitro methods. *Free Radic. Res.* **2002**, *36*, 177–187. [CrossRef] [PubMed]

26. Gawron-Skarbek, A.; Chrzczanowicz, J.; Kostka, J.; Nowak, D.; Drygas, W.; Jegier, A.; Kostka, T. Cardiovascular risk factors and total serum antioxidant capacity in healthy men and in men with coronary heart disease. *Biomed. Res. Int.* **2014**, *2014*, 216964. [CrossRef] [PubMed]

27. Navazesh, M. Methods for collecting saliva. *Ann. N. Y. Acad. Sci.* **1993**, *694*, 72–77. [CrossRef] [PubMed]

28. Giebułtowicz, J.; Wroczyński, P.; Samolczyk-Wanyura, D. Comparison of antioxidant enzymes activity and the concentration of uric acid in the saliva of patients with oral cavity cancer, odontogenic cysts and healthy subjects. *J. Oral. Pathol. Med.* **2011**, *40*, 726–730. [CrossRef] [PubMed]

29. Chard, T. *An Introduction to Radioimmunoassay and Related Techniques*; Elsevier Science: Amsterdam, The Netherlands, 1995; Volume 6.

30. Jarosz, M. *Standards of Human Nutrition*; National Food and Nutrition Institute: Warsaw, Poland, 2012.

31. Kunachowicz, H.; Nadolna, I.; Przygoda, B.; Iwanow, K. *Charts of Nutritive Values of Products and Foods*; PZWL: Warsaw, Poland, 2005.

32. Zhao, L.G.; Shu, X.O.; Li, H.L.; Zhang, W.; Gao, J.; Sun, J.W.; Zheng, W.; Xiang, Y.B. Dietary antioxidant vitamins intake and mortality: A report from two cohort studies of chinese adults in Shanghai. *J. Epidemiol.* **2017**, *27*, 89–97. [CrossRef] [PubMed]

33. Kim, K.; Vance, T.M.; Chun, O.K. Greater total antioxidant capacity from diet and supplements is associated with a less atherogenic blood profile in U.S. Adults. *Nutrients* **2016**, *8*, 15. [CrossRef] [PubMed]

34. Mazidi, M.; Kengne, A.P.; Mikhailidis, D.P.; Cicero, A.F.; Banach, M. Effects of selected dietary constituents on high-sensitivity c-reactive protein levels in U.S. Adults. *Ann. Med.* **2017**. [CrossRef] [PubMed]

35. Kelley, N.S.; Yoshida, Y.; Erickson, K.L. Do *n*-3 polyunsaturated fatty acids increase or decrease lipid peroxidation in humans? *Metab. Syndr. Relat. Disord.* **2014**, *12*, 403–415. [CrossRef] [PubMed]

36. Nanri, A.; Moore, M.A.; Kono, S. Impact of c-reactive protein on disease risk and its relation to dietary factors. *Asian Pac. J. Cancer Prev.* **2007**, *8*, 167–177. [PubMed]

37. Valtueña, S.; Pellegrini, N.; Franzini, L.; Bianchi, M.A.; Ardigò, D.; Del Rio, D.; Piatti, P.; Scazzina, F.; Zavaroni, I.; Brighenti, F. Food selection based on total antioxidant capacity can modify antioxidant intake, systemic inflammation, and liver function without altering markers of oxidative stress. *Am. J. Clin. Nutr.* **2008**, *87*, 1290–1297. [PubMed]

38. Stringa, N.; Brahimaj, A.; Zaciragic, A.; Dehghan, A.; Ikram, M.A.; Hofman, A.; Muka, T.; Kiefte-de Jong, J.C.; Franco, O.H. Relation of antioxidant capacity of diet and markers of oxidative status with c-reactive protein and adipocytokines: A prospective study. *Metabolism* **2017**, *71*, 171–181. [CrossRef] [PubMed]

39. Zhang, M.; Gao, Y.; Wang, X.; Liu, W.; Zhang, Y.; Huang, G. Comparison of the effect of high fruit and soybean products diet and standard diet interventions on serum uric acid in asymptomatic hyperuricemia adults: An open randomized controlled trial. *Int. J. Food Sci. Nutr.* **2016**, *67*, 335–343. [CrossRef] [PubMed]

40. Stedile, N.; Canuto, R.; Col, C.D.; Sene, J.S.; Stolfo, A.; Wisintainer, G.N.; Henriques, J.A.; Salvador, M. Dietary total antioxidant capacity is associated with plasmatic antioxidant capacity, nutrient intake and lipid and dna damage in healthy women. *Int. J. Food Sci. Nutr.* **2016**, *67*, 479–488. [CrossRef] [PubMed]

41. Mejean, C.; Morzel, M.; Neyraud, E.; Issanchou, S.; Martin, C.; Bozonnet, S.; Urbano, C.; Schlich, P.; Hercberg, S.; Peneau, S.; et al. Salivary composition is associated with liking and usual nutrient intake. *PLoS ONE* **2015**, *10*, e0137473. [CrossRef] [PubMed]

42. Zare Javid, A.; Seal, C.J.; Heasman, P.; Moynihan, P.J. Impact of a customised dietary intervention on antioxidant status, dietary intakes and periodontal indices in patients with adult periodontitis. *J. Hum. Nutr. Diet.* **2014**, *27*, 523–532. [CrossRef] [PubMed]

43. Carrión-García, C.J.; Guerra-Hernández, E.J.; García-Villanova, B.; Molina-Montes, E. Non-enzymatic antioxidant capacity (NEAC) estimated by two different dietary assessment methods and its relationship with neac plasma levels. *Eur. J. Nutr.* **2016**, *56*, 1561–1576. [CrossRef] [PubMed]

44. Sinha, R.; Block, G.; Taylor, P.R. Determinants of plasma ascorbic acid in a healthy male population. *Cancer Epidemiol. Biomark. Prev.* **1992**, *1*, 297–302.

45. Kyllästinen, M.J.; Elfving, S.M.; Gref, C.G.; Aro, A. Dietary vitamin C supplementation and common laboratory values in the elderly. *Arch. Gerontol. Geriatr.* **1990**, *10*, 297–301. [CrossRef]

46. Prymont-Przyminska, A.; Bialasiewicz, P.; Zwolinska, A.; Sarniak, A.; Wlodarczyk, A.; Markowski, J.; Rutkowski, K.P.; Nowak, D. Addition of strawberries to the usual diet increases postprandial but not fasting non-urate plasma antioxidant activity in healthy subjects. *J. Clin. Biochem. Nutr.* **2016**, *59*, 191–198. [CrossRef] [PubMed]

47. Gawron-Skarbek, A.; Prymont-Przymińska, A.; Sobczak, A.; Guligowska, A.; Kostka, T.; Nowak; Dariusz; Szatko, F. Comparison of native and non-urate total antioxidant capacity of fasting plasma and saliva among middle-aged and older subjects. *Redox Rep.* **2017**. submitted.

nutrients

MDPI

Article

Lack of Additive Effects of Resveratrol and Energy Restriction in the Treatment of Hepatic Steatosis in Rats

Iñaki Milton-Laskibar [1,2], Leixuri Aguirre [1,2], Alfredo Fernández-Quintela [1,2], Anabela P. Rolo [3], João Soeiro Teodoro [3], Carlos M. Palmeira [3] and María P. Portillo [1,2,*]

[1] Nutrition and Obesity Group, Department of Nutrition and Food Science, University of the Basque Country (UPV/EHU) and Lucio Lascaray Research Institute, Facultad de Farmacia, Vitoria 01006, Spain; inaki.milton@ehu.eus (I.M.-L.); leixuri.aguirre@ehu.eus (L.A.); alfredo.fernandez@ehu.eus (A.F.-Q.)

[2] CIBERobn Physiopathology of Obesity and Nutrition, Institute of Health Carlos III, Vitoria 01006, Spain

[3] Department of Life Sciences and Center for Neurosciences and Cell Biology, University of Coimbra, Coimbra 3004-517, Portugal; anpiro@ci.uc.pt (A.P.R.); jteodoro@ci.uc.pt (J.S.T.); palmeira@ci.uc.pt (C.M.P.)

* Correspondence: mariapuy.portillo@ehu.eus; Tel.: +34-945-013067; Fax: +34-945-013014

Received: 30 May 2017; Accepted: 5 July 2017; Published: 11 July 2017

Abstract: The aims of the present study were to analyze the effect of resveratrol on liver steatosis in obese rats, to compare the effects induced by resveratrol and energy restriction and to research potential additive effects. Rats were initially fed a high-fat high-sucrose diet for six weeks and then allocated in four experimental groups fed a standard diet: a control group, a resveratrol-treated group, an energy restricted group and a group submitted to energy restriction and treated with resveratrol. We measured liver triacylglycerols, transaminases, FAS, MTP, CPT1a, CS, COX, SDH and ATP synthase activities, FATP2/FATP5, DGAT2, PPARα, SIRT1, UCP2 protein expressions, ACC and AMPK phosphorylation and PGC1α deacetylation. Resveratrol reduced triacylglycerols compared with the controls, although this reduction was lower than that induced by energy restriction. The mechanisms of action were different. Both decreased protein expression of fatty acid transporters, thus suggesting reduced fatty acid uptake from blood stream and liver triacylglycerol delivery, but only energy restriction reduced the assembly. These results show that resveratrol is useful for liver steatosis treatment within a balanced diet, although its effectiveness is lower than that of energy restriction. However, resveratrol is unable to increase the reduction in triacylglycerol content induced by energy restriction.

Keywords: resveratrol; energy restriction; liver steatosis; rat

1. Introduction

Excessive fat accumulation in the liver is known as simple hepatic steatosis, which is the most benign form of non-alcoholic fatty liver disease (NAFLD). It is a major cause of chronic liver disease in western societies, and this burden is expected to grow with the increasing incidence of obesity and metabolic syndrome, which are both closely associated with it [1,2]. Energy restriction is a commonly used method for fatty liver treatment [3,4]. In fact, this method has been proved to induce a decrease in intrahepatic fat content in overweight and obese subjects [5,6].

A great deal of attention has been paid by the scientific community in recent years to bioactive molecules present in foods and plants, such as phenolic compounds, which could represent new complementary tools for liver steatosis management. One of the most widely studied molecules is resveratrol (*trans*-3,5,4′-trihydroxystilbene), a phytoalexin occurring naturally in grapes, berries and peanuts [7,8]. Numerous studies have been carried out using resveratrol and different models of liver steatosis in mice and rats [9,10]. The vast majority of these studies have demonstrated that resveratrol is able to prevent liver triacylglycerol accumulation induced by overfeeding conditions. With regard to human beings, its positive effects on liver steatosis have been observed in studies carried out by its administration at doses in the range of 150–500 mg/day for 4–12 weeks [9,11–13]. Nevertheless, it is important to point out that other authors have not observed this beneficial effect [14].

Furthermore, it has been proposed that resveratrol may mimic energy restriction in rodent models [15–18]. Thus, this compound could bring about the benefits of energy restriction without an actual reduction in calorie intake.

Taking all of the information above into account, the aims of the present study were (a) to analyze the effect of resveratrol on liver steatosis previously induced by a high-fat high-sucrose diet in obese rats; (b) to compare the effects induced by resveratrol and energy restriction and (c) to research potential additive effects between resveratrol and energy restriction. Our initial hypothesis is that resveratrol can show a delipidating effect in the liver similar to that induced by a mild energy restriction, and that the combination of both strategies can increase treatment effectiveness.

2. Material and Methods

2.1. Animals, Diets and Experimental Design

The experiment was conducted with forty five 6-week-old male Wistar rats from Harlan Ibérica (Barcelona, Spain) and performed in accordance with the institution guide for the care and use of laboratory animals (M20_2016_039).

The rats were individually housed in polycarbonate metabolic cages (Tecniplast Gazzada, Buguggiate, Italy) and placed in an air-conditioned room (22 ± 2 °C) with a 12-h light-dark cycle. After a 6-day adaptation period, all rats were fed a high-fat high-sucrose (HFHS) diet (OpenSource Diets, Lynge, Denmark; Ref. D12451), for six weeks. This diet provided 45% of the energy as fat, 20% as protein and 35% as carbohydrates (4.7 kcal/g diet). After this period, nine rats (HFHS group) were sacrificed to check whether liver steatosis was induced by comparing their liver lipid content with that of a matched group of rats fed a standard diet for six weeks (normal rats; N group). The remaining animals fed the high-fat high-sucrose diet for six weeks were randomly divided into four experimental groups ($n = 9$): the control group (C), the resveratrol group treated with resveratrol (RSV), the restricted group submitted to a moderate energy restriction (R), and the group both treated with resveratrol as well as submitted to energy restriction (RR). In all cases, the diet was a semi-purified standard diet (OpenSource Diets, Lynge, Denmark; D10012G), and the additional treatment period was six weeks. This semi-purified standard diet provided 16% of the energy as fat, 20% as protein and 64% as carbohydrates (3.9 kcal/g diet). Rats from C and RSV groups had free access to food, and rats from R and RR groups were subjected to a 15% energy restriction. This percentage, that was selected according to previous studies from our laboratory, is below the percentage commonly used in energy restricted diets in humans. The diet amount provided to the rats on the restricted groups was calculated based on the spontaneous food intake in C group. In the RSV and RR groups, resveratrol was added to the diet as previously reported [8] to ensure a dose of 30 mg/kg body weight/day.

At the end of the total experimental period (12 weeks), rats from the four experimental groups were sacrificed after 8–12 h of fasting, under anesthesia (chloral hydrate), by cardiac exsanguination. Livers were dissected, weighed and immediately frozen in liquid nitrogen. Serum was obtained from blood samples after centrifugation ($1000 \times g$ for 10 min, at 4 °C). All samples were stored at -80 °C until analysis.

2.2. Liver Triacylglycerol Content and Serum Transaminases

Total liver lipids were extracted according to the method described by Folch et al. [19]. The lipid extract was dissolved in isopropanol, and the triacylglycerol content was measured using a commercial kit (Spinreact, Barcelona, Spain). Commercial kits were also used for the analysis of serum transaminases aspartate aminotransferase (AST) and alanine aminotransferase (ALT) (Spinreact, Barcelona, Spain).

2.3. Enzyme Activities

The activity of the lipogenic enzyme fatty acid synthase (FAS) was measured by spectrophotometry, as previously described [20]. Briefly, liver samples (0.5 g) were homogenized in 5 mL of buffer (150 mM KCl, 1 mM MgCl$_2$, 10 mM N-acetyl cysteine and 0.5 mM dithiothreitol) and centrifugated to 100,000× g for 40 min at 4 °C. The supernatant fraction was used for FAS activity determination, as the rate of malonyl CoA dependent NADH oxidation [21]. Results were expressed as nanomoles of reduced nicotinamide adenine dinucleotide phosphate (NADPH) consumed per minute per milligram of protein.

In order to assess the assembly and secretion of very low density lipoproteins by the liver, microsomal triglyceride transfer protein (MTP) activity was determined fluorimetrically by using a commercial kit (Sigma-Aldrich, St. Louis, MI, USA). MTP activity was expressed as percentage of transference.

As far as oxidative enzymes are concerned, carnitine palmitoyltransferase-1a (CPT-1a) activity was measured spectrophotometrically in the mitochondrial fraction as previously described [22]. The activity was expressed as nanomoles of coenzyme A formed per minute per milligram of protein. Citrate synthase (CS) activity was assessed spectrophotometrically following the Srere method [23], by measuring the appearance of free CoA. Briefly, frozen liver samples were homogenized in 25 vol (wt/vol) of 0.1 M Tris-HCl buffer (pH 8.0). Homogenates were incubated for 2 min at 30 °C with 0.1 M Tris-HCl buffer containing 0.1 mM DTNB, 0.25 Triton X-100, 0.5 mM oxalacetate and 0.3 mM acetyl CoA, and readings were taken at 412 nm. Then, the homogenates were re-incubated for 5 min and readings were taken at the same wavelength. CS activity was expressed as CoA nanomoles formed per minute, per milligram of protein. The protein content of the samples was determined by the [24], using bovine serum albumine as standard.

For succinate dehydrogenase (SDH), cytochrome c oxidase (COX) and mitochondrial ATP synthase activity determinations, liver samples were powdered with liquid nitrogen, using a mortar and a pestle, and homogenized with homogenization buffer (250 mM sucrose, 10 mM HEPES (pH 7.4), 0.5 mM EGTA and 0.1% fat-free bovine serum albumin) using a Ystral D-79282 homogenizer (Ystral, Ballrechten-Dottingen, Germany). The protein content of the samples was determined using the Biuret method [25], and calibrated with bovine serum albumin. SDH activity was determined polarographically as previously described [26]. Briefly, liver homogenates (2 mg of protein) were suspended under constant magnetic stirring at 25 °C, in 1.4 mL of standard respiratory medium (130 mM sucrose, 50 mM KCl, 5 mM MgCl$_2$, 5 mM KH$_2$PO$_4$, 50 μM EDTA and 5 mM HEPES (pH 7.4) supplemented with 5 mM succinate, 2 μM rotenone, 0.1 μg Antimycin A, 1 mM KCN and 0.3 mg Triton X-100. The reaction was initiated by the addition of 1 mM phenazine methosulfate (PMS). In the case of COX, the activity was also measured polarographically, as previously described [27]. The reaction was carried out at 25 °C in 1.4 mL of standard respiratory medium, supplemented with 2 μM rotenone, 10 μM oxidized Cytochrome c and 0.3 mg Triton X-100. After the addition of 2 mg of liver homogenate protein, the reaction was initiated by adding 5 mM ascorbate plus 0.25 mM tetra methylphenylene-diamine (TMPD). Finally, the activity of ATP Synthase was determined spectrophotometrically at a wavelength of 660 nm, in association with ATP hydrolysis as previously mentioned [28]. Briefly, 2 mg of liver homogenate protein were incubated with 2 mL of reaction medium (125 mM sucrose, 65 mM KCl, 2.5 mM MgCl$_2$ and 0.5 mM HEPES, pH 7.4) at 37 °C. The reaction was initiated by adding 2 mM Mg^{2+}-ATP in the presence or absence of oligomicyn (1 μg/mg protein), and stopped after 3 min by adding 1 mL of 40% trichloroacetic acid. The samples were then centrifugated for 5 min at 3000 rpm, and 1 mL of the supernatant was mixed with 2 mL of H$_2$O and 2 mL of ammonium molybdate. The ATP synthase activity was calculated as the difference in total absorbance and absorbance in the presence of oligomycin.

2.4. Western Blot

For Acetyl CoA carboxylase (ACC), AMP activated protein kinase (AMPK α), sirtuin 1 (SIRT1), fatty acid transport protein 2 (FATP2), uncoupling protein 2 (UCP2), diacylglycerol acyltransferase 2 (DGAT2), fatty acid transport protein 5 (FATP5) and β-actin protein quantification, liver samples of 100 mg were homogenated in 1000 μL of cellular PBS (pH 7.4), containing protease inhibitors (100 mM phenylmethylsulfonyl fluoride and 100 mM iodoacetamide). Homogenates were centrifuged at $800 \times g$ for 10 min at 4 °C. Protein concentration in homogenates was measured by the Bradford method [24] using bovine serum albumin as standard. In the case of peroxisome proliferator-activator receptor alpha (PPARα), and peroxisome proliferator-activated receptor gamma coactivator 1-alpha (PGC1α), nuclear protein extraction was carried out with 100 mg of liver tissue, as previously described [29].

Immunoblot analyses were performed using 60 μg of protein from total or nuclear liver extracts separated by electrophoresis in 7.5% or 10% SDS-polyacrylamide gels and transferred to PVDF membranes. The membranes were then blocked with 5% casein PBS-Tween buffer for 2 h at room temperature. Subsequently, they were blotted with the appropriate antibodies overnight at 4 °C. Protein levels were detected via specific antibodies for ACC (1:1000), AMPK α (1:1000) (Cell Signaling Technology, Danvers, MA, USA), SIRT1 (1:1000), FATP2 (1:1000), UCP2 (1:500), DGAT2 (1:500) (Santa Cruz Biotech, Dallas, TX, USA) FATP5 (1:500), (LifeSpan BioScience, Seattle, WA, USA), PGC1α (1:1000), PPARα (1:500), (Abcam, Cambridge, UK) and β-actin (1:5000) (Sigma, St. Louis, MO, USA). Afterward, polyclonal anti-mouse for β-actin, anti-rabbit for ACC, AMPK, SIRT1, DGAT2, FATP5, PGC1α and PPARα, and anti-goat for FATP2 and UCP2 (1:5000) were incubated for 2 h at room temperature, and ACC, AMPK, SIRT1, FATP2, UCP2, DGAT2, PPARα, FATP5, PGC1α and β-actin were measured. After antibody stripping, the membranes were blocked, and then incubated with phosphorylated ACC (serine 79, 1:1000), phosphorylated AMPK (threonine 172, 1:500) and acetylated lysine (1:1000) (Cell Signaling Technology, Danvers, MA, USA) antibodies. The bound antibodies were visualized by an ECL system (Thermo Fisher Scientific Inc., Rockford, IL, USA) and quantified by a ChemiDoc MP Imaging System (Bio-Rad, Hercules, CA, USA). The measurements were normalized by β-actin in total protein extractions and in the case of the nuclear extraction, equal loading of proteins was confirmed by staining the membranes with Comassie Blue.

2.5. Statistical Analysis

Results are presented as mean $\pm$ SEM. Statistical analysis was performed using SPSS 24.0 (SPSS, Chicago, IL, USA). All the parameters are normally-distributed according to the Shapiro-Wilks test. Data were analyzed by one-way ANOVA followed by Newman-Keuls post-hoc test. Significance was assessed at the $p < 0.05$ level.

3. Results

3.1. Body Weight Gain, Liver Weight, Liver Triacylglycerol Amounts and Serum Transaminases

As explained in the Results section, after six weeks of high-fat high-sucrose feeding, rats (HFHS group) showed significantly increased amounts of triacylglycerols in their livers than rats fed a standard diet for six weeks (N group) (53.6 ± 1.9 mg/g tissue vs. 32.6 ± 4.1 mg/g tissue; $p < 0.050$), indicating that liver steatosis was induced. These results were paralleled by the induction of insulin resistance, as observed in a previous study from our laboratory carried out in this cohort of rats [30].

Body weight gain was similar in C and RSV groups and lower in both restricted groups when compared with the C group ($p < 0.0003$ in R group and $p < 0.0001$ in RR group), with no difference between them. In spite of this difference between restricted and non restricted groups, no differences were observed in liver weight among the four experimental groups (Table 1).

Lower values of triacylglycerol content were found in the three treated groups in comparison with the C group ($p < 0.03$ in RSV group, $p < 0.0002$ in R group and $p < 0.0004$ in RR group). In the case of the groups submitted to a mild energy restriction (R and RR), lower values were found compared with the RSV group ($p < 0.003$ in R group and $p < 0.005$ in RR group), with no differences between them (Table 1).

As far as serum parameters are concerned, triacylglycerols were not modified in resveratrol-treated rats when compared with control animals. By contrast, restricted rats (R and RR groups) showed significantly lower values without differences between them. No changes in serum transaminase concentrations were observed among experimental groups (Table 1).

Table 1. Body weight gain, liver weight, hepatic triacylglycerol (TG) content, liver cholesterol (Chol) content and serum triacylglycerol, alanine aminotransferase (ALT) and aspartate aminotransferase (AST) concentrations of rats fed on the experimental diets for six weeks, and then fed a standard diet (C), or a standard diet supplemented with resveratrol (RSV), or submitted to energy restriction and fed a standard diet (R) or submitted to energy restriction and fed a standard diet supplemented with resveratrol (RR) ($n = 9$/group) for additional six weeks.

	C	RSV	R	RR	ANOVA
Body weight gain (g)	40 ± 4 [a]	46 ± 5 [a]	18 ± 4 [b]	16 ± 2 [b]	$p < 0.001$
Liver weight (g)	10.6 ± 0.2	11.4 ± 0.4	10.7 ± 0.4	11.0 ± 0.3	NS
Hepatic TG (mg/g tissue)	42.6 ± 4.7 [a]	32.4 ± 3.5 [b]	18.5 ± 2.5 [c]	19.7 ± 1.8 [c]	$p < 0.05$
Hepatic Chol (mg/g tissue)	5.3 ± 0.3 [a]	4.2 ± 0.3 [bc]	3.5 ± 0.5 [c]	4.6 ± 0.3 [ab]	$p < 0.05$
Serum TG (mg/dL)	68.2 ± 13.3 [a]	56.7 ± 11.0 [a]	39.6 ± 8.6 [b]	43.8 ± 4.9 [b]	$p < 0.05$
ALT (U/L)	31.2 ± 3.0	31.5 ± 6.6	24.0 ± 2.7	32.7 ± 5.4	NS
AST (U/L)	51.5 ± 3.1	57.6 ± 7.1	47.7 ± 8.1	49.0 ± 15.9	NS

Values are mean $\pm$ SEM. Differences among groups were determined by using one-way ANOVA followed by Newman Keuls post-hoc test. Values not sharing a common letter are significantly different ($p < 0.05$). NS: Not significant.

3.2. Enzyme Activities

No differences in FAS activity were found between the control and each treated group (Figure 1A). On the other hand, MTP activity was greater in the three treated groups when compared with the C group ($p < 0.016$ in RSV group, $p < 0.05$ in R group and $p < 0.0016$ in RR group), without significant differences among the three (Figure 2B).

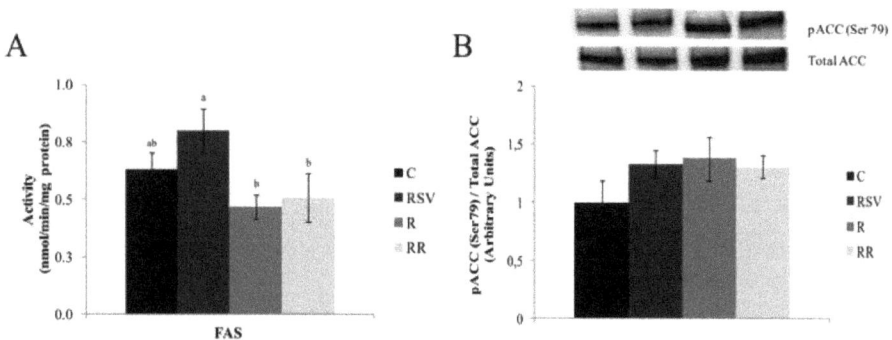

Figure 1. FAS activity (**A**) and phosphorylated ACC (serine 79)/Total ACC ratio (**B**) in liver from rats fed an obesogenic diet for six weeks, and then fed a standard diet (C), or a standard diet supplemented with resveratrol (RSV), or submitted to energy restriction and fed a standard diet (R) or submitted to energy restriction and fed a standard diet supplemented with resveratrol (RR) ($n = 9$/group) for additional six weeks. Values are mean $\pm$ SEM. Differences among groups were determined by using one-way ANOVA followed by Newman Keuls post-hoc test. Values not sharing a common letter are significantly different ($p < 0.05$). FAS: fatty acid synthase, ACC: acetyl CoA carboxylase.

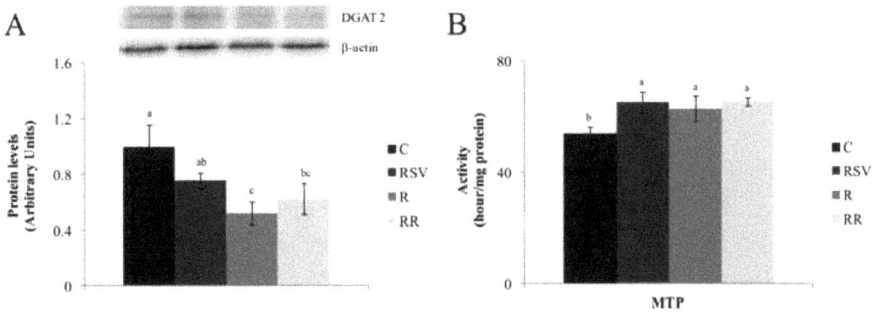

Figure 2. DGAT2 (**A**) protein expression and MTP (**B**) activity in liver from rats fed an obesogenic diet for six weeks, and then fed a standard diet (C), or a standard diet supplemented with resveratrol (RSV), or submitted to energy restriction and fed a standard diet (R) or submitted to energy restriction and fed a standard diet supplemented with resveratrol (RR) (n = 9/group) for additional six weeks. Values are mean $\pm$ SEM. Differences among groups were determined by using one-way ANOVA followed by Newman Keuls post-hoc test. Values not sharing a common letter are significantly different ($p < 0.05$). DGAT2: diacylglycerol acyltransferase 2, MTP: microsomal triglyceride transfer protein.

With regard to oxidative enzymes, the activity of CPT1a was increased in the groups supplemented with resveratrol when compared with the C group ($p < 0.002$ in RSV group and $p < 0.05$ in RR group), with no difference between them. A significantly higher enzyme activity was also observed in the RSV ($p < 0.01$) group when compared with the R group (Figure 3). In the case of the CS activity, the RSV and RR groups showed greater activity when compared with the C group ($p < 0.03$ and $p < 0.003$ respectively), with no differences between them (Figure 3). Moreover, no differences were observed in SDH, (also known as respiratory Complex II) or ATP synthase among experimental groups (Figure 3). Finally, the activity of mitochondrial Complex IV (COX) was significantly increased in both restricted groups ($p < 0.01$ and $p < 0.01$ in R and RR groups respectively), with no differences between them. Its activity in resveratrol-treated rats remained unchanged when compared with the control group (Figure 3).

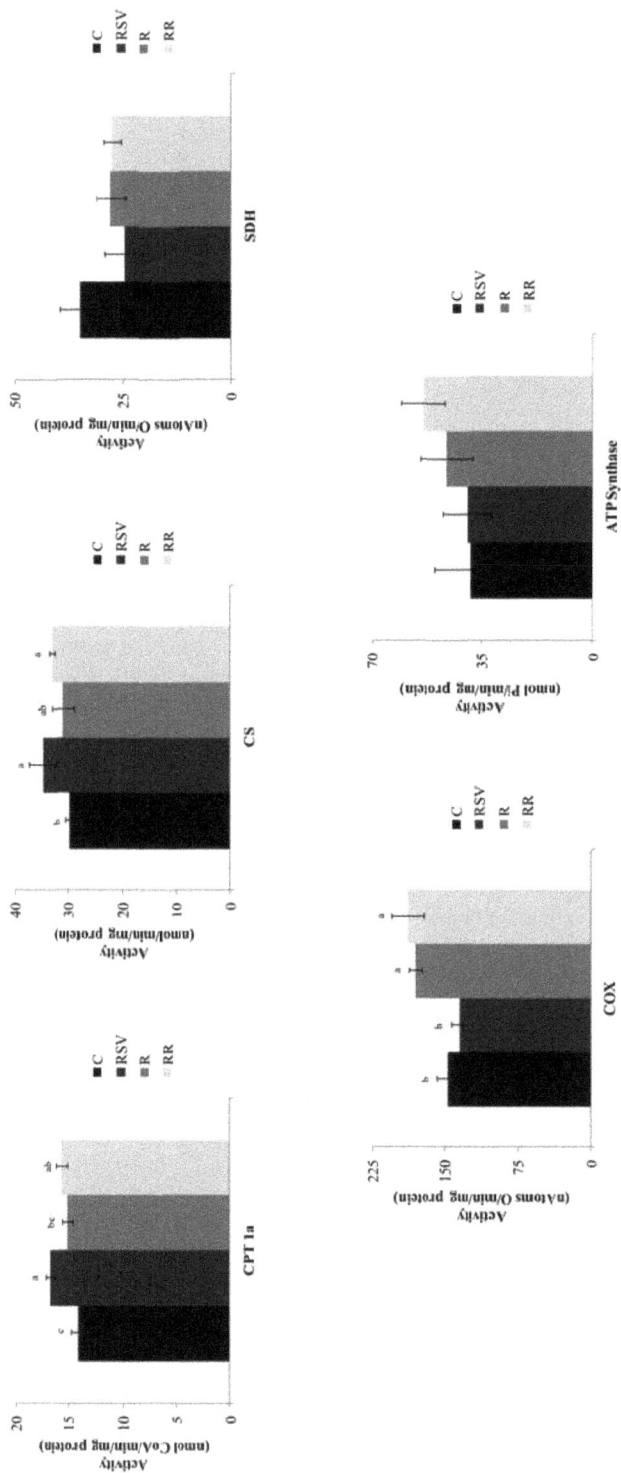

Figure 3. CPT1 and CS, SDH, COX and ATP Synthase activities in liver from rats fed an obesogenic diet for six weeks, and then fed a standard diet (C), or a standard diet supplemented with resveratrol (RSV), or submitted to energy restriction and fed a standard diet (R) or submitted to energy restriction and fed a standard diet supplemented with resveratrol (RR) ($n = 9$/group) for additional six weeks. Values are mean ± SEM. Differences among groups were determined by using one-way ANOVA followed by Newman Keuls post-hoc test. Values not sharing a common letter are significantly different ($p < 0.05$). CPT1a: carnitine palmitoyltransferase-1a, CS: citrate synthase, SDH: succinate dehydrogenase, COX: cytochrome *c* oxidase.

3.3. Western Blot Analysis

The ratio pACC (Ser 79)/Total ACC was used as an index of ACC activity. High values of this ratio were found in treated groups when compared with the controls (+33% in RSV group, +37% in R group and +30% in RR group). These differences showed a statistical trend ($p = 0.08$) (Figure 1B). In the case of pAMPKα (Thr 172)/Total AMPKα ratio, which shows the activation of this enzyme, the three treated groups showed greater phosphorylation ($p < 0.05$ in RSV group, $p < 0.005$ in R group and $p < 0.01$ in RR group), which is to say activation, when compared with C group, with no differences among the three (Figure 4).

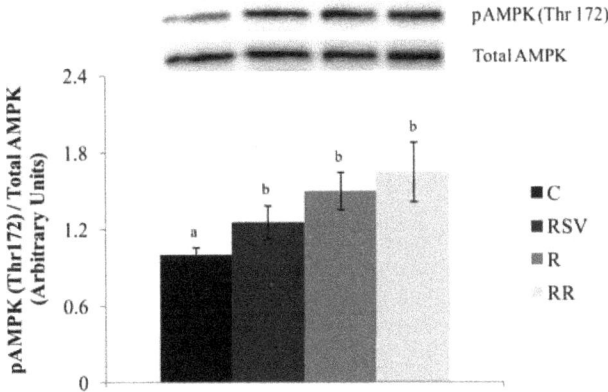

Figure 4. Phosphorylated AMPK (threonine 172)/Total AMPK ratio in liver from rats fed an obesogenic diet for six weeks, and then fed a standard diet (C), or a standard diet supplemented with resveratrol (RSV), or submitted to energy restriction and fed a standard diet (R) or submitted to energy restriction and fed a standard diet supplemented with resveratrol (RR) ($n = 9$/group) for additional six weeks. Values are mean ± SEM. Differences among groups were determined by using one-way ANOVA followed by Newman Keuls post-hoc test. Values not sharing a common letter are significantly different ($p < 0.05$). AMPK: AMP activated protein kinase.

DGAT2 was also measured and lower protein expression was showed by rats from the restricted groups when compared with the C group ($p < 0.01$ in R group and $p < 0.04$ in RR group), with no differences between them (Figure 2B). As far as FATP2 protein is concerned, the groups submitted to a mild energy restriction showed the lowest values, in comparison with the C group ($p < 0.003$ in R group and $p < 0.003$ in RR group), with no difference between them (Figure 5A). On the other hand, in all the treated groups FATP5 protein expression was lower than that in the C group ($p < 0.03$ in RSV group, $p < 0.0003$ in R group and $p < 0.0004$ in RR group) (Figure 5B).

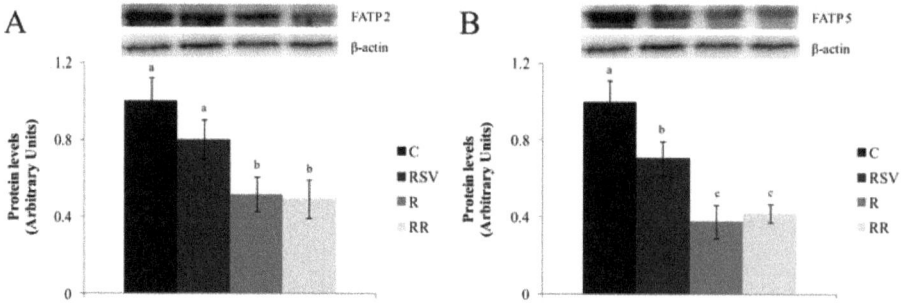

Figure 5. FATP2 (**A**) and FATP5 (**B**) protein expression in liver from rats fed an obesogenic diet for six weeks, and then fed a standard diet (C), or a standard diet supplemented with resveratrol (RSV), or submitted to energy restriction and fed a standard diet (R) or submitted to energy restriction and fed a standard diet supplemented with resveratrol (RR) (n = 9/group) for additional six weeks. Values are mean ± SEM. Differences among groups were determined by using one-way ANOVA followed by Newman Keuls post-hoc test. Values not sharing a common letter are significantly different ($p < 0.05$). FATP2: fatty acid transport protein 2, FATP5: fatty acid transport protein 5.

Regarding proteins related to fatty acid oxidation, no significant changes were induced by experimental treatments in the expression of PPARα (Figure 6A). In the case of PGC-1α acetylation, reduced levels were observed in all treated groups ($p < 0.01$ in RSV group, $p < 0.01$ in R group and $p < 0.008$ in RR group), with no differences among them (Figure 6B). Finally, when the protein expression of SIRT1 and UCP2 were studied, no changes were observed among the different groups (Figure 7).

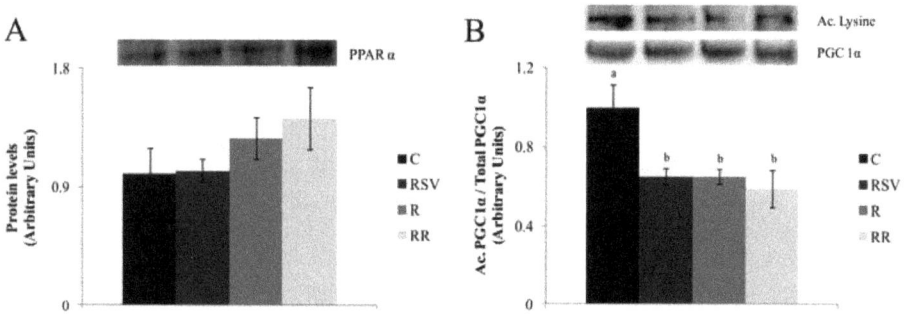

Figure 6. PPARα protein expression (**A**) and Acetylated PGC1α/Total PGC1α (**B**) in liver from rats fed an obesogenic diet for six weeks, and then fed a standard diet (C), or a standard diet supplemented with resveratrol (RSV), or submitted to energy restriction and fed a standard diet (R) or submitted to energy restriction and fed a standard diet supplemented with resveratrol (RR) (n = 9/group) for additional six weeks. Values are mean ± SEM. Differences among groups were determined by using one-way ANOVA followed by Newman Keuls post-hoc test. Values not sharing a common letter are significantly different ($p < 0.05$). PPARα: peroxisome proliferator-activator receptor alpha, PGC1α: peroxisome proliferator-activated receptor gamma coactivator 1-alpha.

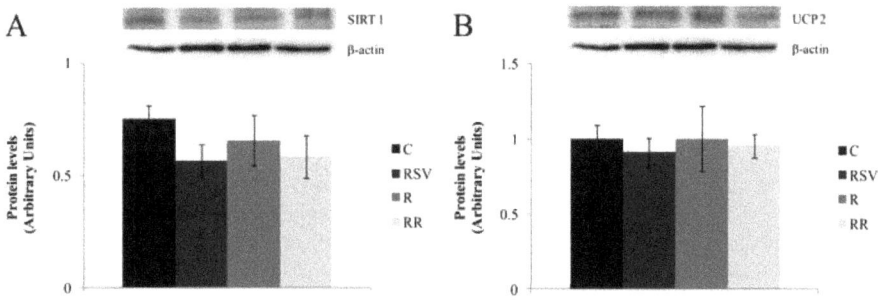

Figure 7. SIRT1 (**A**) and UCP2 (**B**) protein expression in liver from rats fed an obesogenic diet for six weeks and then fed a standard diet (C), or a standard diet supplemented with resveratrol (RSV), or submitted to energy restriction and fed a standard diet (R) or submitted to energy restriction and fed a standard diet supplemented with resveratrol (RR) (n = 9/group) for additional six weeks. Values are mean ± SEM. Differences among groups were determined by using one-way ANOVA followed by Newman Keuls post-hoc test. Values not sharing a common letter are significantly different ($p < 0.05$). SIRT1: sirtuin 1, UCP2: uncoupling protein 2.

4. Discussion

The effectiveness of resveratrol in the reduction of hepatic lipid accumulation, when administered under overfeeding conditions and concurrent with an obesogenic diet, has been largely reported in rodents in the prevention of steatosis [31–40]. Indeed, resveratrol is able to partially prevent liver steatosis associated with overfeeding. However, much less abundant information is available concerning its effects on previously developed liver steatosis reduction [41]. Bearing this in mind, and taking into account that it has been proposed that resveratrol mimics energy restriction [11,16,17,42], which is a common dietary strategy for steatosis treatment, the first aim of the present study was to analyze the effect of this compound on liver steatosis. This had been previously induced by an obesogenic diet when it was added to a standard diet. In the present study, a dose of 30 mg resveratrol/kg body weight/day was used because in a previous study we observed that it was an effective dose in reducing liver triacylglycerol amount in an overfeeding model [35].

For this purpose, the C and RSV groups were compared. The lower hepatic triacylglycerol content observed in the rats from the RSV group (−23.4%) showed that resveratrol was indeed effective, not only in preventing steatosis, as widely described in literature, but also in reducing fat accumulation previously induced by a high-fat high-sucrose feeding. When we compare the percentage of triacylglycerol reduction obtained in this study (−23.4%) with that found in a previous study from our group that was devoted to analyzing the preventive effect of resveratrol on liver steatosis and carried out with the same dose of resveratrol and the same experimental period length (−23.0%) [35], it can be observed that the effectiveness of resveratrol as a preventative molecule was only slightly higher than it was as a therapeutic one. This conclusion is not in good accordance with that obtained by Heebøll et al., who found that the preventive effect of resveratrol was superior to its therapeutic effect. This discrepancy may be due to differences in the experimental design (mainly animal species and resveratrol dose). Surprisingly, serum transaminases were not reduced. This lack of effect may have been due to their being in the range of physiological values [43] after six weeks of obesogenic feeding, as a consequence of the development of a mild degree of steatosis.

Insulin resistance is closely related to liver steatosis. This alteration in glucose homeostasis was studied in this cohort of rats in a previous paper [30], by measuring serum insulin and glucose, HOMA-IR and by carrying out a glucose tolerance test. We observed that resveratrol induced a mild improvement in glycemic control, which fits well with the reduction observed in liver steatosis in these rats.

The amount of triacylglycerols accumulated in hepatocytes is regulated by various metabolic processes: fatty acid uptake, fatty acid synthesis and triacylglycerol esterification on the one hand ("input"), and fatty acid oxidation and triacylglycerol export on the other hand ("output"). Steatosis occurs when "input" exceeds "output" [44,45]. In order to analyze the mechanisms underlying the delipidating action of resveratrol, we assessed its effects on several parameters related to the previously mentioned processes.

As far as de novo lipogenesis in concerned, although FAS activity remained unchanged, a sharp increase in the activity of ACC, the limiting enzyme of this process, was observed in resveratrol-treated rats. Consequently, it can be proposed that this metabolic pathway was likely somehow inhibited by this polyphenol, and thus this could contribute to the reduction in triacylglycerol content. Moreover, FATP5 protein expression was reduced in the RSV group, suggesting a decrease in fatty acid uptake, which could also contribute to the reduction in triacylglycerol content. In fact, the relationship between FATP5 and NAFLD development has been studied in rodents [46] and in humans [47]. As far as fatty acid oxidation is concerned, its involvement in liver delipidation is not clear. Thus, the activities of CPT1a, the enzyme that allows long chain fatty acids to enter into mitochondria, and CS, a marker of mitochondria density, were significantly increased due to resveratrol treatment; this was also the case for the deacetylation level of PGC-1α, the transcription factor co-activator that regulates mitochondria number and function [48,49]. By contrast, the activities of enzymes participating in the respiratory electron transport chain, SDH, COX, and ATP synthase remained unchanged.

DGAT2, the enzyme that catalyzes the binding between diacylglycerol and a long chain fatty acyl-CoA, was not modified by resveratrol treatment. This suggests that the synthesis of triacylglycerols could be reduced by a decrease in fatty acid availability, but not by the inhibition of the assembly process. Moreover, increased MTP activity suggests enhanced delivery of triacylglycerols from liver to plasma. In spite of this effect, serum triacylglycerol concentration was not increased. In order to explain this fact, it is important to remember that this parameter depends not only on triacylglycerol delivery to blood, but also on triacylglycerol clearance from tissues. Thus, increased triacylglycerol clearance in skeletal muscle via lipoprotein lipase cannot be discarded. Although there are no reports in the literature showing the effect of resveratrol on skeletal muscle LPL, our hypothesis stems from the fact that Timmers et al., [11] proposed that resveratrol mimics the effects of training in skeletal muscle, and by the reported increase in LPL expression induced by training in skeletal muscle [50,51].

Taken together, these results suggest that the reduction in hepatic triacylglycerols induced by resveratrol is mainly justified by decreased fatty acid availability for triacylglycerol synthesis, due to reduced de novo synthesis and uptake and increased oxidation, and to the increase in triacylglycerol delivery to blood.

The role of UCP2 in NAFLD development has been intensively studied, but reported studies are controversial [52]. Some studies have shown that hepatocellular UCP2 expression is increased in NAFLD, indicating its potential role in disease development [53–56]. However, other studies have demonstrated that UCP2 deficiency caused diminished hepatic utilization and fatty acid clearance and thus may lead to liver steatosis [57]. Moreover, it has been reported that obesity-related fatty liver is unchanged in UCP2 mitochondrial-deficient mice [55]. Thus, in the present study we analyzed UCP2 protein expression in order to gain more insight concerning this issue. Unfortunately, no changes were observed after resveratrol treatment, meaning that irrespective of the positive or negative effect of UCP2 on steatosis, the delipidating effect of this phenolic compound was not mediated by this uncoupling protein. This result agrees with that reported by Heebøll et al. in mice [41].

Resveratrol has been identified as a potent activator for both SIRT1 and AMPK, two critical signalling molecules regulating the pathways of hepatic lipid metabolism [58]. In the present study, AMPK phosphorylation was increased in the RSV group, meaning that this enzyme was activated by the polyphenol treatment. As far as SIRT1 is concerned, although its protein expression was not modified, the increased deacetylation level of PGC-1α, one of its main targets, suggests that this deacetylase was activated by resveratrol. Consequently, it can be stated that, under our experimental conditions, the activation of the axis SIRT1/AMPK was also involved in resveratrol-induced effects.

Although, as stated in this paper's discussion section, resveratrol is considered an energy restriction mimetic, several authors who have analyzed actions of this polyphenol other than on fatty liver have proposed that the mechanisms underlying the effects of resveratrol and energy restriction are not always the same [17,30,59,60]. In this context, a second aim of the present study was to compare the effects of a mild energy restriction and resveratrol on liver steatosis. Rats from the R group showed a significant reduction in hepatic triacylglycerol when compared with the control group (−56.3%). De novo lipogenesis seems to be reduced in the restricted group because the activity of ACC was decreased by 37%. Furthermore, fatty acid uptake was reduced, as shown by the decrease in FATP2 and FATP5. With regard to the potential contribution of fatty acid oxidation pathway, the results show that energy restriction increased activation of PGC-1α and the activity of COX, with no changes in the rest of oxidative parameters. These results are not surprising because Nisoli et al. [61] reported that a 30% calorie restriction on mice for three months resulted in greatly increased liver mitochondria, evidenced by increases in the proteins cytochrome *c* and cytochrome oxidase subunit IV, and the mRNA levels of PGC-1α, among others. These findings have led to the general acceptance, and have led to incorporation of the concept that energy restriction induces mitochondrial biogenesis. However, Hancock et al. [62] did not find any change in mitochondrial markers in the liver after 14 weeks of 30% energy restriction.

Moreover, the reduced amount of DGAT2 in the R group suggests a decrease in triacylglycerol assembly. These results show that a decrease in triacylglycerol synthesis, due to reduced availability of one of the substrates (fatty acids) and the inhibition of the assembly process, contributed to the reduction in hepatic triacylglycerol content induced by energy restriction. Finally, increased MTP activity indicates enhanced triacylglycerol delivery from liver to plasma. In spite of this effect, serum concentration of triacylglycerols was lower in the R and C group. As in the case of resveratrol treated-rats, it can be argued that due to energy restriction, other tissues can increase the uptake of this lipid species via lipoprotein lipase [63].

As expected, AMPK was phosphorylated and thus, activated. On the other hand, protein expression of SIRT1 was not modified. However, the deacetylation status of PGC-1α suggests its activation. Consequently, it can be stated that under the activation of the axis SIRT1/AMPK was involved in the delipidating effect induced by a mild energy restriction effects.

By comparing the RSV and R groups it can be observed that hepatic fat reduction induced by energy restriction was greater than that induced by resveratrol treatment, meaning that a mild energy restriction (−15%) was more efficient than resveratrol administration. Similarly, the improvement in glycemic control observed in this cohort of rats in our previous paper mentioned before in this paper's discussion section, was greater than that observed in rats treated with resveratrol [30]. In addition, the mechanisms of action of resveratrol and energy restriction were not exactly the same. Both treatment strategies decreased de novo lipogenesis, fatty acid uptake from blood stream and increased fatty acid oxidation and liver triacylglycerol delivery, but only energy restriction reduced triacylglycerol assembly. These results are in good accordance with those reported by Tauriainen et al. [33] when they analyzed the preventive effects of resveratrol and energy restriction on liver steatosis under overfeeding conditions. These authors observed that whereas energy restriction (−30%) totally prevented liver steatosis associated to obesogenic feeding, resveratrol only prevented it partially.

Finally, a third aim of the present study was to seek the effects of resveratrol under energy restriction conditions, and to search for potential additive effects between both treatments. This being the case, the administration or resveratrol together with a restricted diet would increase the effectiveness of this dietary treatment. At this point, it is important to emphasize that although in the vast majority of the reported studies the energy restriction ranges from 20% to 40%, in this case, a lower degree of restriction was chosen (15%) was chosen in the present study. The reason for this was based on a previous study from our group [64]. In that study, we also looked for additive anti-obesity and anti-diabetic effects between resveratrol, at a dose of 30 mg/kg of body weight/day, and 25% energy restriction. We observed that the addition of resveratrol to the restricted diet did not lead to additional reductions in fat mass or in serum insulin concentrations with regard to those produced by energy restriction alone. We believed that one of the reasons that could explain this situation was that the effects caused by energy restriction were strong enough to mask the potential positive effects ascribed to resveratrol. Consequently, a lower degree of energy restriction was preferred in the present study.

In the present study, when the effects observed in both restricted groups (R and RR) were compared, no significant differences were appreciated between them. This suggests that resveratrol is not effective in reducing liver triacylglycerols when it is administered together with a restricted diet. Similarly, no differences in the improvement of glycemic control were observed between both experimental groups, as previously reported by our group. It is interesting to point out that resveratrol behaviour is different depending on the feeding pattern, because, as it has been widely reported, this polyphenol is effective in terms of liver triacylglycerol reduction when administered in a scenario of overfeeding. "On the other hand, an important message is that resveratrol is not able to increase the effects induced by energy restriction, and consequently no additive effects were found".

In conclusion, the present results show that resveratrol administration is useful for liver steatosis treatment in the framework of a balanced diet, although its effectiveness is lower than that of a mild energy restriction. By contrast, resveratrol is not able to increase the reduction in hepatic triacylglycerol content induced by energy restriction. Consequently, our initial hypothesis was not confirmed. The mechanisms of action mediating the effects of these two treatment strategies are very similar but not exactly the same.

Acknowledgments: This research has been supported by MINECO (AGL-2015-65719-FEDER-UE), University of the Basque Country (ELDUNANOTEK UFI11/32), Instituto de Salud Carlos III (CIBERobn) and Basque Government (IT-572-13). Iñaki Milton is a recipient of a doctoral fellowship from the Gobierno Vasco. João Soeiro Teodoro is a recipient of a post-doc grant from the Portuguese Fundação para a Ciência e a Tecnologia, ref. SFRH/BPD/94036/2013.

Author Contributions: I.M.-L., L.A. and A.F.-Q. revised the literature. I.M.-L. and L.A. carried out the Western blot analysis in in vivo samples. I.M.-L., J.S.T., A.P.R. and C.M.P. measured the enzyme activities. L.A. and M.P.P. designed the experiment. M.P.P. wrote the manuscript. All the authors revised and approved the final manuscript.

Conflicts of Interest: The authors declare no conflicts of interest.

References

1. Dongiovanni, P.; Lanti, C.; Riso, P.; Valenti, L. Nutritional therapy for nonalcoholic fatty liver disease. *J. Nutr. Biochem.* **2016**, *29*, 1–11. [CrossRef] [PubMed]
2. Day, C.P.; James, O.F. Steatohepatitis: A tale of two "Hits"? *Gastroenterology* **1998**, *114*, 842–845. [CrossRef]
3. Zivkovic, A.M.; German, J.B.; Sanyal, A.J. Comparative review of diets for the metabolic syndrome: Implications for nonalcoholic fatty liver disease. *Am. J. Clin. Nutr.* **2007**, *86*, 285–300. [PubMed]
4. Trepanowski, J.F.; Canale, R.E.; Marshall, K.E.; Kabir, M.M.; Bloomer, R.J. Impact of caloric and dietary restriction regimens on markers of health and longevity in humans and animals: A summary of available findings. *Nutr. J.* **2011**, *10*, 107. [CrossRef] [PubMed]
5. Shah, K.; Stufflebam, A.; Hilton, T.N.; Sinacore, D.R.; Klein, S.; Villareal, D.T. Diet and exercise interventions reduce intrahepatic fat content and improve insulin sensitivity in obese older adults. *Obesity* **2009**, *17*, 2162–2168. [CrossRef] [PubMed]
6. Larson-Meyer, D.E.; Heilbronn, L.K.; Redman, L.M.; Newcomer, B.R.; Frisard, M.I.; Anton, S.; Smith, S.R.; Alfonso, A.; Ravussin, E. Effect of calorie restriction with or without exercise on insulin sensitivity, beta-cell function, fat cell size, and ectopic lipid in overweight subjects. *Diabetes Care* **2006**, *29*, 1337–1344. [CrossRef] [PubMed]
7. Langcake, P.; Pryce, R.J. The production of resveratrol by vitis vinifera and other members of the vitaceae as a response to infection or injury. *Physiol. Plant Pathol.* **1976**, *9*, 77–86. [CrossRef]

8. Macarulla, M.T.; Alberdi, G.; Gómez, S.; Tueros, I.; Bald, C.; Rodríguez, V.M.; Martínez, J.A.; Portillo, M.P. Effects of different doses of resveratrol on body fat and serum parameters in rats fed a hypercaloric diet. *J. Physiol. Biochem.* **2009**, *65*, 369–376. [CrossRef] [PubMed]

9. Faghihzadeh, F.; Hekmatdoost, A.; Adibi, P. Resveratrol and liver: A systematic review. *J. Res. Med. Sci.* **2015**, *20*, 797–810. [PubMed]

10. Aguirre, L.; Portillo, M.P.; Hijona, E.; Bujanda, L. Effects of resveratrol and other polyphenols in hepatic steatosis. *World J. Gastroenterol.* **2014**, *20*, 7366–7380. [CrossRef] [PubMed]

11. Timmers, S.; Konings, E.; Bilet, L.; Houtkooper, R.H.; van de Weijer, T.; Goossens, G.H.; Hoeks, J.; van der Krieken, S.; Ryu, D.; Kersten, S.; et al. Calorie restriction-like effects of 30 days of resveratrol supplementation on energy metabolism and metabolic profile in obese humans. *Cell Metab.* **2011**, *14*, 612–622. [CrossRef] [PubMed]

12. Faghihzadeh, F.; Adibi, P.; Rafiei, R.; Hekmatdoost, A. Resveratrol supplementation improves inflammatory biomarkers in patients with nonalcoholic fatty liver disease. *Nutr. Res.* **2014**, *34*, 837–843. [CrossRef] [PubMed]

13. Chen, S.; Zhao, X.; Ran, L.; Wan, J.; Wang, X.; Qin, Y.; Shu, F.; Gao, Y.; Yuan, L.; Zhang, Q.; et al. Resveratrol improves insulin resistance, glucose and lipid metabolism in patients with non-alcoholic fatty liver disease: A randomized controlled trial. *Dig. Liver Dis.* **2015**, *47*, 226–232. [CrossRef] [PubMed]

14. Chachay, V.S.; Macdonald, G.A.; Martin, J.H.; Whitehead, J.P.; O'Moore-Sullivan, T.M.; Lee, P.; Franklin, M.; Klein, K.; Taylor, P.J.; Ferguson, M.; et al. Resveratrol does not benefit patients with nonalcoholic fatty liver disease. *Clin. Gastroenterol. Hepatol.* **2014**, *12*, 2092–2103. [CrossRef] [PubMed]

15. Pearson, K.J.; Baur, J.A.; Lewis, K.N.; Peshkin, L.; Price, N.L.; Labinskyy, N.; Swindell, W.R.; Kamara, D.; Minor, R.K.; Perez, E.; et al. Resveratrol delays age-related deterioration and mimics transcriptional aspects of dietary restriction without extending life span. *Cell Metab.* **2008**, *8*, 157–168. [CrossRef] [PubMed]

16. Barger, J.L.; Kayo, T.; Vann, J.M.; Arias, E.B.; Wang, J.; Hacker, T.A.; Wang, Y.; Raederstorff, D.; Morrow, J.D.; Leeuwenburgh, C.; et al. A low dose of dietary resveratrol partially mimics caloric restriction and retards aging parameters in mice. *PLoS ONE* **2008**, *3*, e2264. [CrossRef]

17. Baur, J.A. Resveratrol, sirtuins, and the promise of a dr mimetic. *Mech. Ageing Dev.* **2010**, *131*, 261–269. [CrossRef] [PubMed]

18. Mercken, E.M.; Carboneau, B.A.; Krzysik-Walker, S.M.; de Cabo, R. Of mice and men: The benefits of caloric restriction, exercise, and mimetics. *Ageing Res. Rev.* **2012**, *11*, 390–398. [CrossRef] [PubMed]

19. Folch, J.; Lees, M.; Sloane Stanley, G.H. A simple method for the isolation and purification of total lipides from animal tissues. *J. Biol. Chem. FIELD* **1957**, *226*, 497–509.

20. Zabala, A.; Churruca, I.; Macarulla, M.T.; Rodríguez, V.M.; Fernández-Quintela, A.; Martínez, J.A.; Portillo, M.P. The trans-10,cis-12 isomer of conjugated linoleic acid reduces hepatic triacylglycerol content without affecting lipogenic enzymes in hamsters. *Br. J. Nutr.* **2004**, *92*, 383–389. [CrossRef] [PubMed]

21. Lynen, F. Yeast fatty acid synthase. *Methods Enzymol.* **1969**, *14*, 17–33.

22. Miranda, J.; Fernández-Quintela, A.; Macarulla, M.; Churruca, I.; García, C.; Rodríguez, V.; Simón, E.; Portillo, M. A comparison between clna and cla effects on body fat, serum parameters and liver composition. *J. Physiol. Biochem.* **2009**, *65*, 25–32. [CrossRef] [PubMed]

23. Srere, P. Citrate synthase. *Methods Enzymol.* **1969**, *3*, 3–11.

24. Bradford, M.M. A rapid and sensitive method for the quantitation of microgram quantities of protein utilizing the principle of protein-dye binding. *Anal. Biochem.* **1976**, *72*, 248–254. [CrossRef]

25. Gornall, A.G.; Bardawill, C.J.; David, M.M. Determination of serum proteins by means of the biuret reaction. *J. Biol. Chem.* **1949**, *177*, 751–766. [PubMed]

26. Singer, T.P. Determination of the activity of succinate, NADH, choline, and alpha-glycerophosphate dehydrogenases. *Methods Biochem. Anal.* **1974**, *22*, 123–175. [PubMed]

27. Brautigan, D.L.; Ferguson-Miller, S.; Margoliash, E. Mitochondrial cytochrome c: Preparation and activity of native and chemically modified cytochromes c. *Methods Enzymol.* **1978**, *53*, 128–164. [PubMed]

28. Teodoro, J.S.; Rolo, A.P.; Duarte, F.V.; Simões, A.M.; Palmeira, C.M. Differential alterations in mitochondrial function induced by a choline-deficient diet: Understanding fatty liver disease progression. *Mitochondrion* **2008**, *8*, 367–376. [CrossRef] [PubMed]

29. Aguirre, L.; Hijona, E.; Macarulla, M.T.; Gracia, A.; Larrechi, I.; Bujanda, L.; Hijona, L.; Portillo, M.P. Several statins increase body and liver fat accumulation in a model of metabolic syndrome. *J. Physiol. Pharmacol.* **2013**, *64*, 281–288. [PubMed]

30. Milton-Laskibar, I.; Aguirre, L.; Macarulla, M.T.; Etxeberria, U.; Milagro, F.I.; Martínez, J.A.; Contreras, J.; Portillo, M.P. Comparative effects of energy restriction and resveratrol intake on glycemic control improvement. *Biofactors* **2017**, *43*, 371–378. [CrossRef] [PubMed]

31. Baur, J.A.; Pearson, K.J.; Price, N.L.; Jamieson, H.A.; Lerin, C.; Kalra, A.; Prabhu, V.V.; Allard, J.S.; Lopez-Lluch, G.; Lewis, K.; et al. Resveratrol improves health and survival of mice on a high-calorie diet. *Nature* **2006**, *444*, 337–342. [CrossRef] [PubMed]

32. Shang, J.; Chen, L.L.; Xiao, F.X.; Sun, H.; Ding, H.C.; Xiao, H. Resveratrol improves non-alcoholic fatty liver disease by activating amp-activated protein kinase. *Acta Pharmacol. Sin.* **2008**, *29*, 698–706. [CrossRef] [PubMed]

33. Tauriainen, E.; Luostarinen, M.; Martonen, E.; Finckenberg, P.; Kovalainen, M.; Huotari, A.; Herzig, K.H.; Lecklin, A.; Mervaala, E. Distinct effects of calorie restriction and resveratrol on diet-induced obesity and fatty liver formation. *J. Nutr. Metab.* **2011**, *2011*, 525094. [CrossRef] [PubMed]

34. Poulsen, M.M.; Larsen, J.; Hamilton-Dutoit, S.; Clasen, B.F.; Jessen, N.; Paulsen, S.K.; Kjær, T.N.; Richelsen, B.; Pedersen, S.B. Resveratrol up-regulates hepatic uncoupling protein 2 and prevents development of nonalcoholic fatty liver disease in rats fed a high-fat diet. *Nutr. Res.* **2012**, *32*, 701–708. [CrossRef] [PubMed]

35. Alberdi, G.; Rodríguez, V.M.; Macarulla, M.T.; Miranda, J.; Churruca, I.; Portillo, M.P. Hepatic lipid metabolic pathways modified by resveratrol in rats fed an obesogenic diet. *Nutrition* **2013**, *29*, 562–567. [CrossRef] [PubMed]

36. Xin, P.; Han, H.; Gao, D.; Cui, W.; Yang, X.; Ying, C.; Sun, X.; Hao, L. Alleviative effects of resveratrol on nonalcoholic fatty liver disease are associated with up regulation of hepatic low density lipoprotein receptor and scavenger receptor class b type I gene expressions in rats. *Food Chem. Toxicol.* **2013**, *52*, 12–18. [CrossRef] [PubMed]

37. Andrade, J.M.; Frade, A.C.; Guimarães, J.B.; Freitas, K.M.; Lopes, M.T.; Guimarães, A.L.; de Paula, A.M.; Coimbra, C.C.; Santos, S.H. Resveratrol increases brown adipose tissue thermogenesis markers by increasing sirt1 and energy expenditure and decreasing fat accumulation in adipose tissue of mice fed a standard diet. *Eur. J. Nutr.* **2014**, *53*, 1503–1510. [CrossRef] [PubMed]

38. Choi, Y.J.; Suh, H.R.; Yoon, Y.; Lee, K.J.; Kim, D.G.; Kim, S.; Lee, B.H. Protective effect of resveratrol derivatives on high-fat diet induced fatty liver by activating amp-activated protein kinase. *Arch. Pharm. Res.* **2014**, *37*, 1169–1176. [CrossRef] [PubMed]

39. Pan, Q.R.; Ren, Y.L.; Liu, W.X.; Hu, Y.J.; Zheng, J.S.; Xu, Y.; Wang, G. Resveratrol prevents hepatic steatosis and endoplasmic reticulum stress and regulates the expression of genes involved in lipid metabolism, insulin resistance, and inflammation in rats. *Nutr. Res.* **2015**, *35*, 576–584. [CrossRef] [PubMed]

40. Nishikawa, K.; Iwaya, K.; Kinoshita, M.; Fujiwara, Y.; Akao, M.; Sonoda, M.; Thiruppathi, S.; Suzuki, T.; Hiroi, S.; Seki, S.; et al. Resveratrol increases cd68+ kupffer cells colocalized with adipose differentiation-related protein and ameliorates high-fat-diet-induced fatty liver in mice. *Mol. Nutr. Food Res.* **2015**, *59*, 1155–1170. [CrossRef] [PubMed]

41. Heebøll, S.; Kreuzfeldt, M.; Hamilton-Dutoit, S.; Kjær Poulsen, M.; Stødkilde-Jørgensen, H.; Møller, H.J.; Jessen, N.; Thorsen, K.; Kristina Hellberg, Y.; Bønløkke Pedersen, S.; et al. Placebo-controlled, randomised clinical trial: High-dose resveratrol treatment for non-alcoholic fatty liver disease. *Scand. J. Gastroenterol.* **2016**, *51*, 456–464. [CrossRef] [PubMed]

42. Lam, Y.Y.; Peterson, C.M.; Ravussin, E. Resveratrol vs. Calorie restriction: Data from rodents to humans. *Exp. Gerontol.* **2013**, *48*, 1018–1024. [CrossRef] [PubMed]

43. Boehm, O.; Zur, B.; Koch, A.; Tran, N.; Freyenhagen, R.; Hartmann, M.; Zacharowski, K. Clinical chemistry reference database for wistar rats and C57/BL6 mice. *Biol. Chem.* **2007**, *388*, 547–554. [CrossRef] [PubMed]

44. Fabbrini, E.; Sullivan, S.; Klein, S. Obesity and nonalcoholic fatty liver disease: Biochemical, metabolic, and clinical implications. *Hepatology* **2010**, *51*, 679–689. [CrossRef] [PubMed]

45. Den Boer, M.; Voshol, P.J.; Kuipers, F.; Havekes, L.M.; Romijn, J.A. Hepatic steatosis: A mediator of the metabolic syndrome. Lessons from animal models. *Arterioscler. Thromb. Vasc. Biol.* **2004**, *24*, 644–649. [CrossRef] [PubMed]

46. Doege, H.; Grimm, D.; Falcon, A.; Tsang, B.; Storm, T.A.; Xu, H.; Ortegon, A.M.; Kazantzis, M.; Kay, M.A.; Stahl, A. Silencing of hepatic fatty acid transporter protein 5 in vivo reverses diet-induced non-alcoholic fatty liver disease and improves hyperglycemia. *J. Biol. Chem.* **2008**, *283*, 22186–22192. [CrossRef] [PubMed]

47. Mitsuyoshi, H.; Yasui, K.; Harano, Y.; Endo, M.; Tsuji, K.; Minami, M.; Itoh, Y.; Okanoue, T.; Yoshikawa, T. Analysis of hepatic genes involved in the metabolism of fatty acids and iron in nonalcoholic fatty liver disease. *Hepatol. Res.* **2009**, *39*, 366–373. [CrossRef] [PubMed]

48. Cantó, C.; Auwerx, J. Pgc-1alpha, sirt1 and ampk, an energy sensing network that controls energy expenditure. *Curr. Opin. Lipidol.* **2009**, *20*, 98–105. [CrossRef] [PubMed]

49. Lagouge, M.; Argmann, C.; Gerhart-Hines, Z.; Meziane, H.; Lerin, C.; Daussin, F.; Messadeq, N.; Milne, J.; Lambert, P.; Elliott, P.; et al. Resveratrol improves mitochondrial function and protects against metabolic disease by activating sirt1 and pgc-1alpha. *Cell* **2006**, *127*, 1109–1122. [CrossRef] [PubMed]

50. Seip, R.L.; Semenkovich, C.F. Skeletal muscle lipoprotein lipase: Molecular regulation and physiological effects in relation to exercise. *Exerc. Sport Sci. Rev.* **1998**, *26*, 191–218. [CrossRef] [PubMed]

51. Hildebrandt, A.L.; Pilegaard, H.; Neufer, P.D. Differential transcriptional activation of select metabolic genes in response to variations in exercise intensity and duration. *Am. J. Physiol. Endocrinol. Metab.* **2003**, *285*, E1021–E1027. [CrossRef] [PubMed]

52. Jin, X.; Xiang, Z.; Chen, Y.P.; Ma, K.F.; Ye, Y.F.; Li, Y.M. Uncoupling protein and nonalcoholic fatty liver disease. *Chin. Med. J.* **2013**, *126*, 3151–3155. [PubMed]

53. Chavin, K.D.; Yang, S.; Lin, H.Z.; Chatham, J.; Chacko, V.P.; Hoek, J.B.; Walajtys-Rode, E.; Rashid, A.; Chen, C.H.; Huang, C.C.; et al. Obesity induces expression of uncoupling protein-2 in hepatocytes and promotes liver atp depletion. *J. Biol. Chem.* **1999**, *274*, 5692–5700. [CrossRef] [PubMed]

54. Rashid, A.; Wu, T.C.; Huang, C.C.; Chen, C.H.; Lin, H.Z.; Yang, S.Q.; Lee, F.Y.; Diehl, A.M. Mitochondrial proteins that regulate apoptosis and necrosis are induced in mouse fatty liver. *Hepatology* **1999**, *29*, 1131–1138. [CrossRef] [PubMed]

55. Baffy, G.; Zhang, C.Y.; Glickman, J.N.; Lowell, B.B. Obesity-related fatty liver is unchanged in mice deficient for mitochondrial uncoupling protein 2. *Hepatology* **2002**, *35*, 753–761. [CrossRef] [PubMed]

56. Stärkel, P.; Sempoux, C.; Leclercq, I.; Herin, M.; Deby, C.; Desager, J.P.; Horsmans, Y. Oxidative stress, klf6 and transforming growth factor-beta up-regulation differentiate non-alcoholic steatohepatitis progressing to fibrosis from uncomplicated steatosis in rats. *J. Hepatol.* **2003**, *39*, 538–546. [CrossRef]

57. Sheets, A.R.; Fülöp, P.; Derdák, Z.; Kassai, A.; Sabo, E.; Mark, N.M.; Paragh, G.; Wands, J.R.; Baffy, G. Uncoupling protein-2 modulates the lipid metabolic response to fasting in mice. *Am. J. Physiol. Gastrointest. Liver Physiol.* **2008**, *294*, G1017–G1024. [CrossRef] [PubMed]

58. Ajmo, J.M.; Liang, X.; Rogers, C.Q.; Pennock, B.; You, M. Resveratrol alleviates alcoholic fatty liver in mice. *Am. J. Physiol. Gastrointest. Liver Physiol.* **2008**, *295*, G833–G842. [CrossRef] [PubMed]

59. Marchal, J.; Blanc, S.; Epelbaum, J.; Aujard, F.; Pifferi, F. Effects of chronic calorie restriction or dietary resveratrol supplementation on insulin sensitivity markers in a primate, microcebus murinus. *PLoS ONE* **2012**, *7*, e34289. [CrossRef] [PubMed]

60. Barger, J.L. An adipocentric perspective of resveratrol as a calorie restriction mimetic. *Ann. N. Y. Acad. Sci.* **2013**, *1290*, 122–129. [CrossRef] [PubMed]

61. Nisoli, E.; Tonello, C.; Cardile, A.; Cozzi, V.; Bracale, R.; Tedesco, L.; Falcone, S.; Valerio, A.; Cantoni, O.; Clementi, E.; et al. Calorie restriction promotes mitochondrial biogenesis by inducing the expression of enos. *Science* **2005**, *310*, 314–317. [CrossRef] [PubMed]

62. Hancock, C.R.; Han, D.H.; Higashida, K.; Kim, S.H.; Holloszy, J.O. Does calorie restriction induce mitochondrial biogenesis? A reevaluation. *FASEB J. Fed. Am. Soc. Exp. Biol.* **2011**, *25*, 785–791. [CrossRef] [PubMed]

63. Wang, H.; Eckel, R. Lipoprotein lipase: From gene to obesity. *Am. J. Physiol. Endocrinol. Metab.* **2009**, *297*, E271–E288. [CrossRef] [PubMed]

64. Alberdi, G.; Macarulla, M.T.; Portillo, M.P.; Rodríguez, V.M. Resveratrol does not increase body fat loss induced by energy restriction. *J. Physiol. Biochem.* **2014**, *70*, 639–646. [CrossRef] [PubMed]

nutrients

MDPI

Article

Quercetin and Green Tea Extract Supplementation Downregulates Genes Related to Tissue Inflammatory Responses to a 12-Week High Fat-Diet in Mice

Lynn Cialdella-Kam [1], Sujoy Ghosh [2] , Mary Pat Meaney [3], Amy M. Knab [4], R. Andrew Shanely [5] and David C. Nieman [6,*]

[1] Department of Nutrition, School of Medicine—WG 48, Case Western Reserve University, 10900 Euclid Avenue, Cleveland, OH 44106, USA; lynn.kam@case.edu
[2] Program in Cardiovascular & Metabolic Diseases and Center for Computational Biology, Duke NUS Medical School, 8 College Road, Singapore 169857, Singapore; sujoy.ghosh@duke-nus.edu.sg
[3] Department of Exercise Physiology, School of Health Sciences, Winston-Salem State University, 601 S. Martin Luther King Jr. Drive, Winston-Salem, NC 27110, USA; meaneyMP@wssu.edu
[4] Levine Center for Health and Wellness, Queens University of Charlotte, 1900 Selwyn Avenue, Charlotte, NC 28274, USA; knaba@queens.edu
[5] Department of Health & Exercise Science, Appalachian State University, ASU Box 32071, 111 Rivers Street, 050 Convocation Center, Boone, NC 28608, USA; shanelyra@appstate.edu
[6] Human Performance Laboratory, North Carolina Research Campus, Appalachian State University, 600 Laureate Way, Kannapolis, NC 28081, USA
* Correspondence: niemandc@appstate.edu; Tel.: +1-828-773-0056; Fax: +1-704-250-5409

Received: 19 June 2017; Accepted: 13 July 2017; Published: 19 July 2017

Abstract: Quercetin (Q) and green tea extract (E) are reported to counter insulin resistance and inflammation and favorably alter fat metabolism. We investigated whether a mixture of E + Q (EQ) could synergistically influence metabolic and inflammation endpoints in a high-fat diet (HFD) fed to mice. Male C57BL/6 mice ($n = 40$) were put on HFD (fat = 60%kcal) for 12 weeks and randomly assigned to Q (25 mg/kg of body weight (BW)/day), E (3 mg of epigallocatechin gallate/kg BW/day), EQ, or control groups for four weeks. At 16 weeks, insulin sensitivity was measured via the glucose tolerance test (GTT), followed by area-under-the-curve (AUC) estimations. Plasma cytokines and quercetin were also measured, along with whole genome transcriptome analysis and real-time polymerase chain reaction (qPCR) on adipose, liver, and skeletal muscle tissues. Univariate analyses were conducted via analysis of variance (ANOVA), and whole-genome expression profiles were examined via gene set enrichment. At 16 weeks, plasma quercetin levels were higher in Q and EQ groups vs. the control and E groups ($p < 0.05$). Plasma cytokines were similar among groups ($p > 0.05$). AUC estimations for GTT was 14% lower for Q vs. E ($p = 0.0311$), but non-significant from control ($p = 0.0809$). Genes for cholesterol metabolism and immune and inflammatory response were downregulated in Q and EQ groups vs. control in adipose tissue and soleus muscle tissue. These data support an anti-inflammatory role for Q and EQ, a result best captured when measured with tissue gene downregulation in comparison to changes in plasma cytokine levels.

Keywords: cytokines; fat metabolism; flavonoids; inflammation; insulin resistance; immune function; obesity; metabolic syndrome; phytochemicals

---

1. Introduction

High-fat Western diets are associated with insulin resistance, inflammation, and de novo lipogenesis [1,2], which are factors that contribute to the development of metabolic syndrome. Flavonoid ingestion has the potential to partially offset these effects. In particular, quercetin and epigallocatechin gallate (EGCG) from green tea have been reported to attenuate insulin resistance, counter inflammation, and favorably alter fat metabolism [2–5]. However, the effect of a mixture of quercetin and EGCG has been examined in only a few studies.

Quercetin is a flavonoid that is found in many plant and foods such as onions, green tea, apples, peppers, and berries [6]. Both in vitro and rodent models provide evidence that quercetin supplementation reduces various measures related to metabolic syndrome [2,3,7]. Specifically, quercetin has been reported to blunt pro-inflammatory signaling via regulation of NF-κβ-associated mechanisms in adipocytes, macrophages, and other cell lines [8–13], decrease insulin intolerance in primary human adipocytes and 3T3-L1 cells [8,14], and inhibit adipogenesis in 3T3-L1 cells [14–16] and lipid body formation in macrophages [17]. In rodents, quercetin has been reported to lower levels of circulating inflammatory-related plasma cytokines [18], inhibit pro-inflammatory signals [11,19–21], and improve insulin sensitivity [20–27] and dyslipidemia [20,21,24,26–28]. Very few human studies have examined the relationship between quercetin supplementation and metabolic syndrome risk factors in overweight adults. In a double-blinded, placebo-controlled study, Egert et al. [29] reported that six weeks of supplementation of quercetin at 150 mg/day reduced systolic blood pressure and plasma oxidized low-density lipoprotein (LDL) concentrations in overweight adults (n = 93; mean age = 45.1 years), but had no effect on inflammation. However, the effect of quercetin supplementation on lipid markers appears to vary based on apolipoprotein (APOE) genotype. Similarly, six weeks of onion-extract supplementation (quercetin of 162 mg/day) was associated with a reduction in 24-h ambulatory blood pressure in overweight/obese adults (n = 68, mean age = 47.4 years) with central obesity and pre-hypertension [30]. However, quercetin supplementation had no impact on endothelial function, inflammation, oxidative stress, and lipid and glucose metabolism in these individuals [30]. In large community studies including both normal weight and overweight female adults, quercetin supplementation at 500 mg/day or 1000 mg/day for 12 weeks was reported to have no influence on innate immune function or inflammation [31], body composition [32], or disease risk factors [33]. Quercetin supplementation was, however, associated with a reduction in the severity and number of sick days associated with upper respiratory tract infections (URTI) in adults [34]. To our knowledge, only two studies have examined the influence of quercetin supplementation on insulin sensitivity. In one study, a 17.5% improvement in the homeostatic model assessment of insulin resistance (HOMA-IR) was reported in women with polycystic ovary syndrome (PCOS; n = 82, age = ~30 years) after 12 weeks of quercetin supplementation (1000 mg/day) [35]. In contrast, four weeks of quercetin supplementation (500 mg/day) had no impact on fasting blood glucose levels in healthy males (n = 22, age = 29.9 years) [36].

EGCG, a catechin, is the most abundant flavonoid found in green tea [6] and has been reported to have anti-obesity, anti-diabetic, and anti-inflammatory properties [2,3,37]. Notably, in vitro studies indicate that EGCG suppressed insulin resistance [38,39] and promoted glucose uptake via enhanced GLUT4 translocation [39,40] in skeletal muscle cells, attenuated β-cell release of insulin from mouse and human islet cells [39], and improved insulin sensitivity in human hepatocytes (HepG2 cells) [41]. Furthermore, EGCG was associated with decreased glucose uptake [42], lipid accumulation [43–45], adipogenesis [46], and adipocyte differentiation [44] in 3T3-L1 adipocytes, and reduced inflammation by reactive oxygen species generation in macrophages [47]. In rodents, EGCG and green tea extract have been shown in most studies to reduce total body and adipose tissue weights [37,48,49], decrease blood/plasma glucose and insulin levels [37,48,50], improve insulin sensitivity [37,48], blood pressure, and lipid profile [37,48,51], and reduce unfavorable obesity-associated changes in gut microbiota [52]. Epidemiological research and meta-analyses in general support the anti-obesity and health effects of EGCG [53]. In randomized controlled studies in humans, three studies found a small but significant decrease in body weight, waist circumference, and body fat with green tea supplementation [54–56], while two studies found no change [57,58]. Several meta-analyses of randomized controlled trials with green tea indicate a possible reduction in blood pressure [59–61], total and low-density lipoprotein cholesterol [60,62,63], and fasting blood glucose and insulin insensitivity [64].

Given the independent effects of quercetin and EGCG on metabolic syndrome, we aimed to elucidate whether the combined effort of quercetin and green tea extract supplementation would improve blood glucose tolerance, decrease inflammation, and favorably alter metabolism in mice fed a high-fat diet. Previous studies by our research group suggest that ingestion of both quercetin and EGCG-enriched green tea extract have a greater anti-inflammatory effect than quercetin alone [65–68]. We utilized whole genome transcriptome and real-time polymerase chain reaction (qPCR) analysis of adipose, liver, and skeletal muscle tissues in mice fed high-fat diets to improve our ability to measure potential metabolic and anti-inflammatory effects related to flavonoid ingestion.

2. Materials and Methods

2.1. Animals and Experimental Design

Forty C57BL/6 mice (male, 5 weeks old, $n = 44$), purchased from a commercial vendor (Jackson Laboratory, Bar Harbor, ME, USA), were provided ad libitum access to a high-fat diet (HFD, fat = 60% kcal; BioServ, Frenchtown, NJ, USA) and water and maintained in 12 h light/dark cycle for the first 12 weeks at the animal facility of the North Carolina Research Campus. The experimental design is depicted in Figure 1. After 12 weeks on HFD, the four mice with the least weight gain were excluded from the second phase of the study, and the remaining mice ($n = 40$) were randomly assigned to one of four treatment groups ($n = 10$ per group): quercetin only (Q, 25 mg/kg of body weight (BW)/day of quercetin), green tea extract only (E; 3 mg/kg BW/day of EGCG), quercetin and green tea extract (EQ; 25 mg/kg BW of quercetin plus 3 mg/kg of EGCG), or control. All mice were maintained on HFD and with the exception of the control group were also supplemented with Q, E, or both for four weeks. Body weight was monitored weekly. At 16 weeks, mice underwent a glucose tolerance test and then were sacrificed. Tissue and plasma samples were collected for further analysis (Figure 1). All protocols utilized were approved by The Institutional Animal Care and Use Committee (IACUC) of the North Carolina Research Campus.

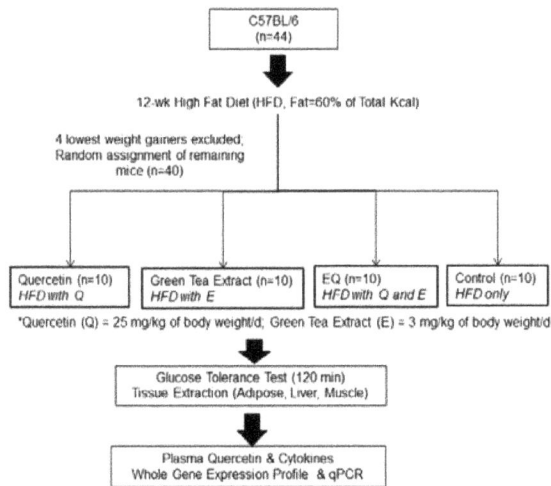

Figure 1. Study Design: C57BL/6 mice ($n = 40$) were placed on a high-fat diet (fat = 60% of total kcal) for 12 weeks and then randomly assigned to a diet supplemented with quercetin only (Q), green tea extract only (E), quercetin + green tea extract (EQ), or control (i.e., high fat diet only) for four weeks. The quercetin dosage was 25 mg of quercetin/kg of body weight (BW) per day, and green tea extract dosage was 3 mg of epigallocatechin gallate/kg BW per day.

2.2. Glucose Tolerance Test and Blood and Tissue Collection

Following the four-week treatment period, mice fasted for 14 h and then were anesthetized and placed on a warming blanket. Next, mice were injected intraperitoneally with 2 g of glucose/kg BW. Blood (~3 μL) was collected from the tail vein, and blood glucose levels were measured at 0, 15, 30, 60 and 120 min using OneTouch Ultra® blood glucometer (LifeScan, Johnson & Johnson, Chesterbrook, PA, USA).

Upon completion of the glucose tolerance test, mice were sacrificed, and whole blood was collected by cardiac puncture and centrifuged at $1000 \times g$ for 10 min at 4 °C. Plasma samples were aliquoted, snap frozen in liquid nitrogen, and stored at −80 °C for later analysis. The following tissue was harvested from the mice: left lobes of kidney and liver, pancreas, visceral adipose, subcutaneous adipose, and skeletal muscle tissue (soleus, gastrocnemius, plantaris, EDL, and quadriceps). All tissue was weighed. Tissue was either stored in RNAlater™ (ThermoFischer Scientific, Waltham, MA, USA) per manufacturer's instructions for genomics or frozen in liquid nitrogen and stored at −80 °C for later analysis.

2.3. Biochemical Assays

Plasma samples were pooled to assess quercetin, which was measured following solid-phase extraction via reversed-phase high-performance liquid chromatography with UV detection as previously described [65–68]. Plasma cytokines (IFN-γ, IL-1β, IL-6, IL-10, KC/GRO/CINC, and TNF-α) were measured using Mouse ProInflammatory 7-Plex Base Kit (Meso Scale Discovery, Rockville, MD, USA) per manufacturer's instructions.

2.4. Genomic Analysis

Whole genome expression profiling was conducted with total RNA isolated from adipose, liver and skeletal muscle from mice in the Q, EQ and control groups. RNA was isolated and quantified, and quality control (QC) was performed on all samples. Expression profiling was performed on Mouse ST 1.1 PEG array (Affymetrix, ThermoFischer Scientific, Waltham, MA, USA) as per the manufacturer's instructions. Signal extraction and background was subtracted for normalization utilizing Robust Multichip Average [69]. Samples that were considered outliers were excluded based on the QC report and scatter plots. Both the mean signal per treatment group and fold-change (log ratio) were calculated. CyberT was used to identify differentially expressed genes [70]. Pathways affected by each treatment relative to the control was determined using overrepresentation analysis via Ingenuity Pathway Analysis (IPA) software (Qiagen, Redwood City, CA, USA).

To quantify the expression of individual genes ($n = 27$), qPCR was performed in tissue samples from fat, liver, and soleus for the four experimental groups using Applied Biosystems™ TaqMan® Gene Expression Assays (ThermoFischer Scientific, Waltham, MA, USA) as per the manufacturer's instructions. Genes examined include those involved in cholesterol regulation (Abca1, Apoa1, Cyp3a41a, Srebf1, and Srebf2), fatty acid metabolism (Lpl, Ppara, Pparag. and Scd1), inflammatory and immune response (Cc12, Cd68, Ikbkb, Il1r1, Nfkb1, and Nr1h3), adipokines (Adipoq and Lep), oxidative stress (Ppargc1a), stress response (Hspa1a, Hspa2, Mapk8, and Sirt1), transcription (Atf2 and Nfact3), and xenobiotics (Cyp2e1).

2.5. Statistical Analysis

Data was summarized using means and standard error. To detect significant differences between groups, one-way ANOVA (time × treatment) was used for blood analysis and gene expression. Whole-genome expression profiles were examined via gene-set enrichment analysis (GSEA) [71]. A *p*-value was set at <0.05 for significance. Analysis was conducted using SAS 9.3 (SAS Institute, Cary, NC, USA).

3. Results

3.1. Body Mass and Biochemical Analysis

At the beginning of the study, the body mass for all mice was 20.0 ± 0.0 g with no differences among groups ($p > 0.05$). Body mass was also similar among groups at 12 weeks ($Q = 47.3 \pm 0.7$ g, $E = 47.1 \pm 0.8$ g, $EQ = 47.1 \pm 0.8$ g, and control $= 47.1 \pm 1.0$ g; $p > 0.05$) and at 16 weeks (i.e., after four weeks of supplementation ($Q = 51.1 \pm 0.6$ g, $E = 50.6 \pm 0.8$ g, $EQ = 50.5 \pm 0.5$ g, and control $= 50.2 \pm 0.7$ g; $p > 0.05$). At 16 weeks, pooled plasma quercetin levels were ~fivefold higher in Q and twofold higher in the EQ group compared to the control group (Figure 2). Glucose tolerance test (GTT) results are presented in Figure 2. Area-under-the-curve (AUC) estimations for plasma glucose were 14% lower for Q vs. EQ ($p = 0.031$) and trended 11% lower than control, but did not reach significance ($p = 0.081$). Plasma glucose was lower for Q vs. control at 60 min ($p = 0.032$; Figure 3). No other differences among groups were detected ($p > 0.05$; Figure 3). Plasma cytokines levels were also similar among groups ($p > 0.05$, Figure 4).

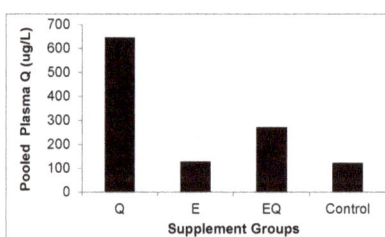

Figure 2. Pooled plasma quercetin at 16 weeks by experimental groups. C57BL/6 mice ($n = 40$) were placed on a high-fat diet (fat = 60% of total kcal) for 12 weeks and then randomly assigned to a diet supplemented with quercetin only (Q), green tea extract only (E), quercetin + green tea extract (EQ), or control (i.e., high-fat diet only) for four weeks. The dosage for quercetin was 25 mg of quercetin/kg of body weight (BW) per day and green tea extract dosage was 3 mg of epigallocatechin gallate/kg BW per day. Plasma samples were pooled for each group and analyzed for quercetin. At 16 weeks, plasma quercetin levels were 525% higher in Q, and 225% higher in EQ compared to control.

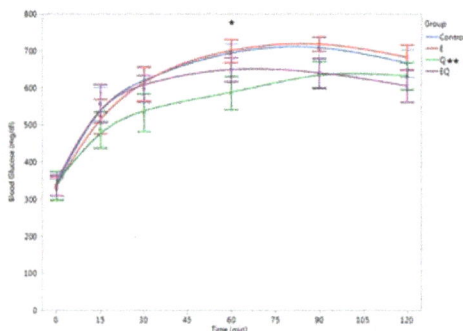

Figure 3. Glucose tolerance curve at 16 weeks by supplement groups. C57BL/6 mice ($n = 40$) were placed on a high-fat diet (fat = 60% of total kcal) for 12 weeks and then randomly assigned to a diet supplemented with quercetin only (Q), green tea extract only (E), quercetin + green tea extract (EQ), or control (i.e., high-fat diet only) for four weeks. The dosage for quercetin was 25 mg of quercetin/kg of body weight (BW) per day and green tea extract dosage was 3 mg of epigallocatechin gallate/kg BW per day. * Q lower than control at 60-min ($p < 0.05$). ** Area-under-the-curve (AUC) estimations lower for Q vs. EQ ($p < 0.05$).

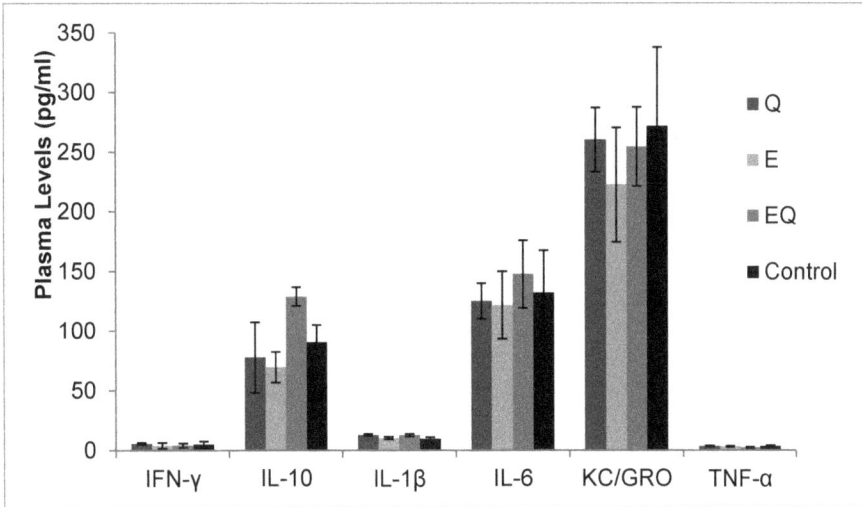

Figure 4. Plasma cytokine levels at 16 weeks by supplement groups. C57BL/6 mice (*n* = 40) were placed on a high-fat diet (fat = 60% of total kcal) for 12 weeks and then randomly assigned to a diet supplemented with quercetin only (Q), green tea extract only (E), quercetin + green tea extract (EQ), or control (i.e., high fat diet only) for four weeks. The dosage for quercetin was 25 mg of quercetin/kg of body weight (BW) per day and green tea extract dosage was 3 mg of epigallocatechin gallate/kg BW per day. Plasma cytokine levels did not differ between supplement groups and control ($p > 0.05$).

3.2. Genomic Analysis

Both microarray and IPA analysis revealed downregulation of genes associated with cholesterol metabolism and immune/inflammation in adipose tissue and soleus muscle tissue, fatty acid metabolism in soleus muscle tissue, and CYP450 metabolism in the liver. EQ resulted in downregulation of over 100 genes in adipose tissue compared to both control and quercetin alone ($p < 0.01$; Figure 5). The specific pathways downregulated by EQ and Q are depicted in Table 1. In skeletal muscle, protein ubiquination, the pathway responsible for marking proteins for degradation, was upregulated by Q treatment relative to the control.

In Table 2, gene expression changes are presented related to the plasma cytokines assessed. Of these, KC/GRO (i.e., Cxcl1) gene was expressed in adipose tissue and liver with an upregulation of KC/GRO detected in the liver of the EQ group compared to control (Table 2). The Il-1β gene was also expressed in the liver, but no difference was found among treatments (Table 2). Q was associated with the downregulation of the Il-1β receptor gene in adipose (Table 2) vs. control, and a downregulation trend was observed for other cytokine receptors genes in adipose tissue and soleus muscle tissue ($p > 0.05$, Table 1). For the EQ treatment, the IL-10 receptor gene was downregulated while the TNF-α receptor gene was upregulated in comparison to the control (Table 2). No differences were detected between Q and EQ groups for the genes presented in Table 2.

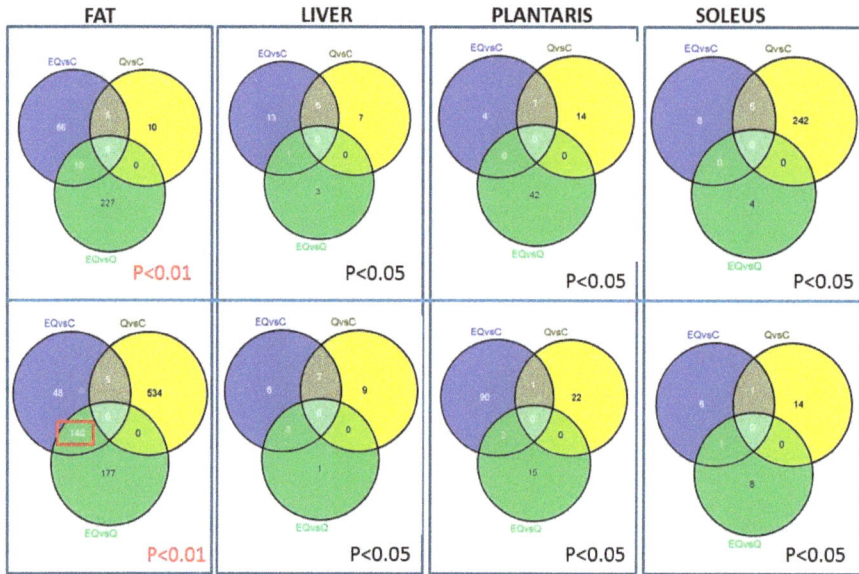

Figure 5. Overlap among differential expressed genes by tissue. Top panel shows the number of downregulated genes and the bottom panel shows upregulated genes. C57BL/6 mice (*n* = 40) were placed on a high-fat diet (fat = 60% of total kcal) for 12 weeks and then randomly assigned to a diet supplemented with quercetin only (Q), green tea extract only (E), quercetin + green tea extract (EQ), or control (i.e., high fat diet only) for four weeks. The dosage for quercetin was 25 mg of quercetin/kg of body weight (BW) per day, and green tea extract dosage was 3 mg of epigallocatechin gallate/kg BW per day. EQ treatment result in the upregulation of 140 genes compared to the control and Q groups.

Table 1. Top canonical pathways altered by four-week supplementation vs. control as identified by Ingenuity Pathway Analysis (IPA) [1].

Downregulated Pathways	Fat	Liver	Muscle	Comments
Steroid Biosynthesis		Q; EQ		Target of Statins
Phagocytosis/ leukocyte extravasation	EQ			Innate Immune Response
EIF2 signaling			Q; EQ	Stress Response
Mitochondrial dysfunction			Q	Associated with disease
eIF4/p70S6K signaling			Q; EQ	Insulin Signaling
Oxidative phosphorylation			Q; EQ	Energy Production
PPARα/RXRα activation			Q	Gene Expression

[1] C57BL/6 mice (*n* = 40) were placed on a high-fat diet (fat = 60% of total kcal) for 12 weeks and then randomly assigned to a diet supplemented with quercetin only (Q), green tea extract only (E), quercetin + green tea extract (EQ), or control (i.e., high fat diet only) for four weeks. The dosage for quercetin was 25 mg of quercetin/kg of body weight (BW) per day and green tea extract dosage was 3 mg of epigallocatechin gallate/kg BW per day. IPA analysis was only conducted on tissue collected from the EQ, Q and control groups.

Table 2. Fold change in genes associated with cytokines assessed in plasma vs. control, based on microarray analysis [1].

Description	Q Change	p	EQ Change	p	Description
Adipose					
Ifngr1	−0.28	0.340	−0.18	0.659	interferon gamma receptor 1
Ifngr2	−0.14	0.820	−0.54	0.090	interferon gamma receptor 2
Il10ra	−0.36	0.550	−0.63	0.194	interleukin 10 receptor, alpha
Il10rb	−0.29	0.550	−0.73	0.047	interleukin 10 receptor, beta
Il1r1	−060	0.037	−0.53	0.087	interleukin 1 receptor, type I
Il1rap	−0.10	0.908	0.10	0.916	interleukin 1 receptor accessory protein
Il1rn	−0.31	0.783	−1.22	0.052	interleukin 1 receptor antagonist
Il6ra	−0.29	0.466	0.32	0.428	interleukin 6 receptor, alpha
Il6st	−0.18	0.645	−0.04	0.977	interleukin 6 signal transducer
Cxcl1	−0.17	0.679	−0.11	0.873	chemokine (C-X-C motif) ligand 1
Tnfrsf1a	−0.47	0.196	0.99	0.188	tumor necrosis factor receptor superfamily, member 1a
Tnfrsf1b	−0.33	0.565	0.20	0.046	tumor necrosis factor receptor superfamily, member 1b
Soleus					
Ifngr1	−0.27	0.423	−0.15	0.775	interferon gamma receptor 1
Il10rb	−0.10	0.876	0.10	0.879	interleukin 10 receptor, beta
Il6ra	−0.18	0.692	−0.12	0.845	interleukin 6 receptor, alpha
Il6st	−0.31	0.151	−0.16	0.602	interleukin 6 signal transducer
Tnfrsf1a	−0.42	0.162	0.21	0.968	tumor necrosis factor receptor superfamily, member 1a
Liver					
Ifngr1	−0.07	0.935	−0.13	0.805	interferon gamma receptor 1
Ifngr2	0.15	0.678	−0.05	0.961	interferon gamma receptor 2
Il10rb	0.00	0.999	−0.06	0.964	interleukin 10 receptor, beta
Il1b	0.07	0.985	0.07	0.986	interleukin 1 beta
Il1r1	0.31	0.799	0.79	0.243	interleukin 1 receptor, type I
Il1rap	−0.01	0.998	0.00	0.999	interleukin 1 receptor accessory protein
Il1rn	0.06	0.971	0.32	0.413	interleukin 1 receptor antagonist
Il6ra	−0.24	0.550	0.55	0.055	interleukin 6 receptor, alpha
Il6st	−0.07	0.929	0.12	0.801	interleukin 6 signal transducer
Cxcl1	0.72	0.222	1.15	0.030	chemokine (C-X-C motif) ligand 1
Tnfrsf1a	−0.42	0.162	0.21	0.968	tumor necrosis factor receptor superfamily, member 1a

[1] C57BL/6 mice ($n = 40$) were placed on a high-fat diet (fat = 60% of total kcal) for 12 weeks and then randomly assigned to a diet supplemented with quercetin only (Q), green tea extract only (E), quercetin + green tea extract (EQ), or control (i.e., high fat diet only) for four weeks. The dosage for quercetin was 25 mg of quercetin/kg of body weight (BW) per day and green tea extract dosage was 3 mg of epigallocatechin gallate/kg BW per day. Individual genes ($n = 27$) were assessed in soleus, liver, and fat. * Significantly different than the control group ($p < 0.05$).

Of the 27 individual genes evaluated in adipose, soleus, and liver via qPCR, Q was associated with downregulation of three genes in adipose tissue, and no gene changes in the soleus or liver tissue compared to the control group (Table 3). In the soleus tissue, EQ and Q were associated with the downregulation of genes (4 and 2 genes, respectively) in the soleus vs. control with no other changes observed in adipose or liver tissue (Table 3).

Table 3. Fold change in genes downregulated in adipose and soleus tissue compared to control by supplement groups as assessed via real-time quantitative polymerase chain reaction (qPCR) analysis.

Description	Q	E	EQ	Pathways
Adipose				
Srebf2	0.44 *	0.57	0.82	Sterol biosynthesis
Atf2	0.51 *	0.83	1.10	Transcriptional activator
Sirt1	0.40 *	0.92	0.74	Stress response
Soleus				
Srebf2	0.60	0.54 *	0.62 *	Sterol biosynthesis
Pparag	1.04	0.71	0.69 *	Fatty acid storage and Glucose metabolism
Scd1	0.44	0.97	0.40 *	Fatty Acid metabolism
Cd68	0.88	0.71	0.57 *	Promote phagocytosis and activation of macrophages
Atf2	1.23	0.62 *	0.85	Transcriptional activator

[1] C57BL/6 mice ($n = 40$) were placed on a high-fat diet (fat = 60% of total kcal) for 12-weeks and then randomly assigned to a diet supplemented with quercetin only (Q), green tea extract only (E), quercetin + green tea extract (EQ), or control (i.e., high fat diet only) for four weeks. The dosage for quercetin was 25 mg of quercetin/kg of body weight (BW) per day and green tea extract dosage was 3 mg of epigallocatechin gallate/kg BW per day. Individual genes ($n = 27$) were assessed in soleus, liver, and fat tissue. * Significantly different than the control group ($p < 0.05$).

4. Discussion

In mice on a 12-week HFD, four weeks of EQ supplementation were associated with the downregulation of over 100 genes in adipose tissue, including those involved in phagocytosis and leukocyte extravasation or trafficking pathways. Recruitment of leukocytes, specifically neutrophils, to adipose has been implicated in chronic inflammation in adipose tissue [72,73] and has been linked to insulin resistance in mice on HFD [73]. Traditional biomarkers for inflammation and glucose tolerance, however, were not different between EQ and control groups, but a mild improvement in blood glucose tolerance was detected with the Q treatment. In adipose and muscle tissue, EQ was associated with a downregulation of cholesterol metabolism compared to control. Cholesterol accumulation in adipose and muscle tissue have been associated with obesity and sarcopenia [74,75]. Genes associated with drug metabolism were also downregulated in EQ vs. control in the liver. The implications, however, are unclear, as changes in drug metabolism vary by metabolic and excretion pathways in obese individuals [76]. Thus, four weeks of EQ supplementation in mice on a 12-week HFD resulted in changes in tissue gene expression suggestive of reduced inflammation and cholesterol metabolism, while blood markers of glucose tolerance and inflammation were largely unaltered.

In the EQ group, the changes in tissue gene expression are indicative of reduced inflammation and leukocyte trafficking, which has been examined as a treatment target for inflammatory diseases [77]. Cytokine levels in the present study were not different among the experimental groups. Our findings in mice (~age in human = 50 years) [78] parallel previous studies in middle-aged humans. In overweight and obese women ($n = 48$, age = 56 years), 10 weeks of supplementation with mixed flavonoid-nutrient-fish oil supplement (Q-mix; 1000 mg quercetin, 400 mg isoquercetin, 120 mg EGCG, 220 mg EPA, and 180 mg DHA, 1000 mg vitamin C, 40 mg niacinamide, and 800 µg folic acid) did not alter biomarkers of inflammation, oxidative stress, and blood lipid levels, but was associated with gene alterations suggestive of enhanced antiviral defense and decreased leukocyte trafficking [79]. Similarly, in a randomized, double-blinded, crossover study in overweight men ($n = 26$, age = 46 years), Bakker et al. [78] reported no change in traditional biomarkers, but did report a shift in nutrigenomic profiles, which was associated with a reduction in inflammation after a five-week, anti-inflammatory dietary mix supplementation (AIDM, 6.3 mg resveratrol, 3.75 mg lycopene, ~38 mg EGCG, 300 mg EPA, 260 mg DHA, 125 mg vitamin C, and 90.7 mg α-tocopherol) vs. placebo [80]. The relative dose of both quercetin and ECGC was higher in the present study compared to the human studies ($Q = 25$ mg/kg BW vs. AIDM = 0 mg/kg BW and Qmix = ~15 mg/kg BW; and $E = 3$ mg/kg BW vs. AIDM = ~0.4 mg/kg BW and Qmix = ~1.3 mg/kg BW). In addition, the respective duration of the supplementation was longer in the present study (i.e., four weeks of supplementation in mice is equivalent to ~10 years in human). Finally, the supplements in the human studies also contained fish oil, which has been associated with improvement in inflammatory biomarkers [81,82]. Taken together, these studies provide evidence that a mixture of flavonoids may be a promising treatment for reducing inflammation in overweight/obese individuals. Further research is needed to elucidate the optimal combination of flavonoids and/or whether the inclusion of fish oil in the supplementation provides additive benefits. Part of the challenge is the inclusion of novel outcome measures that capture metabolic and anti-inflammatory effects that are missed by basic plasma inflammation biomarkers [79]. The tissue-specific transcriptomic change observed in the present study may possibly reflect an earlier stage of tissue response to supplementation. Systemic changes may follow, and further research is thus warranted that examines a prolonged period of supplementation and/or supplementation at higher doses.

In the present study, plasma quercetin levels were lower in the EQ vs. Q group, despite the same dosage of quercetin being provided to both groups. Our findings are consistent with a mouse study conducted by Wang et al. [83], in which the authors reported that total quercetin levels in tissue were lower with the co-administration of EGCG and quercetin in mice. In the intestine, quercetin in humans is absorbed via passive diffusion as a primary route, and organic anion transporting polypeptide (OATPs) as a secondary route [84,85]. EGCG has been speculated to interfere with quercetin absorption via OATPs by acting as a non-competitive inhibitor or decreasing the activity of the transporter [83,86]. Given the high dose of Q administered in the present study, decreased absorption via OATPs could explain the lower plasma Q levels observed in the EQ group compared to Q and thus supporting the hypothesis of EGCG's interaction with this transporter.

A mild improvement in blood glucose tolerance was associated with quercetin alone in this study. As previously discussed, very few human studies [35,36] have examined the impact of quercetin on blood glucose control and insulin resistance. Mehta et al. [82] reported that male Swiss albino mice (age not reported) had less stress-induced hyperglycemia and insulin-resistance following three weeks of quercetin supplementation (30 mg/kg) vs. control. Henagan et al. [87] reported that eight weeks of a low dose of quercetin (~1.6 g/kg BW) compared to a placebo resulted in improved insulin tolerance in male C57BL/6J mice (~14 weeks of age at sacrifice) on HFD, while the high dose (20 g/kg BW) did not alter insulin tolerance. The mice in the present study were older and had a higher dosage of quercetin compared to Henagan et al. [87]. In the EQ group, blood glucose levels were similar to both the control and E groups, but were higher than the Q group. As discussed previously, plasma quercetin levels were lower in the EQ vs. Q, which was possibly related to the interference of EGCG on quercetin's absorption. Thus, the observed differences support quercetin's role in improving glucose tolerance. A potential limitation in the current study is that the glucose tolerance test was conducted while the mice were under general anesthesia, which may have caused stress-induced hyperglycemia in all groups [88]. In addition, blood glucose was measured utilizing a glucometer, which has been reported to overestimate blood glucose levels in hyperglycemic states [89]. Thus, the measured blood glucose may have been higher than actual levels. Furthermore, it is difficult to separate hyperglycemia caused by the stress of anesthesia vs. HFD, and this may have confounded the potential impact of flavonoid supplementation on HFD-induced hyperglycemia.

Finally, the metabolism of flavonoids in human and mice differ, and more research is needed to determine the applicability of our results to human populations. The agreement between our prior human trial and the current mouse-based study indicating a downregulation in expression of genes related to leukocyte trafficking following mixed flavonoid supplementation is one indicator of similar responses between species [79]. In humans, polyphenols are transformed into metabolites with diminished biological impact [90–92]. Unabsorbed polyphenols can undergo bacterial bioconversion by gut microbiota into more bioactive forms [90–92]. Mice have different species of gut bacteria compared to humans, which limits the applicability of this model [92]. Humanized mice models have been suggested that utilize human fecal microbiota transplants (FMT) in mice to create a similar gut microbiome [92]. In addition, mice expressing the human drug metabolizing enzymes, cytochromes P450, may also prove to be a useful animal model in examining flavonoids [93]. Despite these differences, the plasma and urine content of quercetin metabolites are similar between humans and rats in type and number [94], and the bioavailability of EGCG have been reported as similar between human and mice [95]. Taken together, future studies on polyphenol mixtures could compare humanized and standard mouse models (e.g., those with FMT) to determine both similarities and differences on metabolic and inflammation outcome measures.

5. Conclusions

Supplementation with EQ for four weeks in mice fed a high fat diet for 12 weeks was associated with tissue gene expression changes suggestive of reduced inflammation and diminished leukocyte cell trafficking, a result we have previously demonstrated in human participants [79]. Traditional inflammatory biomarkers and glucose tolerance were not altered by EQ, but a mild improvement in glucose tolerance was observed with Q only. Future research should consider comparing flavonoid biotransformation in humanized mouse models to standard mouse models. Furthermore, lower doses and different flavonoid mixtures should be examined in both sedentary and physically active rodent models.

Acknowledgments: This work was supported by funding from Quercegen Pharmaceuticals LLC, Marlborough, MA, USA.

Author Contributions: L.C.-K. wrote the first manuscript. L.C.-K., S.G. and M.P.M. implemented and conducted study and collected the data. L.C.-K., S.G. and D.C.N. analyzed data. L.C.-K., S.G., D.C.N., M.P.M., A.M.K. and R.A.S. interpreted the data and gave critical comments.

Conflicts of Interest: The authors declare no conflict of interest.

References

1. Lee, C.Y. The effect of high-fat diet-induced pathophysiological changes in the gut on obesity: What should be the ideal treatment? *Clin. Transl. Gastroenterol.* **2013**, *4*, e39. [CrossRef] [PubMed]
2. Cherniack, E.P. Polyphenols: Planting the seeds of treatment for the metabolic syndrome. *Nutrition* **2011**, *27*, 617–623. [CrossRef] [PubMed]
3. Siriwardhana, N.; Kalupahana, N.S.; Cekanova, M.; LeMieux, M.; Greer, B.; Moustaid-Moussa, N. Modulation of adipose tissue inflammation by bioactive food compounds. *J. Nutr. Biochem.* **2013**, *24*, 613–623. [CrossRef] [PubMed]
4. Lee, S.G.; Parks, J.S.; Kang, H.W. Quercetin, a functional compound of onion peel, remodels white adipocytes to brown-like adipocytes. *J. Nutr. Biochem.* **2017**, *42*, 62–71. [CrossRef] [PubMed]
5. Kim, H.M.; Kim, J. The effects of green tea on obesity and type 2 diabetes. *Diabetes Metab. J.* **2013**, *37*, 173–175. [CrossRef] [PubMed]
6. Bhagwat, S.; Haytowitz, D.B.; Holden, J.M. USDA Database for the Flavonoid Content of Selected Foods. Available online: http://www.ARS.USDA.Gov/nutrientdata/flav (accessed on 15 May 2017).
7. Li, Y.; Yao, J.; Han, C.; Yang, J.; Chaudhry, M.T.; Wang, S.; Liu, H.; Yin, Y. Quercetin, inflammation and immunity. *Nutrients* **2016**, *8*, 167. [CrossRef] [PubMed]
8. Chuang, C.C.; Martinez, K.; Xie, G.; Kennedy, A.; Bumrungpert, A.; Overman, A.; Jia, W.; McIntosh, M.K. Quercetin is equally or more effective than resveratrol in attenuating tumor necrosis factor-α-mediated inflammation and insulin resistance in primary human adipocytes. *Am. J. Clin. Nutr.* **2010**, *92*, 1511–1521. [CrossRef] [PubMed]
9. Overman, A.; Chuang, C.C.; McIntosh, M. Quercetin attenuates inflammation in human macrophages and adipocytes exposed to macrophage-conditioned media. *Int. J. Obes.* **2011**, *35*, 1165–1172. [CrossRef] [PubMed]

10. Comalada, M.; Ballester, I.; Bailon, E.; Sierra, S.; Xaus, J.; Galvez, J.; De Medina, F.S.; Zarzuelo, A. Inhibition of pro-inflammatory markers in primary bone marrow-derived mouse macrophages by naturally occurring flavonoids: Analysis of the structure-activity relationship. *Biochem. Pharmacol.* **2006**, *72*, 1010–1021. [CrossRef] [PubMed]
11. Dias, A.S.; Porawski, M.; Alonso, M.; Marroni, N.; Collado, P.S.; Gonzalez-Gallego, J. Quercetin decreases oxidative stress, NF-kβ activation, and iNOS overexpression in liver of streptozotocin-induced diabetic rats. *J. Nutr.* **2005**, *135*, 2299–2304. [PubMed]
12. Comalada, M.; Camuesco, D.; Sierra, S.; Ballester, I.; Xaus, J.; Galvez, J.; Zarzuelo, A. In vivo quercitrin anti-inflammatory effect involves release of quercetin, which inhibits inflammation through down-regulation of the NF-kβ pathway. *Eur. J. Immunol.* **2005**, *35*, 584–592. [CrossRef] [PubMed]
13. Nair, M.P.; Mahajan, S.; Reynolds, J.L.; Aalinkeel, R.; Nair, H.; Schwartz, S.A.; Kandaswami, C. The flavonoid quercetin inhibits proinflammatory cytokine (tumor necrosis factor alpha) gene expression in normal peripheral blood mononuclear cells via modulation of the NF-kβ system. *Clin. Vaccine Immunol.* **2006**, *13*, 319–328. [CrossRef] [PubMed]
14. Fang, X.K.; Gao, J.; Zhu, D.N. Kaempferol and quercetin isolated from euonymus alatus improve glucose uptake of 3T3-L1 cells without adipogenesis activity. *Life Sci.* **2008**, *82*, 615–622. [CrossRef] [PubMed]
15. Hsu, C.L.; Yen, G.C. Induction of cell apoptosis in 3t3-L1 pre-adipocytes by flavonoids is associated with their antioxidant activity. *Mol. Nutr. Food Res.* **2006**, *50*, 1072–1079. [CrossRef] [PubMed]
16. Ahn, J.; Lee, H.; Kim, S.; Park, J.; Ha, T. The anti-obesity effect of quercetin is mediated by the AMPK and MAPK signaling pathways. *Biochem. Biophys. Res. Commun.* **2008**, *373*, 545–549. [CrossRef] [PubMed]
17. Lara-Guzman, O.J.; Tabares-Guevara, J.H.; Leon-Varela, Y.M.; Alvarez, R.M.; Roldan, M.; Sierra, J.A.; Londono-Londono, J.A.; Ramirez-Pineda, J.R. Proatherogenic macrophage activities are targeted by the flavonoid quercetin. *J. Pharmacol. Exp. Ther.* **2012**, *343*, 296–306. [CrossRef] [PubMed]
18. Stewart, L.K.; Soileau, J.L.; Ribnicky, D.; Wang, Z.Q.; Raskin, I.; Poulev, A.; Majewski, M.; Cefalu, W.T.; Gettys, T.W. Quercetin transiently increases energy expenditure but persistently decreases circulating markers of inflammation in C57BL/6 mice fed a high-fat diet. *Metabolism* **2008**, *57*, S39–S46. [CrossRef] [PubMed]
19. Milenkovic, M.; Arsenovic-Ranin, N.; Stojic-Vukanic, Z.; Bufan, B.; Vucicevic, D.; Jancic, I. Quercetin ameliorates experimental autoimmune myocarditis in rats. *J. Pharm. Pharm. Sci.* **2010**, *13*, 311–319. [CrossRef] [PubMed]
20. Hu, Q.H.; Zhang, X.; Pan, Y.; Li, Y.C.; Kong, L.D. Allopurinol, quercetin and rutin ameliorate renal NLRP3 inflammasome activation and lipid accumulation in fructose-fed rats. *Biochem. Pharmacol.* **2012**, *84*, 113–125. [CrossRef] [PubMed]
21. Rivera, L.; Moron, R.; Sanchez, M.; Zarzuelo, A.; Galisteo, M. Quercetin ameliorates metabolic syndrome and improves the inflammatory status in obese zucker rats. *Obesity* **2008**, *16*, 2081–2087. [CrossRef] [PubMed]
22. Hu, Q.H.; Wang, C.; Li, J.M.; Zhang, D.M.; Kong, L.D. Allopurinol, rutin, and quercetin attenuate hyperuricemia and renal dysfunction in rats induced by fructose intake: Renal organic ion transporter involvement. *Am. J. Physiol. Renal. Physiol.* **2009**, *297*, 1080–1091. [CrossRef] [PubMed]
23. Kannappan, S.; Anuradha, C.V. Insulin sensitizing actions of fenugreek seed polyphenols, quercetin & metformin in a rat model. *Indian J. Med. Res.* **2009**, *129*, 401–408. [PubMed]
24. Kobori, M.; Masumoto, S.; Akimoto, Y.; Oike, H. Chronic dietary intake of quercetin alleviates hepatic fat accumulation associated with consumption of a western-style diet in C57/BL6J mice. *Mol. Nutr. Food Res.* **2011**, *55*, 530–540. [CrossRef] [PubMed]
25. Shao, L.; Liu, K.; Huang, F.; Guo, X.; Wang, M.; Liu, B. Opposite effects of quercetin, luteolin, and epigallocatechin gallate on insulin sensitivity under normal and inflammatory conditions in mice. *Inflammation* **2013**, *36*, 1–14. [CrossRef] [PubMed]
26. Zhou, M.; Wang, S.; Zhao, A.; Wang, K.; Fan, Z.; Yang, H.; Liao, W.; Bao, S.; Zhao, L.; Zhang, Y.; et al. Transcriptomic and metabonomic profiling reveal synergistic effects of quercetin and resveratrol supplementation in high fat diet fed mice. *J. Proteome Res.* **2012**, *11*, 4961–4971. [CrossRef] [PubMed]
27. Snyder, S.M.; Zhao, B.; Luo, T.; Kaiser, C.; Cavender, G.; Hamilton-Reeves, J.; Sullivan, D.K.; Shay, N.F. Consumption of quercetin and quercetin-containing apple and cherry extracts affects blood glucose concentration, hepatic metabolism, and gene expression patterns in obese C57/BL6J high fat-fed mice. *J. Nutr.* **2016**, *146*, 1001–1007. [CrossRef] [PubMed]

28. Jung, C.H.; Cho, I.; Ahn, J.; Jeon, T.I.; Ha, T.Y. Quercetin reduces high-fat diet-induced fat accumulation in the liver by regulating lipid metabolism genes. *Phytother. Res.* **2013**, *27*, 139–143. [CrossRef] [PubMed]

29. Egert, S.; Bosy-Westphal, A.; Seiberl, J.; Kurbitz, C.; Settler, U.; Plachta-Danielzik, S.; Wagner, A.E.; Frank, J.; Schrezenmeir, J.; Rimbach, G.; et al. Quercetin reduces systolic blood pressure and plasma oxidised low-density lipoprotein concentrations in overweight subjects with a high-cardiovascular disease risk phenotype: A double-blinded, placebo-controlled cross-over study. *Br. J. Nutr.* **2009**, *102*, 1065–1074. [CrossRef] [PubMed]

30. Brull, V.; Burak, C.; Stoffel-Wagner, B.; Wolffram, S.; Nickenig, G.; Muller, C.; Langguth, P.; Alteheld, B.; Fimmers, R.; Naaf, S.; et al. Effects of a quercetin-rich onion skin extract on 24 h ambulatory blood pressure and endothelial function in overweight-to-obese patients with (pre-)hypertension: A randomised double-blinded placebo-controlled cross-over trial. *Br. J. Nutr.* **2015**, *114*, 1263–1277. [CrossRef] [PubMed]

31. Heinz, S.A.; Henson, D.A.; Nieman, D.C.; Austin, M.D.; Jin, F. A 12-week supplementation with quercetin does not affect natural killer cell activity, granulocyte oxidative burst activity or granulocyte phagocytosis in female human subjects. *Br. J. Nutr.* **2010**, *104*, 849–857. [CrossRef] [PubMed]

32. Knab, A.M.; Shanely, R.A.; Jin, F.; Austin, M.D.; Sha, W.; Nieman, D.C. Quercetin with vitamin C and niacin does not affect body mass or composition. *Appl. Physiol. Nutr. Metab.* **2011**, *36*, 331–338. [CrossRef] [PubMed]

33. Knab, A.M.; Shanely, R.A.; Henson, D.A.; Jin, F.; Heinz, S.A.; Austin, M.D.; Nieman, D.C. Influence of quercetin supplementation on disease risk factors in community-dwelling adults. *J. Am. Diet. Assoc.* **2011**, *111*, 542–549. [CrossRef] [PubMed]

34. Heinz, S.A.; Henson, D.A.; Austin, M.D.; Jin, F.; Nieman, D.C. Quercetin supplementation and upper respiratory tract infection: A randomized community clinical trial. *Pharmacol. Res.* **2010**, *62*, 237–242. [CrossRef] [PubMed]

35. Rezvan, N.; Moini, A.; Janani, L.; Mohammad, K.; Saedisomeolia, A.; Nourbakhsh, M.; Gorgani-Firuzjaee, S.; Mazaherioun, M.; Hosseinzadeh-Attar, M.J. Effects of quercetin on adiponectin-mediated insulin sensitivity in polycystic ovary syndrome: A randomized placebo-controlled double-blind clinical trial. *Horm. Metab. Res.* **2017**, *49*, 115–121. [CrossRef] [PubMed]

36. Shi, Y.; Williamson, G. Quercetin lowers plasma uric acid in pre-hyperuricaemic males: A randomised, double-blinded, placebo-controlled, cross-over trial. *Br. J. Nutr.* **2016**, *115*, 800–806. [CrossRef] [PubMed]

37. Sae-tan, S.; Grove, K.A.; Lambert, J.D. Weight control and prevention of metabolic syndrome by green tea. *Pharmacol. Res.* **2011**, *64*, 146–154. [CrossRef] [PubMed]

38. Deng, Y.T.; Chang, T.W.; Lee, M.S.; Lin, J.K. Suppression of free fatty acid-induced insulin resistance by phytopolyphenols in C2C12 mouse skeletal muscle cells. *J. Agric. Food Chem.* **2012**, *60*, 1059–1066. [CrossRef] [PubMed]

39. Pournourmohammadi, S.; Grimaldi, M.; Stridh, M.H.; Lavallard, V.; Waagepetersen, H.S.; Wollheim, C.B.; Maechler, P. Epigallocatechin-3-gallate (EGCG) activates AMPK through the inhibition of glutamate dehydrogenase in muscle and pancreatic SS-cells: A potential beneficial effect in the pre-diabetic state? *Int. J. Biochem. Cell Biol.* **2017**, *88*, 220–225. [CrossRef] [PubMed]

40. Ueda, M.; Nishiumi, S.; Nagayasu, H.; Fukuda, I.; Yoshida, K.; Ashida, H. Epigallocatechin gallate promotes glut4 translocation in skeletal muscle. *Biochem. Biophys. Res. Commun.* **2008**, *377*, 286–290. [CrossRef] [PubMed]

41. Ma, S.B.; Zhang, R.; Miao, S.; Gao, B.; Lu, Y.; Hui, S.; Li, L.; Shi, X.P.; Wen, A.D. Epigallocatechin-3-gallate ameliorates insulin resistance in hepatocytes. *Mol. Med. Rep.* **2017**, *15*, 3803–3809. [CrossRef] [PubMed]

42. Sung, H.Y.; Hong, C.G.; Suh, Y.S.; Cho, H.C.; Park, J.H.; Bae, J.H.; Park, W.K.; Han, J.; Song, D.K. Role of (−)-epigallocatechin-3-gallate in cell viability, lipogenesis, and retinol-binding protein 4 expression in adipocytes. *Naunyn Schmiedebergs Arch. Pharmacol.* **2010**, *382*, 303–310. [CrossRef] [PubMed]

43. Lee, M.S.; Kim, C.T.; Kim, I.H.; Kim, Y. Inhibitory effects of green tea catechin on the lipid accumulation in 3T3-L1 adipocytes. *Phytother. Res.* **2009**, *23*, 1088–1091. [CrossRef] [PubMed]

44. Moon, H.S.; Chung, C.S.; Lee, H.G.; Kim, T.G.; Choi, Y.J.; Cho, C.S. Inhibitory effect of (−)-epigallocatechin-3-gallate on lipid accumulation of 3T3-L1 cells. *Obesity* **2007**, *15*, 2571–2582. [CrossRef] [PubMed]

45. Sakurai, N.; Mochizuki, K.; Kameji, H.; Shimada, M.; Goda, T. (−)-epigallocatechin gallate enhances the expression of genes related to insulin sensitivity and adipocyte differentiation in 3T3-L1 adipocytes at an early stage of differentiation. *Nutrition* **2009**, *25*, 1047–1056. [CrossRef] [PubMed]

46. Wu, M.; Liu, D.; Zeng, R.; Xian, T.; Lu, Y.; Zeng, G.; Sun, Z.; Huang, B.; Huang, Q. Epigallocatechin-3-gallate inhibits adipogenesis through down-regulation of PPARγ and FAS expression mediated by PI3K-AKT signaling in 3T3-L1. *Eur. J. Pharmacol.* **2017**, *795*, 134–142. [CrossRef] [PubMed]

47. Li, M.; Liu, J.T.; Pang, X.M.; Han, C.J.; Mao, J.J. Epigallocatechin-3-gallate inhibits angiotensin II and interleukin-6-induced C-reactive protein production in macrophages. *Pharmacol. Rep.* **2012**, *64*, 912–918. [CrossRef]

48. Bose, M.; Lambert, J.D.; Ju, J.; Reuhl, K.R.; Shapses, S.A.; Yang, C.S. The major green tea polyphenol, (−)-epigallocatechin-3-gallate, inhibits obesity, metabolic syndrome, and fatty liver disease in high-fat-fed mice. *J. Nutr.* **2008**, *138*, 1677–1683. [PubMed]

49. Cunha, C.A.; Lira, F.S.; Rosa Neto, J.C.; Pimentel, G.D.; Souza, G.I.; Da Silva, C.M.; De Souza, C.T.; Ribeiro, E.B.; Sawaya, A.C.; Oller do Nascimento, C.M.; et al. Green tea extract supplementation induces the lipolytic pathway, attenuates obesity, and reduces low-grade inflammation in mice fed a high-fat diet. *Mediat. Inflamm.* **2013**, *2013*. [CrossRef] [PubMed]

50. Sampath, C.; Rashid, M.R.; Sang, S.; Ahmedna, M. Green tea epigallocatechin 3-gallate alleviates hyperglycemia and reduces advanced glycation end products via NRF2 pathway in mice with high fat diet-induced obesity. *Biomed. Pharmacother.* **2017**, *87*, 73–81. [CrossRef] [PubMed]

51. Szulinska, M.; Stepien, M.; Kregielska-Narozna, M.; Suliburska, J.; Skrypnik, D.; Bak-Sosnowska, M.; Kujawska-Luczak, M.; Grzymislawska, M.; Bogdanski, P. Effects of green tea supplementation on inflammation markers, antioxidant status and blood pressure in NaCl-induced hypertensive rat model. *Food Nutr. Res.* **2017**, *61*. [CrossRef] [PubMed]

52. Remely, M.; Ferk, F.; Sterneder, S.; Setayesh, T.; Roth, S.; Kepcija, T.; Noorizadeh, R.; Rebhan, I.; Greunz, M.; Beckmann, J.; et al. Egcg prevents high fat diet-induced changes in gut microbiota, decreases of DNA strand breaks, and changes in expression and DNA methylation of DNMT1 and MLH1 in C57BL/6J male mice. *Oxidative Med. Cell. Longev.* **2017**, *2017*. [CrossRef] [PubMed]

53. Grove, K.A.; Lambert, J.D. Laboratory, epidemiological, and human intervention studies show that tea (camellia sinensis) may be useful in the prevention of obesity. *J. Nutr.* **2010**, *140*, 446–453. [CrossRef] [PubMed]

54. Nagao, T.; Hase, T.; Tokimitsu, I. A green tea extract high in catechins reduces body fat and cardiovascular risks in humans. *Obesity* **2007**, *15*, 1473–1483. [CrossRef] [PubMed]

55. Hase, T.; Komine, Y.; Meguro, S.; Takeda, Y.; Takahashi, H.; Matsui, Y.; Inaoka, S.; Katsuragi, Y.; Tomitsu, J.; Shimasaki, H.; et al. Anti-obesity effects of tea catechins in humans. *J. Oleo Sci.* **2001**, *50*, 599–605. [CrossRef]

56. Basu, A.; Sanchez, K.; Leyva, M.J.; Wu, M.; Betts, N.M.; Aston, C.E.; Lyons, T.J. Green tea supplementation affects body weight, lipids, and lipid peroxidation in obese subjects with metabolic syndrome. *J. Am. Coll. Nutr.* **2010**, *29*, 31–40. [CrossRef] [PubMed]

57. Basu, A.; Du, M.; Sanchez, K.; Leyva, M.J.; Betts, N.M.; Blevins, S.; Wu, M.; Aston, C.E.; Lyons, T.J. Green tea minimally affects biomarkers of inflammation in obese subjects with metabolic syndrome. *Nutrition* **2011**, *27*, 206–213. [CrossRef] [PubMed]

58. Brown, A.L.; Lane, J.; Coverly, J.; Stocks, J.; Jackson, S.; Stephen, A.; Bluck, L.; Coward, A.; Hendrickx, H. Effects of dietary supplementation with the green tea polyphenol epigallocatechin-3-gallate on insulin resistance and associated metabolic risk factors: Randomized controlled trial. *Br. J. Nutr.* **2009**, *101*, 886–894. [CrossRef] [PubMed]

59. Khalesi, S.; Sun, J.; Buys, N.; Jamshidi, A.; Nikbakht-Nasrabadi, E.; Khosravi-Boroujeni, H. Green tea catechins and blood pressure: A systematic review and meta-analysis of randomised controlled trials. *Eur. J. Nutr.* **2014**, *53*, 1299–1311. [CrossRef] [PubMed]

60. Onakpoya, I.; Spencer, E.; Heneghan, C.; Thompson, M. The effect of green tea on blood pressure and lipid profile: A systematic review and meta-analysis of randomized clinical trials. *Nutr. Metab. Cardiovasc. Dis.* **2014**, *24*, 823–836. [CrossRef] [PubMed]

61. Peng, X.; Zhou, R.; Wang, B.; Yu, X.; Yang, X.; Liu, K.; Mi, M. Effect of green tea consumption on blood pressure: A meta-analysis of 13 randomized controlled trials. *Sci. Rep.* **2014**, *4*. [CrossRef] [PubMed]

62. Kim, A.; Chiu, A.; Barone, M.K.; Avino, D.; Wang, F.; Coleman, C.I.; Phung, O.J. Green tea catechins decrease total and low-density lipoprotein cholesterol: A systematic review and meta-analysis. *J. Am. Diet. Assoc.* **2011**, *111*, 1720–1729. [CrossRef] [PubMed]

63. Zheng, X.X.; Xu, Y.L.; Li, S.H.; Liu, X.X.; Hui, R.; Huang, X.H. Green tea intake lowers fasting serum total and LDL cholesterol in adults: A meta-analysis of 14 randomized controlled trials. *Am. J. Clin. Nutr.* **2011**, *94*, 601–610. [CrossRef] [PubMed]

64. Liu, K.; Zhou, R.; Wang, B.; Chen, K.; Shi, L.Y.; Zhu, J.D.; Mi, M.T. Effect of green tea on glucose control and insulin sensitivity: A meta-analysis of 17 randomized controlled trials. *Am. J. Clin. Nutr.* **2013**, *98*, 340–348. [CrossRef] [PubMed]

65. Nieman, D.C.; Henson, D.A.; Davis, J.M.; Dumke, C.L.; Gross, S.J.; Jenkins, D.P.; Murphy, E.A.; Carmichael, M.D.; Quindry, J.C.; McAnulty, S.R.; et al. Quercetin ingestion does not alter cytokine changes in athletes competing in the western states endurance run. *J. Interferon Cytokine Res.* **2007**, *27*, 1003–1011. [CrossRef] [PubMed]

66. Nieman, D.C.; Henson, D.A.; Davis, J.M.; Angela Murphy, E.; Jenkins, D.P.; Gross, S.J.; Carmichael, M.D.; Quindry, J.C.; Dumke, C.L.; Utter, A.C.; et al. Quercetin's influence on exercise-induced changes in plasma cytokines and muscle and leukocyte cytokine mRNA. *J. Appl. Physiol.* **2007**, *103*, 1728–1735. [CrossRef] [PubMed]

67. Nieman, D.C.; Henson, D.A.; Maxwell, K.R.; Williams, A.S.; McAnulty, S.R.; Jin, F.; Shanely, R.A.; Lines, T.C. Effects of quercetin and EGCG on mitochondrial biogenesis and immunity. *Med. Sci. Sports Exerc.* **2009**, *41*, 1467–1475. [CrossRef] [PubMed]

68. Nieman, D.C.; Henson, D.A.; Gross, S.J.; Jenkins, D.P.; Davis, J.M.; Murphy, E.A.; Carmichael, M.D.; Dumke, C.L.; Utter, A.C.; McAnulty, S.R.; et al. Quercetin reduces illness but not immune perturbations after intensive exercise. *Med. Sci. Sports Exerc.* **2007**, *39*, 1561–1569. [CrossRef] [PubMed]

69. Irizarry, R.A.; Hobbs, B.; Collin, F.; Beazer-Barclay, Y.D.; Antonellis, K.J.; Scherf, U.; Speed, T.P. Exploration, normalization, and summaries of high density oligonucleotide array probe level data. *Biostatistics* **2003**, *4*, 249–264. [CrossRef] [PubMed]

70. Baldi, P.; Long, A.D. A bayesian framework for the analysis of microarray expression data: Regularized t-test and statistical inferences of gene changes. *Bioinformatics* **2001**, *17*, 509–519. [CrossRef] [PubMed]

71. Subramanian, A.; Tamayo, P.; Mootha, V.K.; Mukherjee, S.; Ebert, B.L.; Gillette, M.A.; Paulovich, A.; Pomeroy, S.L.; Golub, T.R.; Lander, E.S.; et al. Gene set enrichment analysis: A knowledge-based approach for interpreting genome-wide expression profiles. *Proc. Natl. Acad. Sci. USA* **2005**, *102*, 15545–15550. [CrossRef] [PubMed]

72. Dam, V.; Sikder, T.; Santosa, S. From neutrophils to macrophages: Differences in regional adipose tissue depots. *Obes. Rev.* **2016**, *17*, 1–17. [CrossRef] [PubMed]

73. Talukdar, S.; Oh, D.Y.; Bandyopadhyay, G.; Li, D.; Xu, J.; McNelis, J.; Lu, M.; Li, P.; Yan, Q.; Zhu, Y.; et al. Neutrophils mediate insulin resistance in mice fed a high-fat diet through secreted elastase. *Nat. Med.* **2012**, *18*, 1407–1412. [CrossRef] [PubMed]

74. Krause, B.R.; Hartman, A.D. Adipose tissue and cholesterol metabolism. *J. Lipid Res.* **1984**, *25*, 97–110. [PubMed]

75. Parr, E.B.; Coffey, V.G.; Hawley, J.A. 'Sarcobesity': A metabolic conundrum. *Maturitas* **2013**, *74*, 109–113. [CrossRef] [PubMed]

76. Brill, M.J.; Diepstraten, J.; Van Rongen, A.; Van Kralingen, S.; Van den Anker, J.N.; Knibbe, C.A. Impact of obesity on drug metabolism and elimination in adults and children. *Clin. Pharmacokinet.* **2012**, *51*, 277–304. [CrossRef] [PubMed]

77. Luster, A.D.; Alon, R.; Von Andrian, U.H. Immune cell migration in inflammation: Present and future therapeutic targets. *Nat. Immunol.* **2005**, *6*, 1182–1190. [CrossRef] [PubMed]

78. Dutta, S.; Sengupta, P. Men and mice: Relating their ages. *Life Sci.* **2016**, *152*, 244–248. [CrossRef] [PubMed]

79. Cialdella-Kam, L.; Nieman, D.C.; Knab, A.M.; Shanely, R.A.; Meaney, M.P.; Jin, F.; Sha, W.; Ghosh, S. A mixed flavonoid-fish oil supplement induces immune-enhancing and anti-inflammatory transcriptomic changes in adult obese and overweight women-a randomized controlled trial. *Nutrients* **2016**, *8*, 277. [CrossRef] [PubMed]

80. Bakker, G.C.; Van Erk, M.J.; Pellis, L.; Wopereis, S.; Rubingh, C.M.; Cnubben, N.H.; Kooistra, T.; Van Ommen, B.; Hendriks, H.F. An antiinflammatory dietary mix modulates inflammation and oxidative and metabolic stress in overweight men: A nutrigenomics approach. *Am. J. Clin. Nutr.* **2010**, *91*, 1044–1059. [CrossRef] [PubMed]

81. Schmidt, S.; Stahl, F.; Mutz, K.O.; Scheper, T.; Hahn, A.; Schuchardt, J.P. Different gene expression profiles in normo- and dyslipidemic men after fish oil supplementation: Results from a randomized controlled trial. *Lipids Health Dis.* **2012**, *11*, 105. [CrossRef] [PubMed]

82. Skulas-Ray, A.C. Omega-3 fatty acids and inflammation: A perspective on the challenges of evaluating efficacy in clinical research. *Prostaglandins Other Lipid Mediat.* **2015**, *116–117*, 104–111. [CrossRef] [PubMed]

83. Wang, P.; Heber, D.; Henning, S.M. Quercetin increased bioavailability and decreased methylation of green tea polyphenols in vitro and in vivo. *Food Funct.* **2012**, *3*, 635–642. [CrossRef] [PubMed]

84. D'Andrea, G. Quercetin: A flavonol with multifaceted therapeutic applications? *Fitoterapia* **2015**, *106*, 256–271. [CrossRef] [PubMed]

85. Glaeser, H.; Bujok, K.; Schmidt, I.; Fromm, M.F.; Mandery, K. Organic anion transporting polypeptides and organic cation transporter 1 contribute to the cellular uptake of the flavonoid quercetin. *Naunyn Schmiedebergs Arch. Pharmacol.* **2014**, *387*, 883–891. [CrossRef] [PubMed]

86. Roth, M.; Timmermann, B.N.; Hagenbuch, B. Interactions of green tea catechins with organic anion-transporting polypeptides. *Drug Metab. Dispos.* **2011**, *39*, 920–926. [CrossRef] [PubMed]

87. Henagan, T.M.; Lenard, N.R.; Gettys, T.W.; Stewart, L.K. Dietary quercetin supplementation in mice increases skeletal muscle PGC1alpha expression, improves mitochondrial function and attenuates insulin resistance in a time-specific manner. *PLoS ONE* **2014**, *9*, e89365. [CrossRef] [PubMed]

88. Palermo, N.E.; Gianchandani, R.Y.; McDonnell, M.E.; Alexanian, S.M. Stress hyperglycemia during surgery and anesthesia: Pathogenesis and clinical implications. *Curr. Diabetes Rep.* **2016**, *16*, 33. [CrossRef] [PubMed]

89. Togashi, Y.; Shirakawa, J.; Okuyama, T.; Yamazaki, S.; Kyohara, M.; Miyazawa, A.; Suzuki, T.; Hamada, M.; Terauchi, Y. Evaluation of the appropriateness of using glucometers for measuring the blood glucose levels in mice. *Sci. Rep.* **2016**, *6*. [CrossRef] [PubMed]

90. Gonzalez-Gallego, J.; Garcia-Mediavilla, M.V.; Sanchez-Campos, S.; Tunon, M.J. Fruit polyphenols, immunity and inflammation. *Br. J. Nutr.* **2010**, *104*, 15–27. [CrossRef] [PubMed]

91. Mereles, D.; Hunstein, W. Epigallocatechin-3-gallate (EGCG) for clinical trials: More pitfalls than promises? *Int. J. Mol. Sci.* **2011**, *12*, 5592–5603. [CrossRef] [PubMed]

92. Van Duynhoven, J.; Vaughan, E.E.; Jacobs, D.M.; Kemperman, R.A.; Van Velzen, E.J.J.; Gross, G.; Roger, L.C.; Possemiers, S.; Smilde, A.K.; Doré, J.; et al. Metabolic fate of polyphenols in the human superorganism. *Proc. Natl. Acad. Sci. USA* **2011**, *108*, 4531–4538. [CrossRef] [PubMed]

93. Gonzalez, F.J. Cytochrome P450 humanised mice. *Hum. Genom.* **2004**, *1*, 300–306. [CrossRef] [PubMed]

94. Graf, B.A.; Ameho, C.; Dolnikowski, G.G.; Milbury, P.E.; Chen, C.Y.; Blumberg, J.B. Rat gastrointestinal tissues metabolize quercetin. *J. Nutr.* **2006**, *136*, 39–44. [PubMed]

95. Lambert, J.D.; Lee, M.J.; Lu, H.; Meng, X.; Hong, J.J.; Seril, D.N.; Sturgill, M.G.; Yang, C.S. Epigallocatechin-3-gallate is absorbed but extensively glucuronidated following oral administration to mice. *J. Nutr.* **2003**, *133*, 4172–4177. [PubMed]

nutrients

MDPI

Review

Food-Derived Antioxidant Polysaccharides and Their Pharmacological Potential in Neurodegenerative Diseases

Haifeng Li [1,2,†], Fei Ding [1,†], Lingyun Xiao [2], Ruona Shi [1], Hongyu Wang [1], Wenjing Han [1] and Zebo Huang [1,*]

[1] Center for Bioresources & Drug Discovery and School of Biosciences & Biopharmaceutics, Guangdong Pharmaceutical University, Guangzhou 510006, China; lihf@gdpu.edu.cn (H.L.); 15692435199@163.com (F.D.); 15521217084@163.com (R.S.); 15626200071@163.com (H.W.); 15622104553@163.com (W.H.)
[2] School of Pharmaceutical Sciences, Wuhan University, Wuhan 430071, China; Mandy.Xiao@infinitus-int.com
*   Correspondence: zebohuang@gdpu.edu.cn
†   These authors contributed equally to this work.

Received: 31 May 2017; Accepted: 13 July 2017; Published: 19 July 2017

Abstract: Oxidative stress is known to impair architecture and function of cells, which may lead to various chronic diseases, and therefore therapeutic and nutritional interventions to reduce oxidative damages represent a viable strategy in the amelioration of oxidative stress-related disorders, including neurodegenerative diseases. Over the past decade, a variety of natural polysaccharides from functional and medicinal foods have attracted great interest due to their antioxidant functions such as scavenging free radicals and reducing oxidative damages. Interestingly, these antioxidant polysaccharides are also found to attenuate neuronal damages and alleviate cognitive and motor decline in a range of neurodegenerative models. It has recently been established that the neuroprotective mechanisms of polysaccharides are related to oxidative stress-related pathways, including mitochondrial function, antioxidant defense system and pathogenic protein aggregation. Here, we first summarize the current status of antioxidant function of food-derived polysaccharides and then attempt to appraise their anti-neurodegeneration activities.

Keywords: polysaccharide; antioxidant; oxidative stress; inflammatory stress; proteotoxic stress; neurodegeneration

---

1. Introduction

Oxygen is essential for normal life of aerobic organisms. Due to its high redox potential, oxygen is inevitably involved in the production of reactive oxygen species (ROS) such as superoxide anion, hydroxyl radical and hydrogen peroxide. ROS are known to play an important role in a variety of cellular functions including signal transduction and regulation of enzyme activity [1,2]. Excessive ROS, on the other hand, can also interact with biological molecules and generate by-products such as peroxides and aldehydes, which can cause damages to architecture and function of cells [3,4]. Under normal circumstances, cells have a set of antioxidant defense system, including enzymatic antioxidants such as superoxide dismutase (SOD), catalase (CAT) and glutathione peroxidase (GPx) and non-enzymatic antioxidants such as glutathione and vitamins, to combat excessive ROS [5,6]. However, when a detrimental stress compromises the antioxidant defense system, a fraction of ROS may escape the intrinsic clearance machinery and induce a state of oxidative stress, leading to cell dysfunction [7,8].

Growing evidence has demonstrated that oxidative stress is implicated in the development and progression of many chronic diseases such as neurodegenerative diseases (NDD) and diabetes [9–11].

NDD, including Alzheimer's disease (AD), Parkinson's disease (PD) and Huntington's disease (HD), are a group of chronic disorders pathologically characterized by selective and progressive loss of neurons [12]. Clinical evidence has shown that NDD patients display an oxidative stress-related manifestation, including increases of ROS level, lipid peroxidation and protein oxidation [13,14]. Recent studies have revealed that ROS-induced peroxidation products, such as the lipid peroxidation product malondialdehyde (MDA) and the protein oxidation product carbonyl groups, can damage other cellular components and exacerbate neuronal dysfunction, further demonstrating the detrimental consequence of oxidative stress in neurodegeneration [4,15,16]. Therefore, strategies to reduce oxidative damages are shown to be beneficial to alleviate neurodegenerative symptoms. For example, intake of foods rich in antioxidant ingredients has shown potentials to prevent oxidative stress-related conditions, including NDD [17–19]. Among the reported ingredients, polysaccharides, an important type of natural polymers consisting of monosaccharide units that contain multiple free hydroxyl groups, are shown to have both in vitro and in vivo antioxidant activities [20,21]. Here, we first review the antioxidant effects of food-derived polysaccharides and then focus on their protective function against neurodegeneration.

2. Reduction of Oxidative Stress by Food-Derived Polysaccharides

During the last decade, a large body of evidence has shown that polysaccharides and glycoconjugates from a variety of natural sources, including bacteria, fungi, algae, plants and animals, have antioxidant potentials [20–23]. In particular, polysaccharides isolated from functional and medicinal foods as well as from common foods have drawn great attention in antioxidant studies. Here, we attempt to summarize recent studies of antioxidant polysaccharides from food resources, including vegetables, fruits, cereals, beans, mushrooms, tea, milk products and meat (Table 1) [24–73].

Table 1. The antioxidant activities and mechanisms of food-derived polysaccharides.

Source	Polysaccharide	Test Model	Protective Effect	Potential Mechanism	Ref.
Vegetables					
Zizania latifolia	ZLPs-W	In vitro assays	Scavenging activity against DPPH and ·OH		[24]
Daucus carota	CWSP	In vitro assays	Scavenging activity against DPPH, reducing power, prevention of β-carotene bleaching	Ferrous chelating ability	[25]
Cucurbita maxima Duchesne	WSP	In vitro assays	Scavenging activity against DPPH, inhibition of ascorbic acid oxidation	SOD-like activity	[26]
Solanum tuberosum	PPPWs	In vitro assays	Scavenging activity against DPPH and ABTS, reducing power, total antioxidant capacity		[27]
Potentilla anserine	PAP	H_2O_2-exposed murine splenic lymphocytes	Apoptosis rate↓		[28]
Psidium guajava	PS-PGL	In vitro assays; H_2O_2-exposed Vero cells and zebrafish	Scavenging activity against DPPH, ·OH and alkyl radicals in vitro; Cell viability↑, DNA fragmentation↓, nuclear condensation and morphological disruption↓ in Vero cells; Survival↑, heart-beating rate↓, cell death↓ in zebrafish embryos	ROS level↓ in Vero cells; ROS level↓, MDA content↓ in zebrafish embryos	[29]
Fruits					
Malus pumila	APPS	In vitro assays	Scavenging activity against DPPH, O_2^-· and ·OH, reducing power		[30]
Diospyros kaki L.	PFP	In vitro assays	Scavenging activity against DPPH, O_2^-· and ·OH, reducing power		[31]
Seed watermelon	SWP	H_2O_2-exposed PC12 cells	Cell viability↑, LDH release↓	ROS level↓, 8-OHdG content↓, caspase-3 and caspase-9 activities↓, MMP↑	[32]
Cereals and Beans					
Rice bran	RBP2	In vitro assays	Scavenging activity against DPPH, O_2^-·, ·OH and ABTS, reducing power	Ferrous chelating ability	[33]
Wheat bran	Feruloyl oligosaccharides	AAPH-exposed human erythrocytes	Erythrocyte hemolysis↓	GSH level↓, MDA content↓, PCG level↓	[34]
Glycine max (L.) Merr.	MSF	In vitro assays	Scavenging activity against ABTS, reducing power		[35]
Cicer arietinum L. hull	CHPS	In vitro assays; H_2O_2-exposed PC12 cells	Scavenging activity against ABTS, DPPH O_2^-·; reducing power in vitro; Cell viability↑		[36]

Table 1. Cont.

Source	Polysaccharide	Test Model	Protective Effect	Potential Mechanism	Ref.
Herbs					
Dioscorea opposita	Yam polysaccharide	In vitro assays	Scavenging activity against $O_2^{\cdot-}$ and ·OH		[37]
Epimedium brevicornum Maxim.	EbPS-A1	In vitro assays; PQ-exposed C. elegans	Scavenging activity against DPPH and ·OH in vitro; Survival rate↑ in C. elegans	ROS level↓, MDA content↓, SOD and CAT activities↑ in C. elegans	[38]
Chuanminshen violaceum	CVPS	In vitro assays; D-Gal-treated ICR mice	Scavenging activity against DPPH, $O_2^{\cdot-}$ and ·OH in vitro; Body weights and spleen indices↑ in mice	Activities and mRNA levels of Mn-SOD, Cu/Zn-SOD, GPx and CAT↑, MDA content↓ in mouse liver, heart and brain	[39]
Radix Rehmanniae	RRPs	UV-irradiated mice		GSH level↑, SOD, CAT and GPx activities↑, MDA content↓, IL-2, IL-4 and IL-10 levels↑	[40]
Lycium barbarum	LBPs	H_2O_2-exposed SRA01/04 cells	Cell viability↑, apoptotic rate↓, ratio of ageing cells↓, G0/G1 cell cycle phase arrest↓	ROS level↓, MMP↑, Bcl-2 protein level↑, Bax protein level↓, MDA content↓, SOD activity↑, GSH level↑	[41]
Angelica sinensis	ASP	H_2O_2-exposed PC12 cells; SD rats with middle cerebral artery occlusion	Cell viability↑, apoptosis rate↓ in PC12 cells; Number of microvessels in rat brain↑	ROS level↓, MMP↑ in PC12 cells; SOD and GPx activities↑ in rat cortex	[42]
Sophora subprosrate	SSP	PCV-2 infection RAW264.7 cells		Activities of Total-SOD, Cu/Zn-SOD and Mn-SOD↑, mRNA levels of Mn-SOD↑ and NOX2↓, NOX2 protein level↓, MMP↑	[43]
Cynomorium songaricum Rupr.	CSP	H_2O_2-exposed PC12 cells	Cell viability↑, ratio of sub G1and S phase↓, ratio of G2/M phase↑, apoptosis rate↓, LDH release↓	ROS level↓, MDA content↓, 8-OHdG content↓, SOD and GPx activities↑, capase-3 and capase-9 activities↓	[44]
Tea					
Black tea	BTPS	In vitro assays	Scavenging activity against DPPH and ·OH		[45]
Green tea	TPS1	In vitro assays	Scavenging activity against DPPH, $O_2^{\cdot-}$ and ·OH, ferrous chelating ability, reducing power, total antioxidant capacity, inhibition of lipid hydroperoxide		[46]
Gynostemma pentaphyllum Makino	GPMMP	Cyclophosphamide-treated C57BL/6 mice	Spleen and thymus indices↑, CD4+ T lymphocyte counts↑, total antioxidant capacity↑	CAT, SOD and GPx activities↑, MDA content↓, GSH level↑, IL-2 level in sera and spleen↑	[47]

Table 1. *Cont.*

Source	Polysaccharide	Test Model	Protective Effect	Potential Mechanism	Ref.
Nuts					
Juglans regia L.	SJP	In vitro assays	Scavenging activity against DPPH, ·OH and ABTS, reducing power		[48]
Ginkgo biloba L.	GNP	In vitro assays; Hyperlipemia mice	Scavenging activity against DPPH, $O_2{}^{\cdot-}$ and ·OH in vitro	CAT, SOD and GPx activities↑, MDA content↓ in mouse serum and liver	[49]
Other Plants					
Zizyphus jujuba Mill	ZJPa	In vitro assays	Scavenging activity against $O_2{}^{\cdot-}$ and ·OH	Ferrous chelating ability	[50]
Aloe barbadensis Miller	GAPS-1 and SAPS-1	In vitro assays	Scavenging activity against $O_2{}^{\cdot-}$, ·OH and H_2O_2, reducing power, MDA content↓	Ferrous chelating ability	[51]
Anoectochilus roxburghii	ARPT	CCl_4-treated Kunming mice	Hepatocyte necrosis↓, serum alanine transaminase and aspartate transaminase activities↓	MDA level↓, SOD, CAT and GPx activities↑, GSH level↓, mRNA levels of TNF-α, IL-6 and Bax↓, protein levels of TNF-α, IL-6, NF-κB and cleaved-caspase 3↓ in liver	[52]
Opuntia dillenii Haw	CP	H_2O_2-exposed PC12 cells	Cell viability↑, LDH release↓, apoptosis rate↓	ROS level↓, ratio of Bax/Bcl-2 mRNA level↑	[53]
Camellia oleifera Abel	SCP1	In vitro assays; PQ-exposed *C. elegans*	Scavenging activity against $O_2{}^{\cdot-}$ and ·OH in vitro; Survival rate↑ in *C. elegans*	Ferric chelating ability in vitro; SOD, CAT and GPx activities↑, MDA content↓ in *C. elegans*	[54]
Taraxacum officinale	TOP2	LPS or *t*-BHP-exposed RAW 264.7 cells	NO production↓ in LPS-exposed cells; Cell viability↑ in *t*-BHP-exposed cells	Protein levels of TNF-α, p-IκBα, p-p65, p-Akt, iNOS and heme oxygenase 1↓	[55]

Table 1. *Cont.*

Source	Polysaccharide	Test Model	Protective Effect	Potential Mechanism	Ref.
Mushrooms					
Ganoderma lucidum	*G. lucidum* polysaccharide	Isoproterenol-treated albino rats	Creatinine kinase and LDH activities↓ in serum, cardiac muscle fibers with mild hyalinization	ROS level↓, MDA content↓, SOD and GPx activities↑, GSH level↑, activities of Krebs cycle dehydrogenases and mitochondrial complexes↑, MMP↑	[56]
Lentinus edodes, Ganoderma applanatum, Trametes versicolor	Mushroom polysaccharides	In vitro assays	Scavenging activity against DPPH, reducing power, inhibition of linoleic acid peroxidation	Ferric chelating ability	[57]
Dictyophora indusiata	DIPS	PQ-exposed *C. elegans*	Survival rate↑	ROS level↓, SOD activity↑, MDA content↓, MMP↑, ATP content↑, DAF-16 activation↑	[58]
Other Fungi					
Auricularia auricula	AAP1	In vitro assays; PQ or H_2O_2-exposed *C. elegans*	Scavenging activity against DPPH, $O_2 \cdot^-$ and $\cdot OH$, reducing power in vitro; Survival rate↑ in *C. elegans*	Ferric chelating ability in vitro; ROS level↓, SOD and CAT activities↑ in *C. elegans*	[59]
Tremella fuciformis	TP	UV-irradiated SD rats	Water and collagen content↑, glycosaminoglycan↓, endogenous collagen breakdown↓, ratio of type I/III collagen↑ in rat skin	SOD, GPx and CAT activities↑	[60]
Algae					
Porphyra haitanesis	*P. haitanesis* polysaccharide	In vitro assays; H_2O_2-exposed rat erythrocytes and liver microsome	Scavenging activity against $O_2 \cdot^-$ and $\cdot OH$ in vitro; Erythrocyte hemolysis↓; lipid peroxidation of rat liver microsome↓		[61]
Laminaria japonica	LJPA-P3	In vitro assays	Oxygen radical absorbance capacity, scavenging activity against ABTS		[62]
Fucus vesiculosus	*F. vesiculosus* polysaccharide	In vitro assays	Ferric reducing antioxidant power		[63]
Ulva pertusa	*U. pertusa* polysaccharide	In vitro assays	Scavenging activity against $O_2 \cdot^-$ and $\cdot OH$, reducing power	Ferric chelating ability	[64]
Brown seaweed	Fucoidan	UV-irradiated HS68 cells		ROS level↓, MDA content↓, GSH level↑	[65]
Nostoc commune	*Nostoc* polysaccharide	In vitro assays; PQ-exposed *C. elegans*	Scavenging activity against $O_2 \cdot^-$ and $\cdot OH$ in vitro; Survival rate↑ in *C. elegans*	SOD, CAT and GPx activities↑, MDA content↓ in *C. elegans*	[66]

Table 1. *Cont.*

Source	Polysaccharide	Test Model	Protective Effect	Potential Mechanism	Ref.
Milkproducts					
Milk fermented with lactic acid bacteria	Exopolysaccharides	UV-irradiated hairless mice	Erythema formation, dryness and epidermal proliferation, cyclobutane pyrimidine dimers↓ in mouse skin	mRNA levels of xeroderma pigmentosum complementation group A↑, ratio of mRNA levels of IL10/IL12α and IL10/IFN-γ↓ in mouse skin	[67]
Wine					
Red wine	PS-SI	In vitro assays	Scavenging activity against ·OH, oxygen radical absorbance capacity		[68]
Probiotics					
Bifidobacterium animalis RH	EPS	In vitro assays; D-Gal-treated Kunming mice	Inhibition of linoleic acid peroxidation, total antioxidant capacity, scavenging activity against DPPH, $O_2{}^{\cdot-}$ and ·OH in vitro	Total antioxidant capacity, SOD, CAT and GPx activities↑, MDA content↓ in serum, GST activity and MDA content↓ in liver, MAO activity and lipofuscin level↓ in brain	[69]
Bifidobacterium bifidum WBIN03, Lactobacillus plantarum R31	B-EPS and L-EPS	In vitro assays; H_2O_2-exposed rat erythrocytes	Scavenging activity against DPPH, $O_2{}^{\cdot-}$ and ·OH, inhibition of lipid peroxidation in vitro; Erythrocyte hemolysis↓		[70]
Meat					
Haliotis discus hannai Ino	ASP-1	In vitro assays	Scavenging activity against $O_2{}^{\cdot-}$		[71]
Crassostrea hongkongensis	CHPs	In vitro assays	Scavenging activity against DPPH, ·OH and ABTS, inhibition of linoleic acid peroxidation		[72]
Mytilus coruscus	MP-1	CCl$_4$-treated Kunming mice	Serum alanine transaminase and aspartate transaminase levels↓, necrosis of liver cells↓, immigration of inflammatory cells↓	MDA content↓, SOD activity↑ in liver	[73]

AAPH, 2,2′-Azobis(2-amidinopropane) dihydrochloride; ABTS, 2,2′-Azino-bis(3-ethylbenzothiazoline-6-sulfonic acid); CAT, catalase; CCl$_4$, carbon tetrachloride; Cu/Zn-SOD, copper-zinc superoxide dismutase; D-Gal, D-galactose; DPPH, 2,2-diphenyl-1-picrylhydrazyl radical; GPx, glutathione peroxidase; GSH, glutathione; GST, glutathione S-transferase; H_2O_2, hydrogen peroxide; HS68 cells, human foreskin fibroblast line; IFN-γ, interferon-γ; IκBα, NF-κB inhibitor α; ILs, interleukins; iNOS, inducible nitric oxide synthase; LDH, lactate dehydrogenases; LPS, lipopolysaccharide; MAO, monoamine oxidase; MDA, malondialdehyde; MMP, mitochondrial membrane potential; Mn-SOD, manganese superoxide dismutase; NF-κB, nuclear factor-κB; NO, nitric oxide; NOX2, cytochrome b-245β chain; $O_2{}^{\cdot-}$, superoxide anion; ·OH, hydroxyl radical; RAW 264.7 cells, murine macrophage cell line; PC12 cells, rat pheochromocytoma cell line; PCC, protein carbonyl group; PCV-2, porcine circovirus type 2; PQ, paraquat; ROS, reactive oxygen species; SOD, superoxide dismutase; SRA01/04 cells, SV40 T-antigen-transformed human lens epithelial cell line; t-BHP, tert-Butyl hydroperoxide; TNF-α, tumor necrosis factor α; UV, ultraviolet; 8-OHdG, 8-hydroxy-2′-deoxyguanosine.

2.1. Reduction of Free Radical and Peroxidation Product Levels

Many food-derived polysaccharides are reported to have potent reducing power and free radical scavenging ability in vitro. For example, we have previously isolated a polysaccharide from *Nostoc commune*, a widespread microalga with a long history as food and medicine, and found that the polysaccharide is capable of scavenging both superoxide anion and hydroxyl radicals in vitro [66]. The antioxidant capability of polysaccharides is shown to be related with their functional groups such as hydroxyl, amino, carbonyl and carboxyl groups, e.g., the scavenging capacity of chitosan against superoxide radicals is correlated with its number of hydroxyl and amino groups [74]. The polysaccharide fractions from *Zizyphus jujuba* with higher uronic acid content exhibit stronger free radical scavenging activities than other polysaccharide fractions from the same species containing no uronic acid [75]. These functional groups in polysaccharides can donate hydrogen to electron-deficient free radicals to generate alkoxyl products, which accelerate intramolecular hydrogen abstraction and further induce spirocyclization reaction to prevent radical chain reaction [22,76]. Interestingly, free radicals are usually generated via transition metal ions in in vitro antioxidant assays. In Fenton reaction, for instance, ferrous ion is used to catalyze superoxide or hydrogen peroxide to generate hydroxyl radicals [77]. Therefore, the direct scavenging effect of polysaccharides against free radicals may also be through chelating ions. For example, the polysaccharide fraction GAPS-1 isolated from *Aloe barbadensis* has a higher chelating ability against ferrous ion and meanwhile exhibits stronger scavenging effect against hydroxyl radicals as compared to SAPS-1, another polysaccharide fraction isolated from the same species [51]. Moreover, monosaccharide composition and substitution groups of polysaccharides are reported to play important roles in their chelating capacity, e.g., the chelating ability of the polysaccharides from *Zizyphus jujuba*, a well-known traditional food, against ferrous ion is positively correlated with their galacturonic acid contents [50].

In addition to scavenge free radicals in vitro, antioxidant polysaccharides are also shown to reduce the levels of ROS and associated peroxidation products in cellular and animal models under oxidative stress. For instance, a polysaccharide from the common fungus *Auricularia auricular* is capable of increasing the survival rate and reducing the ROS level in hydrogen peroxide-stressed *Caenorhabditis elegans* [59], while the wheat bran-derived feruloyl oligosaccharides can reduce MDA content and suppress protein carbonyl formation in human erythrocytes exposed to 2,2′-Azobis(2-amidinopropane) dihydrochloride, a potent free radical generator [34]. It is well established that peroxidation products can modify cellular components, leading to cell damages. For instance, MDA interacts with proteins and DNA to generate covalent adducts with mutagenic and carcinogenic effects [3], while protein carbonyl groups can cause rapid degradation of proteins [78]. Therefore, reduction of peroxidation product contents may contribute to the protective effects of feruloyl oligosaccharides against oxidative stress.

Mitochondria are the main source of ROS and energy production in cells. However, mitochondrial dysfunction, including mitochondrial membrane potential (MMP) decline, respiratory chain malfunction and calcium dysregulation, can accelerate ROS generation and reduce ATP generation, leading to oxidative damage and energy deficiency [79,80]. In a vicious cycle, excessive ROS further impair mitochondrial components such as membrane lipids and DNA, resulting in a secondary mitochondrial dysfunction that amplifies oxidative stress [81,82]. Therefore, restoring mitochondrial function is a beneficial strategy to reduce oxidative impairment. Interestingly, recent reports have revealed that the antioxidant function of food-derived polysaccharides is associated with the alleviation of mitochondrial dysfunction. For example, we have recently shown that the polysaccharide DiPS from *Dictyophora indusiata*, an edible mushroom traditionally used for inflammatory and neural diseases, can reduce paraquat-mediated increase of ROS level through elevating MMP in *C. elegans* [58]. A polysaccharide from *Ganoderma lucidum*, a well-known mushroom traditionally used to delay ageing and enhance immune function, is able to attenuate isoproterenol-induced cardiotoxicity via increasing MMP and mitochondrial complex activity in rats [56]. In addition to mitochondria, several other biochemical pathways such as NADPH oxidase also contribute to ROS production [9]. Interestingly, a recent study has found that a polysaccharide from *Sophora subprosrate*, a medicinal food used for inflammatory disorders, can reduce superoxide anion in porcine circovirus type 2-infected murine macrophage RAW264.7 cells by inhibiting the expression of NADPH oxidase, which is a major enzyme responsible for generating superoxide anion in phagocytes [43]. Together, these findings demonstrate that antioxidant polysaccharides can inhibit cellular ROS generation through multiple pathways.

2.2. Improvement of the Antioxidant Defense System

A number of studies have revealed that food-derived polysaccharides can reduce oxidative stress and associated damages through modulation of antioxidant enzymes in experimental models. For example, we have recently found that the acidic polysaccharide EbPS-A1 from *Epimedium brevicornum*, a functional food used for a variety of medical conditions including neurological disorders, can increase oxidative survival and reduce ROS level and MDA content of both wild-type and polyglutamine (polyQ) *C. elegans* under paraquat-induced oxidative stress. The protective effect of EbPS-A1 against paraquat toxicity is shown to be related with increasing SOD and CAT activities [38]. Interestingly, the polysaccharides isolated from the tonic food *Chuanminshen violaceum* are also shown to up-regulate the mRNA expression levels of SOD isoforms and CAT and enhance the activities of these antioxidant enzymes in mice injected with D-galactose [39], an ageing-promoting agent that induces cognitive and motor performance deterioration similar to AD symptoms via oxidative stress and mitochondrial dysfunction [83].

In addition to their effect on antioxidant enzymes, several food-derived polysaccharides are also reported to have modulatory function on non-enzyme components of the cellular antioxidant system. For example, a polysaccharide from *Anoectochilus roxburghii*, a medicinal food used to treat a variety of chronic diseases such as hepatitis and diabetes, is shown to attenuate oxidative stress by increasing glutathione level as well as antioxidant enzyme activities in the hepatic tissue of mice injected with carbon tetrachloride, an organic chemical that can induce hepatotoxicity through increased oxidative stress [84]. Interestingly, *A. roxburghii* polysaccharide is also shown to reduce the mRNA levels of inflammation-related genes including tumor necrosis factor alpha (TNF-α) and interleukin-6 (IL-6) [52]. Oxidative stress is known to increase the expression of TNF-α, a key cytokine that promotes inflammation, while elevated TNF-α level can activate NADPH oxidase, ultimately leading to ROS overproduction [85,86].

2.3. Regulation of Oxidative Stress-Related Signaling

A number of signaling pathways, such as those involving nuclear factor erythroid 2-related factor 2/antioxidant response element (Nrf2/ARE), mitogen-activated protein kinases (MAPKs), phosphoinositide 3 kinase/Akt (PI3K/Akt) and insulin/insulin-like growth factor-1 signaling (IIS), are known to be associated with cellular responses to multiple stresses including oxidative stress [87–89]. For instance, Nrf2, a basic region leucine-zipper transcription factor, plays an important role in cellular antioxidant response. When Nrf2 is activated, it translocates into nucleus and binds to ARE, leading to up-regulation of genes involved in cellular antioxidant and anti-inflammatory defense as well as mitochondrial protection [87]. Interestingly, some food-derived polysaccharides are recently reported to exert their antioxidant activity via Nrf2/ARE pathway in cellular and animal models. For instance, a polysaccharide from *Lycium barbarum*, a medicinal food traditionally used to retard ageing and improve neuronal function, is shown to attenuate ultraviolet B-induced cell viability decrease and ROS level increase in human keratinocytes HaCaT cells by promoting the nuclear translocation of Nrf2 and the expression of Nrf2-dependent ARE target genes [90]. This protective effect of *L. barbarum* polysaccharide can be neutralized by siRNA-mediated Nrf2 silencing, indicating an involvement of Nrf2/ARE pathway in the antioxidant effect of the polysaccharide [90]. Intriguingly, however, the above-mentioned polysaccharide DiPS is shown to increase oxidative survival through promoting nuclear translocation of transcription factor DAF-16/FOXO transcription factor but not SKN-1 (worm homologue of Nrf2) in wild-type *C. elegans* under paraquat exposure, demonstrating the antioxidant activity of the polysaccharide is associated with IIS, an evolutionarily conserved pathway that regulates organismal metabolism and lifespan, as DAF-16 is a key regulator in IIS [58].

Several signaling pathways related to cell death and survival are also involved in the antioxidant effect of food-derived polysaccharides. For example, hydrogen peroxide can induce apoptosis of rat pheochromocytoma PC12 cells via activation of p38 MAPK, while a polysaccharide from the fruiting bodies of the edible mushroom *Morchella importuna* increases the viability of hydrogen peroxide-exposed PC12 cells by inhibiting p38 MAPK phosphorylation [91]. In addition, hydrogen peroxide can inhibit the activation of PI3K/Akt signaling in human neuroblastoma SH-SY5Y cells, while sulfated polysaccharides prepared from fucoidan are able to increase the phosphorylation of PI3K/Akt and inhibit cell apoptosis [92]. Interestingly, the PI3K inhibitor LY294002 can partially prevent the beneficial role of the polysaccharide, demonstrating that modulation of PI3K/Akt pathway contributes to the protective effect of the sulfated polysaccharides against hydrogen peroxide cytotoxicity [92].

Recent studies provide clear evidence for the protective effects of food-derived polysaccharides against oxidative stress. Many polysaccharides exhibit potent reducing power, total antioxidant capacity and scavenging ability against free radicals in vitro. Moreover, some polysaccharides can decrease ROS and peroxidation product levels, improve antioxidant defense system and regulate stress-related signaling events to attenuate oxidative damage in cellular and animal models exposed to a variety of external stimuli, such as hydrogen peroxide, paraquat, ultraviolet radiation and virus. Together, these findings suggest a potential of these dietary polysaccharides to maintain health and prevent oxidative stress-related disorders.

3. Alleviation of Neurodegeneration by Food-Derived Antioxidant Polysaccharides

It is known that oxidative stress and chronic inflammation are two intertwined pathological events in NDD [85]. Excessive ROS can modulate inflammatory signaling to up-regulate the expression of pro-inflammatory factors such as cytokines, which act as potent stimuli in brain inflammation [93,94]. In turn, elevated inflammatory stress further provokes ROS generation via multiple pathways such as nuclear factor κB (NF-κB) signaling [85]. On the other hand, abnormal protein aggregation is known to be a common pathological hallmark of late-onset NDD. These protein aggregates, including amyloid-β peptide (Aβ) aggregates in AD and polyQ aggregates in HD, can induce neuronal damages through induction of oxidative stress, inflammation and mitochondrial dysfunction [95–97]. Oxidative stress can also promote the aggregation of pathogenic proteins as ROS modified-proteins tend to form aggregates [98]. In addition, a variety of chemical interventions, including excitatory amino acids such as glutamate, *N*-methyl-D-aspartate (NMDA) and kainic acid; neurotoxins such as 1-methyl-4-phenyl-1,2,3,6-tetrahydropyridine (MPTP) and 6-hydroxydopamine (6-OHDA); and ageing-promoting agents such as D-galactose, are also shown to induce neurodegenerative symptoms via oxidative and inflammatory stresses [83,99,100]. As oxidative stress plays a pivotal role in neurodegeneration, antioxidant strategies, including food-derived antioxidant polysaccharides, are shown to attenuate neuronal damage and improve cognitive and motor functions in a range of neurodegenerative models (Table 2) [101–142].

Table 2. Protective effects and mechanisms of food-derived antioxidant polysaccharides in neurodegeneration models.

Source	Polysaccharide	Test Model	Protective Effect	Potential Mechanism	Ref.
Ganoderma lucidum	GLP	APP/PS1 transgenic mice	Learning and memory in MWM↑, neural progenitor cell proliferation↑	Aβ deposits↓, protein levels of p-FGFR1, p-ERK and p-Akt↑	[101]
Marine red algae	KCP	Aβ(25–35)-exposed SH-SY5Y cells	Cell viability↑, apoptosis rate↓	Protein level of cleavage caspase 3↓, JNK signaling activation↓	[102]
Undaria pinnatifida sporophylls	Fucoidan	Aβ(25–35) and D-Gal-exposed PC12 cells; D-Gal treated ICR mice	Cell viability↑, apoptosis rate↓ in PC12 cells; Learning and memory in MWM↑	Protein levels of cleaved caspase-3, caspase-8 and caspase-9↓, cytochrome c release↓, SOD activity↑, GSH level↑ in PC12 cells; Aβ deposits in hippocampus↓, SOD activity and GSH level↑ in serum, Ach content↑, ChAT activity↑ and AChE activity↓ in brain	[103]
Laminaria japonica Aresch.	Fucoidan	Aβ40-treated SD rats	Learning and memory in MWM, single-trial passive avoidance and eight-arm radial maze task↑	Ach content↑, ChAT activity↑, AChE activity↓, SOD and GPx activities↑, MDA content↓, Bax/Bcl-2 protein level ratio↓, cleaved caspase-3 protein level↓ in hippocampus	[104]
Polygonatum sibiricum	PS-WNP	Aβ(25–35)-exposed PC12 cells	Cell viability↑, apoptosis rate↓	Bax/Bcl-2 protein level ratio↓, MMP↑, cytochrome c release↓, cleaved caspase-3 protein level↓, caspase-3 activity↓, p-Akt protein level↑	[105]
Lonicera japonica Thunb.	LJW0F2	Aβ42-exposed SH-SY5Y cells	Cell viability↑	Aβ42 aggregates↓	[106]
Ecklonia Kurome Okam.	AOSC	Aβ(25–35)-exposed SH-SY5Y cells	Cell viability↑, apoptosis rate↓, activation of astrocytes↓, cell redox activity↑	ROS level↓, TNF-α and IL-6 level↓, calcium influx in astrocytes↓	[107]
Angelica sinensis	AS	Aβ(25–35)-exposed Neuro 2A cells	Cell viability↑	ROS level↓, GSH level↑, MMP↑, mitochondria mass↑, TBARS content↓, autophagosomes or residual bodies↓	[108]
Lycium barbarum	L. barbarum polysaccharide	APP/PS1 transgenic mice	Learning and memory in MWM↑	Aβ deposits in hippocampus↓	[109]
Lycium barbarum	LBP-III	Aβ(25–35)-exposed rat primary cortical neurons	Maintain neurite fasciculation and neuron integrity	Caspase-3 and caspase-2 activities↓, p-PKR protein level↓	[110]
Ganoderma lucidum	GLA	Aβ(25–35) or Aβ42-exposed rat primary cortical neurons	Apoptosis rate↓, synaptophysin immunoreactivity↑	DEVD-cleavage activity↓, protein levels of p-JNK, p-c-Jun, and p-p38↓	[111]
Rubia cordifolia L.	PS5	T-REx293 cells	Cell viability↑	Aβ42-EGFP aggregates↓	[112]
Dictyophora indusiata	DiPS	C. elegans CL2355	Survival rate↑, chemotaxis index↑	ROS level↓	[58]

Nutrients **2017**, 9, 778

Table 2. *Cont.*

Source	Polysaccharide	Test Model	Protective Effect	Potential Mechanism	Ref.
Gynostemma pentaphyllum Makino	GPP1	Aβ(25–35)-exposed PC12 cells	Cell viability↑, LDH release↓, DNA fragmentation↓	ROS level↓, MDA content↓, SOD activity↑, GSH level↑, Calcium overload↓, MMP↑, Bcl-2 protein level↑, protein levels of Bax, cytochrome c and cleaved caspase-3↓	[113]
Lycium barbarum L.	LBP	6-OHDA-exposed PC12 cells	Cell viability↑, nuclear morphology changes↓, apoptosis rate↓	ROS and NO levels↓, calcium overload↓, protein-bound 3-nitrotyrosine level↓, protein levels of nNOS, iNOS and cleaved caspase-3↓	[114]
Gynostemma pentaphyllum Makino	GP	MPP⁺-exposed PC12 cells	Cell viability↑, LDH release↓, apoptosis rate↓	Cytochrome c release↓, caspase-3 and caspase-9 activities↓, Bax/Bcl-2 protein level ratio↓, protein levels of cleaved caspase-3 and poly (ADP-ribose) polymerase↓	[115]
Spirulina platensis	PSP	MPTP-treated C57BL/6J mice	Number of TH-immunoreactive neurons and DAT binding ratio in the substantia nigra pars compacta↑	TH and DAT mRNA levels in substantia nigra↑, SOD and GPx activity↑ in serum and midbrain	[116]
Chlorella pyrenoidosa	CPS	MPTP-treated C57BL/6J mice	Body weight↑, movement in pole test and gait test↑	Contents of DA, DOPAC and HVA↑, ratio of DOPAC and HVA to DA↓, TH mRNA level↑, striatal Emr1 mRNA level↓, TNF-α, IL-1β and IL-6 levels in serum↓, D-amino acid oxidase and secretory immunoglobulin A levels↑	[117]
Gracilaria cornea J. Agardh	SA-Gc	6-OHDA-treated Wistar rats	Locomotor performance in OFT, rotarod and apomorphine-induced rotation test↑, weight gain↑	DA and DOPAC content↑, NO₂/NO₃ and GSH level↑ in brain, p65, iNOS and IL1β mRNA levels↓, BDNF mRNA level↑	[118]
Stichopus japonicus	SJP	6-OHDA-exposed SH-SY5Y cells	Cell viability↑, apoptosis rate↓, LDH release↓	SOD activity↑, ROS level↓, NO release↓, MDA content↓, MMP↑, cytochrome c release↓, percentage of cells in S phase↑, Bax/Bcl-2 protein level ratio↓, protein levels of Cyclin D3, p-p53, p-p38, p-JNK1/2, p-p65, iNOS and p-IκB↓, cleaved caspase-9/caspase-9 and cleaved caspase-3/caspase-3 protein level ratio↓, p-Akt and IκB protein levels↑	[119]
Hericium erinaceus	EA	MPTP-treated C57BL/6 mice	Apoptosis rate↓, number of normal neurons↑, motor function in RT↑	Nitro-tyrosine and 4-HNE level↓, dopamine, NGF, and GSH level↑, protein levels of Fas, p-JNK1/2, p-p38, DNA damage inducible transcript 3, NF-κB and p65↓	[120]
Epimedium brevicornum Maxim.	EbPS-A1	*C. elegans* HA759	Avoidance index↑	ROS level↓, MDA content↓, SOD and CAT activities↑	[38]
Turbinaria decurrens	TD fucoidan	MPTP-treated C57BL/6 mice	Motor performance in OFT, Narrow beam walking and RT↑, nigral TH immunoreactivity↑	DA, DOPAC, and HVA content↑, TBARS level↓, GSH level↑, SOD and CAT activities↓, GPx activity↑, TH and DAT protein levels↑	[121]

55

Table 2. *Cont.*

Source	Polysaccharide	Test Model	Protective Effect	Potential Mechanism	Ref.
Lycium barbarum	LBP	HEK293-160Q cells; HD-related transgenic mice	Cell viability↑ in HEK293 cells; Survival rate↑, weight gain↑, motor performance in RT↑ in mice	Soluble and aggregated huntingtin levels↓, caspase-3 activity↓, p-Akt/Akt and p-GSK3β/GSK3β protein levels↑ in HEK293 cells; Mutant huntingtin level↓, p-Akt/Akt and p-GSK3β/GSK3β protein levels↑ in mouse brain	[122]
Ganoderma lucidum	GLP	Kainic acid-treated Wistar rats	Frequency of epilepsy↓	CaMK II level↑, ERK1/2 level↓, calcium turnover↓, Caveolin-1 positive cells↑, NF-κB positive cells↓	[123]
Hericium erinaceus	HE	L-Glu-exposed PC12 cells; AlCl3 and D-Gal-treated Balb/c mice	Differentiation rate↑, cell viability↑, apoptosis rate↓, in PC12 cells; learning, memory and locomotor in MWM, Autonomic activities and RT↑	β-tubulin III protein level↑, MMP↑, calcium overload↓, ROS level↓ in PC12 cells; Ach and ChAT contents in mouse serum and hypothalamus↑	[124]
Pleurotus ostreatus	POP	D-Gal and AlCl3-treated Wistar rats	Learning and memory in MWM and SDT↑, hippocampal impairment↓	AchE activity↓, in hippocampus, MDA content↓, SOD, GPx and CAT activities↑ in hippocampus, liver and serum, protein levels of APP, Aβ, BACE1 and p-tau↓, Protein phosphatase 2 protein level↑	[125]
Sargassum fusiforme	SFPS65A	SCO-, ethanol- and sodium nitrite-treated ICR mice	Learning and memory in SDT↑		[126]
Sargassum fusiforme	SFPS	D-Gal-treated ICR mice		CAT and SOD activities↑, MDA content in hearts and MAO in brains↓, protein levels of Nrf2, Bcl-2, p21 and JNK1/2↑, mRNA levels of Nrf2, Cu/Zn-SOD, Mn-SOD, glutamate cysteine ligase and GPX1↑, voltage dependent anion channel 1 protein level↓	[127]
Lycium barbarum	LBA	Homocysteine-exposed cortical neurons	Cell viability↑, apoptosis rate↓	LDH release and caspase-3 activity↓, p-tau-1 protein level↑, cleaved-tau protein level↓, p-ERK1/2 and p-JNK protein levels↓	[128]
Lycium barbarum	LBA	L-Glu- or NMDA-exposed cortical neurons	Cell viability↑, maintained their integrity and fasciculation of neurites	LDH release and caspase-3 activity↓, p-JNK-1/JNK protein level ratio↓	[129]
Saccharomyces cerevisiae	β-glucan	SCO-treated SD rats	Learning, memory, and locomotor in MWM and PTT↑	AChE activity↓	[130]
Flammulina velutipes	FVP	SCO-treated Wistar rats	Learning and memory in MWM and PTT↑	SOD and GPx activities↑, TBARS level↓, Ach, 5-HT, DA and NE content↑, ChAT activity↑, AChE activity↓, connexin 36 and p-CaMK II protein level↑ in hippocampus and cerebral cortex	[131]
Lycium barbarum	LBPs	SCO-treated SD rats	Learning and memory in MWM, NOR and OLR↑, cell proliferation and neuroblast differentiation in dentate gyrus↑	SOD and GPX activities↑, MDA content↓, Bax/Bcl-2 protein level ratio↓ in hippocampus	[132]

Table 2. *Cont.*

Source	Polysaccharide	Test Model	Protective Effect	Potential Mechanism	Ref.
Lycium barbarum	LBP	D-Gal-treated Kunming mice	Weight gain↑, learning and memory in Jumping test↑, thymus and spleen indices↑	Lipid peroxidation, lipofuscin and MAO-B contents↓ in brain	[133]
Polygonatum sibiricum	PSP	SCO-treated Kunming mice	Learning and memory in SDT and Memory test↑	SOD and GPx activities↑, MDA content↓	[134]
Panax ginseng	WGOS	SCO-treated ICR mice	Learning and memory in MWM and NOR↑	mRNA levels of GFAP, IL-1β and IL-6↓ in hippocampus, number of GFAP-positive cells↓ in hippocampal subregions	[135]
Lentinus edodes	LT2	D-Gal-treated Kunming mice	Erythrocyte membrane fluidity↑	SOD and GPx activities↑ in liver, heart and brain	[136]
Angelica sinensis	ASP	D-Gal-treated C57BL/6J mice	Percentage of ageing cells↓	Advanced glycation end-product level in serum↓, ROS level↓, TAOC content↑, 8-OHDG content↓, 4-HNE level↓, protein levels of H2A histone family member X, p16, p21, p53, β-catenin, p-GSK-3β and transcription factor 4↓, mRNA levels of p16, p21 and β-catenin↓, GSK-3β protein level↑	[137]
Tricholoma lobayense	TLH-3	*t*-BHP-exposed HELF cells; D-Gal-treated Kunming mice	Cell viability↑, percentage of ageing cells↓, ratio of G0/G1phase↓, nucleic morphological changes↓, in HELF cells	ROS level↓, in HELF cells; SOD and CAT activities↑, MDA content↓, in mouse liver and serum	[138]
Cuscuta chinensis Lam	PCCL	D-Gal-treated SD rats	Apoptosis rate of cardiomyoctyes↓	Calcium overload↓, Bax/Bcl-2 protein level ratio↓, caspase-3 activity↓, cytochrome c release↓	[139]
Ganoderma atrum	PSG-1	D-Gal-treated Kunming mice	Weight gain↑, lymphocyte proliferation↑	MDA content↓, SOD, CAT and GPx activities↑, GSH level↑, GSSG level↓ in liver, brain and spleen	[140]
Auricularia auricula-judae	APP 1-a	D-Gal-treated Kunming mice	Spleen and thymus indexes↑	MDA content↓, SOD and GPx activities↑ in liver, serum and heart	[141]
Saccharina japonica	DJ0.5	6-OHDA-exposed MES 23.5 cells and SH-SY5Y cells	Cell viability↑		[142]

Aβ, amyloid-β peptide; Ach, acetylcholine; AChE, acetylcholinesterase; APP, amyloid precursor protein; BACE1, β-secretase 1; BDNF, brain-derived neurotrophic factor; CaMK II, calmodulin-dependent protein kinase II; CAT, catalase; ChAT, choline acetyltransferase; CL2355, a nematode that pan-neuronally expresses Aβ42; Cu/Zn-SOD, copper-zinc superoxide dismutase; D-Gal, D-galactose; DA, dopamine; DAT, dopamine transporter; DOPAC, 3,4-Dihydroxyphenylacetic acid; FGFR1, fibroblast growth factor receptor 1; GFAP, glial fibrillary acid protein; GPx, glutathione peroxidase; GSH, glutathione; GSK-3β, glycogen synthase kinase-3β; GSSG, glutathione disulfide; HA759, a nematode that expresses HtnQ150 in ASH neurons; HEK293 cells, human embryonic kidney cell line; HELF cells, human embryonic lung fibroblast line; HVA, homovanillic acid; IκB, NF-κB inhibitor; iNOS, inducible nitric oxide synthase; LDH, lactate dehydrogenases; L-Glu, L-glutamate; MAO, monoamine oxidase; MES 23.5 cells, rodent mesencephalic neuronal cell line; MDA, malondialdehyde; MMP, mitochondrial membrane potential; MPTP, 1-methyl-4-phenyl-1,2,3,6-tetrahydropyridine; MPP⁺, 1-methyl-4-phenylpyridinium; MWM, Morris water maze; NE, norepinephrine; Neuro 2A cells, murine neuroblastoma cell line; NMDA, N-methyl-D-aspartate; nNOS, neuronal nitric oxide synthase; NO, nitric oxide; NOR, novel object recognition; Nrf2, nuclear factor erythroid 2-related factor 2; OLR, object location recognition; OFT, open field test; PC12 cells, rat pheochromocytoma cell line; PS1, presenilin-1; PTT, probe trial test; ROS, reactive oxygen species; RT, Rotarod test; SCO, scopolamine; SDT, step-down test; SH-SY5Y, human neuroblastoma cell line; SOD, superoxide dismutase; TAOC, total antioxidant capacity; TBARS, thiobarbituric acid reactive substances; TH, tyrosine hydroxylase; TNF-α, tumor necrosis factor α; T-REx293, human embryonic kidney cell line transiently transfected with Aβ42-EGFP; *t*-BHP, tert-butylhydroperoxide; 4-HNE, 4-hydroxynonenal; 5-HT, 5-hydroxytryptamine; 6-OHDA, 6-hydroxydopamine; 8-OHDG, 8-hydroxydeoxyguanosine.

3.1. Effects on Alzheimer's Disease

AD is characterized by amyloid plaques and neurofibrillary tangles in the brain, which lead to progressive memory loss and cognitive decline [143]. As global population ages, AD has become a major public health concern. Among therapeutic and nutritional interventions, growing evidence has shown that adequate intake of antioxidants may be helpful to reduce neuronal damages and alleviate AD symptoms [144,145]. For example, dietary intake of α-tocopherol or combined tocopherols shows beneficial effects to alleviate age-related cognitive decline and lower AD risk [146,147].

Antioxidant polysaccharides from various food sources are also found to inhibit Aβ-mediated neurotoxicity in experimental models (Table 2). The polysaccharide PS-WNP from the medicinal food *Polygonatum sibiricum*, for instance, is shown to significantly attenuate Aβ-induced apoptosis of PC12 cells by alleviating mitochondrial dysfunction, regulating apoptosis-related protein Bax and Bcl-2 levels, inhibiting apoptotic executor caspase-3 activation and enhancing Akt phosphorylation [105]. In rats injected with Aβ40 aggregates, fucoidan is shown to attenuate learning and memory deficits by elevating SOD and GPx activities and decreasing MDA content, Bax/Bcl-2 ratio and caspase-3 activity in hippocampal tissue [104]. Using transgenic *C. elegans* models that overexpress Aβ proteins, the *D. indusiata* polysaccharide DiPS is shown to alleviate chemosensory behavior dysfunction, which is associated with reduction of ROS level and MDA content, increase of SOD activity and alleviation of mitochondrial dysfunction [58]. Antioxidant polysaccharides are also shown to modulate pathogenic protein aggregation, e.g., *L. barbarum* polysaccharides can reduce Aβ42 protein level in hippocampal tissue and improve the performance of learning and memory in APP/PS1 mice [109]. Intriguingly, *L. barbarum* polysaccharide is also shown to inhibit the apoptosis and reduce cleaved-tau protein level, the main component of neurofibrillary tangles in AD patients, in rat primary cortical cells exposed to homocysteine, a sulfur-containing amino acid associated with several NDD [128]. Moreover, several studies have uncovered that the regulatory effect of polysaccharides on protein aggregation is through the interaction with aggregation-prone proteins, and this effect is influenced by the chemical structure of polysaccharides. For example, four glycosaminoglycans from different animal tissues are shown to inhibit the neurotoxicity of serum amyloid P component and its interaction with Aβ, and the inhibitory efficacy is correlated with the uronic acid content in glycosaminoglycans [148]. In addition, the well-known glycosaminoglycan heparin is reported to bind with Aβ and promote amyloid fibrillogenesis, while low molecular weight heparin can prevent Aβ aggregation by blocking β-sheet formation and inhibiting fibril formation, suggesting that the molecular weight of polysaccharides may also affect their interaction with proteins [149,150]. Together, these studies demonstrate that the neuroprotective effects of food-derived polysaccharides in AD-like models correlate with their modulation of oxidative and related stresses.

3.2. Effects on Parkinson's Disease

PD is a chronic and progressive NDD characterized by selective loss of dopaminergic neurons in the substantia nigra pars compacta and abnormal accumulation of Lewy bodies in these neurons [151]. The major clinical symptoms of PD include motor symptoms such as tremor and bradykinesia, and neuropsychiatric symptoms such as cognitive decline and anxiety [152]. Current clinic therapy for PD only concentrates on symptomatic management as the available therapeutics do not prevent disease progression [153].

Recent studies have shown that several food-derived antioxidant polysaccharides are capable of inhibiting the neurotoxicity mediated by MPTP and 6-OHDA, which can selectively induce dopaminergic neuron death and cause PD-like motor deficits in experimental models (Table 2). For instance, the polysaccharides from the seaweed *Saccharina japonica* and from the sea cucumber *Stichopus japonicus* can increase 6-OHDA-induced reduction of cell viability in SH-SY5Y cells and murine embryonic stem MES 23.5 cells, respectively [119,142]. The *S. japonicus* polysaccharides are shown to increase SOD activity, regulate the level of apoptosis-related proteins, inhibit NF-κB and p38 MAPK activation and activate PI3K/Akt pathway, indicating the involvement of antioxidant, anti-apoptotic and anti-inflammatory signaling pathways in its neuroprotective effect [119]. Using MPTP-injected mouse models, low molecular weight fucoidan DF and its two fractions DF1 and DF2 are shown to ameliorate dopaminergic neuron injury and prevent dopamine depletion in the substantia nigra through enhancing antioxidant enzyme activities and inhibiting neuronal apoptosis [154]. Interestingly, DF1 exerts better neuroprotective activity than DF and DF2 in general, and their monosaccharide compositions are different: DF1 is a hetero-polysaccharide with low content of fucose and high content of uronic acid and other monosaccharides, while DF2 mainly consists of fucose and galactose, suggesting that chemical composition may play an important role in the neuroprotective activity of fucoidan [154]. In addition, a polysaccharide from the edible microalga *Chlorella pyrenoidosa* is recently shown to reduce bradykinesia, inhibit the loss of striatal dopamine and its metabolites, and increase tyrosine hydroxylase in MPTP-injected mice [117]. The polysaccharide can also elevate the levels of small intestinal secretory immunoglobulin A, a protein that is crucial for the immune function of mucous membranes, in mice serum [117], and has been previously shown to enhance immune function [155]. As immune system dysfunction is known to contribute to PD development and progression [156], immune-related therapies may be a useful strategy to reduce disease risks and retard disease progression [157].

3.3. Effects on Huntington's Disease

HD is an autosomal-dominant neurodegenerative disorder that is clinically manifested by a variety of motor, cognitive and psychiatric deficits [158]. This disease is caused by an abnormal expanded CAG trinucleotide repeat in the huntingtin gene on the short arm of chromosome 4. In normal individuals, the average number of CAG repeats in the huntingtin gene is 17–20; when the number of repeats exceeds 36, the risk of developing HD is significantly increased [159]. The prevalence of HD varies geographically, with the highest rates in Europe (~10–15 per 100,000 individuals) and lower rates in Asia and Africa [160]. Similar with AD and PD, currently there is no efficient treatment for HD.

Among various pharmacological interventions, natural antioxidants such as epigallocatechin gallate and salidroside have been found to alleviate HD-like symptoms in transgenic cellular and animal models [161,162]. Interestingly, several recent studies have uncovered that food-derived antioxidant polysaccharides also have beneficial effects in HD-like animal models (Table 2). For example, the *E. brevicornum* polysaccharide EbPS-A1 can alleviate polyQ-mediated chemosensory dysfunction in transgenic *C. elegans* model HA759 [38], which expresses a polyQ tract of 150 glutamine repeats in amphid sensilla (ASH) neurons, leading to progressive ASH death and chemotactic behavior deficit [163]. EbPS-A1 also reduces ROS level, inhibits lipid peroxidation and enhances antioxidant enzyme activities in HA759 nematodes, indicating that the antioxidant activity of the polysaccharide contributes to its protective effect against polyQ neurotoxicity [38]. Other studies suggest that some antioxidant polysaccharides exert their neuroprotective effects by targeting polyQ aggregate itself, e.g., *L. barbarum* polysaccharide not only increases the viability of HEK293 cells that express mutant-huntingtin containing 160 glutamine repeats but also improves motor behavior and lifespan in HD-related transgenic mice [122]. The neuroprotective effect of *L. barbarum* polysaccharide against mutant-huntingtin toxicity in both cellular and mouse models are shown to be associated with reducing mutant-huntingtin levels and activating AKT [122]. These studies provide an important insight into the therapeutic potential of food-derived antioxidant polysaccharides in HD.

3.4. Effects on Other Neurodegenerative Symptoms

Several recent studies have shown that food-derived antioxidant polysaccharides are capable of inhibiting excitatory amino acid-mediated neurotoxicity, which is implicated in many NDD [99,164]. For instance, *L. barbarum* polysaccharide can increase cell viability and suppress JNK activation in glutamate-exposed rat primary cortical neurons [129], suggesting an involvement of MAPK pathway in the neuroprotective effect of the polysaccharide. Another example is *G. lucidum* polysaccharide, which is shown to alleviate epileptic symptoms and up-regulate the expression of calcium/calmodulin-dependent protein kinase II, a kinase that plays an important role in calcium transfer in neurons, in kainic acid-injected rats [123]. As calcium overload mediates excitatory amino acid-induced neurotoxicity [164], prevention of calcium transporting may contribute to this neuroprotective effect of *G. lucidum* polysaccharide. In addition, other chemicals can also induce cognitive impairment and behavior deficit through increase of oxidative and inflammatory stresses, and the polysaccharides isolated form mushrooms, medicinal herbs and algae are reported to attenuate neurodegenerative symptoms induced by these toxic chemicals. For instance, a polysaccharide from *Pleurotus ostreatus* can decrease escape latency in Morris water maze test and increase passive avoidance latency in step-down test in rats under D-galactose and aluminum chloride challenge [125]. *P. ostreatus* polysaccharide also reduces MDA level and elevates SOD, GPx and CAT activities [125], indicating that the behavior-improving capability of the polysaccharide correlates with reduction of oxidative stress.

A large body of evidence has confirmed that oxidative stress can interact with many other stresses to induce neurodegeneration, indicating its significant role in NDD development. Food-derived antioxidant polysaccharides are recently shown to alleviate neuronal injury, death and dysfunction through modulation of multiple oxidative stress-related pathways, including antioxidant defense system, mitochondrial function, peroxidation products, protein aggregation, inflammation and stress-related signaling (Figure 1), demonstrating their pharmacological potentials in NDD.

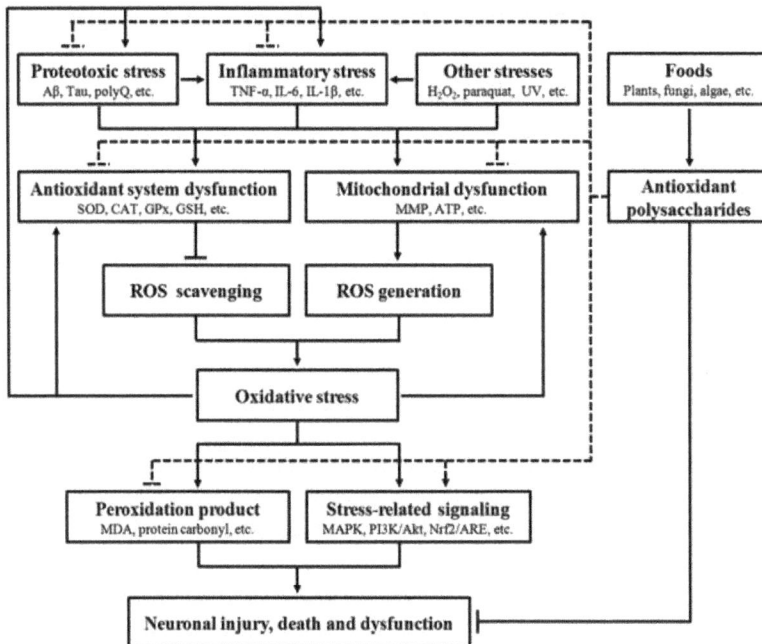

Figure 1. Pharmacological intervention of neurodegeneration by food-derived antioxidant polysaccharides. A number of extrinsic and intrinsic stresses such as proteotoxic stress, inflammatory stress and chemical interruption can stimulate oxidative stress through impairing the function of antioxidant system and mitochondria. Increase of oxidative stress can promote pathogenic protein aggregation and inflammation, eventually leading to neuronal injury, death and dysfunction via multiple biochemical pathways (solid line). However, food-derived antioxidant polysaccharides can exert beneficial effects to suppress neurodegeneration via attenuating oxidative, inflammatory and proteotoxic stresses and regulating stress-related signaling (dashed line).

4. Conclusions

Food-derived polysaccharides have been shown to scavenge free radicals in vitro and reduce oxidative damages in cellular and animal models, and their in vivo antioxidant capacities are related with regulation of peroxidation products, antioxidant defense system and stress-related signaling. As oxidative stress is closely associated with neurodegeneration, some antioxidant polysaccharides are also tested for their anti-NDD activity and found to attenuate neuronal damages and dysfunction in a number of neurodegenerative models. The neuroprotective effects of polysaccharide are associated with alleviation of multiple stresses, including oxidative, inflammatory and proteotoxic stresses (Figure 1). Therefore, consumption of foods rich in antioxidant polysaccharides may not only reduce oxidative damage but also provide protection against oxidative stress-related disorders. It is noted that most recent studies focus on the antioxidant polysaccharides from terrestrial plants and fungi, and relatively less attention is paid to marine organisms although they represent a rich resource of bioactive polysaccharides. In addition, many food-derived antioxidant polysaccharides are shown to have potent immunomodulatory effects, and therefore it would be interesting to explore the involvement of immunomodulation in the neuroprotective effect of antioxidant polysaccharides.

Acknowledgments: This work was supported by the National Natural Science Foundation of China (Grant 81403081), the National High-Tech R & D Program of China (863 Program; Grant 2014AA022001), the International Science & Technology Cooperation Program of China (Grant 2015DFA30280), the Special Funds of the Central Finance to Support the Development of Local Universities and Colleges, and Guangdong Province Department of Education (Grants 2015KGJHZ022 and 2016KQNCX084).

Author Contributions: Zebo Huang and Haifeng Li conceived and designed the review. Haifeng Li, Fei Ding, Lingyun Xiao, Ruona Shi, Hongyu Wang, and Wenjing Han collected the literature, analyzed the data and made figures and tables. Haifeng Li, Fei Ding, and Zebo Huang wrote and revised the manuscript. All authors have read and approved the final version.

Conflicts of Interest: The authors declare that there is no conflict of interests.

References

1.   McCord, J.M. The evolution of free radicals and oxidative stress. *Am. J. Med.* **2000**, *108*, 652–659. [CrossRef]
2.   Reczek, C.R.; Chandel, N.S. ROS-dependent signal transduction. *Curr. Opin. Cell Biol.* **2015**, *33*, 8–13. [CrossRef] [PubMed]
3.   Ayala, A.; Muñoz, M.F.; Argüelles, S. Lipid peroxidation: Production, metabolism, and signaling mechanisms of malondialdehyde and 4-hydroxy-2-nonenal. *Oxid. Med. Cell. Longev.* **2014**, *2014*, 360438. [CrossRef] [PubMed]
4.   Dalle-Donne, I.; Aldini, G.; Carini, M.; Colombo, R.; Rossi, R.; Milzani, A. Protein carbonylation, cellular dysfunction, and disease progression. *J. Cell. Mol. Med.* **2006**, *10*, 389–406. [CrossRef] [PubMed]
5.   Rahal, A.; Kumar, A.; Singh, V.; Yadav, B.; Tiwari, R.; Chakraborty, S.; Dhama, K. Oxidative stress, prooxidants, and antioxidants: The interplay. *BioMed Res. Int.* **2014**, *2014*, 761264. [CrossRef] [PubMed]
6.   Al-Dalaen, S.M.; Al-Qtaitat, A.I. Review article: Oxidative stress versus antioxidants. *J. Biosci. Bioeng.* **2014**, *2*, 60–71. [CrossRef]
7.   Birben, E.; Sahiner, U.M.; Sackesen, C.; Erzurum, S.; Kalayci, O. Oxidative stress and antioxidant defense. *World Allergy Organ. J.* **2012**, *5*, 9–19. [CrossRef] [PubMed]
8.   Gupta, R.K.; Patel, A.K.; Shah, N.; Chaudhary, A.K.; Jha, U.K.; Yadav, U.C.; Gupta, P.K.; Pakuwal, U. Oxidative stress and antioxidants in disease and cancer: A review. *Asian Pac. J. Cancer Prev.* **2014**, *15*, 4405–4409. [CrossRef] [PubMed]
9.   Rahman, T.; Hosen, I.; Towhidul Islam, M.M.; Uddin Shekhar, H. Oxidative stress and human health. *Adv. Biosci. Biotechnol.* **2012**, *3*, 997–1019. [CrossRef]
10. Niedzielska, E.; Smaga, I.; Gawlik, M.; Moniczewski, A.; Stankowicz, P.; Pera, J.; Filip, M. Oxidative stress in neurodegenerative diseases. *Mol. Neurobiol.* **2016**, *53*, 4094–4125. [CrossRef] [PubMed]
11. Newsholme, P.; Cruzat, V.F.; Keane, K.N.; Carlessi, R.; de Bittencourt, P.I., Jr. Molecular mechanisms of ROS production and oxidative stress in diabetes. *Biochem. J.* **2016**, *473*, 4527–4550. [CrossRef] [PubMed]
12. Gammon, K. Neurodegenerative disease: Brain windfall. *Nature* **2014**, *515*, 299–300. [CrossRef] [PubMed]

13. Nikam, S.; Nikam, P.; Ahaley, S.K.; Sontakke, A.V. Oxidative stress in Parkinson's disease. *Indian J. Clin. Biochem.* **2009**, *24*, 98–101. [CrossRef] [PubMed]

14. Chen, Z.; Zhong, C. Oxidative stress in Alzheimer's disease. *Neurosci. Bull.* **2014**, *30*, 271–281. [CrossRef] [PubMed]

15. Melo, A.; Monteiro, L.; Lima, R.M.; Oliveira, D.M.; Cerqueira, M.D.; El-Bachá, R.S. Oxidative stress in neurodegenerative diseases: Mechanisms and therapeutic perspectives. *Oxid. Med. Cell. Longev.* **2011**, *2011*, 467180. [CrossRef] [PubMed]

16. Sultana, R.; Perluigi, M.; Allan Butterfield, D. Lipid peroxidation triggers neurodegeneration: A redox proteomics view into the Alzheimer disease brain. *Free Radic. Biol. Med.* **2013**, *62*, 157–169. [CrossRef] [PubMed]

17. Shahidi, F. Nutraceuticals, functional foods and dietary supplements in health and disease. *J. Food Drug Anal.* **2012**, *20*, 226–230.

18. Herrera, E.; Jiménez, R.; Aruoma, O.I.; Hercberg, S.; Sánchez-García, I.; Fraga, C. Aspects of antioxidant foods and supplements in health and disease. *Nutr. Rev.* **2009**, *67*, S140–144. [CrossRef] [PubMed]

19. Virmani, A.; Pinto, L.; Binienda, Z.; Ali, S. Food, nutrigenomics, and neurodegeneration–neuroprotection by what you eat! *Mol. Neurobiol.* **2013**, *48*, 353–362. [CrossRef] [PubMed]

20. Wang, H.; Liu, Y.M.; Qi, Z.M.; Wang, S.Y.; Liu, S.X.; Li, X.; Wang, H.J.; Xia, X.C. An overview on natural polysaccharides with antioxidant properties. *Curr. Med. Chem.* **2013**, *20*, 2899–2913. [CrossRef] [PubMed]

21. Mei, X.; Yi, C.; Huang, G. The antioxidant activities of polysaccharides and their derivatives. *Curr. Drug Targets.* **2017**, in press. [CrossRef]

22. Wang, J.; Hu, S.; Nie, S.; Yu, Q.; Xie, M. Reviews on mechanisms of in vitro antioxidant activity of polysaccharides. *Oxid. Med. Cell. Longev.* **2016**, *2016*, 5692852. [CrossRef] [PubMed]

23. Wang, Z.J.; Xie, J.H.; Nie, S.P.; Xie, M.Y. Review on cell models to evaluate the potential antioxidant activity of polysaccharides. *Food Funct.* **2017**, *8*, 915–926. [CrossRef] [PubMed]

24. Wang, M.; Zhu, P.; Zhao, S.; Nie, C.; Wang, N.; Du, X.; Zhou, Y. Characterization, antioxidant activity and immunomodulatory activity of polysaccharides from the swollen culms of *Zizania latifolia*. *Int. J. Biol. Macromol.* **2017**, *95*, 809–817. [CrossRef] [PubMed]

25. Ghazala, I.; Sila, A.; Frikha, F.; Driss, D.; Ellouz-Chaabouni, S.; Haddar, A. Antioxidant and antimicrobial properties of water soluble polysaccharide extracted from carrot peels by-products. *J. Food Sci. Technol.* **2015**, *52*, 6953–6965. [CrossRef]

26. Nara, K.; Yamaguchi, A.; Maeda, N.; Koga, H. Antioxidative activity of water soluble polysaccharide in pumpkin fruits (*Cucurbita maxima* Duchesne). *Biosci. Biotechnol. Biochem.* **2009**, *73*, 1416–1418. [CrossRef] [PubMed]

27. Jeddou, K.B.; Chaari, F.; Maktouf, S.; Nouri-Ellouz, O.; Helbert, C.B.; Ghorbel, R.E. Structural, functional, and antioxidant properties of water-soluble polysaccharides from potatoes peels. *Food Chem.* **2016**, *205*, 97–105. [CrossRef] [PubMed]

28. Hu, T.; Wei, X.; Zhang, X.; Cheng, F.; Shuai, X.; Zhang, L.; Kang, L. Protective effect of *Potentilla anserine* polysaccharide (PAP) on hydrogen peroxide induced apoptosis in murine splenic lymphocytes. *Carbohydr. Polym.* **2010**, *79*, 356–361. [CrossRef]

29. Kim, S.Y.; Kim, E.A.; Kim, Y.S.; Yu, S.K.; Choi, C.; Lee, J.S.; Kim, Y.T.; Nah, J.W.; Jeon, Y.J. Protective effects of polysaccharides from *Psidium guajava* leaves against oxidative stresses. *Int. J. Biol. Macromol.* **2016**, *91*, 804–811. [CrossRef] [PubMed]

30. Dou, J.; Guo, Y.; Xue, Z.; Chen, W.; Li, J.; Meng, Y. Purification and antioxidative activity of polysaccharides from cold-extracting apple peel. *Sci. Technol. Food Ind.* **2014**, *1*, 111–115.

31. Zhang, Y.; Lu, X.; Fu, Z.; Wang, Z.; Zhang, J. Sulphated modification of a polysaccharide obtained from fresh persimmon (*Diospyros kaki* L.) fruit and antioxidant activities of the sulphated derivatives. *Food Chem.* **2011**, *127*, 1084–1090. [CrossRef] [PubMed]

32. Liu, Q.; Song, S.; Guo, J.; Luo, S.; Zhang, J. Protective effects of polysaccharide from seed watermelon on the oxidative damage of PC12 cells induced by H_2O_2. *Nat. Prod. Res. Dev.* **2015**, *2*, 338–343.

33. Hefnawy, H.T.; El-Shourbagy, G.A. Chemical analysis and antioxidant activity of polysaccharide extracted from rice bran. *World J. Dairy Food Sci.* **2014**, *9*, 95–104. [CrossRef]

34. Wang, J.; Sun, B.; Cao, Y.; Tian, Y. Protection of wheat bran feruloyl oligosaccharides against free radical-induced oxidative damage in normal human erythrocytes. *Food Chem. Toxicol.* **2009**, *47*, 1591–1599. [CrossRef] [PubMed]

35. Mateos-Aparicio, I.; Mateos-Peinado, C.; Jiménez-Escrig, A.; Rupérez, P. Multifunctional antioxidant activity of polysaccharide fractions from the soybean by product okara. *Carbohydr. Polym.* **2010**, *82*, 245–250. [CrossRef]

36. Ye, Z.; Wang, W.; Yuan, Q.; Ye, H.; Sun, Y.; Zhang, H.; Zeng, X. Box-Behnken design for extraction optimization, characterization and in vitro antioxidant activity of *Cicer arietinum L.* hull polysaccharides. *Carbohydr. Polym.* **2016**, *147*, 354–364. [CrossRef] [PubMed]

37. Yang, W.; Wang, Y.; Li, X.; Yu, P. Purification and structural characterization of Chinese yam polysaccharide and its activities. *Carbohydr. Polym.* **2015**, *117*, 1021–1027. [CrossRef] [PubMed]

38. Xiang, Y.; Zhang, J.; Li, H.; Wang, Q.; Xiao, L.; Weng, H.; Zhou, X.; Ma, C.; Ma, F.; Hu, M.; et al. *Epimedium* polysaccharide alleviates polyglutamine-induced neurotoxicity in *Caenorhabditis elegans* by reducing oxidative stress. *Rejuvenation Res.* **2017**, *20*, 32–41. [CrossRef] [PubMed]

39. Fan, J.; Feng, H.; Yu, Y.; Sun, M.; Liu, Y.; Li, T.; Sun, X.; Liu, S.; Sun, M. Antioxidant activities of the polysaccharides of *Chuanminshen violaceum. Carbohydr. Polym.* **2017**, *157*, 629–636. [CrossRef] [PubMed]

40. Sui, Z.; Li, L.; Liu, B.; Gu, T.; Zhao, Z.; Liu, C.; Shi, C.; Yang, R. Optimum conditions for *Radix Rehmanniae* polysaccharides by RSM and its antioxidant and immunity activity in UVB mice. *Carbohydr. Polym.* **2013**, *92*, 283–288. [CrossRef] [PubMed]

41. Qi, B.; Ji, Q.; Wen, Y.; Liu, L.; Guo, X.; Hou, G.; Wang, G.; Zhong, J. *Lycium barbarum* polysaccharides protect human lens epithelial cells against oxidative stress-induced apoptosis and senescence. *PLoS ONE* **2014**, *9*, e110275. [CrossRef] [PubMed]

42. Lei, T.; Li, H.; Fang, Z.; Lin, J.; Wang, S.; Xiao, L.; Yang, F.; Liu, X.; Zhang, J.; Huang, Z.; et al. Polysaccharides from *Angelica sinensis* alleviate neuronal cell injury caused by oxidative stress. *Neural. Regen. Res.* **2014**, *9*, 260–267. [CrossRef] [PubMed]

43. Su, Z.J.; Yang, J.; Luo, W.J.; Wei, Y.Y.; Shuai, X.H.; Hu, T.J. Inhibitory effect of *Sophora subprosrate* polysaccharide on mitochondria oxidative stress induced by PCV-2 infection in RAW264.7 cells. *Int. J. Biol. Macromol.* **2017**, *95*, 608–617. [CrossRef] [PubMed]

44. Wang, F.; Liu, Q.; Wang, W.; Li, X.; Zhang, J. A polysaccharide isolated from *Cynomorium songaricum Rupr.* protects PC12 cells against H_2O_2-induced injury. *Int. J. Biol. Macromol.* **2016**, *87*, 222–228. [CrossRef] [PubMed]

45. Chen, H.; Qu, Z.; Fu, L.; Dong, P.; Zhang, X. Physicochemical properties and antioxidant capacity of 3 polysaccharides from green tea, oolong tea, and black tea. *J. Food Sci.* **2009**, *74*, C469–474. [CrossRef] [PubMed]

46. Wang, Y.; Yang, Z.; Wei, X. Antioxidant activities potential of tea polysaccharide fractions obtained by ultra filtration. *Int. J. Biol. Macromol.* **2012**, *50*, 558–564. [CrossRef] [PubMed]

47. Shang, X.; Chao, Y.; Zhang, Y.; Lu, C.; Xu, C.; Niu, W. Immunomodulatory and antioxidant effects of polysaccharides from *Gynostemma pentaphyllum* Makino in immunosuppressed mice. *Molecules* **2016**, *21*, 1085. [CrossRef] [PubMed]

48. Ren, X.; He, L.; Wang, Y.; Cheng, J. Optimization extraction, preliminary characterization and antioxidant activities of polysaccharides from *Semen Juglandis. Molecules* **2016**, *21*, 1335. [CrossRef] [PubMed]

49. Yang, Q.; Li, X.; Wang, L.; Lu, F.; Zhang, C.; Zheng, Y.; Xie, H. Physicochemical properties and antioxidant activity of *Ginkgo biloba* L. nut. *Mod. Food Sci. Technol.* **2013**, *10*, 2395–2400.

50. Chang, S.C.; Hsu, B.Y.; Chen, B.H. Structural characterization of polysaccharides from *Zizyphus jujuba* and evaluation of antioxidant activity. *Int. J. Biol. Macromol.* **2010**, *47*, 445–453. [CrossRef] [PubMed]

51. Liu, C.; Wang, C.; Xu, Z.; Wang, Y. Isolation, chemical characterization and antioxidant activities of two polysaccharides from the gel and the skin of *Aloe barbadensis* Miller irrigated with sea water. *Process Biochem.* **2007**, *42*, 961–970. [CrossRef]

52. Zeng, B.; Su, M.; Chen, Q.; Chang, Q.; Wang, W.; Li, H. Protective effect of a polysaccharide from *Anoectochilus roxburghii* against carbon tetrachloride-induced acute liver injury in mice. *J. Ethnopharmacol.* **2017**, *200*, 124–135. [CrossRef] [PubMed]

53. Huang, X.; Li, Q.; Li, H.; Guo, L. Neuroprotective and antioxidative effect of cactus polysaccharides in vivo and in vitro. *Cell. Mol. Neurobiol.* **2009**, *29*, 1211–1221. [CrossRef] [PubMed]

54. Jin, X.; Ning, Y. Antioxidant and antitumor activities of the polysaccharide from seed cake of *Camellia oleifera* Abel. *Int. J. Biol. Macromol.* **2012**, *51*, 364–368. [CrossRef] [PubMed]
55. Park, C.M.; Cho, C.W.; Song, Y.S. TOP 1 and 2, polysaccharides from *Taraxacum officinale*, inhibit NFκB-mediated inflammation and accelerate Nrf2-induced antioxidative potential through the modulation of PI3K-Akt signaling pathway in RAW 264.7 cells. *Food Chem. Toxicol.* **2014**, *66*, 56–64. [CrossRef] [PubMed]
56. Sudheesh, N.P.; Ajith, T.A.; Janardhanan, K.K. *Ganoderma lucidum* ameliorate mitochondrial damage in isoproterenol-induced myocardial infarction in rats by enhancing the activities of TCA cycle enzymes and respiratory chain complexes. *Int. J. Cardiol.* **2013**, *165*, 117–125. [CrossRef] [PubMed]
57. Kozarski, M.; Klaus, A.; Niks˘ic´, M.; Vrvic´, M.; Todorovic´, N.; Jakovljevic´, D.; Griensven, L. Antioxidative activities and chemical characterization of polysaccharide extracts from the widely used mushrooms *Ganoderma applanatum*, *Ganoderma lucidum*, *Lentinus edodes* and *Trametes versicolor*. *J. Food Compos. Anal.* **2012**, *26*, 144–153. [CrossRef]
58. Zhang, J.; Shi, R.; Li, H.; Xiang, Y.; Xiao, L.; Hu, M.; Ma, F.; Ma, C.; Huang, Z. Antioxidant and neuroprotective effects of *Dictyophora indusiata* polysaccharide in *Caenorhabditis elegans*. *J. Ethnopharmacol.* **2016**, *192*, 413–422. [CrossRef] [PubMed]
59. Xu, S.; Zhang, Y.; Jiang, K. Antioxidant activity in vitro and in vivo of the polysaccharides from different varieties of *Auricularia auricula*. *Food Funct.* **2016**, *7*, 3868–3879. [CrossRef] [PubMed]
60. Wen, L.; Gao, Q.; Ma, C.; Ge, Y.; You, Li.; Liu, R.; Fu, X.; Liu, D. Effect of polysaccharides from *Tremella fuciformis* on UV-induced photoaging. *J. Funct. Foods* **2016**, *20*, 400–410. [CrossRef]
61. Zhang, Q.; Yu, P.; Li, Z.; Zhang, H.; Xu, Z.; Li, P. Antioxidant activities of sulfated polysaccharide fractions from *Porphyra haitanesis*. *J. Appl. Phycol.* **2003**, *15*, 305–310. [CrossRef]
62. Cui, C.; Lu, J.; Sun-Waterhouse, D.; Mu, L.; Sun, W.; Zhao, M.; Zhao, H. Polysaccharides from *Laminaria japonica*: Structural characteristics and antioxidant activity. *Lwt-Food Sci. Technol.* **2016**, *73*, 602–608. [CrossRef]
63. Rupérez, P.; Ahrazem, O.; Leal, J.A. Potential antioxidant capacity of sulfated polysaccharides from the edible marine brown seaweed *Fucus vesiculosus*. *J. Agric. Food Chem.* **2002**, *50*, 840–845. [CrossRef] [PubMed]
64. Qi, H.; Zhang, Q.; Zhao, T.; Hu, R.; Zhang, K.; Li, Z. In vitro antioxidant activity of acetylated and benzoylated derivatives of polysaccharide extracted from *Ulva pertusa* (Chlorophyta). *Bioorg. Med. Chem. Lett.* **2006**, *16*, 2441–2445. [CrossRef] [PubMed]
65. Ku, M.J.; Lee, M.S.; Moon, H.J.; Lee, Y.H. Antioxidation effects of polysaccharide fucoidan extracted from seaweeds in skin photoaging. *FASEB J.* **2008**, *22* 647.1.
66. Li, H.; Xu, J.; Liu, Y.; Ai, S.; Qin, F.; Li, Z.; Zhang, H.; Huang, Z. Antioxidant and moisture-retention activities of the polysaccharide from *Nostoc commune*. *Carbohydr. Polym.* **2011**, *83*, 1821–1827. [CrossRef]
67. Morifuji, M.; Kitade, M.; Fukasawa, T.; Yamaji, T.; Ichihashi, M. Exopolysaccharides isolated from milk fermented with lactic acid bacteria prevent ultraviolet-induced skin damage in hairless mice. *Int. J. Mol. Sci.* **2017**, *18*, 146. [CrossRef] [PubMed]
68. Aguirre, M.J.; Isaacs, M.; Matsuhiro, B.; Mendoza, L.; Zúñiga, E.A. Characterization of a neutral polysaccharide with antioxidant capacity from red wine. *Carbohydr. Res.* **2009**, *344*, 1095–1101. [CrossRef] [PubMed]
69. Xu, R.; Shang, N.; Li, P. In vitro and in vivo antioxidant activity of exopolysaccharide fractions from *Bifidobacteriumanimalis* RH. *Anaerobe* **2011**, *17*, 226–231. [CrossRef] [PubMed]
70. Li, S.; Huang, R.; Shah, N.; Tao, X.; Xiong, Y.; Wei, H. Antioxidant and antibacterial activities of exopolysaccharides from *Bifidobacterium bifidum* WBIN03 and *Lactobacillus plantarum* R315. *J. Dairy Sci.* **2014**, *97*, 7334–7343. [CrossRef] [PubMed]
71. Wang, Z.; Liang, H.; Guo, W.; Peng, Z.; Chen, J.; Zhang, Q. Isolation, identification, and antioxidant activity of polysaccharides from the shell of abalone (*Haliotis discus hannai* Ino). *Genet. Mol. Res.* **2014**, *13*, 4883–4892. [CrossRef] [PubMed]
72. Cai, B.; Pan, J.; Wan, P.; Chen, D.; Long, S.; Sun, H. Ultrasonic-assisted production of antioxidative polysaccharides from *Crassostrea hongkongensis*. *Prep. Biochem. Biotechnol.* **2014**, *44*, 708–724. [CrossRef] [PubMed]
73. Xu, H.; Guo, T.; Guo, Y.; Zhang, J.; Li, Y.; Feng, W.; Jiao, B. Characterization and protection on acute liver injury of a polysaccharide MP-I from *Mytilus coruscus*. *Glycobiology* **2008**, *18*, 97–103. [CrossRef] [PubMed]

74. Guo, Z.; Xing, R.; Liu, S.; Yu, H.; Wang, P.; Li, C.; Li, P. The synthesis and antioxidant activity of the Schiff bases of chitosan and carboxymethyl chitosan. *Bioorg. Med. Chem. Lett.* **2005**, *15*, 4600–4603. [CrossRef] [PubMed]

75. Li, J.; Liu, Y.; Fan, L.; Ai, L.; Shan, L. Antioxidant activities of polysaccharides from the fruiting bodies of *Zizyphus Jujuba* cv. Jinsixiaozao. *Carbohydr. Polym.* **2011**, *84*, 390–394. [CrossRef]

76. Francisco, C.G.; Herrera, A.J.; Suárez, E. Intramolecular hydrogen abstraction reaction promoted by alkoxy radicals in carbohydrates. Synthesis of chiral 2,7-dioxabicyclo[2.2.1]heptane and 6,8-dioxabicyclo[3.2.1]octane ring systems. *J. Org. Chem.* **2002**, *67*, 7439–7445. [CrossRef] [PubMed]

77. Yamauchi, R.; Tatsumi, Y.; Asano, M.; Kato, K.; Ueno, Y. Effect of metal salts and fructose on the autoxidation of methyl linoleate in emulsions. *Agric. Biol. Chem.* **1988**, *52*, 849–850. [CrossRef]

78. Dalle-Donne, I.; Rossi, R.; Giustarini, D.; Milzani, A.; Colombo, R. Protein carbonyl groups as biomarkers of oxidative stress. *Clin. Chim. Acta* **2003**, *329*, 23–38. [CrossRef]

79. Lin, M.T.; Beal, M.F. Mitochondrial dysfunction and oxidative stress in neurodegenerative diseases. *Nature* **2006**, *443*, 787–795. [CrossRef] [PubMed]

80. Bhat, A.H.; Dar, K.B.; Anees, S.; Zargar, M.A.; Masood, A.; Sofi, M.A.; Ganie, S.A. Oxidative stress, mitochondrial dysfunction and neurodegenerative diseases; a mechanistic insight. *Biomed. Pharmacother.* **2015**, *74*, 101–110. [CrossRef] [PubMed]

81. Lu, M.; Gong, X. Upstream reactive oxidative species (ROS) signals in exogenous oxidative stress-induced mitochondrial dysfunction. *Cell Biol. Int.* **2009**, *33*, 658–664. [CrossRef] [PubMed]

82. Federico, A.; Cardaioli, E.; Da Pozzo, P.; Formichi, P.; Gallus, G.N.; Radi, E. Mitochondria, oxidative stress and neurodegeneration. *J. Neurol. Sci.* **2012**, *322*, 254–262. [CrossRef] [PubMed]

83. Kumar, A.; Prakash, A.; Dogra, S. Naringin alleviates cognitive impairment, mitochondrial dysfunction and oxidative stress induced by D-galactose in mice. *Food Chem. Toxicol.* **2010**, *48*, 626–632. [CrossRef] [PubMed]

84. Hafez, M.M.; Al-Shabanah, O.A.; Al-Harbi, N.O.; Al-Harbi, M.M.; Al-Rejaie, S.S.; Alsurayea, S.M.; Sayed-Ahmed, M.M. Association between paraoxonases gene expression and oxidative stress in hepatotoxicity induced by CCl$_4$. *Oxid. Med. Cell. Longev.* **2014**, *2014*, 893212. [CrossRef] [PubMed]

85. Fischer, R.; Maier, O. Interrelation of oxidative stress and inflammation in neurodegenerative disease: Role of TNF. *Oxid. Med. Cell. Longev.* **2015**, *2015*, 610813. [CrossRef] [PubMed]

86. Brandes, R.P.; Weissmann, N.; Schröder, K. Nox family NADPH oxidases: Molecular mechanisms of activation. *Free Radic. Biol. Med.* **2014**, *76*, 208–226. [CrossRef] [PubMed]

87. Kaspar, J.W.; Niture, S.K.; Jaiswal, A.K. Nrf2:INrf2 (Keap1) signaling in oxidative stress. *Free Radic. Biol. Med.* **2009**, *47*, 1304–1309. [CrossRef] [PubMed]

88. Kim, E.K.; Choi, E.J. Compromised MAPK signaling in human diseases: An update. *Arch. Toxicol.* **2015**, *89*, 867–882. [CrossRef] [PubMed]

89. Cohen, E.; Dillin, A. The insulin paradox: Aging, proteotoxicity and neurodegeneration. *Nat. Rev. Neurosci.* **2008**, *9*, 759–767. [CrossRef] [PubMed]

90. Li, H.; Li, Z.; Peng, L.; Jiang, N.; Liu, Q.; Zhang, E.; Liang, B.; Li, R.; Zhu, H. *Lycium barbarum* polysaccharide protects human keratinocytes against UVB-induced photo-damage. *Free Radic. Res.* **2017**, *51*, 200–210. [CrossRef] [PubMed]

91. Xiong, C.; Li, Q.; Chen, C.; Chen, Z.; Huang, W. Neuroprotective effect of crude polysaccharide isolated from the fruiting bodies of *Morchella importuna* against H$_2$O$_2$-induced PC12 cell cytotoxicity by reducing oxidative stress. *Biomed. Pharmacother.* **2016**, *83*, 569–576. [CrossRef] [PubMed]

92. Wang, J.; Liu, H.; Zhang, X.; Li, X.; Geng, L.; Zhang, H.; Zhang, Q. Sulfated hetero-polysaccharides protect SH-SY5Y cells from H$_2$O$_2$-induced apoptosis by affecting the PI3K/Akt signaling pathway. *Mar. Drugs* **2017**, *15*, 110. [CrossRef] [PubMed]

93. Gloire, G.; Legrand-Poels, S.; Piette, J. NF-kappaB activation by reactive oxygen species: Fifteen years later. *Biochem. Pharmacol.* **2006**, *72*, 1493–1505. [CrossRef] [PubMed]

94. Hsieh, H.L.; Yang, C.M. Role of redox signaling in neuroinflammation and neurodegenerative diseases. *BioMed Res. Int.* **2013**, *2013*, 484613. [CrossRef] [PubMed]

95. Mayo, K.J.; Cyr, D.M. Protein aggregation and neurodegeneration. *Methods* **2011**, *53*, 185–186. [CrossRef] [PubMed]

96. Kumar, V.; Sami, N.; Kashav, T.; Islam, A.; Ahmad, F.; Hassan, M.I. Protein aggregation and neurodegenerative diseases: From theory to therapy. *Eur. J. Med. Chem.* **2016**, *124*, 1105–1120. [CrossRef] [PubMed]

97. Readnower, R.D.; Sauerbeck, A.D.; Sullivan, P.G. Mitochondria, amyloid β, and Alzheimer's Disease. *Int. J. Alzheimers Dis.* **2011**, *2011*, 104545. [CrossRef] [PubMed]

98. Fox, J.H.; Connor, T.; Stiles, M.; Kama, J.; Lu, Z.; Dorsey, K.; Liebermann, G.; Sapp, E.; Cherny, R.A.; Banks, M.; et al. Cysteine oxidation within N-terminal mutant huntingtin promotes oligomerization and delays clearance of soluble protein. *J. Biol. Chem.* **2011**, *286*, 18320–18330. [CrossRef] [PubMed]

99. Mehta, A.; Prabhakar, M.; Kumar, P.; Deshmukh, R.; Sharma, P.L. Excitotoxicity: Bridge to various triggers in neurodegenerative disorders. *Eur. J. Pharmacol.* **2013**, *698*, 6–18. [CrossRef] [PubMed]

100. Bové, J.; Perier, C. Neurotoxin-based models of Parkinson's disease. *Neuroscience* **2012**, *211*, 51–76. [CrossRef] [PubMed]

101. Huang, S.; Mao, J.; Ding, K.; Zhou, Y.; Zeng, X.; Yang, W.; Wang, P.; Zhao, C.; Yao, J.; Xia, P.; et al. Polysaccharides from *Ganoderma lucidum* promote cognitive function and neural progenitor proliferation in mouse model of Alzheimer's disease. *Stem Cell Rep.* **2017**, *8*, 84–94. [CrossRef] [PubMed]

102. Liu, Y.; Jiang, L.; Li, X. κ-carrageenan-derived pentasaccharide attenuates Aβ25-35-induced apoptosis in SH-SY5Y cells via suppression of the JNK signaling pathway. *Mol. Med. Rep.* **2017**, *15*, 285–290. [CrossRef] [PubMed]

103. Wei, H.; Gao, Z.; Zheng, L.; Zhang, C.; Liu, Z.; Yang, Y.; Teng, H.; Hou, L.; Yin, Y.; Zou, X. Protective effects of fucoidan on Aβ25–35 and D-Gal-induced neurotoxicity in PC12 cells and D-Gal-induced cognitive dysfunction in mice. *Mar. Drugs* **2017**, *15*, 77. [CrossRef] [PubMed]

104. Gao, Y.; Li, C.; Yin, J.; Shen, J.; Wang, H.; Wu, Y.; Jin, H. Fucoidan, a sulfated polysaccharide from brown algae, improves cognitive impairment induced by infusion of Aβ peptide in rats. *Environ. Toxicol. Pharmacol.* **2012**, *33*, 304–311. [CrossRef] [PubMed]

105. Zhang, H.; Cao, Y.; Chen, L.; Wang, J.; Tian, Q.; Wang, N.; Liu, Z.; Li, J.; Wang, N.; Wang, X.; et al. A polysaccharide from *Polygonatum sibiricum* attenuates amyloid-β-induced neurotoxicity in PC12 cells. *Carbohydr. Polym.* **2015**, *117*, 879–886. [CrossRef] [PubMed]

106. Wang, P.; Liao, W.; Fang, J.; Liu, Q.; Hu, M.; Ding, K. A glucan isolated from flowers of *Lonicera japonica* Thunb. Inhibits aggregation and neurotoxicity of Aβ42. *Carbohydr. Polym.* **2014**, *110*, 142–147. [CrossRef] [PubMed]

107. Wang, S.; Li, J.; Xia, W.; Geng, M. A marine-derived acidic oligosaccharide sugar chain specifically inhibits neuronal cell injury mediated by beta-amyloid-induced astrocyte activation in vitro. *Neurol. Res.* **2007**, *29*, 96–102. [CrossRef] [PubMed]

108. Huang, S.H.; Lin, C.M.; Chiang, B.H. Protective effects of *Angelica sinensis* extract on amyloid β-peptide-induced neurotoxicity. *Phytomedicine* **2008**, *15*, 710–721. [CrossRef] [PubMed]

109. Zhang, Q.; Du, X.; Xu, Y.; Dang, L.; Xiang, L.; Zhang, J. The effects of Gouqi extracts on Morris maze learning in the APP/PS1 double transgenic mouse model of Alzheimer's disease. *Exp. Ther. Med.* **2013**, *5*, 1528–1530. [CrossRef] [PubMed]

110. Yu, M.; Lai, C.; Ho, Y.; Zee, S.; So, K.; Yuen, W.; Chang, R. Characterization of the effects of anti-aging medicine *Fructus lycii* on β-amyloid peptide neurotoxicity. *Int. J. Mol. Med.* **2007**, *20*, 261–268. [CrossRef] [PubMed]

111. Lai, C.; Yu, M.; Yuen, W.; So, K.; Zee, S.; Chang, R. Antagonizing beta-amyloid peptide neurotoxicity of the anti-aging fungus *Ganoderma lucidum*. *Brain Res.* **2008**, *1190*, 215–224. [CrossRef] [PubMed]

112. Chakrabortee, S.; Liu, Y.; Zhang, L.; Matthews, H.; Zhang, H.; Pan, N.; Cheng, C.; Guan, S.; Guo, D.; Huang, Z.; et al. Macromolecular and small-molecule modulation of intracellular Aβ42 aggregation and associated toxicity. *Biochem. J.* **2012**, *442*, 507–515. [CrossRef] [PubMed]

113. Jia, D.; Rao, C.; Xue, S.; Lei, J. Purification, characterization and neuroprotective effects of a polysaccharide from *Gynostemma pentaphyllum*. *Carbohydr. Polym.* **2015**, *122*, 93–100. [CrossRef] [PubMed]

114. Gao, K.; Liu, M.; Cao, J.; Yao, M.; Lu, Y.; Li, J.; Zhu, X.; Yang, Z.; Wen, A. Protective effects of *Lycium barbarum* polysaccharide on 6-OHDA-induced apoptosis in PC12 cells through the ROS-NO pathway. *Molecules* **2014**, *20*, 293–308. [CrossRef] [PubMed]

115. Deng, Q.; Yang, X. Protective effects of *Gynostemma pentaphyllum* polysaccharides on PC12 cells impaired by MPP(+). *Int. J. Biol. Macromol.* **2014**, *69*, 171–175. [CrossRef] [PubMed]

116. Zhang, F.; Lu, J.; Zhang, J.; Xie, J. Protective effects of a polysaccharide from *Spirulina platensis* on dopaminergic neurons in an MPTP-induced Parkinson's disease model in C57BL/6J mice. *Neural. Regen. Res.* **2015**, *10*, 308–313. [CrossRef] [PubMed]
117. Chen, P.; Wang, H.; Liu, Y.; Lin, S.; Chou, H.; Sheen, L. Immunomodulatory activities of polysaccharides from *Chlorella pyrenoidosa* in a mouse model of Parkinson's disease. *J. Funct. Foods* **2014**, *11*, 103–113. [CrossRef]
118. Souza, R.B.; Frota, A.F.; Sousa, R.S.; Cezario, N.A.; Santos, T.B.; Souza, L.M.; Coura, C.O.; Monteiro, V.S.; Cristino Filho, G.; Vasconcelos, S.M.; et al. Neuroprotective effects of sulphated agaran from marine alga *Gracilaria cornea* in rat 6-hydroxydopamine Parkinson's disease model: behavioural, neurochemical and transcriptional alterations. *Basic Clin. Pharmacol. Toxicol.* **2017**, *120*, 159–170. [CrossRef] [PubMed]
119. Cui, C.; Cui, N.; Wang, P.; Song, S.; Liang, H.; Ji, A. Neuroprotective effect of sulfated polysaccharide isolated from sea cucumber *Stichopus japonicus* on 6-OHDA-induced death in SH-SY5Y through inhibition of MAPK and NF-κB and activation of PI3K/Akt signaling pathways. *Biochem. Biophys. Res. Commun.* **2016**, *470*, 375–383. [CrossRef] [PubMed]
120. Kuo, H.; Lu, C.; Shen, C.; Tung, S.; Hsieh, M.; Lee, K.; Lee, L.; Chen, C.; Teng, C.; Huang, W.; et al. *Hericium erinaceus* mycelium and its isolated erinacine A protection from MPTP-induced neurotoxicity through the ER stress, triggering an apoptosis cascade. *J. Transl. Med.* **2016**, *14*, 78. [CrossRef] [PubMed]
121. Meenakshi, S.; Umayaparvathi, S.; Saravanan, R.; Manivasagam, T.; Balasubramanian, T. Neuroprotective effect of fucoidan from *Turbinaria decurrens* in MPTP intoxicated Parkinsonic mice. *Int. J. Biol. Macromol.* **2016**, *86*, 425–433. [CrossRef] [PubMed]
122. Fang, F.; Peng, T.; Yang, S.; Wang, W.; Zhang, Y.; Li, H. *Lycium barbarum* polysaccharide attenuates the cytotoxicity of mutant huntingtin and increases the activity of AKT. *Int. J. Dev. Neurosci.* **2016**, *52*, 66–74. [CrossRef] [PubMed]
123. Zhou, S.; Wang, S.; Sun, C.; Mao, H.; Di, W.; Ma, X.; Liu, L.; Liu, J.; Wang, F.; Kelly, P.; et al. Investigation into anti-epileptic effect and mechanisms of *Ganoderma lucidum* polysaccharides in in vivo and in vitro models. *Proc. Nutr. Soc.* **2015**, *74*, E65. [CrossRef]
124. Zhang, J.; An, S.; Hu, W.; Teng, M.; Wang, X.; Qu, Y.; Liu, Y.; Yuan, Y.; Wang, D. The neuroprotective properties of *Hericium erinaceus* in glutamate-damaged differentiated PC12 cells and an Alzheimer's disease mouse model. *Int. J. Mol. Sci.* **2016**, *17*, 1810. [CrossRef] [PubMed]
125. Zhang, Y.; Yang, X.; Jin, G.; Yang, X.; Zhang, Y. Polysaccharides from *Pleurotus ostreatus* alleviate cognitive impairment in a rat model of Alzheimer's disease. *Int. J. Biol. Macromol.* **2016**, *92*, 935–941. [CrossRef] [PubMed]
126. Hu, P.; Li, Z.; Chen, M.; Sun, Z.; Ling, Y.; Jiang, J.; Huang, C. Structural elucidation and protective role of a polysaccharide from *Sargassum fusiforme* on ameliorating learning and memory deficiencies in mice. *Carbohydr. Polym.* **2016**, *139*, 150–158. [CrossRef] [PubMed]
127. Chen, P.; He, D.; Zhang, Y.; Yang, S.; Chen, L.; Wang, S.; Zou, H.; Liao, Z.; Zhang, X.; Wu, M. *Sargassum fusiforme* polysaccharides activate antioxidant defense by promoting Nrf2-dependent cytoprotection and ameliorate stress insult during aging. *Food Funct.* **2016**, *7*, 4576–4588. [CrossRef] [PubMed]
128. Ho, Y.; Yu, M.; Yang, X.; So, K.; Yuen, W.; Chang, R. Neuroprotective effects of polysaccharides from wolfberry, the fruits of *Lycium barbarum*, against homocysteine-induced toxicity in rat cortical neurons. *J. Alzheimers Dis.* **2010**, *19*, 813–827. [CrossRef] [PubMed]
129. Ho, Y.; Yu, M.; Yik, S.; So, K.; Yuen, W.; Chang, R. Polysaccharides from wolfberry antagonizes glutamate excitotoxicity in rat cortical neurons. *Cell. Mol. Neurobiol.* **2009**, *29*, 1233–1244. [CrossRef] [PubMed]
130. Haider, A.; Inam, W.; Khan, S.A.; Hifza; Mahmood, W.; Abbas, G. β-glucan attenuated scopolamine induced cognitive impairment via hippocampal acetylcholinesterase inhibition in rats. *Brain Res.* **2016**, *1644*, 141–148. [CrossRef] [PubMed]
131. Yang, W.; Yu, J.; Zhao, L.; Ma, N.; Fang, Y.; Pei, F.; Mariga, A.; Hu, Q. Polysaccharides from *Flammulina velutipes* improve scopolamine-induced impairment of learning and memory of rats. *J. Funct. Foods* **2015**, *18*, 411–422. [CrossRef]
132. Chen, W.; Cheng, X.; Chen, J.; Yi, X.; Nie, D.; Sun, X.; Qin, J.; Tian, M.; Jin, G.; Zhang, X. *Lycium barbarum* polysaccharides prevent memory and neurogenesis impairments in scopolamine-treated rats. *PLoS ONE* **2014**, *9*, e88076. [CrossRef] [PubMed]
133. Tang, T.; He, B. Treatment of *d*-galactose induced mouse aging with *Lycium barbarum* polysaccharides and its mechanism study. *Afr. J. Tradit. Complement. Altern. Med.* **2013**, *10*, 12–17. [CrossRef] [PubMed]

134. Zhang, F.; Zhang, J.; Wang, L.; Mao, D. Effects of *polygonatum sibiricum* polysaccharide on learning and memory in a scopolamine-induced mouse model of dementia. *Neural Regen. Res.* **2008**, *1*, 33–36.

135. Xu, T.; Shen, X.; Yu, H.; Sun, L.; Lin, W.; Zhang, C. Water-soluble ginseng oligosaccharides protect against scopolamine-induced cognitive impairment by functioning as an antineuroinflammatory agent. *J Ginseng Res.* **2016**, *40*, 211–219. [CrossRef] [PubMed]

136. You, R.; Wang, K.; Liu, J.; Liu, M.; Luo, L.; Zhang, Y. A comparison study between different molecular weight polysaccharides derived from *Lentinus edodes* and their antioxidant activities in vivo. *Pharm. Biol.* **2011**, *49*, 1298–1305. [CrossRef] [PubMed]

137. Mu, X.; Zhang, Y.; Li, J.; Xia, J.; Chen, X.; Jing, P.; Song, X.; Wang, L.; Wang, Y. *Angelica Sinensis* polysaccharide prevents hematopoietic stem cells senescence in D-galactose-induced aging mouse model. *Stem. Cells Int.* **2017**, *2017*, 3508907. [CrossRef] [PubMed]

138. Ding, Q.; Yang, D.; Zhang, W.; Lu, Y.; Zhang, M.; Wang, L.; Li, X.; Zhou, L.; Wu, Q.; Pan, W.; et al. Antioxidant and anti-aging activities of the polysaccharide TLH-3 from *Tricholoma lobayense*. *Int. J. Biol. Macromol.* **2016**, *85*, 133–140. [CrossRef] [PubMed]

139. Sun, S.; Guo, L.; Ren, Y.; Wang, B.; Li, R.; Qi, Y.; Yu, H.; Chang, N.; Li, M.; Peng, H. Anti-apoptosis effect of polysaccharide isolated from the seeds of *Cuscuta chinensis* Lam on cardiomyocytes in aging rats. *Mol. Biol. Rep.* **2014**, *41*, 6117–6124. [CrossRef] [PubMed]

140. Li, W.; Nie, S.; Peng, X.; Liu, X.; Li, C.; Chen, Y.; Li, J.; Song, W.; Xie, M. *Ganoderma atrum* polysaccharide improves age-related oxidative stress and immune impairment in mice. *J. Agric. Food Chem.* **2012**, *60*, 1413–1418. [CrossRef] [PubMed]

141. Zhang, H.; Wang, Z.; Zhang, Z.; Wang, X. Purified *Auricularia auricular-judae* polysaccharide (AAP I-a) prevents oxidative stress in an ageing mouse model. *Carbohydr. Polym.* **2011**, *84*, 638–648. [CrossRef]

142. Jin, W.; Wang, J.; Jiang, H.; Song, N.; Zhang, W.; Zhang, Q. The neuroprotective activities of heteropolysaccharides extracted from *Saccharina japonica*. *Carbohydr. Polym.* **2013**, *97*, 116–120. [CrossRef] [PubMed]

143. Tsai, L.H.; Madabhushi, R. Alzheimer's disease: A protective factor for the ageing brain. *Nature* **2014**, *507*, 439–440. [CrossRef] [PubMed]

144. Feng, Y.; Wang, X. Antioxidant therapies for Alzheimer's disease. *Oxid. Med. Cell. Longev.* **2012**, *2012*, 472932. [CrossRef] [PubMed]

145. Zhao, Y.; Zhao, B. Natural antioxidants in prevention and management of Alzheimer's disease. *Front. Biosci. (Elite Ed.)* **2012**, *4*, 794–808. [CrossRef] [PubMed]

146. Morris, M.C.; Evans, D.A.; Tangney, C.C.; Bienias, J.L.; Wilson, R.S.; Aggarwal, N.T.; Scherr, P.A. Relation of the tocopherol forms to incident Alzheimer disease and to cognitive change. *Am. J. Clin. Nutr.* **2005**, *81*, 508–514. [PubMed]

147. Mangialasche, F.; Kivipelto, M.; Mecocci, P.; Rizzuto, D.; Palmer, K.; Winblad, B.; Fratiglioni, L. High plasma levels of vitamin E forms and reduced Alzheimer's disease risk in advanced age. *J. Alzheimers Dis.* **2010**, *20*, 1029–1037. [CrossRef] [PubMed]

148. Urbányi, Z.; Forrai, E.; Sárvári, M.; Likó, I.; Illés, J.; Pázmány, T. Glycosaminoglycans inhibit neurodegenerative effects of serum amyloid P component in vitro. *Neurochem. Int.* **2005**, *46*, 471–477. [CrossRef] [PubMed]

149. Zhou, X.; Jin, L. The structure-activity relationship of glycosaminoglycans and their analogues with β-amyloid peptide. *Protein Pept. Lett.* **2016**, *23*, 358–364. [CrossRef] [PubMed]

150. Ariga, T.; Miyatake, T.; Yu, R.K. Role of proteoglycans and glycosaminoglycans in the pathogenesis of Alzheimer's disease and related disorders: Amyloidogenesis and therapeutic strategies—A review. *J. Neurosci. Res.* **2010**, *88*, 2303–2315. [CrossRef] [PubMed]

151. Abeliovich, A.; Gitler, A.D. Defects in trafficking bridge Parkinson's disease pathology and genetics. *Nature* **2016**, *539*, 207–216. [CrossRef] [PubMed]

152. Fernandez, H.H. 2015 Update on Parkinson disease. *Clevel. Clin. J. Med.* **2015**, *82*, 563–568. [CrossRef] [PubMed]

153. Wood, L.D. Clinical review and treatment of select adverse effects of dopamine receptor agonists in Parkinson's disease. *Drugs Aging* **2010**, *27*, 295–310. [CrossRef] [PubMed]

154. Wang, J.; Liu, H.; Jin, W.; Zhang, H.; Zhang, Q. Structure-activity relationship of sulfated hetero/galactofucan polysaccharides on dopaminergic neuron. *Int. J. Biol. Macromol.* **2016**, *82*, 878–883. [CrossRef] [PubMed]

155. Yang, F.; Shi, Y.; Sheng, J.; Hu, Q. In vivo immunomodulatory activity of polysaccharides derived from *Chlorella pyrenoidosa*. *Eur. Food Res. Technol.* **2006**, *224*, 225–228. [CrossRef]
156. Panaro, M.A.; Cianciulli, A. Current opinions and perspectives on the role of immune system in the pathogenesis of Parkinson's disease. *Curr. Pharm. Des.* **2012**, *18*, 200–208. [CrossRef] [PubMed]
157. Kanemaru, K. Immunotherapy targeting misfolded proteins in neurodegenerative disease. *Brain Nerve* **2013**, *65*, 469–474. [PubMed]
158. Walker, F.O. Huntington's disease. *Lancet* **2007**, *369*, 218–228. [CrossRef]
159. Berry-Kravis, E. Huntington's disease: Genetics. *Encycl. Mov. Disord.* **2010**, 31–36. [CrossRef]
160. Dayalu, P.; Albin, R.L. Huntington disease: Pathogenesis and treatment. *Neurol. Clin.* **2015**, *33*, 101–114. [CrossRef] [PubMed]
161. Ehrnhoefer, D.E.; Duennwald, M.; Markovic, P.; Wacker, J.L.; Engemann, S.; Roark, M.; Legleiter, J.; Marsh, J.L.; Thompson, L.M.; Lindquist, S.; et al. Green tea (−)-epigallocatechin-gallate modulates early events in huntingtin misfolding and reduces toxicity in Huntington's disease models. *Hum. Mol. Genet.* **2006**, *15*, 2743–2751. [CrossRef] [PubMed]
162. Xiao, L.; Li, H.; Zhang, J.; Yang, F.; Huang, A.; Deng, J.; Liang, M.; Ma, F.; Hu, M.; Huang, Z. Salidroside protects *Caenorhabditis elegans* neurons from polyglutamine-mediated toxicity by reducing oxidative stress. *Molecules* **2014**, *19*, 7757–7769. [CrossRef] [PubMed]
163. Faber, P.W.; Voisine, C.; King, D.C.; Bates, E.A.; Hart, A.C. Glutamine/proline-rich PQE-1 proteins protect *Caenorhabditis elegans* neurons from huntingtin polyglutamine neurotoxicity. *PNAS* **2002**, *99*, 17131–17136. [CrossRef] [PubMed]
164. Ezza, H.S.A.; Khadrawyb, Y.A. Glutamate excitotoxicity and neurodegeneration. *J. Mol. Genet. Med.* **2014**, *8*, 4. [CrossRef]

nutrients

MDPI

Article

Role of Mitochondria and Endoplasmic Reticulum in Taurine-Deficiency-Mediated Apoptosis

Chian Ju Jong [1], Takashi Ito [2], Howard Prentice [3], Jang-Yen Wu [3] and Stephen W. Schaffer [1,*]

[1] Department of Pharmacology, College of Medicine, University of South Alabama, Mobile, AL 36688, USA; cjjong84@gmail.com
[2] Faculty of Biotechnology, Fukui Prefectural University, Fukui 910-1195, Japan; tito@fpu.ac.jp
[3] Program in Integrative Biology and Center for Complex Systems and Brain Sciences, College of Medicine, Florida Atlantic University, Boca Raton, FL 33431, USA; hprentic@health.fau.edu (H.P.); jwu@health.fau.edu (J.-Y.W.)
*   Correspondence: sschaffe@southalabama.edu; Tel.: +011-251-460-6288

Received: 18 May 2017; Accepted: 19 July 2017; Published: 25 July 2017

Abstract: Taurine is a ubiquitous sulfur-containing amino acid found in high concentration in most tissues. Because of its involvement in fundamental physiological functions, such as regulating respiratory chain activity, modulating cation transport, controlling inflammation, altering protein phosphorylation and prolonging lifespan, taurine is an important nutrient whose deficiency leads to severe pathology and cell death. However, the mechanism by which taurine deficiency causes cell death is inadequately understood. Therefore, the present study examined the hypothesis that overproduction of reactive oxygen species (ROS) by complex I of the respiratory chain triggers mitochondria-dependent apoptosis in hearts of taurine transporter knockout (TauTKO) mice. In support of the hypothesis, a 60% decrease in mitochondrial taurine content of 3-month-old TauTKO hearts was observed, which was associated with diminished complex I activity and the onset of mitochondrial oxidative stress. Oxidative damage to stressed mitochondria led to activation of a caspase cascade, with stimulation of caspases 9 and 3 prevented by treatment of 3-month-old TauTKO mice with the mitochondria specific antioxidant, MitoTempo. In 12 month-old, but not 3-month-old, TauTKO hearts, caspase 12 activation contributes to cell death, revealing a pathological role for endoplasmic reticulum (ER) stress in taurine deficient, aging mice. Thus, taurine is a cytoprotective nutrient that ensures normal mitochondrial and ER function, which is important for the reduction of risk for apoptosis and premature death.

Keywords: oxidative stress; mitochondria; endoplasmic reticulum stress; apoptosis; caspase cascade; respiratory chain; mitochondria encoded proteins; tRNA$^{Leu(UUR)}$

1. Introduction

Taurine is a β-amino acid found in very high concentration in excitable tissues. In certain species, such as the cat and fox, the amino acid is considered an essential nutrient, but in humans it is considered a semi-essential nutrient [1]. Neither humans nor cats readily synthesize taurine, therefore, the primary source of taurine for both species is the diet. For man, meat is a primary source of taurine, with the concentration of taurine being particularly high in seafood [2]. Hayes et al. [3] provided the first evidence that taurine is an essential nutrient for normal function of excitable tissues in cats, which developed a retinopathy when fed a taurine deficient diet. It was subsequently shown that taurine deficient cats also develop dilated cardiomyopathy [4]. In contrast to cats, adult rodents deprived of dietary taurine do not develop overt taurine deficiency, as they readily synthesize taurine in the liver [5]. Nonetheless, rodents become taurine deficient if exposed to a high concentration of a taurine transporter inhibitor or subjected to a genetic alteration that limits the uptake of taurine by excitable tissues. Hence, a taurine deficient cardiomyopathy develops in rodents lacking the myocardial taurine transporter (TauTKO mice) [6].

One of the primary physiological functions of taurine is its antioxidant activity, an action attributed to several mechanisms [7]. First, some investigators have attributed taurine's antioxidant actions to elevations in the activity of antioxidant enzymes, however, by reducing the amount of damaging ROS (reactive oxygen species), taurine could indirectly elevate the activity of the antioxidant defenses. Second, taurine serves as an important anti-inflammatory agent, which involves a myeloperoxidase-catalyzed reaction between taurine and hypochlorous acid to generate an anti-inflammatory product, taurine chloramine. However, through the myeloperoxidase reaction, taurine also reduces the levels of the neutrophil-generated ROS, hypochlorous acid [8]. Third, reductions in intramitochondrial taurine content are associated with elevations in mitochondrial superoxide generation, leading to the suggestion that the mitochondria are the primary source of ROS generated by taurine deficient tissues [9]. While that conclusion may be valid for β-alanine-mediated taurine depletion, it is known that the taurine transport inhibitor, β-alanine, is a naturally occurring substance that exerts other actions within the cell besides reductions in taurine levels. Therefore, further studies were warranted to establish the source of mitochondrial ROS in the taurine deficient cell. Another unanswered question relates to the consequences of taurine deficiency-mediated oxidative stress. It is known that taurine exerts anti-apoptotic activity, however, it is unclear if its anti-apoptotic and antioxidant activities are directly related [10]. Moreover, it remains to be determined whether mitochondrial ROS is the only cause of apoptosis.

The taurine deficient heart is also characterized by diminished activity of the sarcoplasmic reticular (SR) Ca^{2+} ATPase [11], an effect consistent with the reduction in amplitude and prolongation of the relaxation phase of the Ca^{2+} transient, as well as the defect in systolic and diastolic function of the TauTKO heart. Alterations in Ca^{2+} homeostasis, as well as oxidative stress, have been known to also trigger endoplasmic reticulum (ER) stress and initiate an ER stress-mediated quality control process known as the unfolded protein response (UPR) [12]. Three transmembrane sensor proteins, PERK (protein kinase RNA (PKR)-like ER kinase), IRE-1 (inositol-requiring protein-1) and ATF6 (activating transcription factor 6) initiate distinct pathways of the UPR. In unstressed cells, the most abundant ER chaperone, GRP78, binds to PERK and ATF6, maintaining the chaperone in its inactive state. An elevation in unfolded proteins in the ER promotes the release of GRP78 from PERK and ATF6, allowing the two sensor proteins to initiate their respective UPR pathways. The aim of the UPR pathways is to reduce the cellular load of misfolded and unfolded proteins, restore ER function and allow the cell to function despite the unfolded protein load. However, if ER stress is excessive and overwhelms the capacity of the cell to restore normal ER function, two of the UPR pathways promote cell death. Hence, the present study examines the role of both mitochondrial ROS and ER stress in the initiation of apoptosis in TauTKO hearts.

2. Materials and Methods

2.1. Model of Taurine Deficiency (TauTKO)

Wild-type (WT) and homozygous taurine transporter knockout (TauTKO) mice were generated by mating heterozygous (TauTKO$^{+/-}$) C57BL/6 mouse pairs [6]. This study was conducted using either 3- or 12-month-old mice. Animal handling and experimental procedures followed the Animal Welfare Act and the Guide for the Care and Use of Laboratory Animals and were approved by the Animal Care and Use Committee of the University of South Alabama.

2.2. Measurement of Mitochondrial Taurine Content

Mitochondria were prepared according the method described by Grishko et al. [13]. The mitochondrial fraction was re-suspended in mitochondrial buffer and a small aliquot was kept for protein concentration, while the remaining was used for taurine analysis. Isolated mitochondria were homogenized in ice-cold 1 M perchloric acid and 2 mM EDTA and subjected to centrifugation at $10,000 \times g$ for 10 min. The resulting supernatant was used to measure mitochondrial taurine content as determined by changes in absorbance at 355 nm [14].

2.3. Assay of Respiratory Chain Complexes

Complex I activity (NADH dehydrogenase) was evaluated according to the method of Ricci et al. [10]. Mitochondria, prepared according to the method of Grishko et al. [13], were suspended in 10 mM Tris buffer (pH 8.0) and then incubated for 5 min at 37 °C. The reaction was initiated by addition of NADH and was monitored at 340 nm for 3 min, after which 5 μM rotenone was added and changes in absorbance at 340 nm were observed for an additional 2 min. Complex I activity was calculated from the difference between NADH oxidation in the presence and absence of rotenone and expressed as mmol/min/mg protein.

Complex II activity (succinate dehydrogenase) was determined according to the method of Ricci et al. [10]. Isolated mitochondria were suspended in 50 mM potassium phosphate buffer (pH 7.4) containing 20 mM succinate and incubated for 3 min at 37 °C. To the mitochondrial suspension was added 50 mM potassium phosphate buffer containing 500 μM 2,6-dichlorophenolindophenol, 20 mM KCN, 20 μg/mL rotenone and 20 μg/mL antimycin A. The reaction was initiated by addition of 25 μM decylubiquinone and followed for 3 min. Complex II activity (succinate dehydrogenase) was determined from the reduction of 2,6-dichlorophenolindophenol at 600 nm and expressed as μmol/min/mg protein [10].

Complex III was determined using the method of Chen et al. [15]. Isolated mitochondria were suspended in 50 mM Tris buffer (pH 7.4) containing 250 mM sucrose, 1 mM EDTA, 50 μM oxidized cytochrome c, 2 mM KCN and 10 μg/mL rotenone. After 10 min pre-incubation at 37 °C, the reaction was initiated by addition of 10 mM decylubiquinol and monitored at 550 nm for 3 min before addition of 40 μM antimycin, after which the reaction was further monitored for 2 min. Complex III activity (expressed as mmol/min/mg protein) was calculated from the difference in cytochrome c reduction in the presence and absence of antimycin A.

Complex IV was assayed using the method of Ma et al. [16]. Mitochondria were suspended in 10 mM potassium phosphate buffer (pH 7.4). The reaction was initiated by addition of substrate, ferrocytochrome c (50 μM), and monitored at 550nm. Complex IV activity was evaluated from the rate of reduced cytochrome c oxidation in the presence and absence of KCN (2 mM).

Complex V was assayed as described previously by our group [17]. The assay is based on a spectrophotometric assay of ATP combining pyruvate kinase and lactate dehydrogenase and monitoring changes in NADH oxidation at 340 nm. Complex V activity was evaluated from the rate of NADH oxidation in the presence and absence of oligomycin (1 μg/mL).

2.4. Quantitative Real Time PCR Method of Measuring Levels of ND6 mRNA

Total RNA was isolated from the heart using TRIzol LS Reagent (Invitrogen, Carlsbad, CA, USA) according to the protocol. Quantitative real time PCR was then performed using the iScript One-Step RT-PCR Kit with SYBR Green (Bio-Rad, Hercules, CA, USA) according to the protocol. Primer sequences for ND6 were 5′-ataggatcctcccgaatcaaccct-3′ (forward) and 5′-aggattggtgctgtgggtgaaaga-3′ (reverse), and primer sequences used for β-actin were 5′-gtgacgttgacatccgtaaa-3′ (forward) and 5′-ctcaggaggagcaatgatct-3′ (reverse).

2.5. Assay of Aconitase Activity

The mitochondrial fraction was re-suspended in buffer containing 50 mM Tris-HCl, pH 7.4 and 0.2 mM sodium citrate. Aconitase was assayed using the Bioxytech Aconitase-340 kit by monitoring the increase in NADH absorbance at 340 nm. Aconitase activity was normalized relative to succinate dehydrogenase (complex II activity), whose activity is unaffected by oxidative stress.

2.6. Determination of Glutathione Redox State

Hearts were homogenized in ice-cold 50 mM phosphate buffer pH 7.4 and homogenates were then centrifuged at $10,000 \times g$ for 10 min. A small aliquot of the supernatant, which is defined as the total lysate, was kept for protein concentration and the remaining was deproteinized with 1 M perchloric acid and 2 mM EDTA and subjected to centrifugation at $10,000 \times g$ for 10 min. The supernatant was neutralized to pH 6–7 and centrifuged again at $10,000 \times g$ for 10 min. The resulting supernatant was used to assess the glutathione redox state using a glutathione assay kit. The glutathione redox state was determined as the ratio of reduced glutathione (GSH) to oxidized glutathione (GSSG). Both GSH and GSSG content were determined as an increase in absorbance at 405 nm. GSSG content was measured after derivatizing pre-existing GSH in each sample with 2-vinylpyridine.

2.7. Measurement of Protein Carbonylation

Total lysates were prepared by initially homogenizing hearts in radioimmuno-precipitation assay (RIPA) lysis buffer. The homogenates were centrifuged at $10,000 \times g$ for 20 min and the supernatant was finally collected. The protein concentration was measured by BCA assay. Protein (10–20 µg) was assayed for the degree of protein carbonylation using the Oxyblot Protein Oxidation Detection kit (Catalog No S7150, Millipore; Darmstadt, Germany). Briefly, proteins were derivatized with 2,4-dinitrophenylhydrazine (DNPH), separated by polyacrylamide gel electrophoresis and subjected to Western blotting. Carbonylated proteins were detected utilizing the primary antibody, which is specific for the DNP moiety of proteins.

2.8. Determination of Protein Content via Western Blot Analysis

Total lysates were prepared by homogenizing hearts in RIPA lysis buffer (50 mM Tris base, pH 8.0, 150 mM NaCl, 0.5% deoxycholic acid, 1% NP-40, 0.1% sodium dodecyl sulfate). Homogenates were centrifuged at $10,000 \times g$ for 10 min and the resulting supernatants were collected as total lysates. Isolated mitochondria were prepared as described earlier. The mitochondrial fraction was suspended in RIPA lysis buffer. The protein concentration was measured by the bicinchoninic assay (BCA). Protein (20–30 µg) was mixed with an equal volume of $5\times$ sample buffer (1.25 mM Tris HCl, pH 6.8, 1% sodium dodecyl sulfate, 10% glycerol, 5% β-mercaptoethanol) and then boiled for 5 min. Proteins were separated by sodium dodecyl sulfate polyacrylamide gel electrophoresis (SDS-PAGE) and transferred onto a nitrocellulose membrane. The membranes then were blocked in blocking buffer (5% milk in tris buffered saline with Tween 20) and incubated with an appropriate primary antibody overnight at 4 °C. The primary antibodies used were specific for the following proteins: ND1 (sc-20493), ND2 (sc-20496), ND3 (sc-26760), ND4 (sc-20499), ND6 (Molecular Probes A31587), COX1 (Molecular Probes A6403), cytochrome b (sc-11436), SDH (sc-25851), β-actin (sc-130656) caspase 3 (sc-7148), caspase 12 (Cell Signaling #2202), caspase 9 (Cell Signaling #9508), GRP78 (sc-13968), CHOP (Cell Signaling #2895S), phospho-IRE1 (Abcam #48187), PERK (sc-13073), phosphor-PERK (sc-32577), ATF4 (Cell Signaling #11815), XBP-1 (Abcam #37152). PARP (sc-7150). The next day, membranes were washed before being incubated with an appropriate secondary antibody. After washing, western blots were analyzed by enhanced chemiluminescent reagents.

2.9. Evaluation of Mitochondrial Oxidative Stress

WT and TauTKO mice were administered either PBS (vehicle) or MitoTempo (1.4 mg/kg/day) by intraperitoneal injection for 7 consecutive days. On the 7th day, mice were killed and hearts were immediately removed and rapidly frozen with aluminum tongs cooled in liquid nitrogen. Samples were stored at −80 °C.

2.10. Statistical Analyses

All results were reported as means ± SEM. The statistical significance of the data was determined using the Student's *t*-test for comparison within groups or ANOVA followed by the Newman-Keuls test for comparison between groups. Values of $p < 0.05$ were considered statistically significant.

3. Results

3.1. Effect of Taurine on Mitochondrial Function

Ito et al. [6] have shown that global deletion of the taurine transporter gene suppresses taurine uptake by the heart, as shown by the presence of nondetectable levels of taurine in the cytosol of the TauTKO heart vs. nearly 28 mM in the wild-type (WT) heart. However, the mitochondria of the 3-month-old TauTKO heart appear to be resistant to the loss of taurine, as mitochondrial taurine content of the TauTKO heart was 40 nmol/mg protein, which is only 60% less than the taurine content of the age-matched, WT heart (Figure 1).

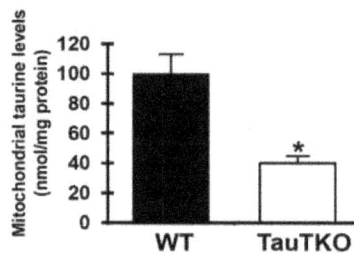

Figure 1. Reduced mitochondrial taurine content of taurine transporter (TauTKO) hearts. Following isolation of the mitochondrial fraction from homogenized hearts, extracts were prepared by precipitating protein. Mitochondrial taurine content was then measured. Values shown represent means ± SEM of 5–7 different hearts. * $p < 0.05$. WT, the wild-type.

It has been previously shown that formation of the mitochondrial taurine conjugate, 5-taurinomethyluridine-tRNA[Leu(UUR)], enhances UUG decoding [18]. Of the three leucine codons (CUN, UUA, UUG) present in the mitochondria, two (UUA and UUG) interact with the anticodon of tRNA[Leu(UUR)] during the course of protein synthesis. However, the interaction of UUG is highly dependent on the presence of the 5-taurinomethyluridine conjugate of tRNA[Leu(UUR)] [19,20]. The most frequently used leucine codon in mitochondrial protein biosynthesis is CUN, with UUA being intermediate and UUG being the least used codon [20]. Because of the importance of the taurine conjugate in the interaction of the UUG codon with the anticodon of tRNA[Leu(UUR)], we tested the hypothesis that taurine deficiency selectively diminishes the expression of UUG-dependent, mitochondria encoded proteins in the heart. As seen in Figure 2, the level of ND6 was significantly reduced in the TauTKO heart relative to that of several mitochondria encoded proteins examined in the WT heart.

To provide insight into the mechanism underlying the decline in ND6 protein content, the mRNA level of ND6 was determined in TauTKO and WT hearts. In contrast to most conditions in which protein expression is altered by taurine deficiency, no significant change in the mRNA content of ND6 was observed (data not shown; $p = 0.15$).

ND6 is not only a subunit of complex I but a facilitator of complex I assembly [20], therefore, we tested whether taurine deficiency specifically decreases the activity of complex I without affecting the activities of the other respiratory chain complexes. In accordance with that hypothesis, we found that the activity of complex I was reduced 60% in the mitochondria of the TauTKO heart while the activities of complexes II–V were unaffected by taurine deficiency (Figure 3).

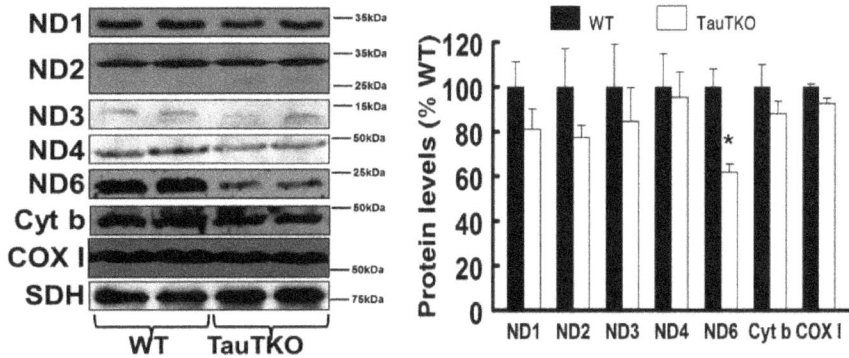

Figure 2. ND6 is reduced in TauTKO hearts. The mitochondrial fraction was isolated from homogenized hearts and then subjected to western blot analyses. The left panel shows representative gels for ND1, ND2, ND3, ND4, ND6, Cyt b (cytochrome b) and COX I (cytochrome c oxidase I), with succinate dehydrogenase (SDH) serving as the loading control. The right panel shows the means $\pm$ SEM of the mitochondrial protein/SDH ratio of 6–9 different hearts. Values are expressed relative to wild-type (WT), where WT is fixed at 100%. * $p < 0.05$.

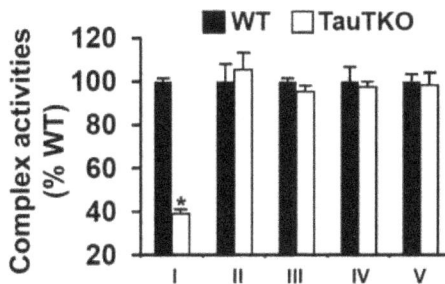

Figure 3. Taurine depletion decreases complex I activity. Isolated mitochondrial fractions of WT and TauTKO hearts were assayed for the activities of complexes I–V. Values of individual complexes shown represent means $\pm$ SEM of 4–6 different hearts. * $p < 0.05$.

Reduced flux of electrons through the respiratory chain is often associated with the diversion of electrons from the respiratory chain to the acceptor, oxygen, forming in the process superoxide anion. To determine if the mitochondria of the TauTKO heart are oxidatively stressed, the glutathione redox ratio (GSH/GSSG) of TauTKO hearts was determined. Supporting the view that the taurine deficient heart is oxidatively stressed, it was shown that the glutathione redox ratio was 2-fold less in the 3-month-old TauTKO heart than in that of the age-matched, WT heart (Figure 4A). However, the levels of ROS differ between various organelles within the cell, with mitochondria being unique because of their ability to not only generate superoxide but also to degrade ROS [21]. Therefore, to determine if taurine deficiency enhances oxidative stress in the mitochondria, the activity of the ROS-sensitive, citric acid cycle enzyme, aconitase, was assayed. As seen in Figure 4B, taurine deficiency is associated with a 30% decrease in aconitase activity. Another marker of oxidative stress is protein carbonylation, in which amino acid residues, such as lysine, arginine, proline and threonine, undergo oxidation to form a protein carbonyl derivative containing an aldehyde or carbonyl group [22]. Figure 4C shows that the amount of carbonylated protein in the cellular lysate and mitochondrial extract of the taurine deficient heart increased 50%. Together, these data show that taurine deficiency is associated with an increase in oxidative stress.

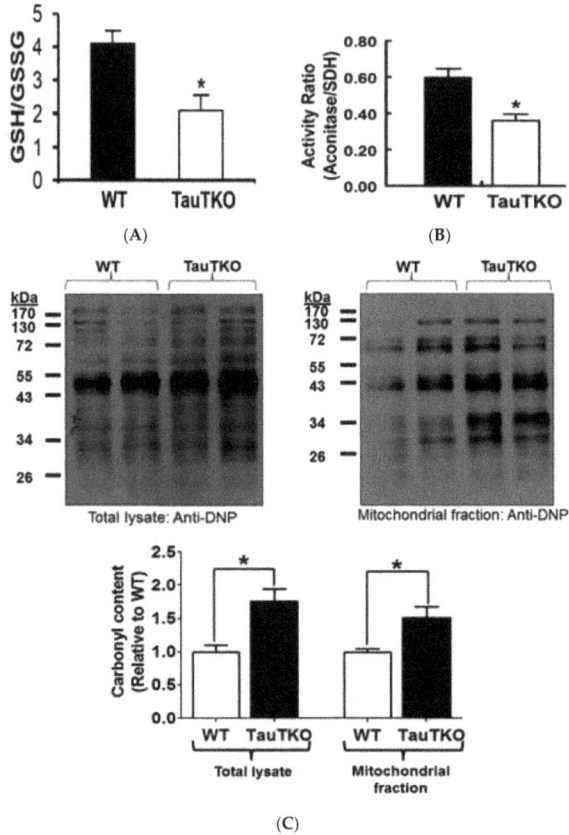

Figure 4. Taurine depletion causes oxidative stress. (**A**) Reduced (GSH) and oxidized (GSSG) glutathione content were determined and the data expressed as the glutathione redox state (GSH/GSSG), with values shown representing means ± SEM of 6–8 hearts. * $p < 0.05$; (**B**) Aconitase activity of WT and TauTKO mitochondria were assayed and normalized relative to SD activity. Values represent means ± SEM of 4–6 hearts. * $p < 0.05$; (**C**) Following preparation of total heart lysates and the mitochondrial fraction, proteins were derivatized with 2,4-dinitrophenylhydrazine and then subjected to western blot analysis of carbonylated proteins. The top panels show representative gels of carbonylated proteins of total lysate and the mitochondrial fraction. Values shown in the bottom panel represent means ± SEM for relative cellular and mitochondrial carbonylated protein content from 4–5 hearts. Values are expressed relative to WT, where WT is fixed at 1.0. * $p < 0.05$.

3.2. Taurine Depletion Induces Cell Death—An Effect Mediated by Oxidative Stress

Excessive oxidative stress commonly triggers mitochondria-dependent cell death via activation of the initiator protease, caspase 9. As shown in Figure 5A, there is a significant increase in the levels of the active, cleaved form of caspase 9 in the 3-month-old TauTKO heart, which resulted in a net increase in the active cleaved/pro-caspase 9 ratio of 60% relative to that of the WT heart (Figure 5A). To determine if mitochondrial oxidative stress contributes to the activation of caspase 9, the cleaved/pro-caspase 9 ratio was examined in 3-month-old TauTKO mice that had been administered MitoTempo (1.4 mg/kg/day) by intraperitoneal injection for 7 consecutive days before removal of the heart. Although treatment with MitoTempo had no effect on the generation of the active form of caspase 9 in the WT heart, it abolished the activation of caspase 9 in the 3-month-old TauTKO heart.

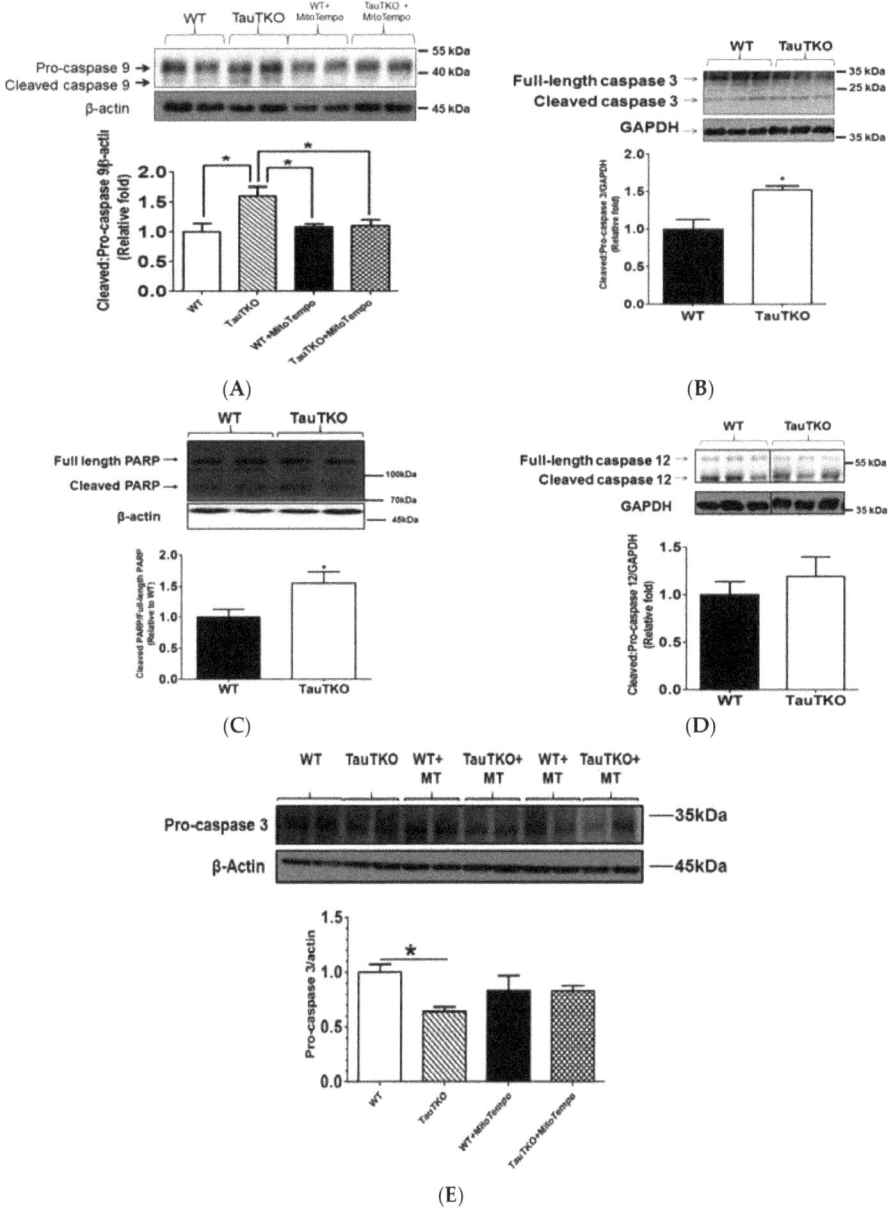

Figure 5. Taurine depletion induces apoptosis. Total heart lysates of 3-month-old TauTKO and WT hearts were subjected to western blot analyses of the active forms of caspase 9, caspase 3 and caspase 12, as well as the inactive pro-caspase forms of the three proteases (**A–E**). In (**A**), the data are expressed as the ratio of active caspase 9/pro-caspase 9 while in (**B**), the data are depicted as the ratio of cleaved caspase 3/pro-caspase 3. In (**C**), the data are expressed as cleaved PARP levels. In (**D**), the ratio of cleaved/pro-caspase 12 is shown. In (**E**), TauTKO and WT mice were treated with the mitochondria-specific antioxidant, MitoTempo, for 7 days and changes in the levels of the inactive pro-caspase 3 zymogen were determined. Each panel contains a representative gel. The representative bands for WT and TauTKO were spliced from one original gel and the splice junction is indicated by the black splicing line. Values shown represent means ± SEM of 6–9 hearts. All values are expressed relative to WT, where WT is fixed at 1.0. * $p < 0.05$.

The active form of caspase 9 is a mitochondrial-localized protease, which initiates an apoptotic cascade by cleaving pro-caspase 3 to its active, cleaved caspase 3 form. Figure 5B shows that the content of the active, cleaved form of caspase 3 is increased in 3-month-old TauTKO mice relative to that of the age-matched control WT mice, causing the ratio of cleaved caspase 3/pro-caspase 3 to increase 50% in young TauTKO hearts relative to that of young WT hearts. Activation of caspase 3 subsequently cleaves poly-ADP (ribose) polymerase (PARP) and induces apoptosis. Indeed, as shown in Figure 5C, a significant 50% increase in the cleaved levels of PARP is observed in TauTKO hearts. In contrast to caspases 3 and 9, the level of the cleaved, active form of caspase 12, which is activated following initiation of UPR by ER stress, was identical in the 3 month- old TauTKO and WT hearts (Figure 5D). Treatment of 3-month-old TauTKO mice with MitoTempo abolishes the significant decrease in pro-caspase 3 content but the mitochondrial antioxidant has no effect on pro-caspase 3 levels in WT mice (Figure 5E).

3.3. Potential Crosstalk between Mitochondria and ER

Crosstalk between mitochondria and the sarcoplasmic reticulum of the heart has been observed in specific mitofusion 2-containing mitochondrial-reticular microdomains [23]. It has been demonstrated that ROS and Ca^{2+} are capable of promoting crosstalk between the two organelles [24,25]. According to Ramila et al. [11], Ca^{2+} handling by the ER (SR) of the TauTKO heart is defective, a change attributed to impaired activity of several protein kinases and a phosphatase. Because high levels of Ca^{2+} are required for normal protein folding by the ER [24], the size of the ER stores and the activity of the SR Ca^{2+} ATPase are determinants of ER stress [26]. Moreover, ROS perturbs the ER, thereby initiating ER stress, in part by disrupting proper disulfide-sulfhydryl interchange and disulfide bridge formation via the oxidoreductases that require an oxidative environment in the ER lumen to promote proper protein folding and ER function [27]. Because of the link between mitochondria and the ER, as well as the importance of ER stress in the initiation of apoptosis, we examined the status of two of the three transmembrane sensor proteins (PERK and IRE1) and their downstream effectors; the two UPR pathways examined are involved in the initiation of apoptosis [28]. As shown in Figure 6A, the content of phosphorylated PERK and ATF4, which are components and biomarkers of the PERK pathway, are reduced 25–35% in the 3-month-old TauTKO heart relative to that of the WT control heart. Similarly, two biomarkers of the IRE1 signaling pathway, phospho-IRE1 and spliced XBP-1, were 20% lower in the 3-month-old TauTKO heart than in the 3-month-old WT heart, an effect that was diminished upon treatment of the animal with MitoTempo (Figure 6B). Common to both the PERK and IRE1 pathways is the chaperone, GRP78, which in normal ER is associated with the transmembrane sensor proteins, maintaining them in an inactive state. A key beginning event in the initiation of UPR is the dissociation of GRP78 from the three transmembrane sensor proteins, freeing them to activate distinct UPR pathways. In most examples of ER stress and UPR signaling, GRP78 is upregulated and UPR signaling is stimulated [29,30]. However, as seen in Figure 6C, levels of GRP78 are reduced in the 3-month-old TauTKO heart, an effect associated with inhibition of UPR signaling. The pattern that develops in the 3-month-old TauTKO heart is the mirror image of that seen normally during initiation of ER stress and activation of UPR signaling, suggesting that UPR is not initiated in the young TauTKO heart.

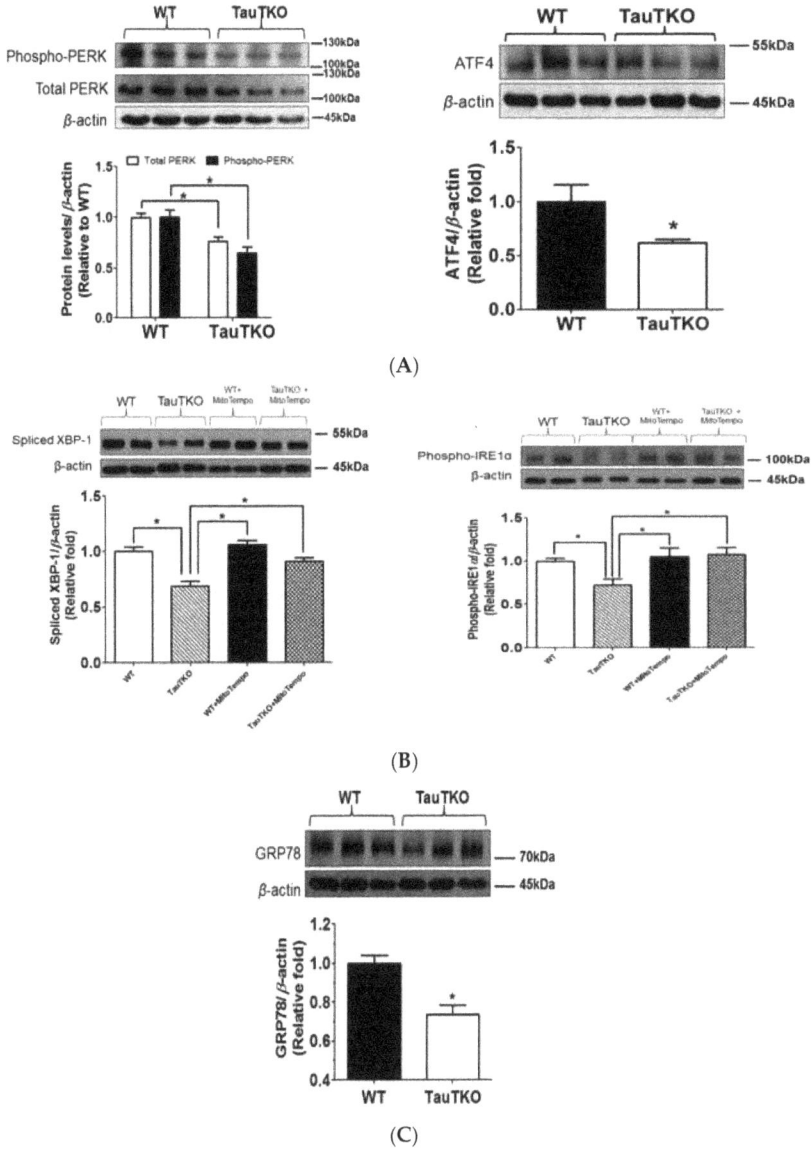

Figure 6. Unfolded protein response (UPR) is not activated in TauTKO hearts at an early age. Total lysates from 3-month-old WT and TauTKO mice treated with or without MitoTempo were subjected to western blot analysis of (A) phosphorylated PERK and ATF4 (B) spliced XBP-1 and phosphorylated-IRE1 and spliced XBP-1 and (C) GRP78. Each panel contains a representative gel and summation data expressed as means ± SEM of 6–9 hearts. Values are expressed relative to WT, where WT is fixed at 1.0. * $p < 0.05$.

To further test the role of UPR in taurine deficiency-mediated apoptosis, the content of the two most important mediators of UPR-mediated apoptosis, CHOP (CCAAT/enhancer binding protein), and the ER localized protease, caspase 12, were determined. Figure 7A shows that CHOP content was significantly elevated in the hearts of the 12-month-old TauTKO mice relative to those of those of year-old WT hearts. Similarly, levels of the active, cleaved form of caspase 12, which is likely activated by calpain and the caspase cascade [31], are nearly 3.0 fold greater in the TauTKO heart than in that of the WT heart (Figure 7B).

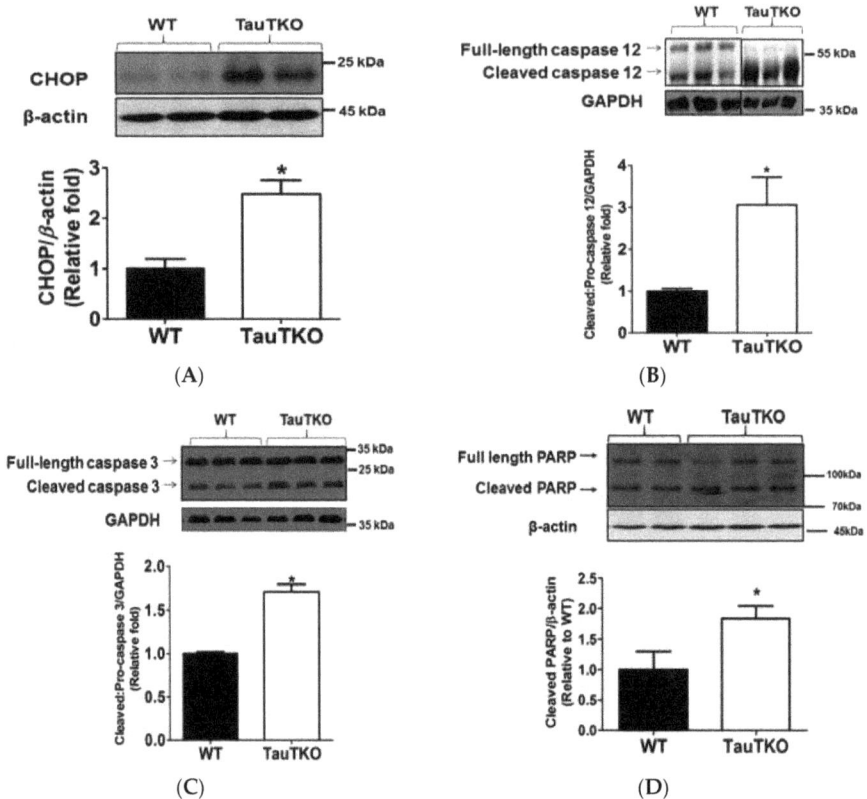

Figure 7. CHOP and UPR caspase cascade are activated in older TauTKO hearts. Total heart lysates from 12-month-old TauTKO and WT mouse hearts were subjected to western blot analysis of (**A**) CHOP, (**B**) caspase 12, (**C**) caspase 3 and (**D**) PARP. Each panel contains a representative gel and summation data expressed as means ± SEM of 6–9 hearts. The representative bands for WT and TauTKO shown in B were spliced from one original gel and the splice junction is indicated by the black splicing line. All values are expressed relative to WT, where WT is fixed at 1.0. * $p < 0.05$.

Following its cleavage, caspase 12 is known to activate caspase 3, an effector protease that induces downstream apoptotic events. In the 3-month old TauTKO mouse heart, the activation of caspase 3 is solely related to caspase 9 cleavage. However, by the time the TauTKO mouse reaches one-year of age, caspase 12 is also capable of activating caspase 3. As seen in Figure 7C, there is a 70% increase in the cleaved/pro-caspase 3 ratio and a significant 70% increase in the cleaved PARP. Because caspases 9 and 12 are more active in the 1 year-old TauTKO mouse than the age-matched WT mouse, we assume that the active state of caspase 3 in the 1 year-old TauTKO is related to the activation of both initiator caspases.

Activation of caspases 9 and 12 cause cardiomyocyte death, which is an important process in the development of cardiomyopathy [6]. In the taurine deficient cat, onset of cardiomyopathy can ultimately be fatal. Therefore, we examined the mortality rates of TauTKO and WT mice over the time period of 3 months-of-age and 12-months-of-age. Beginning with 71, 3-month-old WT mice and 59, 3-month-old TauTKO mice, we found that one WT mouse died between the ages of 3-months and 1 year, which corresponds to a mortality rate over the designated time frame of 1.4%. By comparison, the mortality rate of the TauTKO mice was 3.4%. As the animals aged, the mortality rate increased, particularly in the TauTKO mice [32]. Therefore, the increase in myocardial apoptosis is associated with an elevation in the mortality rate of the taurine deficient animal.

4. Discussion

The present study demonstrates that myocardial deficiency of the natural nutrient, taurine, leads to impaired mitochondrial complex I activity, a defect associated with an increase in mitochondrial oxidative stress and the initiation of mitochondria-mediated apoptosis as early as 3-months-of-age (Figure 8). However, by 12-months-of-age, the ER-localized protease, caspase 12, becomes active and contributes to the initiation of the caspase cascade and the increase in mortality rate (Figure 8).

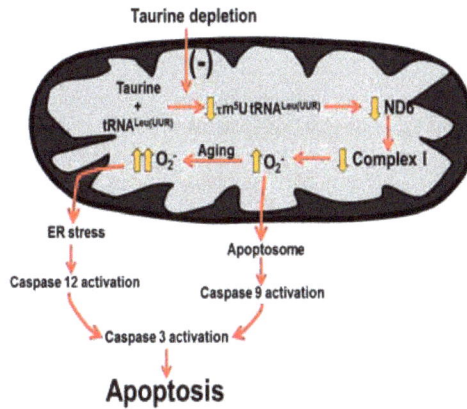

Figure 8. Mechanisms underlying taurine deficiency-mediated myocardial cell death.

4.1. Mitochondrial Actions of Taurine

The animal model utilized in the present study is the taurine transporter knockout mouse (TauTKO), whose cytosolic taurine levels in the heart are too low to detect [6]. By comparison, the mitochondria from 3-month-old TauTKO hearts contain 40% of normal taurine content (Figure 1). Despite the resistance of the mitochondria to taurine loss, the 60% decline in taurine content noted in the present study is associated with significant respiratory chain dysfunction (Figures 2 and 3).

In normal mitochondria, taurine forms a conjugate with a uridine residue located in the anticodon wobble position of mitochondrial tRNA$^{Leu(UUR)}$ [18,20]. This modification strengthens the interaction of the AAU anticodon of tRNA$^{Leu(UUR)}$ with the UUG codon of mitochondrial mRNAs, thereby facilitating UUG decoding and the translation of mitochondria encoded proteins whose mRNA contains multiple UUG codons [19,20]. The present study shows that one of the mitochondria encoded proteins, ND6, is highly sensitive to taurine deficiency, likely related to the high number of UUG codons (8) found in its mRNA and the large number of leucine residues of ND6 that are dependent on the UUG codon (42%) [20,33,34].

The reduction in ND6 levels in the TauTKO heart is not caused by a decrease in ND6 mRNA. Rather, the reduction in ND6 expression appears related to a decrease in 5-taurinomethyluridine-tRNA$^{Leu(UUR)}$ levels, thereby diminishing the interaction between the UUG codons of ND6 mRNA and the AAU* anticodon of tRNA$^{Leu(UUR)}$, where U* represents the modified wobble uridine moiety, 5-taurinomethyluridine and U represents unmodified uridine. Because the mRNA of ND6 contains 8 UUG codons, the weak interaction between the codon UUG and anticodon AAU, would lead to a decrease in ND6 biosynthesis.

ND6 serves an important function in complex I, as it is not only a structural subunit of the complex but also facilitates proper assembly of the complex [35]. Therefore, taurine deficiency is associated with significant decreases in complex I activity and mitochondrial respiration, the latter defect bypassed in taurine deficient mitochondria respiring a complex II substrate, such as succinate [9,20,33].

The bottleneck that develops in complex I of the taurine deficient heart has a dramatic effect on energy metabolism [36]. Not only is flux of electrons through the respiratory chain diminished by impaired NADH dehydrogenase activity (innate complex I enzyme activity), but the increase in the myocardial NADH/NAD$^+$ ratio (resulting from the decrease in NADH dehydrogenase activity) adversely affects energy metabolism by inhibiting key enzymes of glucose and fatty acid metabolism [36]. Thus, the taurine deficient heart becomes energy deficient, an effect that likely contributes to impaired mechanical performance of the heart.

The decline in respiratory chain electron activity, as noted for taurine deficiency, respiratory chain inhibitors, mitochondrial damage and respiratory chain mutations, is associated with enhanced superoxide generation at one of two respiratory chain sites, complex I or complex III [21,37,38]. Since taurine deficiency specifically decreases complex I activity, it is not surprising that it functions like the complex I inhibitor, rotenone, which stimulates the generation of superoxide anion by complex I [9,21]. Other factors that are capable of increasing the generation of superoxide by complex I are increased proton-motive force, reduced CoQ, elevated NADH/NAD$^+$ ratio and increased oxygen concentration [39]. Thus, the most important properties increasing ROS generation by the taurine deficient heart are diminished electron flow and elevated NADH/NAD$^+$ ratio [36].

Superoxide and other ROS generated by complex I are targeted to the mitochondrial matrix, while complex III preferentially targets ROS to the intermembrane space [40]. Thus, the proteins undergoing oxidation by complex I-generated ROS are different from those oxidized by complex III-generated ROS. One of the proteins modified in the taurine deficient heart is aconitase, an enzyme from the citric acid cycle that is located in the matrix. Because aconitase is abundant in the mitochondrial matrix, the observed decrease in activity is unlikely to alter citric acid cycle flux, which is largely diminished in the TauTKO heart because of elevations in the NADH/NAD ratio [36].

On the other hand, the oxidation of specific mitochondrial proteins can lead to mitochondrial outer membrane permeabilization (MOMP) and the release of cytochrome c from the mitochondria. Normally, most cytochrome c is associated with the inner mitochondrial membrane through weak electrostatic interactions with acidic phospholipids although a stronger interaction develops between cytochrome c and the phospholipid, cardiolipin [41]. ROS generated by complexes I and III are capable of oxidizing cardiolipin, breaking the interaction with cytochrome c. Therefore, massive amounts of cytochrome c become available for release from the mitochondria upon MOMP. The prominent mechanism for initiation of MOMP is oligomerization of pro-apoptotic Bcl-2 family members, Bax and Bak. However, MOMP can also be initiated at the level of the mitochondrial inner membrane upon formation of the mitochondrial permeability transition pore, which allows the transfer of solutes with a molecular weight of 1500 Da or less. Massive matrix swelling can cause MOM rupture, releasing cytochrome c [42]. In the cytosol, cytochrome c combines with ATP and APAF-1 to form a large complex, referred to as the apoptosome. According to Hu et al. [43], activation of the zymogen of caspase 9 by the apoptosome leads to a 2–3 order of magnitude increase in proteolytic activity of caspase 9. In turn, the active form of caspase 9 enhances the proteolytic activity of the effector, caspase 3. Several lines of evidence indicate that this sequence of events is responsible for the death of cardiomyocytes in the 3-month-old taurine deficient heart. First, the increase in ROS in taurine deficient cells is restricted to the mitochondria [9,33]. Second, treatment of TauTKO mice with the mitochondria-specific antioxidant, MitoTempo, diminishes the rise in mitochondrial ROS and prevents the activation of caspases 9 and 3. Third, β-alanine-mediated taurine deficiency leads to mitochondrial fragmentation and oxidative stress, effects reversed by restoration of normal taurine content [9]. Fourth, Montessuit et al. [44] reported that mitochondrial fragmentation facilitates Bax oligomerization and apoptosis. Although it was suggested that the factor that mediates mitochondrial fragmentation, Drp1, may directly participate in Bax oligomerization, Kashnareva et al. [45] found no Drp1 in MOM, ruling out Drp1 in MOMP. Recently, Shetewy et al. [9] showed that taurine deficient cells undergo mitochondrial fragmentation, which may increase the susceptibility of taurine deficient cells to Bax-induced oligomerization and MOMP. Further study of this hypothesis is warranted. Fifth, hearts of TauTKO mice, but not those of WT mice, undergo cleavage of the natural caspase 3 substrate, PARP.

4.2. Role of ER Stress in TauTKO Hearts

The 3-month-old TauTKO mouse exhibits no evidence of ER stress and UPR signaling despite severe mitochondrial oxidative stress, diminished energy metabolism, accumulation of ubiquitinated and carbonylated proteins and even impaired Ca^{2+} handling by the SR [11,33,36,46]. The reason for the lack of UPR signaling in 3-month-old taurine deficient mice is unclear but it may be related to the low levels of GRP78, which are below the threshold required to activate the protein sensors. Nonetheless, the findings clearly show that the oxidant status of the mitochondria influences ER function and that the TauTKO heart does not develop overt ER stress until the animal is older. As expected, downstream pro-apoptotic effectors, CHOP and caspase 12, remain within the normal range in the 3-month-old TauTKO heart.

We have previously shown that UPR signaling is enhanced in skeletal muscle of older TauTKO mice [32]. Although the status of caspase 12 and CHOP was not examined in our earlier study, we did observe an increase in GRP78 levels, as well as enhanced ATF6 and IRE1 signaling (ATF6 and spliced XBP1 content). Since the activation of caspase 12 is associated with IRE1 signaling, the TauTKO heart appears to be well equipped to activate caspase 12. In agreement with that notion, we found that caspase 12, as well as CHOP, are upregulated in the 12-month-old TauTKO heart.

It has previously been shown that taurine treatment downregulates CHOP and decreases caspase 12 activity in models of severe stress, thereby diminishing cell death [47–49]. In the case of glutamate excitotoxicity and stroke, taurine treatment suppresses ATF6 and IRE1 signaling [49], the same pathways affected by taurine deficiency. Moreover, both taurine treatment and taurine deficiency alter similar biochemical processes, Ca^{2+} transport and oxidative stress [11,33,49]. Thus, taurine treatment and taurine deficiency appear to affect caspase 12 by modulating IRE1 signaling.

Many of the effects of aging have been attributed to oxidative stress [50]. However, unlike taurine deficiency, complex III is a primary source of mitochondrial ROS generation in aging. Because complex III targets ROS to the intermembrane space while taurine deficiency in young rodents or isolated cells is associated with matrix oxidative stress [9,32,39], oxidative stress is more widely distributed in the 12-month-old TauTKO heart than in the young TauTKO heart. These differences in ROS compartmentalization provide a logical explanation for the differences in UPR signaling between the older and younger TauTKO hearts. One of the UPR pathways that should be sensitive to these differences is IRE1, which is a determinant of caspase 12 activation state. The regulation of IRE1 signaling is extremely complicated, involving ROS, protein phosphorylation, CREB, TRAF2, Bcl-2 family of proteins, and a scaffold [51]. Nonetheless, the data support a role for ER stress and IRE1 signaling in the activation of caspase 12 in 12-month-old TauTKO hearts. These alterations likely contribute to the severity of the cardiomyopathy that develops in TauTKO mice.

5. Conclusions

The present study outlines a novel mechanism for initiation of taurine deficiency-mediated apoptosis (Figure 8). We propose that declines in mitochondrial taurine content lead to a decrease in the formation of the taurine conjugate, 5-taurinomethyluridine-tRNA (Leu-UUR), which in turn alters the binding of the anticodon of the tRNA to the UUG codon of mitochondrial mRNAs. Consequently, the biosynthesis of ND6, whose mRNA contains 8 UUG codons, significantly declines. Because

ND6 serves as a subunit of complex I and facilitates assembly of the complex, activity of complex I also declines in taurine deficiency. The resulting disruption of electron flow through the respiratory chain leads to the diversion of electrons to oxygen, increasing the formation of superoxide and other damaging oxidants. Mitochondrial damage leads to increased membrane permeability, releasing cytochrome c that contributes to the formation of the apoptosome and the activation of caspases 9 and 3, the latter an effector caspase that promotes apoptosis.

In aging, oxidative stress initiates ER stress, which upon stimulation of the UPR pathways, activates caspase 12, an initiating caspase that further stimulates caspase 3 and apoptosis. The rise in cardiac apoptosis promotes the development of cardiomyopathy, increasing mortality in taurine deficiency. Thus, taurine is an important nutrient that is required for normal mitochondrial function. Hence, taurine deficiency reduces lifespan by promoting mitochondrial-dependent and ER stress-mediated apoptosis.

Author Contributions: Chian Ju Jong performed most experiments and assisted in the writing of the manuscript. Takashi Ito provided the TauTKO mice and information on mortality rate. Howard Prentice and Jang-Yen Wu provided assistance with ER stress and finances. Stephen W. Schaffer supervised the study and helped with the writing of the manuscript.

Conflicts of Interest: The authors declare no conflict of interest.

References

1. Gaull, G.E. Taurine as a conditionally essential nutrient in man. *J. Am. Coll. Nutr.* **1986**, *5*, 121–125. [CrossRef] [PubMed]
2. Yamori, Y.; Taguchi, T.; Hamada, A.; Kunimasa, K.; Mori, H.; Mori, M. Taurine in health and diseases: Consistent evidence from experimental and epidemiological studies. *J. Biomed. Sci.* **2010**, *17*, S6. [CrossRef] [PubMed]
3. Hayes, K.C.; Carey, R.E.; Schmidt, S.Y. Retinal degeneration associated with taurine deficiency in the cat. *Science* **1975**, *188*, 949–951. [CrossRef] [PubMed]
4. Pion, P.D.; Kittleson, M.D.; Rogers, Q.R.; Morris, J.G. Myocardial failure in cats associated with low plasma taurine: A reversible cardiomyopathy. *Science* **1987**, *237*, 764–768. [CrossRef] [PubMed]
5. Hayes, K.C. Taurine requirement in primates. *Nutr. Rev.* **1985**, *43*, 65–70. [CrossRef] [PubMed]
6. Ito, T.; Kimura, Y.; Uozumi, Y.; Takai, M.; Muraoka, S.; Matsuda, T.; Uyeki, K.; Yoshiyama, M.; Ikawa, M.; Okabe, M.; et al. Taurine depletion caused by knocking out the taurine transporter gene leads to cardiomyopathy with cardiac atrophy. *J. Mol. Cell. Cardiol.* **2008**, *44*, 927–937. [CrossRef] [PubMed]
7. Shimada, K.; Jong, C.J.; Takahashi, K.; Schaffer, S.W. Role of ROS production and turnover in the antioxidant activity of taurine. *Adv. Exp. Med. Biol.* **2015**, *803*, 581–596. [PubMed]
8. Marcinkiewicz, J.; Kontny, E. Taurine and inflammatory diseases. *Amino Acids* **2014**, *46*, 7–20. [CrossRef] [PubMed]
9. Shetewy, A.; Shimada-Takaura, K.; Warner, D.; Jong, C.J.; Mehdi, A.B.; Alexeyev, M.; Takahashi, K.; Schaffer, S.W. Mitochondrial defects associated with β-alanine toxicity: Relevance to hyper-beta-alaninemia. *Mol. Cell. Biochem.* **2016**, *416*, 11–22. [CrossRef] [PubMed]
10. Ricci, C.; Pastukh, V.; Leonard, J.; Turrens, J.; Wilson, G.; Schaffer, D.; Schaffer, S.W. Mitochondrial DNA damage triggers mitochondrial superoxide generation and apoptosis. *Am. J. Physiol.* **2008**, *294*, C413–C422. [CrossRef] [PubMed]
11. Ramila, K.C.; Jong, C.J.; Pastukh, V.; Ito, T.; Azuma, J.; Schaffer, S.W. Role of protein phosphorylation in excitation-contraction coupling in taurine deficient hearts. *Am. J. Physiol.* **2015**, *308*, H232–H239. [CrossRef] [PubMed]
12. Sano, R.; Reed, J.C. ER stress-induced cell death mechanisms. *Biochim. Biophys. Acta* **2013**, *1833*, 3460–3470. [CrossRef] [PubMed]
13. Grishko, V.; Pastukh, V.; Solodushko, V.; Gillespie, M.; Azuma, J.; Schaffer, S. Apoptotic cascade initiated by angiotensin II in neonatal cardiomyocytes: Role of DNA damage. *Am. J. Physiol.* **2003**, *285*, H2364–H2372. [CrossRef] [PubMed]

14. Shaffer, J.E.; Kocsis, J.J. Taurine mobilizing effects of beta alanine and other inhibitors of taurine transport. *Life Sci.* **1981**, *28*, 2727–2736. [CrossRef]
15. Chen, Q.; Vazquez, E.J.; Moghaddas, S.; Hoppel, C.L.; Lesnefsky, E.J. Production of reactive oxygen species by mitochondria: Central role of coomplex III. *J. Biol. Chem.* **2003**, *278*, 36927–36931. [CrossRef] [PubMed]
16. Ma, Y.Y.; Zhang, X.L.; Wu, T.F.; Liu, Y.P.; Wang, Q.; Zhang, Y.; Song, J.Q.; Wang, Y.J.; Yang, Y.L. Analysis of the mitochondrial complexes I–V enzyme activities of peripheral leukocytes in oxidative phosphorylation disorder. *J. Child Neurol.* **2011**, *26*, 974–979. [CrossRef] [PubMed]
17. Schaffer, S.W.; Seyed-Mozaffari, M.; Cutcliff, C.R.; Wilson, G.L. Postreceptor myocardial defect in a rat model of non-insulin-dependent diabetes mellitus. *Diabetes* **1986**, *55*, 593–597. [CrossRef]
18. Suzuki, T.; Suzuki, T.; Wada, T.; Saigo, K.; Watanabe, K. Taurine as a constituent of mitochondrial tRNAs: New insights into the functions of taurine and human mitochondrial diseases. *EMBO J.* **2002**, *21*, 6581–6589. [CrossRef] [PubMed]
19. Kurata, S.; Ohtsuki, T.; Wada, T.; Kirino, Y.; Takai, K.; Sigo, K.; Watanabe, K.; Suzuki, T. Decoding property of C5 uridine modification at the wobble position of tRNA anticodon. *Nucleic Acids Res.* **2003**, *3*, 245–246. [CrossRef]
20. Schaffer, S.W.; Jong, C.J.; Ito, T.; Azuma, J. Role of taurine in the pathogenesis of MELAS and MERRF. *Amino Acids* **2014**, *46*, 47–56. [CrossRef] [PubMed]
21. Turrens, J.F. Mitochondrial formation of reactive oxygen species. *J. Physiol.* **2003**, *552*, 335–344. [CrossRef] [PubMed]
22. Dalle-Donne, I.; Aldini, G.; Carini, M.; Colombo, R.; Rossi, R.; Milzani, A. Protein carbonylation, cellular dysfunction, and disease progression. *J. Cell. Mol. Med.* **2006**, *10*, 389–406. [CrossRef] [PubMed]
23. Chen, Y.; Csordas, G.; Jowdy, C.; Schneider, T.; Csordas, N.; Wang, W.; Liu, Y.; Kohlhaas, M.; Meiser, M.; Bergem, S.; et al. Mitofusin 2-containing mitochondrial-reticular microdomains direct rapid cardiomyocyte bioenergetic responses via interorganelle Ca^{2+} crosstalk. *Circ. Res.* **2012**, *111*, 863–875. [CrossRef] [PubMed]
24. Malhotra, J.D.; Kaufman, R.J. ER stress and its functional link to mitochondria: Role in cell survival and death. *Cold Spring Harb. Perspect. Biol.* **2011**, *3*, 1004424. [CrossRef] [PubMed]
25. Bhandary, B.; Marahatta, A.; Kim, H.R.; Chae, H.J. An involvement of oxidative stress in endoplasmic reticulum stress and its associated diseases. *Int. J. Mol. Sci.* **2013**, *14*, 434–456. [CrossRef] [PubMed]
26. Younce, C.W.; Burmeister, M.A.; Ayala, J.E. Exendin-4 attenuates high glucose-induced cardiomyocyte apoptosis via inhibition of endoplasmic reticulum stress and activation of SERCA2a. *Am. J. Physiol.* **2013**, *304*, C508–C518. [CrossRef] [PubMed]
27. Groenendyk, J.; Sreenivasaiah, P.K.; Kim, D.H.; Agellon, L.B.; Michalak, M. Biology of endoplasmic reticulum stress in the heart. *Circ. Res.* **2010**, *107*, 1185–1197. [CrossRef] [PubMed]
28. Harding, H.P.; Novoa, I.; Zhang, Y.; Zeng, H.; Wek, R.; Schapira, M.; Ron, D. Regulated translation initiation controls stress-induced gene expression in mammalian cells. *Mol. Cell* **2000**, *6*, 1099–1108. [CrossRef]
29. Liu, X.R.; Cao, L.; Li, T.; Chen, L.L.; Yu, Y.Y.; Huang, W.J.; Liu, L.; Tan, X.Q. Propofol attenuates H_2O_2 induced oxidative stress and apoptosis via the mitochondria- and ER-medicated pathways in neonatal rat cardiomyocytes. *Apoptosis* **2017**, *22*, 639–646. [CrossRef] [PubMed]
30. Hulmi, J.J.; Hentila, J.; DeRuisseau, K.C.; Oliveira, B.M.; Papaioannou, K.G.; Autio, R.; Kujala, U.M.; Ritvos, O.; Kainulainen, H.; Korkmaz, A.; et al. Effects of muscular dystrophy, exercise and blocking activin receptor IIB ligands on the unfolded response and oxidative stress. *Free Radic. Biol. Med.* **2016**, *99*, 308–322. [CrossRef] [PubMed]
31. Martinez, J.A.; Zhang, Z.; Svetlov, S.I.; Hayes, R.L.; Wang, K.K.; Larner, S.F. Calpain and caspase processing of caspase-12 contribute to the ER stress-induced cell death pathway in differentiated PC12 cells. *Apoptosis* **2010**, *15*, 1480–1493. [CrossRef] [PubMed]
32. Ito, T.; Yoshikawa, N.; Inui, T.; Miyazaki, N.; Schaffer, S.W.; Azuma, J. Tissue depletion of taurine accelerates skeletal muscle senescence and leads to early death in mice. *PLoS ONE* **2014**, *9*, e107409. [CrossRef] [PubMed]
33. Jong, C.J.; Azuma, J.; Schaffer, S. Mechanism underlying the antioxidant activity of taurine: Prevention of mitochondrial oxidant production. *Amino Acids* **2012**, *42*, 2223–2232. [CrossRef] [PubMed]
34. Jong, C.J.; Ito, T.; Mozaffari, M.; Azuma, J.; Schaffer, S. Effect of beta-alanine treatment on mitochondrial tuarine level and 5-taurinomethyluridine content. *J. Biomed. Sci.* **2010**, *17*, S25. [CrossRef] [PubMed]
35. Vartak, R.; Deng, J.; Fang, H.; Bai, Y. Redefining the roles of mitochondrial DNA-encoded subunits in respiratory complex I assembly. *Biochim. Biophys. Acta* **2015**, *1852*, 1531–1539. [CrossRef] [PubMed]

36. Schaffer, S.W.; Shimada-Takaura, K.; Jong, C.J.; Ito, T.; Takahashi, K. Impaired energy metabolism of the taurine-deficient heart. *Amino Acids* **2016**, *48*, 549–558. [CrossRef] [PubMed]

37. Scheubel, R.J.; Tostlebe, M.; Simm, A.; Rohrbach, S.; Prondzinsky, R.; Gellerich, F.N.; Silber, R.E.; Holtz, J. Dysfunction of mitochondrial respiratory chain complex I in human failing myocardium is not due to disturbed mitochondrial gene expression. *J. Am. Coll. Cardiol.* **2002**, *40*, 2174–2181. [CrossRef]

38. Sanders, L.H.; Greenamyre, J.T. Oxidative damage to macromolecules in human Parkinson disease and the rotenone model. *Free Radic. Biol. Med.* **2013**, *62*, 111–120. [CrossRef] [PubMed]

39. Murphy, M.P. How mitochondria produce reactive oxygen species. *Biochem. J.* **2009**, *417*, 1–13. [CrossRef] [PubMed]

40. Bleier, L.; Wittig, I.; Heide, H.; Steger, M.; Brandt, U.; Drose, S. Generator-specific targets of mitochondrial reactive oxygen species. *Free Radic. Biol. Med.* **2015**, *78*, 1–10. [CrossRef] [PubMed]

41. Garrido, C.; Galluzzi, L.; Brunet, M.; Puiq, P.E.; Didelot, C.; Kroemer, G. Mechanisms of cytochrome c release from mitochondria. *Cell Death Differ.* **2006**, *13*, 1423–1433. [CrossRef] [PubMed]

42. Kuwana, T.; Newmeyer, D.D. Bcl-2 family proteins and the role of mitochondria in apoptosis. *Curr. Opin. Cell Biol.* **2003**, *15*, 691–699. [CrossRef] [PubMed]

43. Hu, Q.; Wu, D.; Chen, W.; Yan, Z.; Shi, Y. Proteolytic processing of the caspase-9 zymogen is required for apoptosome-mediated activation of caspase-9. *J. Biol. Chem.* **2013**, *288*, 15142–15147. [CrossRef] [PubMed]

44. Montessuit, S.; Somasekharan, S.P.; Terrones, O.; Lucken-Ardjomande, S.; Herzig, S.; Schwarzenbacher, R.; Manstein, D.J.; Bossy-Wetzel, E.; Basanez, G.; Meda, P.; et al. Membrane remodeling induced by dynamin-related protein Drp1 stimulates Bax oligomerization. *Cell* **2010**, *142*, 889–901. [CrossRef] [PubMed]

45. Kashnareva, Y.; Andreyev, A.Y.; Kuwana, T.; Newmeyer, D.D. Bax activation initiates the assembly of a multimeric catalyst that facilitates Bax pore formation in mitochondrial outer membranes. *PLoS Biol.* **2012**, *10*, e1001394. [CrossRef] [PubMed]

46. Jong, C.J.; Ito, T.; Schaffer, S.W. The ubiquitin-proteasome system and autophagy are defective in the taurine-deficient heart. *Amino Acids* **2015**, *47*, 2609–2622. [CrossRef] [PubMed]

47. Pan, C.; Prentice, H.; Price, A.L.; Wu, J.Y. Beneficial effect of taurine on hypoxia- and glutamate-induced endoplasmic reticulum stress pathways in primary neuronal culture. *Amino Acids* **2012**, *43*, 845–855. [CrossRef] [PubMed]

48. Chowdhury, S.; Sinha, K.; Banerjee, S.; Sil, P.C. Taurine protects cisplatin-induced cardiotoxicity by modulating inflammatory and endoplasmic reticulum stress responses. *Biofactors* **2016**, *42*, 647–664. [CrossRef] [PubMed]

49. Prentice, H.; Modi, J.P.; Wu, J.-Y. Mechanisms of neuronal protection against excitotoxicity, endoplasmic reticulum stress, and mitochondrial dysfunction in stroke and neurodegenerative diseases. *Oxid. Med. Cell. Longev.* **2015**, *2015*, 964518. [CrossRef] [PubMed]

50. Lesnefsky, E.J.; Chen, Q.; Hoppel, C.L. Mitochondrial metabolism in aging heart. *Circ. Res.* **2016**, *118*, 1593–1611. [CrossRef] [PubMed]

51. Hetz, C.; Martinon, F.; Rodriguez, D.; Glimcher, L.H. The unfolded protein response: Integrating stress signals through the stress sensor IRE1α. *Physiol. Rev.* **2011**, *91*, 1219–1243. [CrossRef] [PubMed]

nutrients

MDPI

Article

Effects of Pomegranate Juice Supplementation on Oxidative Stress Biomarkers Following Weightlifting Exercise

Achraf Ammar [1,2,*], Mouna Turki [3], Omar Hammouda [1,4], Hamdi Chtourou [1], Khaled Trabelsi [1], Mohamed Bouaziz [5], Osama Abdelkarim [2], Anita Hoekelmann [2], Fatma Ayadi [3], Nizar Souissi [6], Stephen J. Bailey [7], Tarak Driss [4] and Sourour Yaich [8]

[1] Research Unit: Education, Motricity, Sport and health, UR15JS01, High Institute of Sport and Physical Education of Sfax, Sfax University, Sfax 3000, Tunisia; hammouda.o@parisnanterre.fr (O.H.); h_chtourou@yahoo.fr (H.C.); trabelsikhaled@gmail.com (K.T.)
[2] Institute of Sport Science, Otto-von-Guericke-University, Magdeburg 39106, Germany; osamahalim@ymail.com (O.A.); anita.hoekelmann@ovgu.de (A.H.)
[3] Laboratory of Biochemistry, CHU Habib Bourguiba, Sfax University, Sfax 3000, Tunisia; mouna.turki@gmail.com (M.T.); ayadi_fatma@medecinesfax.org (F.A.)
[4] Research Center on Sport and Movement (Centre de Recherches sur le Sport et le Mouvement, CeRSM), UPL, Univ Paris Nanterre, UFR STAPS, Nanterre F-92000, France; tarak.driss@parisnanterre.fr
[5] High Institute of Biotechnology, Sfax University, Sfax 3000, Tunisia; mohamed.bouaziz@cbs.rnrt.tn
[6] National Observatory of Sport, Tunis 1003, Tunisia; n_souissi@yahoo.fr
[7] School of Sport, Exercise and Health Sciences, Loughborough University, Loughborough LE11 3AJ, UK; S.Bailey2@lboro.ac.uk
[8] Department of Community Medicine and Epidemiology, Hédi Chaker Hospital, Sfax University, Sfax 3000, Tunisia; kammoun.sourour@laposte.net
* Correspondence: ammar.achraf@ymail.com; Tel.: +216-2042-0326

Received: 9 June 2017; Accepted: 21 July 2017; Published: 29 July 2017

Abstract: The aim of this study was to test the hypothesis that pomegranate juice supplementation would blunt acute and delayed oxidative stress responses after a weightlifting training session. Nine elite weightlifters (21.0 ± 1 years) performed two Olympic-Weightlifting sessions after ingesting either the placebo or pomegranate juice supplements. Venous blood samples were collected at rest and 3 min and 48 h after each session. Compared to the placebo condition, pomegranate juice supplementation attenuated the increase in malondialdehyde (-12.5%; $p < 0.01$) and enhanced the enzymatic ($+8.6\%$ for catalase and $+6.8\%$ for glutathione peroxidase; $p < 0.05$) and non-enzymatic ($+12.6\%$ for uric acid and $+5.7\%$ for total bilirubin; $p < 0.01$) antioxidant responses shortly (3 min) after completion of the training session. Additionally, during the 48 h recovery period, pomegranate juice supplementation accelerated ($p < 0.05$) the recovery kinetics of the malondialdehyde (5.6%) and the enzymatic antioxidant defenses compared to the placebo condition (9 to 10%). In conclusion, supplementation with pomegranate juice has the potential to attenuate oxidative stress by enhancing antioxidant responses assessed acutely and up to 48 h following an intensive weightlifting training session. Therefore, elite weightlifters might benefit from blunted oxidative stress responses following intensive weightlifting sessions, which could have implications for recovery between training sessions.

Keywords: lipid peroxidation; power training; polyphenol; antioxidant

1. Introduction

Oxidative stress reflects an imbalance between oxidant production and antioxidant responses where the former exceeds the latter. It is well documented that strenuous exercise acutely increases oxidative stress biomarkers and is accompanied by a prolonged pro-oxidant redox status following such exercise [1–4]. Indeed, it has been reported that lipid peroxidation markers are increased immediately following exercise performed near the anaerobic threshold intensity [5], and during short-term maximal efforts such as sprint [6,7] and strength exercises [3,4,8], with such redox perturbations maintained for up to 48 h following high intensity exercise [2–4]. These changes in oxidative stress biomarkers are also accompanied by an increase in antioxidant responses following exercise. Specifically, intensive physical efforts (i.e., sprint, swimming, strength, and Wingate exercises) have been shown to increase the content of glutathione peroxidase (GPX), catalase (CAT), uric acid (UA), and total bilirubin (Tbil) immediately post exercise [3,9–11] and to increase the content of CAT and GPX up to 48 h after weightlifting exercise [2–4].

On the other hand, there is emerging evidence that supplementation with pomegranate juice (POMj), which is a rich source of polyphenols and other biologically active compounds (e.g., flavonols, flavanoids, gallicacid, ellagic acid, quercetin, ellagitannins, and nitrate) [12,13], confers several health benefits during stressful situations [14]. Indeed, POMj has been shown to prevent oxidative stress by enhancing antioxidant status (+130%) [15] and reducing oxidative stress biomarkers including, lipid peroxidation (−65%) and low-density lipoprotein oxidation (−90%) [16]. Additionally, POMj has been reported to exhibit the most powerful antioxidant effect compared to other fruit and vegetable juices [17,18]. Indeed, after a comparative spectrophotometric study, POMj was shown to be the most effective in reducing low-density lipoprotein oxidation and inhibiting cellular oxidative stress in macrophages compared to green tea, red wine, and orange, blueberry and cranberry juices [17]. Similarly, using an in vitro comparative assay, POMj manifested the highest capacity to neutralize free radicals with an antioxidant activity three times higher than red wine and green tea (Trolox equivalent antioxidant capacity = 18–20 vs. 6–8) [18]. The underlying mechanisms mediating the potent antioxidant properties of POMj are not yet clear, but its effectiveness has been attributed to enhanced polyphenol bioavailability compared to other foods rich in polyphenols [17].

Although it is established that polyphenol-rich POMj can improve numerous physiological processes in individuals placed under stress [14–16] and that physical exercise is a considerable physiological stressor [1–4,19], relatively few studies have evaluated the antioxidant potential of POMj supplementation during and following intense physical activity [20–22]. Moreover, despite evidence that Olympic weightlifting exercises are often accompanied by an acute and delayed increase in oxidative stress biomarkers [2–4,8], only one recent study has evaluated the efficacy of POMj supplementation to modulate biochemical responses (i.e., inflammatory and muscle damage parameters) during a weightlifting exercise session [23]. However, biomarkers of antioxidant capacity and oxidative stress, which are considered as key components of muscle fatigue and overtraining syndrome [1], were not measured in this study. Therefore, further research is required to assess the potential biochemical basis for POMj as a nutritional recovery strategy post intensive weightlifting exercise.

The aim of the present study was to assess whether consumption of natural POMj could reduce the immediate increase of oxidative stress responses and accelerate the recovery of such responses following a weightlifting training session. We hypothesized that the consumption of natural POMj could blunt acute and delayed oxidative stress responses following a weightlifting training session, which may help in recovering the resting levels of redox parameters at 48 h post exercises.

2. Materials and Methods

2.1. Participants

Nine elite male weightlifters (21 ± 1 years, 80 ± 10 kg, 1.75 ± 0.08 m (mean $\pm$ standard deviation (SD))) volunteered to participate in this study. The participants were recruited on the basis of: (i) they trained at least five sessions per week, (ii) they had at least 3 years' experience of Olympic weightlifting, (iii) they did not have any injuries, and (iv) they did not use any antioxidant or anti-inflammatory drugs during the experimental period nor for one month before commencement of the study [23]. Additionally, participants were instructed to avoid the consumption of creatine and the consumption of large amounts of foods rich in antioxidants or polyphenols (e.g., blueberries, coffee, tea, grape, cherry, curcuma, red wine, and dark chocolate [24]) and they self-reported that they adhered to this requirement. After receiving a thorough explanation of the possible risks and discomforts associated with the experimental procedures, each participant provided their written informed consent to take part in the experiment. The study was conducted according to the Declaration of Helsinki. The protocol and the consent form were fully approved (identification code: 8/16) by the review board "Local committee of the Laboratory of Biochemistry, CHU Habib Bourguiba, Sfax, Tunisia" before the commencement of the assessments. Additionally, all ongoing and related trials for this intervention are registered in Clinical Trials.gov (identification code: NCT02697903).

2.2. Experimental Design

One week before the start of the experimental period, the heaviest weight lifted in a single repetition (1-RM) was assessed for each participant in each Olympic movement (Figure 1). Thereafter, participants performed, as part of their habitual training-program from 08:00 to 09:45, two training sessions (Figure 1) after ingesting the placebo (PLA) (before session 1) and POMj (before session 2) supplements. A recovery period of 48 h separated the PLA and POMj training sessions. Each training session comprised three Olympic Weightlifting exercises (snatch, clean and jerk, and squat) with five sets for each exercise. Specifically, participants completed 2 sets of 3 repetitions at 85% of 1-RM and 3 sets of 2 repetitions at 90% of 1-RM [3,4,23,25]. The PLA and POMj supplements were administered in 250 mL doses and ingested three times daily during the 48 h preceding these two training sessions. Moreover, to allow sufficient time for circulating polyphenols to become elevated and to fully exercise their antioxidant and ergogenic effects [20–23], participants consumed an additional 500mL of PLA or POMj 60 min before the training sessions (Figure 1) [23]. Before and after each training session, fasting blood samples (blood samples 2–5, Figure 1) were collected. Additionally, to assess the recovery kinetics of the biological parameters, blood samples were collected in a resting recovered state (i.e., after 10 days of recovery (blood sample 6) and immediately (3 min) after the training session which preceded the PLA session (blood sample 1) [23].

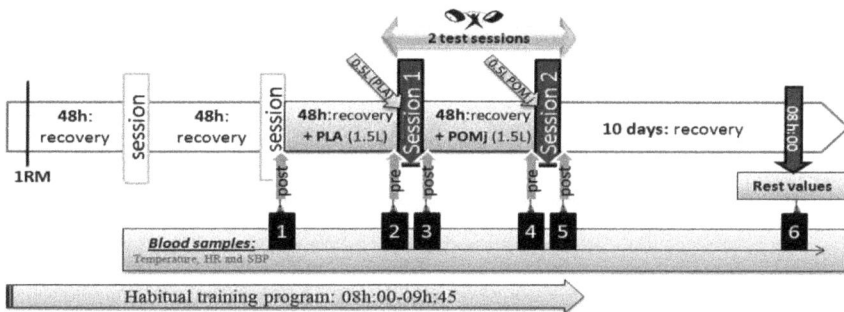

Figure 1. Experimental design.1-RM, One-repetition maximum; PLA, placebo; POMj, pomegranate juice.

Before test sessions, participants underwent an overnight fast and were only permitted to drink one glass of water (15–20 cL) to avoid the potential confounding influence of postprandial thermogenesis [23,26]. Additionally, given that randomly assigning the supplements would have resulted in some participants consuming the POMj supplement before the PLA supplementation with only a 48 h wash-out period, and given that the beneficial effects of POMj could persist for up to three weeks after consumption [27], we elected to avoid any alteration in the biochemical blood levels at the beginning and during the PLA administration period by avoiding the consumption of POMj before PLA. Therefore, as previously described by Ammar et al. [23], PLA was administered before receiving the POMj supplement.

2.3. Pomegranate Juice and Placebo Supplementations

The natural POMj was prepared from a fresh pomegranate fruit 48 h before the beginning of the experimentation and was frozen and stored at −4 °C. No additional chemical products were added to the natural POMj. Each 500 mL of POMj contained 2.56 g of total polyphenols, 1.08 g of orthodiphenols, 292.6 mg of flavonoids, and 46.75 mg of flavonols [23]. PLA juice consisted of a pomegranate-flavored commercial drink containing water, citric acid, natural flavor and natural identical flavor (pomegranate), sweeteners (aspartame × (0.3 g/L), acesulfameK (0.16 g/L), stabilizers (Arabic-gum), and did not contain antioxidants, vitamins, nor polyphenols [23].

2.4. Phenolic Compounds

2.4.1. Extraction of Phenolic Fraction

The phenolic extracts were obtained following the procedure of Chtourou et al. [28] with some modifications. Firstly, the POMj sample (4 g) was added to 2 mL of *n*-hexane and 4 mL of a methanol/water (60:40, *v*/*v*) mixture in a 20 mL centrifuge tube. After vigorous mixing, they were centrifuged for 3 min. The hydroalcoholic phase was collected, and the hexane phase was re-extracted twice with 4 mL of the methanol/water (60:40, *v*/*v*) solution each time. Finally, the hydro alcoholic fractions were combined, washed with 4 mL of *n*-hexane to remove the residual POMj, then concentrated and dried by evaporative centrifuge in vacuum at 35 °C.

2.4.2. Determination of the Total Phenol and O-Diphenol Contents

The determination of the total phenolic compounds was performed by means of the Folin-Ciocalteau reagent using the method described by Gargouri et al. [29]. The total phenolic content was expressed as milligrams of gallic acid (GA) equivalent per kilogram of pomegranate (POM) (y = 0.011x, R^2 = 0.990). The optical density (OD) was measured at λ = 765 nm using a spectrophotometer (Shimadzu UV-1800 PC, Shimadzu, Kyoto, Japan). The concentration of *o*-diphenolic compounds in the methanolic extract was determined by the method of Dridi-Gargouri et al. [30]. The total *o*-diphenolic content was expressed as milligrams of GA equivalent per kilogram of POM (y = 1.144x, R^2 = 0.999). The OD was measured at λ = 370 nm, using the same spectrophotometer.

2.4.3. Determination of Total Flavonoids

Total flavonoids were measured by a colorimetric assay developed by Gargouri et al. [29]; 1 mL aliquot of appropriately diluted sample or standard solutions of catechin (20, 40, 60, 80, and 100 mg/L) was added to a 10mL volumetric flask containing 4 mL double-distillate H_2O. At zero time, 0.30 mL 5% Sodium nitrite ($NaNO_2$) was added to the flask. After 5 min, 0.30 mL 10% Aluminium chloride ($AlCl_3$) was added. At 6 min, 2 mL (1 $mol \cdot L^{-1}$) NaOH was added to the mixture. Immediately, the reaction flask was diluted to volume with the addition of 2.40 mL of double-distillate H_2O and thoroughly mixed. Concerning the absorbance of the mixture, pink in color, it was determined at 510 nm versus prepared water blank. As for the total flavonoids of fruits, they were expressed on a fresh weight basis as mg/100 g catechin equivalents. It is worth noting that the samples were analyzed in triplicate.

2.5. Blood Sampling and Analysis

Blood samples (7 mL) were collected from a forearm vein. Samples were placed in an ice bath and immediately centrifuged at 2500 rpm and 4 °C for 10 min. Aliquots of the resulting plasma were stored at −80 °C until analysis. To eliminate inter-assay-variance, all samples were analyzed in the same assay run. All assays were performed in-duplicate in the same laboratory with simultaneous use of a control serum from Randox. UA and Tbil were determined spectrophotometrically (Architect Ci-4100-ABBOTT, Abbott Deutschland, Wiesbaden, Germany) using uricase and diazonium methods, respectively, and using commercial kits (Ref AU: 3P39-21, Ref Tbil: 6L45, Abbott Deutschland, Wiesbaden, Germany) with an intra-assay-coefficient of variation of 0.5% [3]. CAT activity was measured by assessing the decrease in H_2O_2 concentration [31]. When H_2O_2 was added at low concentration (0.2 M) to a sample with CAT, this enzyme catalyzed the transformation of this substrate to oxygen and water. To check the activity, a kinetic curve had to be measured for 30 s at λ = 240 nm using a molar extinction coefficient of 43.6 $cm^{-1}M^{-1}$ to know the amount of H_2O_2 eliminated [31]. To determine GPX activity, the continuous decrease in nicotinamide adenine dinucleotide phosphate (NADPH) concentration was measured, while glutathione (GSH) levels were maintained, following the methods described by Flohe and Gunzler [32]. This method is based on the rise of the absorbance, during 3 min at λ = 340 nm, because of the oxidation of NADPH in presence of GSH, t-butyl hydroperoxide, glutathione reductase (GR), and the sample. The molar extinction coefficient used for the calculations was ε = 6.22 $\times$ 10^3 $cm^{-1}M^{-1}$. One CAT or GPX enzymatic unit (units) was defined as the amount of the enzyme that catalyzes the conversion of one micromole of substrate per minute. Malonaldehyde (MDA) was measured according to procedures described by Wong et al. [33]. Plasma proteins were precipitated with methanol and removed from the reaction mixture by centrifugation. The protein-free extract was fractionated by high pressure liquid chromatography (HPLC) on a column of octadecyl silica gel, to separate the MDA-thiobarbituric acid (TBA) adduct from interfering chromogens. The MDA-TBA adduct was eluted from the column with methanol/phosphate buffer and quantified spectrophotometrically at λ = 532 nm. Plasma lipoperoxide concentrations were computed by reference to a calibration curve prepared by assays of tetraethoxypropane.

2.6. Statistical Analyses

All statistical analyses were performed using STATISTICA 10.0 Software (StatSoft, Maisons-Alfort, France). Normality of the data distribution was confirmed using the Shapiro-Wilks-W-test. To analyze the effect of POMj supplementation on the biological responses during training sessions (pre-post values), a two-way analysis of variance ANOVA (2 levels (supplementation (PLA and POMj)) $\times$ 2 levels (training-session (Pre and Post)) with repeated measures) was employed. To analyze the effect of POMj supplementation on the recovery kinetics of the selected parameters, a one-way ANOVA was utilized. When significant main effects were observed, Tukey's honest-significance-difference (HSD) post-hoc tests were conducted. Effect sizes were calculated as partial eta-squared (ηp^2) for the ANOVA analysis to assess the potential practical significance of the findings. Pearson correlation was used to assess the correlation between the responses of MDA and the antioxidant markers. Statistical significance was set at $p < 0.05$ and data are presented as mean $\pm$ SD unless otherwise stated.

3. Results

3.1. Acute Effect of POMj on Oxidative Stress Biomarkers Following a Weightlifting Training Session

The oxidative stress responses immediately (3 min) after the weightlifting training sessions are presented in Table 1. The lipid peroxidation biomarker, MDA, increased pre-post training sessions during both PLA (p = 0.0002) and POMj conditions (p = 0.0006). However, the rate of increase was lower (-12.47%) after POMj supplementation compared to that of PLA supplementation ($p < 0.01$). Similarly, biomarkers of enzymatic and non-enzymatic antioxidants increased pre to post training session completed with both the PLA (p = 0.04 for CAT, and GPX and p = 0.02 for UA and Tbil) and POMj (p = 0.0004 for GPX, p = 0.0003 for CAT, and Tbil and p = 0.0001 for UA) supplements. However, the rates of increase pre/post training session were enhanced after POMj supplementation (+8.59%, +6.76%, +12.63%, and +5.68% for CAT, GPX, UA, and Tbil, respectively). For all these parameters (i.e., MDA, CAT, GPX, UA, and Tbil), a significant training-session (p = 0.003, p = 0.009, p = 0.008, p = 0.005, and p = 0.006, respectively) and POMj (p = 0.004, p = 0.008, p = 0.03, p = 0.006, and p = 0.04, respectively) effects were observed. Additionally, significant interactions (Training session $\times$ supplementation condition) were observed for MDA (p = 0.02), CAT (p = 0.04), and UA (p = 0.03).

Table 1. Acute oxidative stress responses to weightlifting training session following pomegranate juice (POMj) and placebo (PLA) supplementation.

Variables	Placebo		% of Change	Pomegranate		% of Change	Δ (POMj-PLA) in %	ANOVA		
	Pre	Post		Pre	Post			Pomegranate Effect	Training Effect	Interaction
Biomarkers of lipid peroxidation										
MDA (μmol/L)	1.82 ± 0.22	2.44 ± 0.18 *	+34.07%	1.76 ± 0.24	2.15 ± 0.26 *	+22.03%	−12.47%	$F_{(1,8)} = 15.9$ $p = 0.004$ $\eta p^2 = 0.3$	$F_{(1,8)} = 36.8$ $p = 0.0003$ $\eta p^2 = 0.6$	$F_{(1,8)} = 8.4$ $p = 0.02$ $\eta p^2 = 0.6$
Biomarkers of enzymatic antioxidant system										
CAT (Units)	15.31 ± 1.51	18.51 ± 1.37 *	+21.02%	15.19 ± 1.32	19.68 ± 1.42 *	+29.61%	+08.59%	$F_{(1,8)} = 12.3$ $p = 0.008$ $\eta p^2 = 0.5$	$F_{(1,8)} = 26.2$ $p = 0.0009$ $\eta p^2 = 0.6$	$F_{(1,8)} = 6.0$ $p = 0.04$ $\eta p^2 = 0.4$
GPX (Units)	0.87 ± 0.08	1.04 ± 0.08 *	+20.46%	0.86 ± 0.09	1.09 ± 0.08 *	+27.13%	+06.76%	$F_{(1,8)} = 6.9$ $p = 0.03$ $\eta p^2 = 0.4$	$F_{(1,8)} = 27.3$ $p = 0.0008$ $\eta p^2 = 0.5$	$F_{(1,8)} = 4.3$ $p = 0.07$ $\eta p^2 = 0.2$
Biomarkers of non-enzymatic antioxidant system										
UA (μmol/L)	321.4 ± 19.6	402.3 ± 22.6 *	+25.16%	299.6 ± 16.9	412.8 ± 18.9 *	+37.79%	+12.63%	$F_{(1,8)} = 13.72$ $p = 0.006$ $\eta p^2 = 0.5$	$F_{(1,8)} = 31.6$ $p = 0.0005$ $\eta p^2 = 0.6$	$F_{(1,8)} = 6.9$ $p = 0.03$ $\eta p^2 = 0.6$
Tbil (μmol/L)	12.30 ± 1.88	15.28 ± 2.32 *	+24.21%	12.01 ± 2.01	15.58 ± 2.43 *	+29.87%	+05.68%	$F_{(1,8)} = 5.9$ $p = 0.04$ $\eta p^2 = 0.4$	$F_{(1,8)} = 29.8$ $p = 0.0006$ $\eta p^2 = 0.7$	$F_{(1,8)} = 3.9$ $p = 0.08$ $\eta p^2 = 0.1$

* Significant differences between pre-post training session. MDA, malonaldehyde; CAT, catalase; GPX, glutathione peroxidase; UA, uric acid; Tbil = total bilirubin; ANOVA, analysis of variance.

3.2. Delayed Effect of POMj on the Recovery Kinetics of the Oxidative Stress Parameters

Table 2 shows the values of the oxidative stress parameters immediately (3 min), 48 h, and 10 days (resting values) after the training sessions during the PLA and POMj trials. From 3 min to 48 h after the training session, the markers of lipid peroxidation as well as the markers of antioxidant responses decreased significantly in both PLA ($p = 0.004$ for MDA, $p = 0.008$ for CAT and GPX, $p = 0.005$ for UA, and $p = 0.007$ for Tbil) and POMj ($p = 0.002$ for MDA, $p = 0.004$ for CAT and GPX, and $p = 0.003$ for UA and Tbil) conditions. However, the rates of decrease were higher ($p < 0.05$) following POMj supplementation compared to that of PLA supplementation (i.e., Δ rate of decrease = 5.63%, 8.94%, 10.21%, 3.57%, and 7.42% for MDA, CAT, GPX, UA, and Tbil, respectively).

Table 2. Recovery kinetics of the oxidative stress responses after placebo and pomegranate supplementation.

Variables	3 min Post Training	48 h Recovery	Rest Values (10 days)	Δ 48 h-3 min in %	ANOVA
		PLA			
Biomarkers of lipid peroxidation					
MDA (μmol/L)	2.44 ± 0.18	1.76 ± 0.24 [a,b]	1.41 ± 0.20	−27.86%	$F_{(2,16)} = 12.2, p = 0.0006, \eta p^2 = 0.4$
Biomarkers of enzymatic antioxidant system					
CAT (Units)	18.51 ± 1.37	15.19 ± 1.32 [a,b]	13.4 ± 1.14	−17.94%	$F_{(2,16)} = 13.3, p = 0.0004, \eta p^2 = 0.4$
GPX (Units)	1.04 ± 0.08	0.86 ± 0.09 [a,b]	0.71 ± 0.08	−17.31%	$F_{(2,16)} = 12.7, p = 0.0005, \eta p^2 = 0.4$
Biomarkers of non-enzymatic antioxidant system					
UA (μmol/L)	402.3 ± 22.6	299.6 ± 16.9 [a]	291.9 ± 18.2	−25.62%	$F_{(2,16)} = 5.1, p = 0.02, \eta p^2 = 0.5$
Tbil (μmol/L)	15.28 ± 2.32	12.01 ± 2.01 [a]	10.93 ± 1.74	−21.40%	$F_{(2,16)} = 4.0, p = 0.04, \eta p^2 = 0.5$
		POMj			
Biomarkers of lipid peroxidation					
MDA (μmol/L)	2.15 ± 0.26	1.43 ± 0.23 [a]	1.41 ± 0.20	−33.49%	$F_{(2,16)} = 6.1, p = 0.01, \eta p^2 = 0.4$
Biomarkers of enzymatic antioxidant system					
CAT (Units)	19.68 ± 1.42	14.39 ± 1.44 [a]	13.4 ± 1.14	−26.88%	$F_{(2,16)} = 4.2, p = 0.03, \eta p^2 = 0.5$
GPX (Units)	1.09 ± 0.08	0.79 ± 0.07 [a]	0.71 ± 0.08	−27.52%	$F_{(2,16)} = 4.4, p = 0.03, \eta p^2 = 0.4$
Biomarkers of non-enzymatic antioxidant system					
UA (μmol/L)	412.8 ± 18.9	292.3 ± 20.6 [a]	291.9 ± 18.2	−29.19%	$F_{(2,16)} = 3.9, p = 0.04, \eta p^2 = 0.6$
Tbil (μmol/L)	15.58 ± 2.43	11.09 ± 1.87 [a]	10.93 ± 1.74	−28.82%	$F_{(2,16)} = 3.8, p = 0.04, \eta p^2 = 0.5$

[a]: Significant differences between 48 h and 3 min post training session. [b]: Significant difference between 48 h recovery and rest values.

Moreover, after supplementing with POMj, the 48 h recovery period was sufficient to recover the resting values of all parameters (i.e., no significant differences were observed between the values of 48 h recovery and the rest values, $p > 0.05$). However, following PLA supplementation, the resting values of the non-enzymatic antioxidants (i.e., UA and Tbil) were recovered after 48 h, but the values of MDA and the enzymatic antioxidants (i.e., CAT and GPX) remained elevated after the same period ($p = 0.03$ for MDA and $p = 0.02$ for CAT and GPX, between the values at 48 h and at rest).

3.3. Relationship between the Responses of Lipid Peroxidation and Antioxidant System

Table 3 shows the relationship between the responses (Δ% rate of change from PLA to POMj conditions) of MDA and the antioxidant parameters. Regression lines for significant correlations were plotted and added to the data supplement (Figure S1 and S2). The acute MDA content following weightlifting exercise was significantly correlated (Table 3, Figure S1) to the enzymatic ($r = 0.6, p = 0.005$ with CAT and $r = 0.5, p = 0.03$ with GPX) and non-enzymatic antioxidant markers with the highest correlation registered between MDA and UA ($r = 0.7, p = 0.0008$). With regard to the delayed response of MDA during the 48 h recovery period, the change in the rate of lipid peroxidation was only correlated (Table 3, Figure S2) to the enzymatic antioxidants CAT ($r = 0.6$, $p = 0.006$) and GPX ($r = 0.5$, $p = 0.02$).

Table 3. Relationship between the lipid peroxidation and antioxidants measures (Δ rate of change % (POMj-PLA)) following a weightlifting training session.

Variables	Acute Response	Delayed Response
Relationship between MDA and the enzymatic antioxidant system		
MDA—CAT	$r = 0.6$, $p = 0.005$	$r = 0.6$, $p = 0.006$
MDA—GPX	$r = 0.5$, $p = 0.03$	$r = 0.5$, $p = 0.02$
Relationship between MDA and the non-enzymatic antioxidant system		
MDA—UA	$r = 0.7$, $p = 0.0008$	$p > 0.05$
MDA—Tbil	$r = 0.5$, $p = 0.04$	$p > 0.05$

4. Discussion and conclusions

The aim of the present study was to investigate the effect of supplementation with a natural POMj on the acute and delayed oxidative stress responses following a weightlifting training session. The main results showed that, compared to PLA, the consumption of POMj 48 h before and during the training session enhanced the recovery kinetics of the acute and delayed oxidative stress responses. Specifically, there was: (i) a reduction in the immediate increase of the MDA (-12.47%), (ii) an increase in acute antioxidant responses (e.g., $+12.63\%$ and $+8.59\%$ for UA and CAT, respectively), and (iii) an acceleration in the delayed recovery kinetics of MDA (5.63%) and the antioxidant markers (e.g., 8.94% for CAT and 10.21% for GPX) with POMj compared to that of PLA supplementation.

4.1. Effect of Weightlifting Training Sessions on Oxidative Stress Responses

The present study reported a significant training-session effect on the oxidative stress parameters (i.e., lipid peroxidation and antioxidant markers) with higher values following a training session compared to pre-training session values in both experimental conditions. These findings are in line with previous studies reporting increased oxidative stress biomarkers following weightlifting exercises [3,4,8]. Exercise-induced oxidative stress has been attributed to reactive oxygen species (ROS) generated through the enzymes xanthine oxidase, NADPH oxidase, and phospholipase A2 [19]. For example, xanthine oxidase has been implicated as an important contributor to oxidative injury to tissues [34], especially during exercise [1] and after ischemic insults [35]. Moreover, increases in hydrogen ions (H^+), catecholamine production, as well as muscle damage, and inflammation post-intensive exercise have been also suggested to contribute to the increased oxidative stress [1,3,7,34]. These effects might be linked to leucocyte adhesion to the endothelial wall [36]; excessive nitric oxide production in activated macrophages [36], which would be expected to facilitate the production of the potent oxidizing and nitrating species peroxynitrite in an environment of increased superoxide production; and iron liberation from haemoglobin or ferritin, promoting Fenton chemistry and the generation of the potent oxidizing hydroxyl radicals [37,38]. These factors are likely to have interacted to evoke the exercise-induced oxidative stress observed in the current study, albeit not directly measured currently.

4.2. Acute Effect of POMj on Oxidative Stress Responses Following theTraining Session.

In this study, POMj supplementation showed a significant impact on the acute post-exercise MDA and enzymatic and non-enzymatic antioxidant responses. Specifically, a lower rate of increase in MDA and greater rate of increase in antioxidant markers were observed immediately post training in the POMj trial compared to the PLA trial. Previous studies in sedentary participants have demonstrated a potent antioxidant effect of POMj on neutralizing reactive oxygen species (ROS) [15–18,39]. However, the results of the current study extend these observations to trained subjects and confirm a beneficial effect in attenuating oxidative stress responses provoked by exhaustive physical exercise. These findings are in line with previous results by Mazani et al. [21] who showed that 240mL of POMj consumed daily for 14 days prior to treadmill running exercise (70% max heart rate) significantly increased the activity of enzymatic antioxidants (i.e., higher pre-post exercise change for GPX and superoxide dismutase (SOD) and attenuated markers of lipid peroxidation post exercise. Similarly, these findings are consistent with those of Tsang et al. [22] who showed that one week of POMj consumption (500 mL/day containing 1.69 g total phenolics/L) significantly decreased urinary levels of lipid peroxidation in the POMj group 30 min after treadmill exercise sessions (50% Wmax). The acute attenuation in post-exercise lipid peroxidation responses after POMj supplementation might be linked to the antioxidant properties of POMj [21,22] leading to reduced acute oxidative stress, lipid peroxidation, tissue edema and/or metabolic by-product accumulation [23,40]. Indeed, UA and CAT responses were increased using POMj (12.63% and 8.59%, respectively) and showed the strongest correlation to the acute MDA response (r = 0.7 and 0.6, respectively) during the training session in the present study. Therefore, it could be suggested that the antioxidant activities of UA and CAT have the highest effect on reducing the acute MDA response immediately following intensive physical exercise completed after POMj supplementation. The apparent efficacy of POMj as an antioxidant might be attributed to its effectiveness as a polyphenol donor [41].

4.3. Delayed Effect of POMj on the Recovery Kinetics of the Oxidative Stress Parameters

After PLA ingestion, only UA and Tbil returned to resting levels 48 h following training with MDA, CAT, and GPX remaining elevated compared to the resting values. These observations confirm previous results showing that a 48 h recovery period lowers MDA levels and antioxidant responses compared to values observed immediately post-training, but was not sufficient to completely recover lipid peroxidation and enzymatic antioxidant parameters to resting levels [3,4]. However, given that the enzymatic (CAT and GPX) but not the non-enzymatic (UA and Tbil) antioxidants remained increased concomitantly with increased markers of lipid peroxidation at 48 h post exercise, and that CAT and GPX correlated with the delayed response of MDA (r = 0.6 and r = 0.5, respectively), it seems that enzymatic antioxidants have a greater potential to mitigate the delayed oxidizing effects of ROS exposure. In contrast, results from the present study indicate that POMj supplementation during the recovery period facilitated the return of the post-training-session values of all parameters to their resting concentration levels. Indeed, the results showed that the rates of decrease of MDA, CAT, and GPX were higher (5.63%, 8.94%, and 10.21%, respectively) with POMj compared to that of PLA supplementation. The more rapid recovery kinetics of lipid peroxidation and antioxidant markers following high-intensity resistance exercise with POMj supplementation confirms the potent antioxidant properties of polyphenol-rich POMj reported previously [20,27]. Indeed, in healthy non active individuals, 15 days of POMj consumption was shown to enhance antioxidant responses, as evidenced by increased erythrocyte glutathione and serum SOD and GPX content, and reduced MDA, protein carbonyls, and matrix metalloproteinases 2 and 9 levels [21]. Importantly, similar responses (lower MDA and protein carbonyl levels) have also been reported following aerobic-based-exercises in adult endurance athletes [20].

Although the exact mechanisms underlying the antioxidant effects of POMj are not entirely understood [20,21], these effects are likely mediated, at least in part, through the high polyphenol content of POMj [39,42] since polyphenols confer antioxidant effects [41]. Indeed. polyphenols can abate the oxidative damage of proteins, lipids, and other cell constituents by the rapid donation of an electron (accompanied by a hydrogen nucleus) to a free radical (e.g., superoxides, peroxynitrite) from hydroxyl (–OH) groups attached to their phenolic rings [36,43]. Through this scavenging process, polyphenols can chemically reduce and stabilize or inactivate free radical species, thereby inhibiting lipid peroxidation and other oxidative modifications (e.g., low density lipoprotein (LDL) oxidation). Furthermore, the antioxidant effect of polyphenols has been attributed to the suppression of free radical formation by modulating antioxidant enzymes and chelating metal ions (Fe^{2+}, Cu^{2+}) involved in free radical production [44,45]. Other possible mechanisms that might underpin the antioxidant properties of polyphenols include the inhibition of leucocyte immobilization and xanthine oxidase activity [36], enhanced endothelial function, lowering of the absorption of pro-oxidant nutrients such as iron [46], the modulation of enzymatic activities (i.e., enhancement of glutathione peroxidase, catalase, NADPH-quinoneoxidoreductase, glutathione S-transferase, and/or cytochrome P450 enzyme) [47], and recycling of antioxidant and reducing agents (e.g., vitamin E and C) [41,45]. However, while the present findings offer some insight into some of the underlying antioxidant mechanisms of POMj, more studies are needed to address the mechanisms for elevated UA and CAT post POMj ingestion, how this links to oxidative stress, and what are the tissues (or cells) of action.

Despite our findings suggesting that POMj supplementation has the potential to blunt oxidative stress responses following weightlifting exercise, it has been previously reported that the acute oxidative stress response following such exercise sessions is greater in the morning compared to the evening session (+12%) [3]. Therefore, while the present results showed that POMj supplementation could blunt the morning acute oxidative responses by 12.5% compared to the PLA condition, further research is required to assess whether the effects can be reproduced at different times of day.

In conclusion, the results of the present study extend on previous studies reporting improved protection against oxidative stress in healthy and diseased conditions after POMj consumption by indicating that POMj supplementation can reduce the acute oxidative stress response to an intensive session of resistance exercises by enhancing antioxidant responses and accelerating the recovery kinetics of oxidative stress markers. These findings about reduced oxidative stress following POMj supplementation might have implications for recovery of performance following intensive resistance training.

5. Limitations and Perspectives

The present study is the first to demonstrate a potential protective effect of polyphenol-rich POMj supplementation on the degree of lipid peroxidation induced by intense weightlifting exercises, as inferred from a lower circulating MDA content. However, it should be acknowledged that since exercise has been shown to provoke oxidative modifications to several biological components (e.g., proteins, lipids, and DNA) [48], and since MDA is just one product of lipid peroxidation, two or more biomarkers are recommended to more accurately infer oxidative damage [49,50]. Accordingly, further studies using multiple redox-related biomarkers are required to confirm the potential positive effects of POMj supplementation on blunting lipid damage and oxidative stress following physical exercise and to specify the target tissues and the exact antioxidant mechanisms of POMj.

Using the same protocol of the present this study, a recent study reported an increase in the total (+8.3%) and maximal (+3.3%) load lifted during a weightlifting training session, and a lowering in markers of muscle damage and the delayed onset of muscle soreness 48 h after the weightlifting training session after POMj supplementation compared to that of placebo supplementation [23]. Therefore, the findings of the current study might improve understanding of the mechanisms for blunted muscle damage following weightlifting training. However, while consumption of polyphenol-rich beverages can blunt post exercise redox perturbation and muscle damage, and can accelerate the recovery of skeletal muscle force production post strenuous exercise, polyphenol consumption during a training intervention has been reported to blunt some of the physiological adaptations elicited by the training program [51]. Therefore, athletes wishing to use POMj as a potential recovery aid during intensified training periods, where the degree of oxidative stress, inflammation, and muscle damage will be greater, need to balance the aim of promoting recovery from training sessions with the potential to attenuate the exercise-induced redox signaling that provokes physiological adaptations to exercise training. Further research is required to optimize the POMj supplementation guidelines.

Supplementary Materials: The following are available online at www.mdpi.com/2072-6643/9/8/819/s1. File S1: Receipt of Clinical trials.gov registry, Figure S1: Regression line for the significant relationships between the lipid peroxidation and antioxidants acute measures (Δ rate of change % (POMj-PLA)) following weightlifting training session, Figure S2: Regression line for the significant relationships between the lipid peroxidation and antioxidants delayed measures (Δ rate of change % (POMj-PLA)) following weightlifting training session.

Acknowledgments: The authors wish to thank all the participants for their maximal effort and cooperation. The authors report no funding source for this work.

Author Contributions: A.A., H.C., K.T., and N.S. conceived and designed the experiments; A.A., H.C., K.T., and S.Y. performed the experiments; A.A., T.D., and S.Y. analyzed the data; A.A., M.T., M.B., and F.A. contributed reagents/materials/analysis tools; A.A., H.C., O.H., K.T., O.A., A.H., S.J.B., T.D., and N.S. wrote the paper.

Conflicts of Interest: The authors declare no conflict of interest.

References

1. Finaud, J.; Lac, G.; Filaire, E. Oxidative stress: Relationship with exercise and training. *Sports Med.* **2006**, *36*, 327–358. [CrossRef] [PubMed]
2. Ammar, A.; Chtourou, H.; Souissi, N. Effect of time-of-day on biochemical markers in response to physical exercise. *J. Strength Cond. Res.* **2017**, *31*, 272–282. [CrossRef] [PubMed]
3. Ammar, A.; Chtourou, H.; Hammouda, O.; Trabelsi, K.; Chiboub, J.; Turki, M.; AbdelKarim, O.; El Abed, K.; Ben Ali, M.; Hoekelmann, A.; et al. Acute and delayed responses of C-reactive protein, malondialdehyde and antioxidant markers after resistance training session in elite weightlifters: Effect of time of day. *Chronobiol. Int.* **2015**, *32*, 1211–1222. [CrossRef] [PubMed]
4. Ammar, A.; Chtourou, H.; Hammouda, O.; Turki, M.; Ayedi, F.; Kallel, C.; AbdelKarim, O.; Hoekelmann, A.; Souissi, N. Relationship between biomarkers of muscle damage and redox status in response to a weightlifting training session: Effect of time-of-day. *Physiol. Int.* **2016**, *103*, 243–261. [PubMed]
5. Goldfarb, A.H.; McKenzie, M.J.; Bloomer, R.J. Gender comparisons of exercise-induced oxidative stress: Influence of antioxidant supplementation. *Appl. Physiol. Nutr. Metab.* **2007**, *32*, 1124–1131. [CrossRef] [PubMed]
6. Baker, J.S.; Bailey, D.M.; Hullin, D.; Young, I.; Davies, B. Metabolic implications of resistive force selection for oxidative stress and markers of muscle damage during 30 s of high-intensity exercise. *Eur. J. Appl. Physiol.* **2004**, *92*, 321–327. [CrossRef] [PubMed]
7. Kayatekin, B.; Gönenç, S.; Açikgöz, O.; Uysal, N.; Dayi, A. Effects of sprint exercise on oxidative stress in skeletal muscle and liver. *Eur. J. Appl. Pshysiol.* **2002**, *87*, 141–144. [CrossRef] [PubMed]
8. Liu, J.F.; Chang, W.Y.; Chan, K.H.; Tsai, W.Y.; Lin, C.L.; Hsu, M.C. Blood lipid peroxides and muscle damage increased following intensive resistance training of female weightlifters. *Ann. N. Y. Acad. Sci.* **2005**, *1042*, 255–261. [CrossRef] [PubMed]
9. Groussard, C.; Rannou-Bekono, F.; Machefer, G.; Chevanne, M.; Vincent, S.; Sergent, O.; Cillard, J.; Gratas-Delamarche, A. Changes in blood lipid peroxidation markers and antioxidants after a single sprint anaerobic exercise. *Eur. J. Appl. Pshysiol.* **2003**, *89*, 14–20. [CrossRef] [PubMed]
10. Hammouda, O.; Chtourou, H.; Chahed, H.; Ferchichi, S.; Chaouachi, A.; Kallel, C.; Miled, A.; Chamari, K.; Souissi, N. High intensity exercise affects diurnal variation of some biological markers in trained subjects. *Int. J. Sports Med.* **2012**, *33*, 886–891. [CrossRef] [PubMed]
11. Inal, M.; Akyüz, F.; Turgut, A.; Getsfrid, W.M. Effect of aerobic and anaerobic metabolism on free radical generation swimmers. *Med. Sci. Sports Exerc.* **2001**, *33*, 564–567. [CrossRef] [PubMed]
12. Sreekumar, S.; Sithul, H.; Muraleedharan, P.; Azeez, J.M.; Sreeharshan, S. Pomegranate fruit as a rich source of biologically active compounds. *BioMed Res. Int.* **2014**, *2014*, 686921. [CrossRef] [PubMed]
13. Hord, N.G.; Tang, Y.; Bryan, N.S. Food sources of nitrates and nitrites: The physiologic context for potential health benefits. *Am. J. Clin. Nutr.* **2009**, *90*, 1–10. [CrossRef] [PubMed]
14. Zarfeshany, A.; Asgary, S.; Javanmard, S.H. Potent health effects of pomegranate. *Adv. Biomed. Res.* **2014**, *3*, 100. [PubMed]
15. Aviram, M.; Rosenblat, M.; Gaitini, D.; Nitecki, S.; Hoffman, A.; Dornfeld, L.; Volkova, N.; Presser, D.; Attias, J.; Liker, H. Pomegranate juice consumption for 3 years by patients with carotid artery stenosis reduces common carotid intima-media thickness, blood pressure and LDL oxidation. *Clin. Nutr.* **2004**, *23*, 423–433. [CrossRef] [PubMed]

16. Kelawala, N.; Ananthanarayan, L. Antioxidant activity of selected foodstuffs. *Int. J. Food Sci. Nutr.* **2004**, *55*, 511–516. [CrossRef] [PubMed]

17. Azadzoi, K.M.; Schulman, R.N.; Aviram, M.; Siroky, M.B. Oxidative stress in arteriogenic erectile dysfunction: Prophylactic role of antioxidants. *J. Urol.* **2005**, *174*, 386–393. [CrossRef] [PubMed]

18. Seeram, N.P.; Aviram, M.; Zhang, Y.; Henning, S.M.; Feng, L.; Dreher, M.; Heber, D. Comparison of antioxidant potency of commonly consumed polyphenol-rich beverages in the united states. *J. Agric. Food Chem.* **2008**, *56*, 1415–1422. [CrossRef] [PubMed]

19. Powers, S.K.; Jackson, M.J. Exercise-induced oxidative stress: Cellular mechanisms and impact on muscle force production. *Physiol. Rev.* **2008**, *88*, 1243–1276. [CrossRef] [PubMed]

20. Fuster-Muñoz, E.; Roche, E.; Funes, L.; Martínez-Peinado, P.; Sempere, J.; Vicente-Salar, N. Effects of pomegranate juice in circulating parameters, cytokines, and oxidative stress markers in endurance-based athletes: A randomized controlled trial. *Nutrition* **2016**, *32*, 539–545. [CrossRef] [PubMed]

21. Mazani, M.; Fard, A.S.; Baghi, A.N.; Nemati, A.; Mogadam, R.A. Effect of pomegranate juice supplementation on matrix metalloproteinases 2 and 9 following exhaustive exercise in young healthy males. *J. Pak. Med. Assoc.* **2014**, *64*, 785–790. [PubMed]

22. Tsang, C.; Wood, G.; Al-Dujaili, E. Pomegranate juice consumption influences urinary glucocorticoids, attenuates blood pressure and exercise-induced oxidative stress in healthy volunteers. In *Endocrine Abstracts, Presented at Society for Endocrinology BES Meeting, Birmingham, UK, 11–14 April 2011*; British Society of Endocrinology: Bristol, UK, 2011.

23. Ammar, A.; Turki, M.; Chtourou, H.; Hammouda, O.; Trabelsi, K.; Kallel, C.; Abdelkarim, O.; Hoekelmann, A.; Bouaziz, M.; Ayadi, F.; et al. Pomegranate supplementation accelerates recovery of muscle damage and soreness and inflammatory markers after a weightlifting training session. *PLoS ONE.* **2016**, *11*, e0160305. [CrossRef] [PubMed]

24. Pérez-Jiménez, J.; Neveu, V.; Vos, F.; Scalbert, A. Identification of the 100 richest dietary sources of polyphenols: An application of the Phenol-Explorer database. *Eur. J. Clin. Nutr.* **2010**, *64*, 112–120. [CrossRef] [PubMed]

25. Ammar, A.; Chtourou, H.; Trabelsi, K.; Padulo, J.; Turki, M.; El Abed, K.; Hoekelamann, A.; Hakim, A. Temporal specificity of training: intra-day effects on biochemical responses and Olympic-Weightlifting performances. *J. Sports Sci.* **2015**, *33*, 358–368. [CrossRef] [PubMed]

26. Ammar, A.; Chtourou, H.; Turki, M.; Hammouda, O.; Chaari, A.; Boudaya, M.; Driss, T.; Ayadi, F.; Souissi, N. Acute and delayed responses of steroidal hormones, blood lactate and biomarkers of muscle damage after a resistance training session: Time-of-day effects. *J. Sports Med. Phys. Fit.* **2017**. [CrossRef]

27. Matthaiou, C.M.; Goutzourelas, N.; Stagos, D.; Sarafoglou, E.; Jamurtas, A.; Koulocheri, S.D.; Haroutounian, S.A.; Tsatsakis, A.M.; Kouretas, D. Pomegranate juice consumption increases gsh levels and reduces lipid and protein oxidation in human blood. *Food Chem. Toxicol.* **2014**, *73*, 1–6. [CrossRef] [PubMed]

28. Chtourou, M.; Gargouri, B.; Jaber, H.; Bouaziz, M. Comparative Study of olive quality from *ChemlaliSfaxArbequina* cultivated in Tunisia. *Eur. J. Lipid Sci.* **2013**, *115*, 631–640. [CrossRef]

29. Gargouri, B.; Ammar, S.; Zribi, A.; Ben Mansour, A.; Bouaziz, M. Effect of growing region on quality characteristics and phenolic compounds of chemlali extra-virgin olive oils. *Acta Physiol. Plant.* **2013**, *35*, 2801–2812. [CrossRef]

30. Dridi-Gargouri, O.; Kallel-Trabelsi, S.; Bouaziz, M.; Abdelhèdi, R. Synthesis of 3-O-methylgallic acid a powerful antioxidant by electrochemical conversion of syringic acid. *Biochim. Biophys. Acta* **2013**, *1830*, 3643–3649. [CrossRef] [PubMed]

31. Wheeler, C.R.; Salzman, J.A.; Elsayed, N.M.; Omaye, S.T.; Korte, D.W. Automated assays for superoxide dismutase, catalase, glutathione peroxidase, and glutathione reductase activity. *Anal. Biochem.* **1990**, *184*, 193–199. [CrossRef]

32. Flohé, L.; Günzler, W.A. Assays of glutathione peroxidase. *Methods Enzymol.* **1984**, *105*, 114–120. [PubMed]

33. Wong, S.; Knight, J.; Hopfer, S.; Zaharia, O.; Leach, C.N.; Sunderman, F. Lipoperoxides in plasma as measured by liquid-chromatographic separation of malondialdehyde-thiobarbituric acid adduct. *Clin. Chem.* **1987**, *33*, 214–220. [PubMed]

34. Sahlin, K.; Cizinsky, S.; Warholm, M.; Höberg, J. Repetitive static muscle contractions in humans—A trigger of metabolic and oxidative stress? *Eur. J. Appl. Pshysiol. Occup. Physiol.* **1992**, *64*, 228–236. [CrossRef]

35. Sanhueza, J.; Valdes, J.; Campos, R.; Garrido, A.; Valenzuela, A. Changes in the xanthine dehydrogenase/xanthine oxidase ratio in the rat kidney subjected to ischemia-reperfusion stress: Preventive effect of some flavonoids. *Res. Commun. Chem. Pathol. Pharmacol.* **1992**, *78*, 211–218. [PubMed]

36. Nijveldt, R.J.; Van Nood, E.; Van Hoorn, D.E.; Boelens, P.G.; Van Norren, K.; Van Leeuwen, P.A. Flavonoids: A review of probable mechanisms of action and potential applications. *Am. J. Clin. Nutr.* **2001**, *74*, 418–425. [PubMed]

37. Childs, A.; Jacobs, C.; Kaminski, T.; Halliwell, B.; Leeuwenburgh, C. Supplementation with vitamin C and n-acetyl-cysteine increases oxidative stress in humans after an acute muscle injury induced by eccentric exercise. *Free Radic. Biol. Med.* **2001**, *31*, 745–753. [CrossRef]

38. Nelson, C.W.; Wei, E.P.; Povlishock, J.T.; Kontos, H.A.; Moskowitz, M.A. Oxygen radicals in cerebral ischemia. *Am. J. Physiol.* **1992**, *263*, H1356–H1362. [PubMed]

39. Gil, M.I.; Tomás-Barberán, F.A.; Hess-Pierce, B.; Holcroft, D.M.; Kader, A.A. Antioxidant activity of pomegranate juice and its relationship with phenolic composition and processing. *J. Agric. Food Chem.* **2000**, *48*, 4581–4589. [CrossRef] [PubMed]

40. Fenercioglu, A.K.; Saler, T.; Genc, E.; Sabuncu, H.; Altuntas, Y. The effects of polyphenol-containing antioxidants on oxidative stress and lipid peroxidation in type 2 diabetes mellitus without complications. *J. Endocrinol. Investig.* **2010**, *33*, 118–124. [CrossRef]

41. Perron, N.R.; Brumaghim, J.L. A review of the antioxidant mechanisms of polyphenol compounds related to iron binding. *Cell Biochem. Biophys.* **2009**, *53*, 75–100. [CrossRef] [PubMed]

42. Halliwell, B.; Rafter, J.; Jenner, A. Health promotion by flavonoids, tocopherols, tocotrienols, and other phenols: Direct or indirect effects? Antioxidant or not? *Am. J. Clin. Nutr.* **2005**, *81*, 268–276.

43. Castaneda-Ovando, A.; de Lourdes Pacheco-Hernández, M.; Páez-Hernández, M.E.; Rodríguez, J.A.; Galán-Vidal, C.A. Chemical studies of anthocyanins: A review. *Food Chem.* **2009**, *113*, 859–871. [CrossRef]

44. Korkina, L.G.; Afanas' Ev, I.B. Antioxidant and chelating properties of flavonoids. *Adv. Pharmacol.* **1996**, *38*, 151–163.

45. Fraga, C.G.; Galleano, M.; Verstraeten, S.V.; Oteiza, P.I. Basic biochemical mechanisms behind the health benefits of polyphenols. *Mol. Asp. Med.* **2010**, *31*, 435–445. [CrossRef] [PubMed]

46. Scalbert, A.; Manach, C.; Morand, C.; Rémésy, C.; Jiménez, L. Dietary polyphenols and the prevention of diseases. *Crit. Rev. Food Sci. Nutr.* **2005**, *45*, 287–306. [CrossRef] [PubMed]

47. Vauzour, D.; Rodriguez-Mateos, A.; Corona, G.; Oruna-Concha, M.J.; Spencer, J.P. Polyphenols and human health: Prevention of disease and mechanisms of action. *Nutrients.* **2010**, *2*, 1106–1131. [CrossRef] [PubMed]

48. Radak, Z.; Zhao, Z.; Koltai, E.; Ohno, H.; Atalay, M. Oxygen consumption and usage during physical exercise: The balance between oxidative stress and ROS-dependent adaptive signaling. *Antioxid. Redox Signal.* **2013**, *18*, 1208–1246. [CrossRef] [PubMed]

49. Halliwell, B.; Whiteman, M. Measuring reactive species and oxidative damage in vivo and in cell culture: How should you do it and what do the results mean? *Br. J. Pharmacol.* **2004**, *142*, 231–255. [CrossRef] [PubMed]

50. Cobley, J.N.; Close, G.L.; Bailey, D.M.; Davison, G.W. Exercise redox biochemistry: Conceptual, methodological and technical recommendations. *Redox Biol.* **2017**, *12*, 540–548. [CrossRef] [PubMed]

51. Gliemann, L.; Schmidt, J.F.; Olesen, J.; Biensø, R.S.; Peronard, S.L.; Grandjean, S.U.; Mortensen, S.P.; Nyberg, M.; Bangsbo, J.; Pilegaard, H.; et al. Resveratrol blunts the positive effects of exercise training on cardiovascular health in aged men. *J. Physiol.* **2013**, *591*, 5047–5059. [CrossRef] [PubMed]

nutrients

MDPI

Article

Comparison of Australian Recommended Food Score (ARFS) and Plasma Carotenoid Concentrations: A Validation Study in Adults

Lee Ashton [1,2], Rebecca Williams [1,2], Lisa Wood [3], Tracy Schumacher [1,2,4], Tracy Burrows [1,2], Megan Rollo [1,2], Kristine Pezdirc [1,2], Robin Callister [2,3] and Clare Collins [1,2,*]

[1] Faculty of Health and Medicine, School of Health Sciences, University of Newcastle, Callaghan, NSW 2308, Australia; lee.ashton@newcastle.edu.au (L.A.); rebecca.williams@newcastle.edu.au (R.W.); tracy.schumacher@newcastle.edu.au (T.S.); Tracy.Burrows@newcastle.edu.au (T.B.); Megan.rollo@newcastle.edu.au (M.R.); Kristine.pezdirc@newcastle.edu.au (K.P.)
[2] Priority Research Centre in Physical Activity and Nutrition, University of Newcastle, Callaghan, NSW 2308, Australia; Robin.Callister@newcastle.edu.au
[3] Priority Research Centre in Physical Activity and Nutrition, Faculty of Health and Medicine, School of Biomedical Sciences and Pharmacy, University of Newcastle, Callaghan, NSW 2308, Australia; lisa.wood@newcastle.edu.au
[4] Gomeroi gaaynggal Centre, Department of Rural Health, University of Newcastle, Tamworth, NSW 2340, Australia
* Correspondence: clare.collins@newcastle.edu.au; Tel.: +61-02-4921-5646

Received: 24 July 2017; Accepted: 15 August 2017; Published: 17 August 2017

Abstract: Diet quality indices can predict nutritional adequacy of usual intake, but validity should be determined. The aim was to assess the validity of total and sub-scale score within the Australian Recommended Food Score (ARFS), in relation to fasting plasma carotenoid concentrations. Diet quality and fasting plasma carotenoid concentrations were assessed in 99 overweight and obese adults (49.5% female, aged 44.6 ± 9.9 years) at baseline and after three months (198 paired observations). Associations were assessed using Spearman's correlation coefficients and regression analysis, and agreement using weighted kappa (K_w). Small, significantly positive correlations were found between total ARFS and plasma concentrations of total carotenoids ($r = 0.17$, $p < 0.05$), β-cryptoxanthin ($r = 0.18$, $p < 0.05$), β-carotene ($r = 0.20$, $p < 0.01$), and α-carotene ($r = 0.19$, $p < 0.01$). Significant agreement between ARFS categories and plasma carotenoid concentrations was found for total carotenoids (K_w 0.12, $p = 0.02$), β-carotene (K_w 0.14, $p < 0.01$), and α-carotene (K_w 0.13, $p < 0.01$). In fully-adjusted regression models the only signification association with ARFS total score was for α-carotene ($\beta = 0.19$, $p < 0.01$), while ARFS meat and fruit sub-scales demonstrated significant relationships with α-carotene, β-carotene, and total carotenoids ($p < 0.05$). The weak associations highlight the issues with self-reporting dietary intakes in overweight and obese populations. Further research is required to evaluate the use of the ARFS in more diverse populations.

Keywords: validation; dietary methods; diet quality index; food frequency questionnaire; comparative validity

1. Introduction

Optimal diet quality can be described as alignment of an individual's usual dietary intake with National Dietary Guidelines and includes the concepts of nutrient intake adequacy and food variety within key healthy food groups [1]. Poor diet quality, characterized by lower intakes of nutrient dense foods and higher intakes of energy-dense, nutrient-poor foods increases the risk of obesity, hypertension, hyperlipidaemia, type 2 diabetes, cardiovascular disease (CVD), and all-cause morbidity and mortality [1–3]. While collection of accurate dietary intake and overall diet quality is challenging [4], valid assessment of food and nutrient intakes is important in understanding their relationship with health and the development or prevention of disease.

Traditional methods of dietary assessment can be burdensome [5]. Prospective methods, including weighed or estimated food records require recording all food and drinks consumed within a defined time period (usually three or seven days). Food records require a high level of literacy and motivation to frequently weigh, measure, or estimate foods [4]. These demands can discourage individuals from adhering to the dietary assessment methods and elicit a conscious or unconscious change in usual eating behaviours (e.g., consume less or choose to eat foods that are easier to prepare and/or report to reduce the burden of recording) [6]. Retrospective methods, including 24 h recalls, can have a substantial burden due to the extensive training required for interviewers and those responsible for data entry including coding of reported food and beverage items, processing, analysis, and quality control [7], although the emergence of free web-based software (i.e., ASA-24) has reduced some of this burden [8]. Food frequency questionnaires (FFQs) enable assessment of longer-term dietary intake in a cost-effective and timely manner [9] with lower respondent burden compared to other methods. However, there are considerable limitations including; reporting errors related to incomplete food lists, voluntary or involuntary misreporting, inaccuracies in frequency options, and portion sizes used [10], and the time they take to complete [4].

Another approach is to develop short-form versions of longer FFQ's, which still capture key aspects of usual dietary intake. Brief dietary assessment tools including diet quality and variety scores or indexes have been developed to provide a single continuous variable that can be used as an overall indicator of nutrient quality and these are increasingly used in research as proxies for assessing overall dietary intakes, due to their low cost and lower subject and analytic burden [11]. These brief tools have the potential to be used as stand-alone questionnaires and may be especially useful in large, nationally representative surveys [11,12]. This is the approach used by a number of a priori diet quality scores [1], including the Australian Recommended Food Score (ARFS) [13] and Healthy Eating Index [14]. Importantly, a review of diet quality tools has shown that they are able to quantify the risk of some health outcomes, including biomarkers of disease and risk of CVD, some cancers, and mortality [1].

There is a need for balance between collection of valid and reliable data versus burden on participants and researchers. The ARFS has previously demonstrated acceptable reproducibility (ICC: 0.87, 95% CI 0.83, 0.90) and validity (0.53, 95% CI 0.37–0.67) in adults compared to the nutrient intakes estimated from the FFQ from which it is derived [13]. However, this approach to validity carries the risk of correlated errors [15]. Plasma biomarkers provide objective assessments of nutrients, and independent measures of intake when validating dietary assessment tools [16]. Therefore, the primary aim of the current study was to examine the relationship between the total ARFS and fasting plasma carotenoid concentrations in a sample of adults. The secondary aim was to examine the associations between the ARFS sub-scales and fasting plasma carotenoid concentrations.

2. Materials and Methods

2.1. Participants

Data were obtained from a subset of participants from a previous weight loss RCT (Clinical Trials Registry—ANZCTR, number: ACTRN12610000197033) with methods and primary analysis published in detail elsewhere [17]. Briefly, the population sample included in the current analysis were overweight (BMI 25.0 to 29.9 kg/m^2) or obese (BMI $\geq$ 30.0 kg/m^2) adults, aged 18–60 years, recruited from the Hunter region of New South Wales, Australia from October to December 2009. Assessments were conducted at baseline and three months in the Human Performance Laboratory at the University of Newcastle, Australia, Callaghan campus. Individuals included in the current analysis were those who completed the Australian Eating Survey (AES) from which the ARFS was calculated and provided a blood sample for assessment of plasma carotenoid concentrations at both baseline and after three months' follow-up. Participants were equally selected from quartiles of baseline fruit and vegetable intake to ensure the data had a spread of low to high intakes. This study was conducted according to the guidelines laid down in the Declaration of Helsinki and all procedures involving human subjects/patients were approved by the University of Newcastle Human Research Ethics Committee (approval No. H-2010-1170). Written informed consent was obtained from all study participants.

2.2. Australian Recommended Food Score

The ARFS is derived from the AES FFQ [18] and uses a subset of 70 questions related to core nutrient-dense foods recommended in the Australian Dietary Guidelines [19]. The ARFS is calculated by summing the points within eight sub-scales as shown in Table 1, with 20 questions related directly to vegetable intake, 12 to fruit, 13 to protein foods (seven to meat and six to vegetarian sources of protein), 12 to breads/cereals, 10 to dairy foods, one to water, and two to spreads/sauces. The total score ranges from zero to a maximum of 73 points. Briefly, most foods were awarded one point for a reported consumption of ≥once per week, but differed for some items depending on national dietary guideline recommendations [19] with consideration of the Australian Guide to Healthy Eating [20]. Some of the food items for meat (i.e., beef, lamb) and dairy (i.e., ice-cream, frozen yoghurt) had a limit placed on their score for higher intakes due to higher intakes being associated with potentially higher saturated fat or disease risk. Additional points were awarded for greater consumption of vegetables and healthier choices for bread and milk.

Table 1. Scoring method for items in the ARFS.

Food Group	Items Giving 1 Point	Items Giving More Than 1 Point	ARFS
Vegetables	3–4 nightly meals with vegetables; ≥1 per week of each of the following vegetables: potato, pumpkin, sweet potato, cauliflower, green beans, spinach, cabbage or Brussels sprouts, peas, broccoli, carrots, zucchini or eggplant or squash, capsicum, corn, mushrooms, tomatoes, lettuce, celery or cucumber, avocado, onion or leek or shallots/spring onion.	2 points for ≥5 nightly meals with vegetables	21
Fruit	≥1 piece of fruit per day, ≥1 per week of each of the following fruit: canned fruit, fruit salad, dried fruit, apple or pear, orange or mandarin or grapefruit, banana, peach or nectarine or plum or apricot, mango or paw-paw, pineapple, grapes or strawberries or blueberries, melon (any variety).		12
Protein foods-meat/flesh	≤1 serve of mincemeat per month but greater than never; 1–4 serve per week of: beef or lamb with or without sauce and/or vegetables per week chicken without batter or crumbing but with or without sauce and/or vegetables, pork with or without sauce and/or vegetables; ≥1 per week of fresh fish, canned tuna or salmon or sardines, other seafood (e.g., prawns, lobster).		7
Vegetarian sources of protein	≥1 per week of the following: nuts (e.g., peanuts, almonds), nut butters, eggs, soybeans or tofu, baked beans, other beans or lentils (e.g., chickpeas, split peas).		6
Breads and cereals (grains)	Usual bread choice is 'other' (e.g., rye, high-fiber white); ≥1 per week of the following: muesli, cooked porridge, breakfast cereal (e.g., Weet-bix, Nutri-grain, Cornflakes), bread or pita bread or toast, English muffin or bagel or crumpet, rice, other grains (e.g., couscous, burghul), noodles (e.g., egg noodles, rice noodles), pasta, tacos or burritos or enchiladas, clear soup with rice or noodles.	2 points if usual bread choice is 'brown' (multigrain or whole meal).	13
Dairy	≥2 serves of: milk, yoghurt or cheese per day; ≥1 serve per week but ≤1 serves per day of flavoured milk, ice cream, frozen yoghurt; ≥1 serve per week but ≤4 serves per day of cheese, cheese spread or cream cheese; ≥1 serve per week of plain milk, yoghurt (not frozen), cottage cheese or ricotta.	2 points if usual type of milk is reduced fat milk or skim milk, or soy milk	11
Water	≥4 glasses of water (including tap, unflavoured bottled water, and unflavoured mineral water).		1
Spreads/sauces *	≥1 serve per week of: yeast extract spread; tomato or barbecue sauce		2
Total			73

* Yeast extract spread and tomato ketchup/barbecue sauce were included in the score as they contain a large amount of B-group vitamins or β-carotene respectively [21].

2.3. Plasma Carotenoids

Phlebotomists collected blood samples in EDTA-coated tubes after an overnight fast and samples were processed then stored at −80 °C until thawed for analysis. High-performance liquid chromatography was used to determine plasma carotenoid concentrations of α-carotene, β-carotene, lycopene, β-cryptoxanthin, and lutein/zeaxanthin. To isolate the carotenoids from the plasma, ethanol was added to allow deproteination followed by ethyl acetate containing internal standards (canthaxanthin), then vortexed and centrifuged (3000 rpm. for 5 min at 4 °C). The supernatant was collected in separate tubes and stored on ice. This process was repeated three times, adding ethyl acetate twice, then hexane to the pellet. Milli-Q (Milli-Q Advantage A10, Merck Millipore, Melbourne, VIC, Australia) water was then added to the pooled supernatant and the mixture vortexed and centrifuged. The supernatant was decanted and placed on a nitrogen evaporator until completely evaporated. The dried extract was reconstituted in dichloromethane:methanol (1:2 vol/vol). Chromatography was performed on Agilent 1200 series gas chromatograph (Agilent Technologies, Santa Clara, CA, USA, part No. G1311-90011) including Chemstation (Chemstation OpenLab CDS software, Agilent Technologies, Melbourne, VIC, Australia) data analysis software at a flow rate of 0.3 mL/min. Carotenoids were analysed using a mobile phase of acetonitrile:dichloromethane:methanol 0.05% ammonium acetate (85:10:5), and integrated and analysed at a wavelength of 450 nm. Total carotenoids were calculated by summing β-carotene, lycopene, α-carotene, β-cryptoxanthin, and lutein/zeaxanthin concentrations.

2.4. Statistical Methods

Data were analysed using Stata Version 12 (StataCorp, College Station, TX, USA) using an alpha level of 0.05. The data from each participant at baseline and 3-months were treated as independent variables. The strength of association between the ARFS or its sub-scales and plasma carotenoid concentrations were assessed in three ways:

1. Spearman's correlations coefficients were used due to non-normal distribution of plasma carotenoid concentrations. Correlation strength was described as poor <0.20, moderate 0.2–0.6 and strong >0.6 [22].
2. Linear regression models to standardized variables, with standard errors clustered on unique participant identifiers and 95% confidence intervals to examine the relationship between specific plasma carotenoids and component of ARFS, while adjusting for known influential factors including: baseline values for total energy intake, total fat intake, age, sex, and BMI. R^2 values and coefficients (95% CI) are also reported, with $R^2 \geq 0.26$ considered large, ≥ 0.13–<0.26 medium and ≤ 0.02 small [23].
3. Precision of the agreement of between categorical assessments of ARFS and plasma carotenoid values was tested using weighted kappa (K_w) statistics to assess whether the ARFS correctly classified participants into the same tertiles of intake based on both total ARFS or ARFS sub-scales and plasma concentrations.

3. Results

A total of 99 overweight or obese adults (49.5% female) completed the AES (from which ARFS was calculated) and provided plasma carotenoid concentration assessments at both baseline and three months. The majority were non-smokers (94.9%), with a mean ARFS of 33 ± 9 points from a maximum of 73. Mean energy intake of participants (10,550 kJ/day ± 3581) was comparable to the average Australian adult aged >18 years (9955 kJ/day) [24]; plasma levels of carotenoids were also comparable to the weighted mean of plasma carotenoid concentrations from a systematic review of 142 studies [25] (156 µg/dL vs. 169 µg/dL—converted to micrograms/deciliter for comparison purposes). The highest concentrations were for lutein/zeaxanthin (60 ± 54) and lycopene (44 ± 31). Table 2 summarizes the subject baseline characteristics.

Table 2. Baseline characteristics of participants (n = 99).

Characteristic	Mean ± SD or n (%)
Age (years)	44.6 ± 9.9
Female	49 (49.5%)
Anthropometric	
Weight (kg)	93.2 ± 14.5
Height (cm)	171.0 ± 8.7
BMI (kg/m^2)	31.8 ± 3.8
Overweight *	39 (39.4%)
Obese I *	40 (40.4%)
Obese II *	18 (18.2%)
Obese III *	2 (2.0%)
Smoking status	
Non-smoker	94 (94.9%)
ARFS (total possible score)	
Total Score (73)	33.0 ± 8.8
Vegetables (21)	12.8 ± 4.2
Fruit (12)	4.7 ± 2.9
Protein—Meat (7)	2.7 ± 1.2
Protein—Vegetarian sources (6)	1.7 ± 1.2
Breads/cereals-Grains (13)	5.3 ± 2.0
Dairy (11)	4.3 ± 1.8
Spreads/Sauces (2)	±0.8
Plasma Carotenoid concentrations (µg/dL)	
α-Carotene	7.4 ± 6.1
β-Carotene	29.7 ± 22.7
Lycopene	44.0 ± 31.2
Lutein/zeaxanthin	59.6 ± 53.7
β-Cryptoxanthin	10.5 ± 7.6
Total carotenoids	156.0 ± 82.4

* Defined using World Health Organization cut offs [26]: Overweight: 25.0 to 29.99 kg/m^2, Obese I: 30.0 to 34.99 kg/m^2, Obese II: 35.0 to 39.99 kg/m^2, Obese III: ≥40.0 kg/m^2.

Table 3 summarizes crude Spearman's correlations between the total ARFS or individual ARFS sub-scales and the total or individual plasma carotenoid concentrations. The total ARFS diet quality score was significantly associated with plasma concentrations of total carotenoids (r = 0.17, p < 0.05), β-cryptoxanthin (r = 0.18, p < 0.05), β-carotene (r = 0.20, p < 0.01), and α -carotene (r = 0.19, p < 0.01). Moderate correlations (r > 0.20) were found between the ARFS fruit subscale score and α-carotene, β-carotene, and β-cryptoxanthin, and between the combined fruit and vegetable score and plasma concentrations of α-carotene and β-carotene. In addition, moderate correlations were found between the ARFS meat subscale score and plasma concentrations of β-carotene, β-cryptoxanthin and total carotenoids. There were other statistically significant correlations found, however, these were classified as poor correlations (r < 0.20).

Table 3. Spearman rank correlations between total ARFS and sub-scale scores and plasma carotenoid concentrations (based on 198 paired observations).

	α-Carotene	β-Carotene	Lycopene	Lutein/Zeaxanthin	β-Cryptoxanthin	Total Plasma Carotenoids
Total ARFS	0.192 **	0.195 **	0.041	0.074	0.180 *	0.170 *
ARFS—Vegetables	0.189 **	0.194 *	−0.050	0.002	0.103	0.095
ARFS—Fruit	0.278 ***	0.260 ***	0.038	0.047	0.217 **	0.193 **
ARFS—Fruit and vegetables combined	0.266 ***	0.258 ***	−0.028	0.027	0.158 *	0.157 *
ARFS Meat	0.195 **	0.215 **	0.092	0.002	0.254 ***	0.209
ARFS Vegetarian alternatives	0.026	0.048	−0.070	0.048	0.007	0.029
ARFS Grains	0.012	−0.007	0.098	0.107	0.123	0.072
ARFS Dairy	−0.033	−0.015	0.056	0.084	0.082	0.041
ARFS Spreads/sauces	0.037	−0.006	0.075	−0.064	−0.022	−0.015

* P-value < 0.05; ** p-value < 0.01; *** p-value < 0.001.

Tables 4 and 5 summarize the unadjusted and the adjusted linear regression analyses between total ARFS and sub-scale component scores and total and individual plasma carotenoid concentrations. In the adjusted regression model (adjusted for total energy intake, total fat intake, BMI, age and sex) total ARFS scores significantly explained the variation in α-carotene ($p < 0.01$), and the ARFS fruit sub-scale significantly explained the variation in α-carotene ($p < 0.001$) and β-carotene ($p < 0.05$), while the ARFS combined fruit and vegetables score explained the variation in α-carotene ($p < 0.001$). The ARFS meat sub-scale explained the variation in plasma concentrations of α-carotene, β-carotene and total carotenoids (all $p < 0.05$).

Table 6 summarizes the analyses examining the extent of agreement between tertiles of ARFS score and tertiles of specific plasma carotenoid concentrations using kappa statistics. Level of agreement indicated by kappa statistics was significant for total ARFS score when compared to plasma concentrations of α-carotene (K_W 0.13, $p < 0.01$), β-carotene (K_W 0.14, $p < 0.01$), and total carotenoids (K_W 0.12, $p = 0.02$). So too was ARFS fruit sub-scale with plasma concentrations of α-carotene (K_W 0.16, $p < 0.01$), β-carotene (K_W 0.19, $p < 0.001$), β-cryptoxanthin (K_W 0.15, $p < 0.01$) and total carotenoids (K_W 0.16, $p < 0.01$). Significant agreement was also evident for ARFS combined fruit and vegetable score with α-carotene (K_W 0.11, $p = 0.03$) and β-carotene (K_W 0.14, $p < 0.01$). There was significant agreement for ARFS meat sub-scale with plasma concentrations of α-carotene (K_W 0.11, $p = 0.02$), β-carotene (K_W 0.11, $p = 0.03$), β-cryptoxanthin (K_W 0.19, $p < 0.01$) and total carotenoids (K_W 0.12, $p = 0.01$). Additionally, there was significant agreement for ARFS grains sub-scale with plasma concentrations of lutein/zeaxanthin (K_W 0.14, $p < 0.01$), and for ARFS dairy sub-scale with β-cryptoxanthin (K_W 0.09, $p = 0.03$).

Table 4. Unadjusted regression analyses between participant ARFS and plasma carotenoid concentrations (based on 198 paired observations).

Variable	α-Carotene			β-Carotene			Lycopene			Lutein/Zeaxanthin			β-Cryptoxanthin			Total Carotenoids		
	β	95% CI	R²	β	95% CI	R²	β	95% CI	R²	β	95% CI	R²	β	95% CI	R²	β	95% CI	R²
Total ARFS	0.19 ***	0.08, 0.30	0.03	0.86 *	0.11, 1.60	0.02	0.31	−0.20, 0.82	0.02	0.08	−0.94, 1.11	0.00	0.07	−0.04, 0.19	0.01	1.51 *	0.06, 2.96	0.02
ARFS—Vegetables	0.23	−0.00, 0.46	0.01	1.40	−0.39, 3.18	0.01	−0.01	−1.05, 1.03	0.01	0.51	−1.99, 3.00	0.00	0.06	−0.18, 0.31	0.00	2.19	−1.11, 5.49	0.01
ARFS—Fruit	0.87 ***	0.41, 1.34	0.08	3.56 **	0.92, 6.19	0.04	1.05	−0.66, 2.76	0.04	0.46	−2.09, 3.01	0.00	0.28	−0.03, 0.59	0.01	6.22 **	1.61, 10.83	0.03
ARFS—Fruit and vegetables	0.32 ***	0.16, 0.48	0.04	1.54 *	0.17, 2.91	0.03	0.25	−0.49, 0.99	0.03	0.36	−1.18, 1.90	0.00	0.10	−0.06, 0.26	0.01	2.57 *	0.35, 4.79	0.02
ARFS Meat	1.29 **	0.35, 2.24	0.04	8.99 **	3.34, 14.64	0.06	3.64	−0.15, 7.44	0.06	−0.86	−4.80, 3.09	0.00	0.94	−0.11, 1.77	0.03	14.01 ***	5.57, 22.45	0.04
ARFS Vegetarian alternatives	0.54	−0.88, 1.96	0.01	−0.78	−5.73, 4.17	0.00	−2.28	−6.05, 1.48	0.00	1.22	−6.48, 8.92	0.01	−0.30	−1.24, 0.64	0.00	−1.61	−12.56, 9.34	0.00
ARFS grains	0.23	−0.35, 0.81	0.00	−0.13	−3.66, 3.41	0.00	1.57	−1.03, 4.17	0.00	−0.19	−5.14, 4.77	0.01	0.24	−0.37, 0.85	0.00	1.72	−5.66, 9.10	0.00
ARFS Dairy	−0.05	−0.85, 0.75	0.00	0.04	−4.69, 4.78	0.00	0.99	−3.20, 5.17	0.00	−0.83	−5.11, 3.45	0.00	0.15	−0.50, 0.81	0.00	0.30	−9.40, 10.00	0.00
ARFS Spreads/sauces	0.96	−0.72, 2.62	0.00	−1.39	−9.67, 6.89	0.00	5.21	−1.54, 11.97	0.00	−5.16	−14.80, 4.48	0.01	−0.38	−1.75, 1.00	0.01	−0.76	−19.15, 17.63	0.00

β = Regression coefficient. CI = Confidence Interval. R² = Partial Correlation coefficient. ARFS = Australian Recommended Food Score. * p-value < 0.05; ** p-value < 0.01; *** p-value < 0.001.

Table 5. Adjusted regression analyses between participant ARFS and plasma carotenoid concentrations (based on 198 paired observations).

Variable	α-Carotene			β-Carotene			Lycopene			Lutein/Zeaxanthin			β-Cryptoxanthin			Total Carotenoids		
	β	95% CI	R²	β	95% CI	R²	β	95% CI	R²	β	95% CI	R²	β	95% CI	R²	β	95% CI	R²
Total ARFS	0.19 **	0.08, 0.29	0.08	0.43	−0.26, 1.11	0.08	0.38	−0.12, 0.88	0.12	0.11	−1.10, 1.32	0.02	0.05	−0.08, 0.17	0.09	1.14	−0.28, 2.57	0.11
ARFS—Vegetables	0.20	−0.05, 0.44	0.05	0.73	−0.81, 2.27	0.05	0.09	−0.96, 1.14	0.12	0.51	−2.07, 3.09	0.02	0.01	−0.24, 0.27	0.08	1.54	−1.55, 4.64	0.11
ARFS—Fruit	0.90 ***	0.41, 1.39	0.12	2.26 *	0.07, 4.45	0.12	1.41	−0.60, 3.42	0.13	0.50	−2.76, 3.77	0.02	0.13	−0.21, 0.48	0.08	5.21	−0.40, 10.46	0.12
ARFS—Fruit and vegetables	0.31 ***	0.15, 0.48	0.08	0.91	−0.25, 2.06	0.08	0.38	−0.41, 1.17	0.12	0.38	−1.34, 2.11	0.02	0.04	−0.14, 0.22	0.10	2.02	−0.19, 4.22	0.12
ARFS Meat	1.08 *	0.02, 2.13	0.07	6.03 **	1.50, 10.57	0.07	3.07	−0.63, 6.77	0.14	−2.07	−7.31, 3.18	0.02	0.68	−0.20, 1.56	0.10	8.79 *	0.42, 17.16	0.10
ARFS Vegetarian alternatives	0.60	−0.75, 1.94	0.05	−0.64	−5.16, 3.87	0.05	−1.82	−5.56, 1.93	0.11	2.05	−6.25, 10.34	0.02	−0.12	−1.01, 0.80	0.08	0.07	−10.19, 10.33	0.10
ARFS grains	0.13	−0.54, 0.80	0.05	−1.85	−6.58, 2.88	0.05	1.49	−1.01, 3.99	0.12	−0.13	−5.54, 5.29	0.02	0.18	−0.39, 0.75	0.08	0.18	−8.54, 8.18	0.10
ARFS Dairy	0.13	−0.94, 0.67	0.05	−0.99	−5.46, 3.49	0.05	1.26	−2.69, 5.22	0.12	−0.36	−5.25, 4.53	0.02	0.21	−0.41, 0.84	0.08	−0.00	−9.26, 9.26	0.10
ARFS Spreads/sauces	0.91	−0.77, 2.59	0.05	−2.65	−11.22, 5.93	0.05	4.75	−1.68, 11.17	0.12	−5.88	−16.44, 4.69	0.03	−0.21	−1.51, 1.09	0.08	−3.08	−20.89, 14.72	0.10

* Models were adjusted for baseline values for energy intake, total fat intake, BMI, sex, and age. β = regression coefficient. CI = confidence interval. R² = partial correlation coefficient. ARFS = Australian Recommended Food Score. * p-value < 0.05; ** p-value < 0.01; *** p-value < 0.001.

Table 6. Extent of the agreement between tertiles of ARFS score with tertiles for plasma carotenoids concentrations (based on 198 paired observations).

Variable	n = 198 (100%)			Kappa (K$_w$)	p-Value
	Same Tertile	Adjacent Tertile	Misclassified		
α-carotene					
Total ARFS	80 (40%)	85 (43%)	33 (17%)	0.13	<0.01
ARFS—vegetables	69 (35%)	96 (48%)	33 (17%)	0.07	0.11
ARFS—fruit	79 (40%)	86 (43%)	33 (17%)	0.16	<0.01
ARFS—fruit and vegetables	73 (37%)	93 (47%)	32 (16%)	0.11	0.03
ARFS Meat	70 (35%)	100 (51%)	28 (14%)	0.11	0.02
ARFS Vegetarian alternatives	66 (33%)	88 (44%)	44 (22%)	0.02	0.36
ARFS grains	65 (33%)	78 (39%)	55 (28%)	−0.02	0.62
ARFS Dairy	62 (31%)	99 (50%)	37 (19%)	−0.03	0.72
ARFS Spreads/sauces	65 (33%)	66 (33%)	67 (34%)	−0.01	0.60
β-carotene					
Total ARFS	77 (39%)	90 (45%)	31 (16%)	0.14	0.01
ARFS—vegetables	72 (36%)	93 (47%)	33 (17%)	0.09	0.06
ARFS—fruit	87 (44%)	75 (38%)	36 (18%)	0.19	<0.001
ARFS—fruit and vegetables	78 (39%)	88 (44%)	32 (16%)	0.14	<0.01
ARFS Meat	73 (37%)	95 (48%)	30 (15%)	0.12	0.02
ARFS Vegetarian alternatives	61 (31%)	93 (47%)	34 (17%)	−0.01	0.54
ARFS grains	58 (29%)	83 (42%)	57 (29%)	−0.06	0.88
ARFS Dairy	79 (40%)	78 (39%)	41 (21%)	0.05	0.13
ARFS Spreads/sauces	66 (33%)	65 (33%)	67 (34%)	−0.01	0.55
Lycopene					
Total ARFS	66 (33%)	92 (46%)	40 (20%)	0.01	0.44
ARFS—vegetables	54 (27%)	99 (50%)	45 (23%)	−0.09	0.95
ARFS—fruit	67 (34%)	85 (43%)	46 (23%)	0.03	0.33
ARFS—fruit and vegetables	61 (31%)	88 (44%)	49 (25%)	−0.06	0.85
ARFS Meat	73 (37%)	91 (46%)	34 (17%)	0.09	0.05
ARFS Vegetarian alternatives	59 (30%)	87 (44%)	52 (26%)	−0.06	0.88
ARFS grains	64 (32%)	85 (43%)	49 (25%)	0.01	0.43
ARFS Dairy	58 (29%)	108 (55%)	32 (16%)	−0.02	0.68
ARFS Spreads/sauces	66 (33%)	67 (34%)	65 (33%)	0.00	0.48
Lutein/Zeaxanthin					
Total ARFS	63 (32%)	100 (51%)	35 (18%)	0.02	0.37
ARFS—vegetables	63 (32%)	97 (49%)	38 (19%)	0.00	0.48
ARFS—fruit	71 (36%)	81 (41%)	46 (23%)	0.05	0.20
ARFS—fruit and vegetables	70 (35%)	86 (43%)	42 (21%)	0.03	0.27
ARFS Meat	59 (30%)	95 (48%)	44 (22%)	−0.05	0.80
ARFS Vegetarian alternatives	70 (35%)	87 (44%)	41 (21%)	0.06	0.14
ARFS grains	78 (39%)	81 (41%)	39 (20%)	0.14	<0.01
ARFS Dairy	63 (32%)	104 (53%)	31 (16%)	0.01	0.39
ARFS Spreads/sauces	59 (30%)	67 (34%)	72 (36%)	−0.07	0.90
β-cryptoxanthin					
Total ARFS	73 (37%)	90 (45%)	35 (18%)	0.08	0.07
ARFS—vegetables	64 (32%)	91 (46%)	43 (22%)	−0.02	0.61
ARFS—fruit	81 (41%)	79 (40%)	38 (19%)	0.15	<0.01
ARFS—fruit and vegetables	68 (34%)	92 (46%)	38 (19%)	0.05	0.19
ARFS Meat	85 (43%)	83 (42%)	30 (15%)	0.19	<0.01
ARFS Vegetarian alternatives	69 (35%)	79 (40%)	50 (25%)	0.01	0.46
ARFS grains	66 (33%)	83 (42%)	49 (25%)	0.02	0.34
ARFS Dairy	77 (39%)	88 (44%)	33 (17%)	0.09	0.03
ARFS Spreads/sauces	64 (32%)	65 (33%)	69 (35%)	−0.03	0.69
Total carotenoids					
Total ARFS	76 (38%)	91 (46%)	31 (16%)	0.12	0.02
ARFS—vegetables	73 (37%)	88 (44%)	37 (19%)	0.06	0.11
ARFS—fruit	75 (38%)	86 (43%)	37 (19%)	0.12	0.02
ARFS—fruit and vegetables	67 (34%)	97 (49%)	34 (17%)	0.06	0.13
ARFS Meat	78 (39%)	86 (43%)	34 (18%)	0.12	0.01
ARFS Vegetarian alternatives	70 (35%)	86 (43%)	42 (21%)	0.06	0.15
ARFS grains	68 (34%)	82 (41%)	48 (24%)	0.04	0.23
ARFS Dairy	66 (33%)	99 (50%)	33 (17%)	0.02	0.31
ARFS Spreads/sauces	60 (30%)	66 (33%)	72 (36%)	−0.06	0.87

4. Discussion

The current study examined the relative validity of using a brief diet quality score using the ARFS, in relation to a range of fasting plasma carotenoid concentrations in a sample of overweight and obese adults. The total ARFS was significantly associated with plasma concentrations of total carotenoids, β-carotene and α-carotene as shown by stronger positive correlations, and a small proportion of values misclassified within tertiles of ARFS and plasma carotenoid concentrations. Only the association between total ARFS and α-carotene remained significant in the fully-adjusted regression analyses, which suggests that the ARFS is influenced by factors such as energy intake, fat intake, BMI, sex, and age, which is similar to other dietary variables [27]. However, the ARFS appears to be able to quantify intakes of food sources of α-carotene, such as orange juice, tangerines, raspberries, and tomato sauce-based dishes (e.g., spaghetti bolognaise) [28,29]. This is important because consumption of diets rich in carotenoids have been epidemiologically associated with a lower risk for several diseases, including cancer [30], with the antioxidant properties of carotenoids playing an important role in this. Specifically for α-carotene, research from a prospective cohort study in >15,000 US adults found that people with higher plasma α-carotene concentrations were at a significantly lower risk of death from CVD ($p = 0.007$), cancer ($p = 0.02$), and all other causes ($p < 0.001$) when compared to those with low levels over the 14-year follow-up period [31].

For the individual ARFS sub-scales, both meat and fruit demonstrated significant associations with some of the plasma carotenoid concentrations as shown by; (i) stronger correlations; and (ii) significant associations in the fully-adjusted regression analyses. Specifically, after adjusted regression analysis, fruit was significantly associated with α-carotene and β-carotene, while meat was significantly associated with total carotenoids, β-carotene, and α-carotene. The association of the fruit subscale with plasma concentrations of α-carotene may be attributed to the high content within orange juice which is considered one of the most highly consumed fruits among Australian adults [32]. Association of the ARFS meat sub-scale with plasma carotenoid concentrations may be driven by foods with high β-carotene and α-carotene content, such as vegetables and tomato-based sauces commonly served with meat (i.e., spaghetti bolognaise). The items that form the scoring for the meat subscale are based on questions from the FFQ which asks the type of meat intake (beef, lamb, chicken, or pork) with (or without) vegetables or sauce. This supports previous evidence which has shown that higher intakes of unprocessed red meat, chicken, and fish are associated with higher intakes of vegetables [33]. The significant association of fruit and meat with plasma carotenoids, which remains after adjusting for confounders suggests that a wider range of plasma carotenoids appear to be better predictors of specific food group intake rather than whole diet. This is likely due to food groups, such as spreads/sauces and vegetarian sources of protein which influence total ARFS to weaken associations.

Although the ARFS vegetable sub-scale demonstrated significant associations with α-carotene and β-carotene using correlation analysis, the significant association no longer remained in the adjusted regression analysis. Plasma concentrations of lutein/zeaxanthin had poor associations with total ARFS and individual subscales of ARFS for all analyses conducted. Lutein/zeaxanthin are most commonly found in green leafy vegetables (e.g., kale, spinach, broccoli, peas, and lettuce) and egg yolks, with the highest content found in kale [34]. There were no questions assessing kale intake in the AES FFQ and thus may explain the poor associations.

In parallel with similar studies, the strongest correlations between diet and plasma carotenoids were for α-carotene, β-carotene and β-cryptoxanthin [25]. This may be attributed to these carotenoids being highly prevalent in commonly consumed fruit such as apples, apricots, oranges, tangerines, and peaches [29,32]. It is widely acknowledged that individuals are more likely to meet fruit targets than vegetable targets [35]. This may also explain the lack of associations of diet with plasma lutein/zeaxanthin as these are abundant in vegetable sources such as dark greens, including spinach, kale, and broccoli [29]. Limited associations with lycopene may be due to the increased variety of foods containing this carotenoid [29], hence, making it difficult to pick up an association, or due to the lack of foods contributing to lycopene within the ARFS score.

A recent systematic review of 142 biomarker studies comparing plasma carotenoid concentrations and dietary carotenoid intakes, determined mean correlation values by meta-analysis of all relevant studies and gave R values classified as moderate, ranging from 0.26 to 0.47 across the different dietary methods [25]. When diet was assessed using an FFQ, associations appeared to be the weakest compared to other methods and is likely due to the limited range of included food items within FFQ food lists. Additionally, FFQs assess food intake for a long period of time (i.e., six months), while the half-life of plasma carotenoids is 26–76 days [36]. This may explain the weaker associations observed in this current study, as diet quality assessed from ARFS was derived from a subset of FFQ questions. Furthermore, the AFRS does not capture the variation of the AES FFQ as it collapses responses down to two categories. Therefore, some of the contribution to carotenoid intake of those consuming frequently (i.e., multiple times per week) is likely to be lost. Furthermore, the FFQ was not able to determine how food was consumed (i.e., raw, cooked with fat, etc.) which may influence the bioavailability of carotenoids from food. Although correlations in the current study were lower than most of the full-scale dietary measures in the review [25], our results did demonstrate a potential role for use of the ARFS, as some associations were within range (albeit at the lower end of the range) of the full assessment, particularly for fruit.

Limitations

Using plasma carotenoids as biomarkers of components of dietary intake has some limitations. For example, plasma carotenoid concentrations can be influenced by a number of dietary, metabolic and lifestyle factors including: the intra and inter variability in individual's digestion and absorption and the amount of fat in the diet [25]. Despite this, adjustments were made in the regression analyses to account for potential confounders. Although supplements were not adjusted for, previous research using this dataset found no influence on results [37]. For most foods, the scoring for ARFS was based on consumption of greater than, or equal to, once per week. To better quantify diet quality, future studies could include a more comprehensive feature of frequency through more response options. Additionally, the ARFS included spreads/sauces as a sub-scale (namely vegemite as a source of B-group vitamins and tomato or barbecue sauce as a source of or β-carotene), which may have influenced the results as these are generally not considered a component of higher diet quality. However, the overall consumption was low so the effect is deemed to be small. The samples used were from overweight and obese individuals. As carotenoids have an antioxidant role in the body, previous research suggested that people who are overweight or obese have lower plasma concentrations of carotenoids than healthy weight people [38]. Additionally, overweight and obese individuals are more likely to over-report healthy foods such a fruit and vegetables which may influence results [39]. The lack of association between the ARFS vegetable sub-scale and plasma carotenoid concentrations, suggests that this tool may not validly predict vegetable intake and/or variety, likely due to misreporting of vegetable intake. Plasma carotenoid concentrations are considered to be the ideal biological markers of fruit and vegetable intake [40] they are invasive, expensive and responsive to short term intakes [41]. This has implications for measuring a person's usual or long term intakes. In particular, issues have been raised regarding single biomarker assessment compared to FFQ analysis which report longer term intakes of food [42]. The AES FFQ used in this current study reflects intake over the previous 6 months, but plasma carotenoids have a short half-life and may be more reflective of shorter term intake. Therefore, other dietary assessment methods such as 24 h recall or diet history are likely to have higher associations than a FFQ [25]. However, in the current study observations from two time-points (baseline and three-months) for plasma carotenoid concentrations were used. Responses from the AES FFQ were self-reported and therefore subject to reporting bias [43]. Finally, this study compared the precision of agreement with weighted kappa (K_w) statistics with data categorized into tertiles. This approach has been implemented in previous diet validation studies [44–46] and indicates to what extent the dietary assessment tool is able to rank participants correctly and this reflects agreement at individual level [47]. However, this approach is limited in that the percent agreement can include chance agreement [48]. It is recommended to undertake multiple statistical tests to evaluate validity of dietary intake assessment methods [46].

5. Conclusions

Consistent with a recent meta-analysis [25], the current study demonstrated that the diet quality of overweight and obese adults, assessed using the Australian Recommended Food Score (ARFS), has a low correlation with plasma carotenoid concentrations. Stronger relationships were evident between total ARFS and ARFS fruit and meat sub-scales with plasma carotenoids, although associations were weaker after adjustment for confounders. Findings further highlight the implications with self-reporting dietary intakes in overweight and obese populations. Further research is required to evaluate use of the ARFS in clinical practice, epidemiologic research and public health interventions, as a brief continuous measure of diet quality, including in more diverse populations.

Acknowledgments: This research project was funded by a Meat and Livestock Australia Human Nutrition Research Program grant. The funding body had no role in the research study and the views expressed in this manuscript are those of the authors. The authors acknowledge the families who participated in the study as well as the student support for data collection and data entry.

Author Contributions: C.C. and T.B. designed the current study. L.A. and C.C. drafted the manuscript. L.A. undertook the statistical analysis with assistance from T.S. and R.W., L.A., R.W., L.W., T.S., T.B., M.R., K.P., R.C. and C.C. contributed to data interpretation, commented on drafts and approved the final manuscript.

Conflicts of Interest: The authors declare no conflict of interest.

References

1. Wirt, A.; Collins, C.E. Diet quality—What is it and does it matter? *Public Health Nutr.* **2009**, *12*, 2473–2492. [CrossRef] [PubMed]
2. Reedy, J.; Krebs-Smith, S.M.; Miller, P.E.; Liese, A.D.; Kahle, L.L.; Park, Y.; Subar, A.F. Higher diet quality is associated with decreased risk of all-cause, cardiovascular disease, and cancer mortality among older adults. *J. Nutr.* **2014**, *144*, 881–889. [CrossRef] [PubMed]
3. Refshauge, A.; Kalisch, D. *Risk Factors Contributing to Chronic Disease*; Australian Institute of Health and Welfare (AIHW): Canberra, Australia, 2012.
4. Thompson, F.E.; Subar, A.F.; Loria, C.M.; Reedy, J.L.; Baranowski, T. Need for technological innovation in dietary assessment. *J. Am. Diet. Assoc.* **2010**, *110*, 48. [CrossRef] [PubMed]
5. Ashman, A.M.; Collins, C.E.; Brown, L.J.; Rae, K.M.; Rollo, M.E. A brief tool to assess image-based dietary records and guide nutrition counselling among pregnant women: An evaluation. *JMIR Mhealth Uhealth* **2016**, *4*, e123. [CrossRef] [PubMed]
6. Rebro, S.M.; Patterson, R.E.; Kristal, A.R.; Cheney, C.L. The effect of keeping food records on eating patterns. *J. Am. Diet. Assoc.* **1998**, *98*, 1163–1165. [CrossRef]
7. Cullen, K.W.; Watson, K.; Himes, J.H.; Baranowski, T.; Rochon, J.; Waclawiw, M.; Sun, W.; Stevens, M.; Slawson, D.L.; Matheson, D. Evaluation of quality control procedures for 24-h dietary recalls: Results from the girls health enrichment multisite studies. *Prev. Med.* **2004**, *38*, 14–23. [CrossRef] [PubMed]
8. Subar, A.F.; Kirkpatrick, S.I.; Mittl, B.; Zimmerman, T.P.; Thompson, F.E.; Bingley, C.; Willis, G.; Islam, N.G.; Baranowski, T.; McNutt, S. The automated self-administered 24-h dietary recall (ASA24): A resource for researchers, clinicians and educators from the national cancer institute. *J. Acad. Nutr. Diet.* **2012**, *112*, 1134. [CrossRef] [PubMed]
9. Shim, J.-S.; Oh, K.; Kim, H.C. Dietary assessment methods in epidemiologic studies. *Epidemiol. Health* **2014**, *36*, e2014009. [CrossRef] [PubMed]
10. Wong, J.E.; Parnell, W.R.; Black, K.E.; Skidmore, P.M. Reliability and relative validity of a food frequency questionnaire to assess food group intakes in New Zealand adolescents. *Nutr. J.* **2012**, *11*, 65. [CrossRef] [PubMed]
11. Marshall, S.; Burrows, T.; Collins, C. Systematic review of diet quality indices and their associations with health-related outcomes in children and adolescents. *J. Hum. Nutr. Diet.* **2014**, *27*, 577–598. [CrossRef] [PubMed]
12. Contento, I.R.; Randell, J.S.; Basch, C.E. Review and analysis of evaluation measures used in nutrition education intervention research. *J. Nutr. Educ. Behav.* **2002**, *34*, 2–25. [CrossRef]
13. Collins, C.E.; Burrows, T.L.; Rollo, M.E.; Boggess, M.M.; Watson, J.F.; Guest, M.; Duncanson, K.; Pezdirc, K.; Hutchesson, M.J. The comparative validity and reproducibility of a diet quality index for adults: The australian recommended food score. *Nutrients* **2015**, *7*, 785–798. [CrossRef] [PubMed]

14. Guenther, P.M.; Casavale, K.O.; Reedy, J.; Kirkpatrick, S.I.; Hiza, H.A.; Kuczynski, K.J.; Kahle, L.L.; Krebs-Smith, S.M. Update of the healthy eating index: Hei-2010. *J. Acad. Nutr. Diet.* **2013**, *113*, 569–580. [CrossRef] [PubMed]

15. Kipnis, V.; Subar, A.F.; Midthune, D.; Freedman, L.S.; Ballard-Barbash, R.; Troiano, R.P.; Bingham, S.; Schoeller, D.A.; Schatzkin, A.; Carroll, R.J. Structure of dietary measurement error: Results of the open biomarker study. *Am. J. Epidemiol.* **2003**, *158*, 14–21. [CrossRef] [PubMed]

16. Kaaks, R. Biochemical markers as additional measurements in studies of the accuracy of dietary questionnaire measurements: Conceptual issues. *Am. J. Clin. Nutr.* **1997**, *65*, 1232S–1239S. [PubMed]

17. Collins, C.E.; Morgan, P.J.; Jones, P.; Fletcher, K.; Martin, J.; Aguiar, E.J.; Lucas, A.; Neve, M.; McElduff, P.; Callister, R. Evaluation of a commercial web-based weight loss and weight loss maintenance program in overweight and obese adults: A randomized controlled trial. *BMC Public Health* **2010**, *10*, 669. [CrossRef] [PubMed]

18. Collins, C.E.; Boggess, M.M.; Watson, J.F.; Guest, M.; Duncanson, K.; Pezdirc, K.; Rollo, M.; Hutchesson, M.J.; Burrows, T.L. Reproducibility and comparative validity of a food frequency questionnaire for australian adults. *Clin. Nutr.* **2014**, *33*, 906–914. [CrossRef] [PubMed]

19. National Health and Medical Research Council (NHMRC). *Eat for Health: Australian Dietary Guidelines*; Department of Health and Ageing: Canberra, Australia, 2013.

20. Smith, A.; Schmerlaib, Y.; Kellett, E. *The Australian Guide to Healthy Eating: Background Information for Nutrition Educators*; Commonwealth Department of Health and Family Services: Canberra, Australia, 1998.

21. Food Standards Australia New Zealand. *Nuttab 2006–Australian Food Composition Tables*; FSANZ: Canberra, Australia, 2006.

22. McNaughton, S.; Hughes, M.; Marks, G. Validation of a ffq to estimate the intake of pufa using plasma phospholipid fatty acids and weighed foods records. *Br. J. Nutr.* **2007**, *97*, 561–568. [CrossRef] [PubMed]

23. Cohen, J. *Statistical Power Analysis for the Behavior Science*; Lawrance Eribaum Association: Hillsdale, NJ, USA, 1988.

24. Australian Bureau of Statistics. 4364.0.55.007—Australian Health Survey: Nutrition First Results—Foods and Nutrients, 2011–2012. Available online: http://www.abs.gov.au/ausstats/abs@.nsf/Lookup/4364.0.55.007main+features12011-12 (accessed on 10 February 2017).

25. Burrows, T.; Williams, R.; Rollo, M.; Wood, L.; Garg, M.; Jensen, M.; Collins, C. Plasma carotenoid levels as biomarkers of dietary carotenoid consumption: A systematic review of the validation studies. *J. Nutr. Intermed. Metab.* **2014**, *1*, 47. [CrossRef]

26. World Health Organization. Obesity: Preventing and Managing the Global Epidemic. Available online: http://www.who.int/nutrition/publications/obesity/WHO_TRS_894/en/ (accessed on 6 June 2017).

27. Thiele, S.; Mensink, G.B.; Beitz, R. Determinants of diet quality. *Public Health Nutr.* **2004**, *7*, 29–37. [CrossRef] [PubMed]

28. Chug-Ahuja, J.K.; Holden, J.M.; Forman, M.R.; Mangels, A.R.; Beecher, G.R.; Lanza, E. The development and application of a carotenoid database for fruits, vegetables, and selected multicomponent foods. *J. Am. Diet. Assoc.* **1993**, *93*, 318–323. [CrossRef]

29. Mangels, A.R.; Holden, J.M.; Beecher, G.R.; Forman, M.R.; Lanza, E. Carotenoid content of fruits and vegetables: An evaluation of analytic data. *J. Am. Diet. Assoc.* **1993**, *93*, 284–296. [CrossRef]

30. Stahl, W.; Sies, H. Bioactivity and protective effects of natural carotenoids. *Biochim. Biophys. Acta* **2005**, *1740*, 101–107. [CrossRef] [PubMed]

31. Li, C.; Ford, E.S.; Zhao, G.; Balluz, L.S.; Giles, W.H.; Liu, S. Serum α-carotene concentrations and risk of death among us adults: The third national health and nutrition examination survey follow-up study. *Arch. Intern. Med.* **2011**, *171*, 507–515. [CrossRef] [PubMed]

32. Australian Bureau of Statistics. 4364.0.55.012—Australian Health Survey: Consumption of Food Groups from the Australian Dietary Guidelines, 2011–2012. Available online: http://www.abs.gov.au/ausstats/abs@.nsf/0/E1BEB9FF17756D25CA257FAF001A3BF4?Opendocument (accessed on 6 June 2017).

33. Jenkins, L.; Mcevoy, M.; Patterson, A.; Sibbritt, D. Higher unprocessed red meat, chicken and fish intake is associated with a higher vegetable intake in mid-age non-vegetarian women. *Nutr. Diet.* **2012**, *69*, 293–299. [CrossRef]

34. Abdel-Aal, E.-S.M.; Akhtar, H.; Zaheer, K.; Ali, R. Dietary sources of lutein and zeaxanthin carotenoids and their role in eye health. *Nutrients* **2013**, *5*, 1169–1185. [CrossRef] [PubMed]

35. Australian Bureau of Statistics. Australian Bureau of Statistics. Australian Health Survey: First Results 2014–2015. Available online: http://www.abs.gov.au/AUSSTATS/abs@.nsf/DetailsPage/4364.0.55.0012014-15?OpenDocument (accessed on 4 January 2017).

36. Burri, B.J.; Neidlinger, T.R.; Clifford, A.J. Serum carotenoid depletion follows first-order kinetics in healthy adult women fed naturally low carotenoid diets. *J. Nutr.* **2001**, *131*, 2096–2100. [PubMed]

37. Williams, R.; Wood, L.; Collins, C.; Callister, R. Comparison of fruit and vegetable intakes during weight loss in males and females. *Eur. J. Clin. Nutr.* **2016**, *70*, 28–34. [CrossRef] [PubMed]

38. Burrows, T.L.; Warren, J.M.; Colyvas, K.; Garg, M.L.; Collins, C.E. Validation of overweight children's fruit and vegetable intake using plasma carotenoids. *Obesity* **2009**, *17*, 162–168. [CrossRef] [PubMed]

39. Macdiarmid, J.; Blundell, J. Assessing dietary intake: Who, what and why of under-reporting. *Nutr. Res. Rev.* **1998**, *11*, 231–253. [CrossRef] [PubMed]

40. Monsen, E.R. Dietary reference intakes for the antioxidant nutrients: Vitamin C, vitamin E, selenium, and carotenoids. *J. Am. Diet. Assoc.* **2000**, *100*, 637–640. [CrossRef]

41. Pollard, J.; Wild, C.; White, K.; Greenwood, D.; Cade, J.; Kirk, S. Comparison of plasma biomarkers with dietary assessment methods for fruit and vegetable intake. *Eur. J. Clin. Nutr.* **2003**, *57*, 988–998. [CrossRef] [PubMed]

42. Boushey, C.J.; Coulston, A.M.; Rock, C.L.; Monsen, E. *Nutrition in the Prevention and Treatment of Disease*; Academic Press: New York, NY, USA, 2001.

43. Calvert, C.; Cade, J.; Barrett, J.; Woodhouse, A.; Group, U.S. Using cross-check questions to address the problem of mis-reporting of specific food groups on food frequency questionnaires. *Eur. J. Clin. Nutr.* **1997**, *51*, 708–712. [CrossRef] [PubMed]

44. Schumacher, T.L.; Burrows, T.L.; Rollo, M.E.; Wood, L.G.; Callister, R.; Collins, C.E. Comparison of fatty acid intakes assessed by a cardiovascular-specific food frequency questionnaire with red blood cell membrane fatty acids in hyperlipidaemic australian adults: A validation study. *Eur. J. Clin. Nutr.* **2016**, *70*, 1433–1438. [CrossRef] [PubMed]

45. Pauwels, S.; Doperé, I.; Huybrechts, I.; Godderis, L.; Koppen, G.; Vansant, G. Reproducibility and validity of an ffq to assess usual intake of methyl-group donors. *Public Health Nutr.* **2015**, *18*, 2530–2539. [CrossRef] [PubMed]

46. Lombard, M.J.; Steyn, N.P.; Charlton, K.E.; Senekal, M. Application and interpretation of multiple statistical tests to evaluate validity of dietary intake assessment methods. *Nutr. J.* **2015**, *14*, 40. [CrossRef] [PubMed]

47. Flood, V.M.; Smith, W.T.; Webb, K.L.; Mitchell, P. Issues in assessing the validity of nutrient data obtained from a food-frequency questionnaire: Folate and vitamin B 12 examples. *Public Health Nutr.* **2004**, *7*, 751–756. [CrossRef] [PubMed]

48. Gibson, R.S. *Principles of Nutritional Assessment*; Oxford University Press: New York, NY, USA, 2005.

nutrients

MDPI

Review

The Role of MicroRNAs in the Chemopreventive Activity of Sulforaphane from Cruciferous Vegetables

Christopher Dacosta and Yongping Bao *

Norwich Medical School, University of East Anglia, Norwich NR4 7UQ, UK; Christopher.A.Dacosta@gmail.com
* Correspondence: Y.Bao@uea.ac.uk; Tel.: +44-1603-591778

Received: 13 July 2017; Accepted: 15 August 2017; Published: 19 August 2017

Abstract: Colorectal cancer is an increasingly significant cause of mortality whose risk is linked to diet and inversely correlated with cruciferous vegetable consumption. This is likely to be partly attributable to the isothiocyanates derived from eating these vegetables, such as sulforaphane, which is extensively characterised for cytoprotective and tumour-suppressing activities. However, its bioactivities are likely to extend in complexity beyond those currently known; further insight into these bioactivities could aid the development of sulforaphane-based chemopreventive or chemotherapeutic strategies. Evidence suggests that sulforaphane modulates the expression of microRNAs, many of which are known to regulate genes involved at various stages of colorectal carcinogenesis. Based upon existing knowledge, there exist many plausible mechanisms by which sulforaphane may regulate microRNAs. Thus, there is a strong case for the further investigation of the roles of microRNAs in the anti-cancer effects of sulforaphane. There are several different types of approach to the wide-scale profiling of microRNA differential expression. Array-based methods may involve the use of RT-qPCR or complementary hybridisation probe chips, and tend to be relatively fast and economical. Cloning and deep sequencing approaches are more expensive and labour-intensive, but are worth considering where viable, for their greater sensitivity and ability to detect novel microRNAs.

Keywords: broccoli; cancer; cruciferous vegetable; microRNA; sulforaphane; isothiocyanate

---

1. Colorectal Cancer

Colorectal cancer is a major and increasingly common cause of morbidity and premature death. Its global incidence was 1.4 million in 2012, according to the World Health Organization, and is rising, making it the third most commonly diagnosed cancer [1]. In 2015, 774,000 mortalities were directly attributable to colorectal cancer—a figure 58% higher than in 1990 [2]. It is rare for tangible symptoms of colorectal cancer to present until it has already progressed to an advanced (usually terminal) stage, at which currently available treatments are unable to provide a cure [3]. Therefore, strategies for limiting the disease burden from colorectal cancers must include improvements to early-stage diagnosis in asymptomatic individuals, as well as new preventive initiatives.

Reported incidence correlates positively with economic development and/or the adoption of "Western" dietary patterns [4]. Increasing rates of chronic diseases such as cancers are inevitable in countries experiencing rising living standards, partly due to reduced rates of premature death from communicable diseases. Economic development also tends to facilitate improvements to public and/or affordable private healthcare, leading to improved diagnosis. Nevertheless, there is compelling evidence to suggest that commonly associated dietary and lifestyle changes have a significant impact upon the development of colorectal cancer [5]. Particularly, one's risk of developing colorectal cancer is believed to be increased by obesity and a high intake of red meat and/or alcohol and reduced by a fibre-rich diet abundant in fruits and vegetables and a physically active lifestyle, according to guidelines published by Bowel Cancer UK [6]. The apparently strong connections between diet and colorectal cancer risk are unsurprising, given the inevitable exposure of the colorectal tissues to ingested compounds and products of the gut microbiota. The apparently protective effects of diets rich in plant-based foods are believed to be largely attributable to the phytochemicals found in them, many families of which have been studied regarding their direct biological (often cytoprotective) activity—both in vitro and in vivo—since the middle of the twentieth century.

2. Cruciferous Vegetables

Epidemiological studies have revealed a particular inverse correlation between the intake of cruciferous vegetables and colorectal cancer risk; one stronger than that between the latter and the intake of other vegetables [7,8]. Cruciferous vegetables refer to those of the Brassicaceae family and include broccoli, cabbage, and Brussel sprouts. Particular to this plant family are glucosinolates—a group of compounds endogenously synthesised and derived from glucose and amino acid residues. Upon the rupture of plant cells—such as occurs from the consumption of the vegetables or from parasitic attack—the glucosinolates are able to be hydrolysed by endogenous myrosinase enzymes. Intact plant tissue separates glucosinolates from myrosinase enzymes by compartmentalising the former in S-cells [9] and the latter in myrosin cells [10]. Only upon cell rupture are the myrosinase enzymes able to hydrolyse the glucosinolates. Several types of compound are potentially formed, including isothiocyanates, thiocyanates, and nitriles [11].

3. Sulforaphane

Isothiocyanates are to date the most-studied and best-characterised of known glucosinolate-hydrolysis-derived products in terms of their bioactivity. They are believed to play a defensive role in the plants via their cytotoxic effects on microorganisms and small parasitic animals, but to be directly beneficial to human health via broad anti-inflammatory and antioxidant effects, and thus are able to help inhibit the development of cancers [12], cardiovascular diseases [13], and osteoarthritis [14,15]. Broccoli is particularly high in a particular glucosinolate called glucoraphanin, whose myrosinase-mediated hydrolysis generates an isothiocyanate called sulforaphane (SFN, 1-isothiocyanato-4-(methylsulfinyl)butane), the structure of which is depicted in Figure 1.

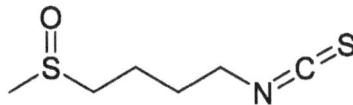

Figure 1. The molecular structure of 1-isothiocyanato-4-(methylsulfinyl)butane), also known as sulforaphane.

A multitude of experiments in vitro and in vivo reportedly demonstrate the ability of SFN to both defend healthy cells against chemical and/or radiation-induced carcinogenesis [16–18] and to inhibit the proliferation, migration, invasive potential and survival of tumour cells [19–21]. It is likely that the former, cytoprotective function of SFN is largely attributable to the induction of nuclear factor (erythroid-derived 2)-like 2 (Nrf2), resulting from the separation of cytoplasmic Nrf2 from Kelch-like ECH-associated protein 1 (Keap1), as illustrated in Figure 2. This allows Nrf2 to enter the nucleus, where it transcriptionally activates various genes, including those coding for antioxidant proteins such as thioredoxin reductase 1 [22] and uridine diphosphate (UDP)-glucuronosyltransferases [23]. These antioxidant proteins act to reduce reactive oxygen species (ROS) levels. ROS can react non-specifically with and thereby damage macromolecules such as lipids, proteins, nucleic acids and carbohydrates, promoting DNA damage–associated mutation and inflammatory signaling linked to age-related decline of function and the pathogeneses of chronic conditions such as atherosclerosis, neurodegenerative diseases and cancers. Therefore, a hypothesis popular in the middle of the twentieth century demonised ROS as toxic metabolic by-products, whose elimination could even halt the "ageing process" [24]. However, it has since been proven that ROS are vital for life and health, as essential components of cell-signaling pathways. For example, the glucose-induced secretion of insulin by β-pancreatic cells is dependent upon glucose-induced ROS generation [25]. Additionally, hydrogen peroxide binds to the regulatory domains of protein kinase C in a manner that promotes cell proliferation and inhibits apoptosis [26].

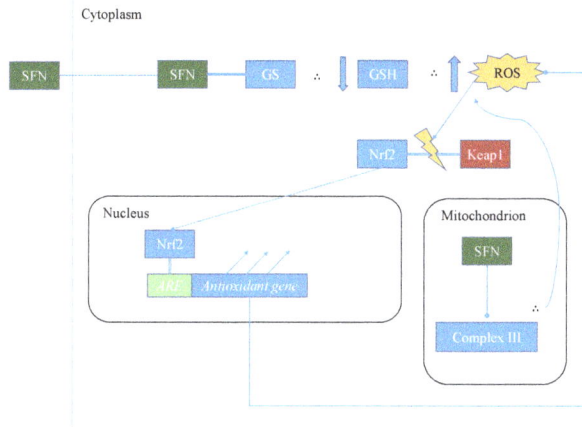

Figure 2. An illustration of the mechanisms by which sulforaphane (SFN) can induce Nrf2. SFN depletes intracellular GSH by forming complexes with it, and also inhibits complex III of the mitochondrial respiratory chain, thus increasing ROS levels. This acute increase in ROS causes Nrf2 to dissociate from Keap1, thereby enabling it to enter the nucleus where it transcriptionally activates antioxidant genes via *ARE* binding.

SFN is commonly referred to as an "antioxidant" based on its widely demonstrated ability to help protect cells against oxidative stress at low-to-moderate doses [27]. However, SFN actually has an acute pro-oxidant effect in cells, largely by depleting intracellular glutathione due to the formation and export of SFN-glutathione complexes [28]. SFN can also increase mitochondrial ROS generation by inhibiting complex III of the mitochondrial respiratory chain, which causes the accumulation of ubisemiquionine, from which molecular oxygen receives electrons, resulting in the formation of superoxide and hydrogen peroxide [29]. SFN-induced acute oxidative stress is widely believed to be a significant driver of SFN-mediated Nrf2 induction, in addition to the SFN-mediated inhibition of p38 mitogen-activated protein kinase (MAPK), whose phosphorylation of Nrf2 inhibits Nrf2-Keap1 dissociation [30]. At low-to-moderate doses, the ensuing antioxidant response tends to outweigh those of the initial oxidative stress in redox terms, leading to a net protection against oxidative stress [31]. This is believed to be largely responsible for SFN's cytoprotective potential in healthy cells. However, very high doses of SFN can be cytotoxic if the pro-oxidant effects induce significant macromolecular damage and/or ROS-mediated apoptosis before the mounting and execution of a sufficient antioxidant response, as illustrated by the sketch in Figure 3. This probably underlies the toxicity of SFN towards microorganisms and parasitic insects, as well as its observed abilities to inhibit tumour cell survival and metastasis [32]. The redox-modulating effects of SFN can thus be described as an example of hormesis. Hormesis is an ancient concept long characterised in various literature for medicinal and/or poisonous herbs—and more recently in scientific literature for nutrients, phytochemicals, and pharmaceutical drugs. In hormesis, low dose-exposure to a particular substance or stimulus has a net beneficial impact (in the case of sulforaphane, oxidative stress-inhibiting) upon the host that are consequential to the protective response it initiates. On the other hand, the net effects of higher dose-exposure are opposite and adverse (in the case of sulforaphane, oxidative stress-inducing) [33]. Doses obtainable from the consumption of cruciferous vegetables by humans are far below the cytotoxic threshold, thus tend to confer either neutral or cytoprotective effects. For example, one study has demonstrated plasma SFN concentrations to reach 10μM following the consumption of broccoli sprouts by study participants [34], which is a concentration demonstrated as non-cytotoxic, Nrf2-inducing and cytoprotective in various cell lines and systems.

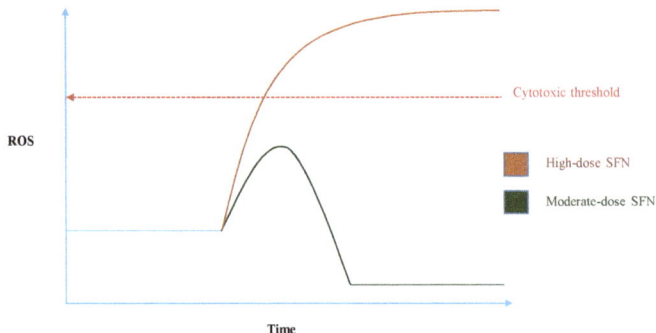

Figure 3. A sketch illustrating the hormetic effects of sulforaphane on cells with regards to oxidative stress, in that net reduction of oxidative stress is observed upon exposure to moderate doses, whereas increased oxidative stress and/or cytotoxicity occurs from high exposure.

Clearly, the induction of Nrf2 in healthy cells tends to be desirable from an anti-cancer perspective due to the inhibition of potentially carcinogenic ROS-induced mutation. In tumour cells, however, it can enhance their survival and proliferation, and make them more resistant to cytotoxicity-dependent anti-cancer therapies [35]. Nrf2 hyperactivation is in fact a marked feature of some cancers and a significant contributor to their aggressiveness. Thus, it has even been speculated that cytoprotective doses of SFN might be detrimental to patients undergoing treatment for advanced cancers, due to the induction of Nrf2 in tumour cells [36]. However, it is important to note that Nrf2 induction does not necessarily boost cell proliferation—particularly in early-stage tumour cells—and has in fact been reported to repress the proliferation of lung-cancer cells by inducing the breakdown of polyamines [37], whilst even to be responsible for the anti-proliferative effects of allicin in HCT-116 cells [38]. Nrf2 can also upregulate the surface expression of IL-17D, which in a systemic context could facilitate natural killer cell-mediated cell death in tumours [39]. Therefore, it is apparent that the interactions between SFN and redox status, and between redox status and carcinogenesis, are complicated.

4. MicroRNAs

ROS are not the only potential means by which high-dose SFN could repress the survival, proliferation, and metastatic characteristics of tumour cells. SFN can induce tumour-suppressing epigenetic changes by directly inhibiting histone deacetylases [40], some of which have a tendency to be aberrantly upregulated in colorectal cancer, thereby repressing various tumour suppressor genes at the transcriptional level via chromatin deacetylation [41]. However, another potential and less comprehensively studied means by which SFN may interact with cancers is the modulation of microRNA (miRNA) expression.

MiRNAs are small non-coding RNAs that are typically 18-25 nucleotides in length and that originate from various genetic loci such as the exons or introns of protein-coding genes and long non-coding exonic clusters (arrays). They post-transcriptionally regulate the expression of at least 30% of protein-coding genes in humans [42] by modulating messenger RNA (mRNA) translation, thus playing major roles in human development and health. Unsurprisingly, major roles for miRNAs in carcinogenesis are frequently reported [43], which is interesting in light of the apparent potential for SFN to modulate miRNA expression in several colorectal cell lines [44]. This leads to reasonable assumption that miRNA modulation has a role to play in SFN's complex interactions with colorectal cancer.

4.1. Biogenesis

The canonical pathway of miRNA biogenesis begins with the RNA polymerase II-mediated transcription of a long 5′-capped and 3′-polyadenylated transcript (the pri-miRNA), which is subsequently cleaved by DiGeorge Syndrome Critical Region 8 (DGCR8) to form products with distinctive hairpin-loop structures and 3′ 2-nucleotide overhangs (the pre-miRNAs) (Figure 4). The overhangs are recognised by exportin 5, which subsequently exports the pre-miRNAs to the cytoplasm where the same overhangs are recognised by Dicer [45]. A pre-miRNA may alternatively be formed by the conversion—by the lariat debranching enzyme (Ldbr)—of an intronic tract released upon the maturation of a protein-coding mRNA [46]. Dicer cleaves pre-miRNAs in their loop regions to generate linear duplexes, which are then unwound, and one strand from each of which remains Dicer-bound, then together with Dicer and Argonaute (AGO) proteins becomes part of an RNA-induced silencing complex (RISC) [45]. These single-stranded Dicer-bound RNAs are the mature miRNAs. Mature miRNAs can alternatively be generated non-canonically via the direct processing of a pre-miRNA by AGO2 followed by covalent modification and/or trimming [47].

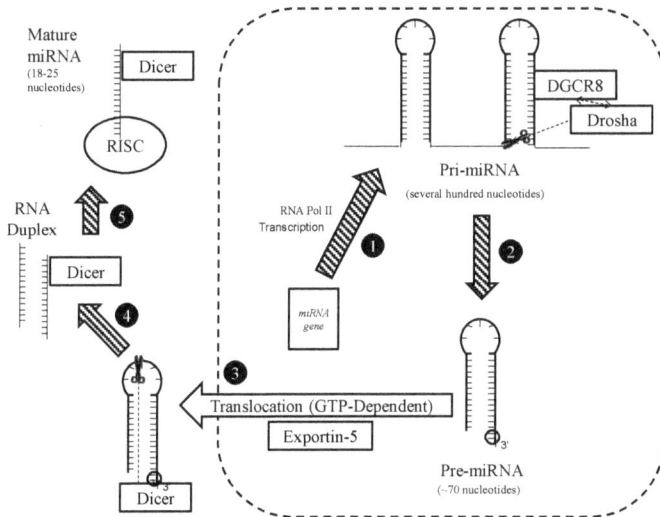

Figure 4. A diagram illustrating the canonical pathway of miRNA expression in animal cells. (1) A genetic locus is transcribed by RNA polymerase II, producing a pri-miRNA, which is several hundred nucleotides long. (2) The pri-miRNA is bound by DGCR8, which recruits Drosha, which cleaves the pri-miRNA into pre-miRNAs that are about 70 nucleotides long and have 2-nucleotide overhangs at their 3′ ends. (3) The 3′ 2-nucleotide overhangs are recognised by Exportin 5, which uses GTP to transport them from the nucleus to the cytoplasm. (4) Dicer recognises the same 3′ overhang and makes a nick in the loop region of the pre-miRNA, generating an imperfectly paired linear RNA duplex, each strand of which bears a 3′ 2-nucleotide overhang. (5) The linear RNA duplex is unwound; one strand remains associated with Dicer as the mature miRNA, and becomes part of a RISC upon association with AGOs.

4.2. Notes on Nomenclature

From each pre-miRNA, two mature miRNAs may be formed; one from each arm of the hairpin-loop motif. It was once assumed that only one of these tended to be functional (typically that with lower stability at the 5′ end), and that that from the other arm was degraded. Therefore—of the two mature miRNAs potentially formed from the processing of pre-miR-29b—the "dominant" product would have been called "miR-29b" whilst the other would have been termed "miR-29b*". However, as evidence against this phenomenon being true in the majority of cases accumulated, miRBase nomenclature was changed such that the "*" suffix was dropped, and mature miRNAs were appended with "-5p" or "-3p" suffixes to denote the pre-miRNA arm of origin. According to this nomenclature, any mature miRNA names lacking a suffix are assumed to denote the only pre-miRNA processing product thus far identified. This updated miRBase nomenclature is used throughout this review.

4.3. Activities

The canonical mechanism of miRNA-mediated repression begins with the interaction of a 6–8 nucleotide seed region at the 5′ end of the miRNA with a locus or loci in the 3′-untranslated region (3′-UTR) of the mRNA, with which it is at least partially complementary in sequence. The RISC represses the translation of and/or degrades the bound mRNA, although the triggering of degradation is believed to be relatively rare in mammals, and to require perfect complementarity between the miRNA seed region and mRNA 3′-UTR [48]. AGO2 is the only AGO in mammals thus far shown to possess endonuclease activity [49]. Since only partial complementarity is required for translational repression, any given miRNA has the potential to target many different mRNA transcripts, whilst any given mRNA is potentially targeted by a multitude of miRNAs.

Although typically characterised as translational repressors, there are contexts in which an miRNA may conversely upregulate the translation of its target. For example, miR-369-3p promotes the translation of its target—tumour necrosis factor-α (TNF-α)—under serum-starved conditions, but represses it under normal conditions [50] as illustrated in Figure 5. This is facilitated by the association of the miRNA with AGO2 specifically, which itself is facilitated by the association of fragile X mental retardation-related protein 1 (FXR1) with AGO2 [50].

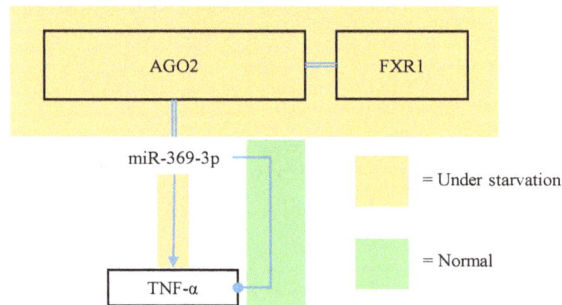

Figure 5. A diagram illustrating the ability of miR-369-3p to upregulate its target TNF-α under starvation, but to conversely repress it under normal conditions.

Hypothetically, serum-starvation could impact the solubility and/or localisation of AGO2-FXR1 complexes in a manner that promotes their association with miR-369-3p. The TNF-α transcript has an AU-rich element toward the 3′ end, which is required for its miRNA-mediated translational upregulation [50]. MiR-10a-5p was demonstrated to bind immediately downstream of 5′-oligopyrimidine motifs present in the 5′-UTRs of ribosomal protein mRNAs, and consequently upregulate their translation under amino acid starvation, in E14 ES mouse embryonic stem cells [51]. It has been proposed that miRNAs have a tendency to repress the translation of their targets in dividing cells, but conversely promote such in quiescent cells [52]. The apparent dependence of miRNA-mediated effects on nutritional status could be important when considering the roles of miRNAs at different stages of carcinogenesis, and in the context of certain therapies. Some miRNAs have been demonstrated to have RISC-independent effects, such as acting as decoys for RNA-binding proteins. For example, miR-328-3p acts as a decoy for the hnRNP E2 RNA-binding protein—which typically represses the transcription of a tumour-suppressing myeloid differentiation factor, CCAAT/enhancer-binding protein α—in chronic myeloid leukaemia [53]. MiR-328-3p thereby de-represses this tumour suppressor gene, and is demonstrably under-expressed in chronic myeloid leukaemia [54].

4.4. IsomiRs

Both at the cleavage of pri-miRNAs to form pre-miRNAs, and of the latter to form mature miRNAs, the Drosha and Dicer enzymes do not necessarily cleave at a precise locus, but have the potential to "slip" by several nucleotides in either direction [55]. Therefore, the mature miRNA sequences catalogued in miRBase do not necessarily represent specific RNAs of fixed sequence, but rather consensus sequences of distributions of isoforms that vary from the consensus in the form of having additional or missing nucleotides at either end. Such variant isoforms are called isomiRs. Perhaps the isomiR phenomenon evolved in cases where several isomiRs of a given miRNA all target a given mRNA in a desirable fashion, but affect different undesirable "off-target" mRNAs—i.e., the presence of multiple isomiRs could provide a mechanism for intensifying the desirable regulation of a specific mRNA target, whilst diffusing undesirable side-effects on other mRNAs [56].

5. Linking Sulforaphane, MicroRNAs, and Colorectal Cancer

A Google Scholar search was performed on 17th February 2017, using the following search string: intitle:microrna | mirna | micrornas | mirnas | "mir-"intitle:colorectal | colon | rectal | bowel intitle:cancer | tumour | tumor | cancers | tumours | tumors | carcinogenesis | tumorigenesis. This revealed existing reports of 144 miRNAs with apparent functions in colorectal cancer—of these, 85 were apparently tumour suppressive, 45 oncogenic, and 14 ambiguous in that reports of both oncogenic and tumour-suppressive function were found. Some examples of each are listed in Table 1, along with reported target genes.

Table 1. Listed examples of reportedly tumour suppressive, oncogenic and ambiguous miRNAs and their reported target genes.

MicroRNA	Reported Role in Colorectal Cancer	Reported Target Genes
let-7a-5p		NIRF [57]
miR-34a-5p		E2F3 [58]
miR-101-3p		PTGS2 [59]
miR-126-3p		PIK3R2 [60]
miR-143-5p	Tumour Suppressive	DNMT3A [61]; ERK5 [62]
miR-195-5p		BCL2 [63]
miR-200a-3p		CTNNB1 [64]
miR-150-5p		MUC4 [65]; MYB [66]
miR-451a		MIF [67]
miR-17-5p		P130 [68]
miR-92a-3p		BIM [69]
miR-23a-3p	Oncogenic	MTSS1 [70]
miR-27a-3p		ZBTB10 [71]
miR-135b-5p		TGFβR2, DAPK1, APC [72]
miR-9-5p	Ambiguous	CDH [73]; TM4SF1 [74]
miR-21-5p		Pdcd4 [75]; CDC25A [76]; TGFβR2 [77]

The apparent involvement of many miRNAs in colorectal cancer is unsurprising, given the major roles of miRNAs in human health and development. There is also existing evidence to suggest that miRNA expression can be modulated in non-cancerous colonic cell lines by SFN. For example, Slaby et al. reported that SFN appeared to upregulate miR-9-3p, miR-23b-3p, miR-27b-5p, miR-27b-3p, miR-30a-3p, miR-135b-3p, miR-145-5p, miR-146a-5p, miR-342-3p, miR-486-5p, miR-505-3p, miR-629-5p, and miR-758-3p, but to downregulate miR-106a-3p, miR-155-5p, and miR-633-3p [44]. Two cell lines—normal derived colon mucosa 460 (NCM460) and normal derived colon mucosa 356 (NCM356)—were used, and TaqMan Low Density qPCR Arrays were used to profile the differential expression of 754 human miRNAs following 48 h SFN treatment [44]. This is a convincing indicator that SFN is able to modulate miRNA expression in colorectal cells, although it is important to note that the miRNAs reported as differentially expressed were not confirmed as so by additional assays. Also, only non-cancerous cell lines and a single time point post-treatment were studied, whereas the effects in cancerous colonic cells might differ substantially from those in their non-cancerous counterparts, and certain miRNAs may be transiently modulated at earlier time points in response to SFN treatment.

Therefore, further experiments to profile the modulation of miRNA expression at different time points, and in a cancerous colorectal cell line, could add significantly to the findings of Slaby et al. as illustrated in Figure 6. It would also be prudent to further examine miRNAs that are reportedly differentially expressed according to the wide-scale profiling process, using single-target assays, in order to rule out the possibility of them being artefacts of the former. Encouragingly, a human study reportedly showed that the consumption of 160g broccoli/day by participants was able to alter blood miRNA expression profiles, thus indicating that systemic regulation of miRNA expression in vivo is possible by doses of SFN obtainable from typical consumption of broccoli, although it cannot be ruled out that the observed modulations were mediated by components of broccoli other than SFN, such as fibre, and selenium and/or other micronutrients [78].

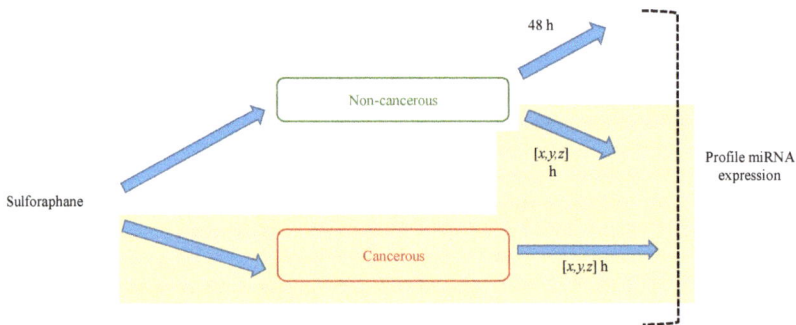

Figure 6. A basic overview of the miRNA-profiling experiments carried out by Slaby et al. and suggested further experiments. Slaby et al. treated non-cancerous colonic cell lines with sulforaphane, then profiled differences in miRNA expression at 48 h. Possible further experiments involving colorectal cancer cell lines and additional time points are illustrated and highlighted in yellow.

5.1. Mechanisms of Sulforaphane-Mediated Modulation

The possible mechanisms by which SFN may modulate miRNA expression are wide-ranging and could involve downstream effects of the SFN-mediated modulation of histone deacetylase (HDAC) activity, redox status, inflammatory signalling, miRNA-processing protein expression, and induction of Nrf2. For example, SFN is known to inhibit the activity of several HDACs [34] including HDAC3, which itself transcriptionally repressed the pro-apoptotic miR-15a-5p/16-1-5p cluster in mantle cell lymphoma cells [79]. SFN (15 µM) was also reported to directly inhibit HDAC3 activity in colorectal cancer HCT-116 cells, thereby inhibiting their proliferation, whilst having no similar effect in non-cancerous colonic CCD-841 cells [41]. SFN is also known to affect redox status—via its acute pro-oxidant and/or Nrf2-inducing effects—which can affect miRNA expression in various ways. For example, oxidative stress has been reported to inhibit Dicer activity in JAR trophoblast cells [80], to inhibit the Drosha partner protein DGCR8 [81], to activate Drosha via glycogen synthase kinase 3β activation [81], to inhibit adenosine diphosphate (ADP)-ribosylation and thus activity of AGO2 [82], and to induce ER stress such that induces an endoribonuclease (RNAse) called inositol-required enzyme 1α that can degrade the pre-miRNAs typically giving rise to miR-17-5p, miR-34a-5p, miR-96-5p and miR-125b-5p [81]. Interestingly, SFN-mediated Nrf2 induction may have additional, redox-independent consequences, since the 5′ flanking regions of certain miRNA genetic loci possess the antioxidant responsive element (ARE), as is found in the promoter regions for antioxidant protein-coding genes [83]. Nrf2 was reported to transcriptionally downregulate miR-29b-3p via ARE binding [83], but to upregulate the transcription of the mir-125b-1 and mir-29b-1 pre-miRNAs in acute myeloid leukaemia cells [84].

The anti-inflammatory effects of SFN are likely to also play a role in SFN-mediated miRNA modulation; an inflammatory medium containing TNF, IL-6, IL-8, and IL-1β was reported to upregulate miR-155-5p in several breast cancer cell lines, and miR-146a-5p in the HCT-15 and HCT-116 colorectal cancer cell lines [85]. Finally, proteins involved in miRNA biogenesis, including RNA polymerase II, Dicer, Drosha, DGCR8, Exportin 5, Ldbr and AGOs, are all potentially susceptible to SFN-mediated modulation. Further complicating the picture are considerations that HDAC activity, redox status, inflammatory signalling, miRNA-processing protein expression and Nrf2 can all cross-interact, as illustrated in Figure 7. For example, ROS can inhibit HDACs and activate histone acetyltransferases [86], whilst reciprocally, the ROS-generating DUOX nicotinamide adenine dinucleotide phosphate (NADPH) oxidases tend to be hypermethylated in lung cancer cell lines [87]. Inflammatory processes tend to reduce extracellular pH [88], which can promote H3 and H4 deacetylation [89]. Conversely, the HDAC inhibitor ITF2357 was reported to inhibit inflammatory cytokine expression in LPS-stimulated peripheral blood mononuclear cells [90].

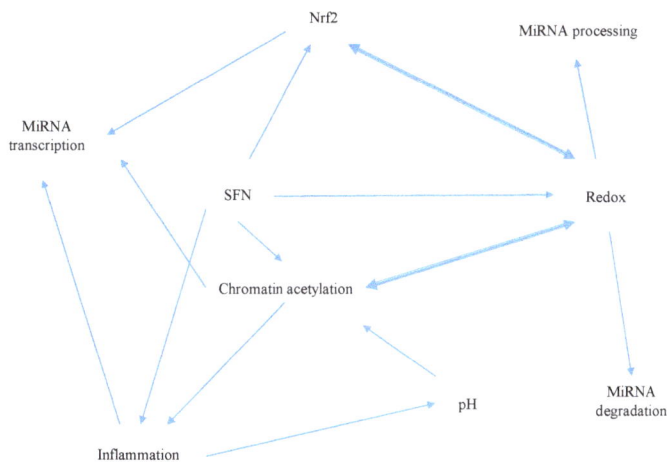

Figure 7. A diagram to illustrate the complexity of the network of cross-interactions between potential mechanisms of SFN-mediated miRNA modulation.

5.2. Interaction of MicroRNAs with Pathogenesis

The potential for miRNAs to interact with colorectal cancer pathogenesis at different stages is vast.

5.2.1. Interaction with Classic Vogelstein-Model Pathogenesis

According to the canonical Vogelstein model of colorectal cancer pathogenesis [91]—which is summarised in Figure 8—tumorigenesis typically begins with a reduction in adenomatous polyposis coli (APC) activity, resulting in increased β-catenin activity. β-catenin facilitates the formation of an aberrant crypt focus (ACF) via the upregulation of cell proliferation and stem cell renewal. There are two miRNAs commonly overexpressed in colorectal cancer specimens—miR-135a-5p and miR-135b-3p—which are both able to repress the translation of the APC protein [92], whilst miR-17-5p has been reported to also promote β-catenin activity [93]. Conversely, miR-320a has been shown to repress β-catenin and thereby inhibit tumour growth [93]. The next stage of pathogenesis according to the Vogelstein model is the upregulation of KRAS activity—believed to typically result from hyperactivating KRAS mutations. This leads to further acceleration of cell proliferation and the progression of the ACF to an early adenoma [91]. MiR-18a-5p, miR-18a-3p, miR-143-3p, and several let-7 miRNAs have all been demonstrated to translationally repress KRAS [92,93]. The following stage of pathogenesis according to the Vogelstein model is the loss of the apoptotic *DCC* gene and the TGF-β-driven tumour suppressor genes SMAD2 and SMAD4, and subsequent progression from an early to a late adenoma [91].

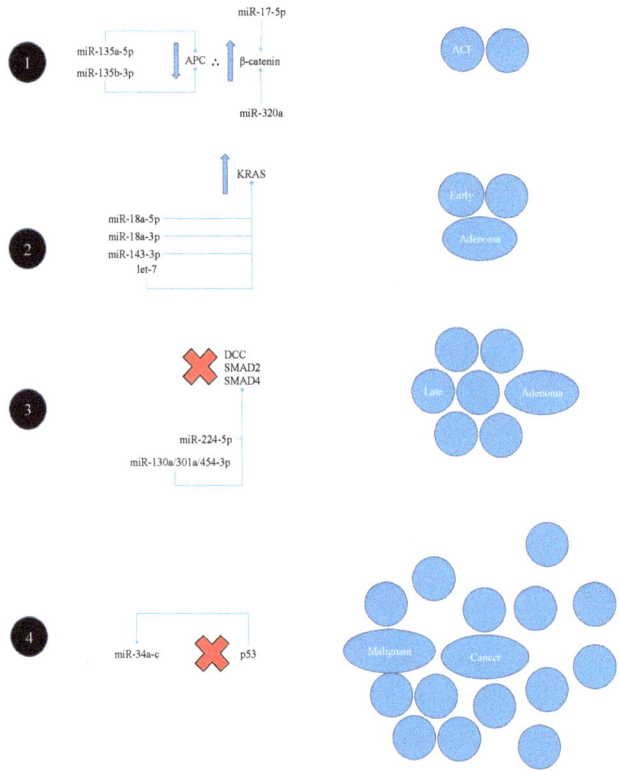

Figure 8. An illustration of the classic Vogelstein model of colorectal carcinogenesis and the reported potential interactions of miRNAs with each stage. (1) A reduction in APC activity derepresses β-catenin, which promotes cell proliferation and stem cell renewal. (2) Increased KRAS activity further promotes proliferation and the formation of an early adenoma. (3) The tumour suppressor genes DCC, SMAD2, and SMAD4 are lost, resulting in progression to a late adenoma. (4) The loss of p53 facilitates the eventual progression of the late adenoma to a malignant cancer.

Interestingly, miR-224-5p and the miR-130a/301a/454-3p family have all been reported to directly repress SMAD4, whilst miR-106a-5p and miR-21-5p have been shown to repress the transforming growth factor (TGF)-β receptor II [93]. The final stage of pathogenesis according to the Vogelstein model, in which the tumour develops into a malignant cancer, is the loss of p53 activity—typically via the mutational inactivation of *TP53* [91]. Tumour suppression by p53 is partly mediated by its ability to induce the miR-34a-c family, including miR-34a-5p, which can repress the HDAC sirtuin 1 [92]. This miRNA is typically induced in response to DNA damage in a largely p53-dependent manner, and—along with other members of its family—tends to be deleted or hypermethylated in colorectal cancer specimens [93].

5.2.2. Alternative Pathogenesis Model Interaction

As illustrated in Figure 9, there are many other miRNAs with the potential to interact with colorectal carcinogenesis additional to those mentioned above. For example, miR-144-3p and miR-25-3p are reported to repress the mTOR and SMAD7 oncogenes, respectively, the latter of which may inhibit TGF-β's tumour suppressive functions [93]. MiR-145-5p is able to inhibit growth factor-induced proliferation by repressing the insulin-like growth factor-1 receptor and insulin receptor substrate, whilst miR-126-3p can repress p85β—a promoter of the oncogenic phosphoinositide 3-kinase (PI3K) signalling pathway [92]. MiR-21-5p can conversely upregulate this oncogenic pathway by repressing phosphate and tensin homolog (PTEN), and miR-103-3p can repress the tumour suppressor gene KLF4 [92]. MiR-26a-5p has been shown to be able to interact with cancer cell metabolism in the form of promoting aerobic glycolysis (i.e., the Warburg effect) by repressing pyruvate dehydrogenase protein X component, and thus the mitochondrial synthesis of acetyl-coenzyme A from pyruvate [94]. The Warburg effect can metabolically enhance the accumulation of intermediates that are involved in macromolecular synthesis and are thus required in abundance for rapid proliferation [95].

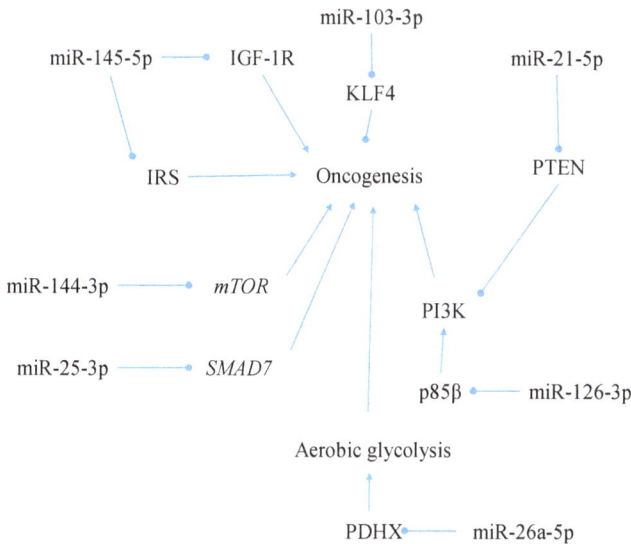

Figure 9. A diagram illustrating the interaction of miRNAs with genes and pathways regulating the pathogenesis of colorectal cancer.

To summarise, there is much published evidence to suggest that miRNAs are involved in the pathogenesis of colorectal cancer, and that SFN is likely to modulate miRNA expression in the colorectum. Based upon existing knowledge regarding SFN's bioactivity, there are many plausible mechanisms by which SFN might modulate miRNA expression. There are also many reports of specific miRNAs regulating tumour suppressor genes and/or oncogenes known to be involved in colorectal carcinogenesis. Together, these premises present a strong case for the further investigation of SFN-mediated miRNA modulation in cancerous and/or non-cancerous cell lines, and its possible implication for SFN's potential to interact with colorectal cancers at different stages and in various contexts.

6. MicroRNA Assay Methods

6.1. Array-Based Methods

There are several types of wide-scale miRNA expression profiling methods that are frequently used, including the miRNA RT-qPCR array technique employed by Slaby et al. for which the TaqMan Low Density Array kit and human Megaplex RT Primer Pool v3.0 were used [44]. Such methods run hundreds of RT-qPCRs in parallel—each specific for a different miRNA—and by nature are relatively fast and economical with regards to required RNA sample input and overall cost. An alternative array-type approach involves the use of a miRNA array hybridisation chip—a medium coated with probes of sequence antisense to those of known miRNAs, to which miRNAs in samples can be hybridised, and which generate signals in response to the quantity of material bound to each specific probe [96]. Similarly to the RT-qPCR array method, this also tends to be relatively fast and economical. However, both of these approaches come with certain caveats. The range of detectable miRNAs is limited to those known at the time of development of the particular assay kit(s) used, which is an issue given that the database of known human miRNAs continues to grow; miRBase v.14 was released at the beginning of 2010 and catalogued 894 human miRNAs, whilst by the release of miRBase v.20 in the middle of 2013 there were 2555, and the current release, v.21, catalogues 2588 [97]. These approaches also come with sensitivity limitations.

6.2. Cloning and Deep Sequencing

An approach different to those described above, and with a number of advantages, is to clone all of the miRNAs present in a given sample into libraries, and then subject the libraries to deep sequencing, the data from which are analysed to evaluate the differential expression of known miRNAs, as well as to identify potentially novel miRNAs [98]. Data can be retained and later re-mapped to updated lists of miRNAs from later releases of miRBase. The cloning-sequencing approach also tends to be more sensitive in terms of profiling the differential expression of known miRNAs. The typical workflow for such experiments tends to begin with the ligation of 5'- and 3'-adapter oligomers to all small RNAs present in samples, followed by the reverse transcription of all adapter-ligated RNA to complementary DNA (cDNA). The cDNA products are then amplified by PCR and then size-separated by PAGE, in order to separate cloned miRNAs from clones of different types of small RNA [99]. Cloned miRNAs are then subject to deep sequencing, and the data generated are normalised and analysed to identify known human miRNAs that are differentially expressed between different conditions.

One challenge presented by this process is the inevitable sequence-dependence bias towards certain miRNAs over others in the adapter-ligation step, which results in the "favoured" miRNAs being more abundantly cloned and thus appearing more abundant upon analysis of deep sequencing data [98]. Such ligation bias should not impact the apparent differential expression of any given miRNA across different samples, given that a bias towards a specific miRNA would occur equally across all samples if the same adapters are used. However, one issue of ligation bias is the potential for certain miRNAs to go undetected, if strongly "disfavoured" by the ligation process. In order to mitigate ligation bias, HD adapter pools were developed by the Dalmay laboratory by adding to then-existing Illumina 5'- and 3'- adapters, four nucleotides at random, at the ligating ends [98]. This means that instead of single 5'- and 3'-adapters being used, pools consisting of 256 variants each are generated—each of which have different miRNA ligation bias profiles. These adapter pools have been demonstrated to dramatically increase miRNA coverage and to enable the detection of novel miRNAs [98]. A method of miRNA library construction using Dalmay HD adapters has been published by Xu et al. [99].

The main disadvantages of miRNA cloning-sequencing methods vs. the previously discussed array-based approaches are the greater cost and time involved, both for the construction and the sequencing of the libraries. However—where viable with regards to time and economics—investment in the cloning-sequencing approach could be worthwhile for the potential additional knowledge regarding differential miRNA expression obtained. It is also conceivable that overall costs may continue to decrease over time as a result of ongoing technological development, as per recently observable trends [100].

6.3. Comparison of Approaches to MicroRNA Profiling

A brief comparison of the above-discussed types of approach to the wide-scale profiling of miRNA expression is provided in Table 2.

Table 2. A brief comparison of array-based and library construction-sequencing-based approaches to wide-scale miRNA expression profiling.

Type of Method	Cost	Relative Time Required	MiRNA Detection
Array	Medium	Low	Limited to miRNAs known at the time of array development.
Library-Sequencing	High	High	Can identify and assay novel miRNAs, and retain data for re-mapping against updated miRNA databases.

7. Summary

Colorectal cancer poses an increasingly important health burden globally, with apparent links to diet that are unsurprising, given the liability of colorectum to be exposed to ingested compounds and products of the gut microbiota. Cruciferous vegetables such as broccoli and cauliflower are inversely correlated with colorectal cancer risk more strongly than other vegetables, and this is believed to be at least partially attributable to the isothiocyanates obtained by consuming these vegetables, such as SFN from broccoli. Isothiocyanates have been widely studied and shown to have anti-inflammatory, antioxidant, and cytoprotective effects through well-studied mechanisms, such as the induction of Nrf2. However, numerous anti-cancer effects have been demonstrated in vitro and in vivo that cannot be solely attributed to Nrf2 induction, particularly those acting to suppress advanced cancer cell proliferation. Since it is apparent that the effects of SFN are wide-ranging and complex—especially so in cancer—it is clear that further knowledge regarding its bioactivities would aid in the development of chemopreventive and/or chemotherapeutic strategies based upon it.

MiRNAs are known to translationally regulate the expression of least 30% of protein-coding genes in humans and to play major roles in health and development, particularly in carcinogenesis. Their biogenesis and activities are highly complex both in their nature and their potential for regulation. There exists ample evidence in the published literature for the involvement of miRNAs in colorectal carcinogenesis, and numerous reports exist of direct miRNA-mediated regulation of specific tumour suppressor genes and/or oncogenes. Based upon existing knowledge regarding SFN's activity, there exist many plausible mechanisms by which SFN might modulate the expression of different miRNAs. Therefore, the case for further investigation of the roles of miRNAs in the anti-cancer effects of SFN is strong.

There are several types of approach to wide-scale miRNA expression profiling, including array-based methods involving RT-qPCR or complementary probe hybridisation. These are relatively fast and economical, but detectable miRNAs are limited to those known at a given point in time. MiRNA cloning-deep sequencing is an alternative approach that can confer greater sensitivity that is not limited to the detection of miRNAs known at a given point in time. The challenge of adapter-ligation bias toward certain miRNAs potentially encountered at the miRNA cloning stage has been addressed by the development of Dalmay HD adapters—a library construction protocol using these adapters has been published. The downsides of the library construction and sequencing approach are its greater cost, time, and/or labour requirements. However, it is likely to be more informative with regard to differential miRNA expression and thus worth considering where viable.

Acknowledgments: The authors would like to thank the Faculty of Medicine and Health Sciences at the University of East Anglia for the provision of a PhD studentship.

Author Contributions: C.D. initially reviewed the literature and produced the first draft of the manuscript. Y.B. reviewed several versions of the manuscript and suggested the inclusion of additional information both in the main text and in the figures that were incorporated into the final manuscript. The content of the final version of the manuscript was agreed upon by both authors.

Conflicts of Interest: The authors declare no conflict of interest.

References

1. World Health Organization: International Agency for Research on Cancer. Colorectal cancer estimated incidence, mortality and prevalence worldwide in 2012. Available online: http://globocan.iarc.fr/Pages/fact_sheets_cancer.aspx (accessed on 17 October 2016).
2. Lozano, R.; Naghavi, M.; Foreman, K.; Lim, S.; Shibuya, K.; Aboyans, V.; Abraham, J.; Adair, T.; Aggarwal, R.; et al. Global and regional mortality from 235 causes of death for 20 age groups in 1990 and 2010: A systematic analysis for the Global Burden of Disease Study 2010. *Lancet* **2012**, *380*, 2095–2128. [CrossRef]
3. Smith, D.; Ballal, M.; Hodder, R.; Soin, G.; Selvachandran, S.N.; Cade, D. Symptomatic presentation of early colorectal cancer. *Ann. R. Coll. Surg. Engl.* **2006**, *88*, 185–190. [CrossRef] [PubMed]
4. Bener, A. Colon cancer in rapidly developing countries: Review of the lifestyle, dietary, consanguinity and hereditary risk factors. *Oncol. Rev.* **2011**, *5*, 5–11. [CrossRef]
5. Phatak, S.; Larson, D.; Hunter, A.; Neal, C.; Contreras, K.; Pontsler, K.; Armbrust, T.; Rodriguez, D.; Abercrombie, B.; Hintze, K.; et al. Ancestral and multi-generational consumption of the total Western diet in mice promotes colitis-associated colorectal cancer in third-generation offspring. *FASEB J.* **2017**, *31*. [CrossRef]
6. Bowel Cancer UK. Understanding Bowel Cancer: Diet & Exercise. Available online: http://www.bowelcanceruk.org.uk/understanding-bowel-cancer/diet-exercise-(1)/ (accessed on 2 September 2016).
7. Tse, G.; Eslick, G.D. Cruciferous vegetables and risk of colorectal neoplasms: a systematic review and meta-analysis. *Nutr. Cancer* **2014**, *66*, 128–139. [CrossRef] [PubMed]
8. Chan, R.; Lok, K.; Woo, J. Prostate cancer and vegetable consumption. *Mol. Nutr. Food Res.* **2009**, *53*, 201–216. [CrossRef] [PubMed]
9. Koroleva, O.A.; Davies, A.; Deeken, R.; Thorpe, M.R.; Tomos, A.D.; Hedrich, R. Identification of a New Glucosinolate-Rich Cell Type in Arabidopsis Flower Stalk. *Plant Physiol.* **2000**, *124*, 599–608. [CrossRef] [PubMed]
10. Andréasson, E.; Jørgensen, L.B.; Höglund, A.; Rask, L.; Meijer, J. Different Myrosinase and Idioblast Distribution in Arabidopsis and *Brassica napus*. *Plant Physiol.* **2001**, *127*, 1750–1763. [CrossRef] [PubMed]
11. Burow, M.; Bergner, A.; Gershenzon, J.; Wittstock, U. Glucosinolate hydrolysis in Lepidium sativum—Identification of the thiocyanate-forming protein. *Plant Mol. Biol.* **2007**, *63*, 49–61. [CrossRef] [PubMed]
12. Clarke, J.D.; Dashwood, R.H.; Ho, E. Multi-targeted prevention of cancer by sulforaphane. *Cancer Lett.* **2008**, *269*, 291–304. [CrossRef] [PubMed]
13. Bai, Y.; Wang, X.; Zhao, S.; Ma, C.; Cui, J.; Zheng, Y. Sulforaphane Protects against Cardiovascular Disease via Nrf2 Activation. *Oxid. Med. Cell Longev.* **2015**, *2015*, 13. [CrossRef] [PubMed]
14. Davidson, R.K.; Jupp, O.; Bao, Y.; MacGregor, A.J.; Donell, S.T.; Cassidy, A.; Clark, I.M. Can sulforaphane prevent the onset or slow the progression of osteoarthritis? *Nutr. Bull.* **2016**, *41*, 175–179. [CrossRef]

15. Davidson, R.; Gardner, S.; Jupp, O.; Bullough, A.; Butters, S.; Watts, L.; Donell, S.; Traka, M.; Saha, S.; Mithen, R.; et al. Isothiocyanates are detected in human synovial fluid following broccoli consumption and can affect the tissues of the knee joint. *Sci. Rep.* **2017**, *7*, 3398. [CrossRef] [PubMed]

16. Ramos-Gomez, M.; Kwak, M.K.; Dolan, P.M.; Itoh, K.; Yamamoto, M.; Talalay, P.; Kensler, T.W. Sensitivity to carcinogenesis is increased and chemoprotective efficacy of enzyme inducers is lost in *nrf2* transcription factor-deficient mice. *Proc. Natl. Acad. Sci. USA* **2001**, *98*, 3410–3415. [CrossRef] [PubMed]

17. Yanaka, A.; Mutoh, M. Su2056 Dietary Intake of Sulforaphane Glucosinolates Inhibits Colon Tumorigenesis in Mice Treated With Azoxymethane and Radioactive Cesium. *Gastroenterology* **2016**, *150*, S623. [CrossRef]

18. Pereira, L.; Silva, P.; Duarte, M.; Rodrigues, L.; Duarte, C.; Albuquerque, C.; Serra, A. Targeting Colorectal Cancer Proliferation, Stemness and Metastatic Potential Using *Brassicaceae* Extracts Enriched in Isothiocyanates: A 3D Cell Model-Based Study. *Nutrients* **2017**, *9*, 368. [CrossRef] [PubMed]

19. Gupta, P.; Kim, B.; Kim, S.H.; Srivastava, S.K. Molecular targets of isothiocyanates in cancer: Recent advances. *Mol. Nutr. Food Res.* **2014**, *58*, 1685–1707. [CrossRef] [PubMed]

20. Liu, K.C.; Shih, T.Y.; Kuo, C.L.; Ma, Y.S.; Yang, J.L.; Wu, P.P.; Huang, Y.P.; Lai, K.C.; Chung, J.G. Sulforaphane Induces Cell Death Through G2/M Phase Arrest and Triggers Apoptosis in HCT 116 Human Colon Cancer Cells. *Am. J. Chin. Med.* **2016**, *44*, 1289–1310. [CrossRef] [PubMed]

21. Samantha, L.M.; Rishabh, K.; Trygve, O.T. Mechanisms for the inhibition of colon cancer cells by sulforaphane through epigenetic modulation of microRNA-21 and human telomerase reverse transcriptase (hTERT) down-regulation. *Curr. Cancer Drug Targets* **2017**, *17*, 1–10.

22. Wang, Y.; Dacosta, C.; Wang, W.; Zhou, Z.; Liu, M.; Bao, Y. Synergy between sulforaphane and selenium in protection against oxidative damage in colonic CCD841 cells. *Nutr. Res.* **2015**, *35*, 610–617. [CrossRef] [PubMed]

23. Yueh, M.F.; Tukey, R.H. Nrf2-Keap1 signaling pathway regulates human UGT1A1 expression in vitro and in transgenic *UGT1* mice. *J. Biol. Chem.* **2007**, *282*, 8749–8758. [CrossRef] [PubMed]

24. Harman, D. Aging: a theory based on free radical and radiation chemistry. *J. Gerontol.* **1956**, *11*, 298–300. [CrossRef] [PubMed]

25. Pi, J.; Zhang, Q.; Fu, J.; Woods, C.G.; Hou, Y.; Corkey, B.E.; Collins, S.; Andersen, M.E. ROS signaling, oxidative stress and Nrf2 in pancreatic beta-cell function. *Toxicol. Appl. Pharmacol.* **2010**, *244*, 77–83. [CrossRef] [PubMed]

26. Gopalakrishna, R.; Gundimeda, U. Antioxidant regulation of protein kinase C in cancer prevention. *J. Nutr.* **2002**, *132*, 3819S–3823S. [PubMed]

27. Guerrero-Beltrán, C.E.; Calderón-Oliver, M.; Pedraza-Chaverri, J.; Chirino, Y.I. Protective effect of sulforaphane against oxidative stress: Recent advances. *Exp. Toxicol. Pathol.* **2012**, *64*, 503–508. [CrossRef] [PubMed]

28. Higgins, L.G.; Kelleher, M.O.; Eggleston, I.M.; Itoh, K.; Yamamoto, M.; Hayes, J.D. Transcription factor Nrf2 mediates an adaptive response to sulforaphane that protects fibroblasts *in vitro* against the cytotoxic effects of electrophiles, peroxides and redox–cycling agents. *Toxicol. Appl. Pharmacol.* **2009**, *237*, 267–280. [CrossRef] [PubMed]

29. Sestili, P.; Paolillo, M.; Lenzi, M.; Colombo, E.; Vallorani, L.; Casadei, L.; Martinelli, C.; Fimognari, C. Sulforaphane induces DNA single strand breaks in cultured human cells. *Mutat. Res.* **2010**, *689*, 65–73. [CrossRef] [PubMed]

30. Keum, Y.S.; Yu, S.; Chang, P.P.; Yuan, X.; Kim, J.H.; Xu, C.; Han, J.; Agarwal, A.; Kong, A.N. Mechanism of Action of Sulforaphane: Inhibition of p38 Mitogen-Activated Protein Kinase Isoforms Contributing to the Induction of Antioxidant Response Element-Mediated Heme Oxygenase-1 in Human Hepatoma HepG2 Cells. *Cancer Res.* **2006**, *66*, 8804–8813. [CrossRef] [PubMed]

31. Pal, S.; Badireenath Konkimalla, V. Hormetic Potential of Sulforaphane (SFN) in Switching Cells' Fate Towards Survival or Death. *Mini. Rev. Med. Chem.* **2016**, *16*, 980–995. [CrossRef] [PubMed]

32. Tierens, K.F.M.J.; Thomma, B.P.H.J.; Brouwer, M.; Schmidt, J.; Kistner, K.; Porzel, A.; Mauch-Mani, B.; Cammue, B.P.A.; Broekaert, W.F. Study of the Role of Antimicrobial Glucosinolate-Derived Isothiocyanates in Resistance of Arabidopsis to Microbial Pathogens. *Plant Physiol.* **2001**, *125*, 1688–1699. [CrossRef] [PubMed]

33. Mattson, M.P.; Calabrese, E.J. *Hormesis*; Mattson, M.P., Calabrese, E.J.B., Eds.; Humana Press: New York, NY, USA, 2010; Volume 1, p. 59.

34. Myzak, M.C.; Hardin, K.; Wang, R.; Dashwood, R.H.; Ho, E. Sulforaphane inhibits histone deacetylase activity in BPH-1, LnCaP and PC-3 prostate epithelial cells. *Carcinogenesis* **2006**, *27*, 811–819. [CrossRef] [PubMed]

35. Lau, A.; Villeneuve, N.F.; Sun, Z.; Wong, P.K.; Zhang, D.D. Dual roles of Nrf2 in cancer. *Pharmacol. Res.* **2008**, *58*, 262–270. [CrossRef] [PubMed]

36. Bao, Y.; Wang, W.; Zhou, Z.; Sun, C. Benefits and Risks of the Hormetic Effects of Dietary Isothiocyanates on Cancer Prevention. *PLoS ONE* **2014**, *9*, e114764. [CrossRef] [PubMed]

37. Murray-Stewart, T.; Hanigan, C.L.; Woster, P.M.; Marton, L.J.; Casero, R.A., Jr. Histone Deacetylase Inhibition Overcomes Drug Resistance through a miRNA-Dependent Mechanism. *Mol. Cancer Ther.* **2013**, *12*, 2088–2099. [CrossRef] [PubMed]

38. Bat-Chen, W.; Golan, T.; Peri, I.; Ludmer, Z.; Schwartz, B. Allicin Purified From Fresh Garlic Cloves Induces Apoptosis in Colon Cancer Cells Via Nrf2. *Nutr. Cancer* **2010**, *62*, 947–957. [CrossRef] [PubMed]

39. Saddawi-Konefka, R.; Seelige, R.; Gross, E.T.; Levy, E.; Searles, S.C.; Washington, A., Jr.; Santosa, E.K.; Liu, B.; O'Sullivan, T.E.; Harismendy, O.; et al. Nrf2 Induces Il-17D to Mediate Tumor and Virus Surveillance. *Cell Rep.* **2016**, *16*, 2348–2358. [CrossRef] [PubMed]

40. Nian, H.; Delage, B.; Ho, E.; Dashwood, R.H. Modulation of histone deacetylase activity by dietary isothiocyanates and allyl sulfides: Studies with sulforaphane and garlic organosulfur compounds. *Environ. Mol. Mutagen* **2009**, *50*, 213–221. [CrossRef] [PubMed]

41. Rajendran, P.; Kidane, A.I.; Yu, T.W.; Dashwood, W.M.; Bisson, W.H.; Lohr, C.V.; Ho, E.; Williams, D.E.; Dashwood, R.H. HDAC turnover, CtIP acetylation and dysregulated DNA damage signaling in colon cancer cells treated with sulforaphane and related dietary isothiocyanates. *Epigenetics* **2013**, *8*, 612–623. [CrossRef] [PubMed]

42. Lewis, B.P.; Burge, C.B.; Bartel, D.P. Conserved seed pairing, often flanked by adenosines, indicates that thousands of human genes are microRNA targets. *Cell* **2005**, *120*, 15–20. [CrossRef] [PubMed]

43. Del Carmen Martínez-Jiménez, V.; Méndez-Mancilla, A.; Portales-Pérez, D.P. miRNAs in nutrition, obesity and cancer: The biology of miRNAs in metabolic disorders and its relationship with cancer development. *Mol. Nutr. Food Res.* **2017**. [CrossRef] [PubMed]

44. Slaby, O.; Sachlova, M.; Brezkova, V.; Hezova, R.; Kovarikova, A.; Bischofova, S.; Sevcikova, S.; Bienertova-Vasku, J.; Vasku, A.; Svoboda, M.; et al. Identification of microRNAs regulated by isothiocyanates and association of polymorphisms inside their target sites with risk of sporadic colorectal cancer. *Nutr. Cancer* **2013**, *65*, 247–254. [CrossRef] [PubMed]

45. Gregory, R.I.; Shiekhattar, R. MicroRNA Biogenesis and Cancer. *Cancer Res.* **2005**, *65*, 3509–3512. [CrossRef] [PubMed]

46. Westholm, J.O.; Lai, E.C. Mirtrons: microRNA biogenesis via splicing. *Biochimie* **2011**, *93*, 1897–1904. [CrossRef] [PubMed]

47. Cifuentes, D.; Xue, H.; Taylor, D.W.; Patnode, H.; Mishima, Y.; Cheloufi, S.; Ma, E.; Mane, S.; Hannon, G.J.; Lawson, N.D.; et al. A Novel miRNA Processing Pathway Independent of Dicer Requires Argonaute2 Catalytic Activity. *Science* **2010**, *328*, 1694–1698. [CrossRef] [PubMed]

48. Bartel, D.P. MicroRNAs: Genomics, Biogenesis, Mechanism, and Function. *Cell* **2004**, *116*, 281–297. [CrossRef]

49. Liu, J.; Rivas, F.V.; Wohlschlegel, J.; Yates, J.R.; Parker, R.; Hannon, G.J. A role for the P-body component GW182 in microRNA function. *Nat. Cell Biol.* **2005**, *7*, 1261–1266. [CrossRef] [PubMed]

50. Vasudevan, S.; Tong, Y.; Steitz, J.A. Switching from Repression to Activation: MicroRNAs Can Up-Regulate Translation. *Science* **2007**, *318*, 1931–1934. [CrossRef] [PubMed]

51. Ørom, U.A.; Nielsen, F.C.; Lund, A.H. MicroRNA-10a Binds the 5′UTR of Ribosomal Protein mRNAs and Enhances Their Translation. *Mol. Cell* **2008**, *30*, 460–471. [CrossRef] [PubMed]

52. Zhao, S.; Liu, M.F. Mechanisms of microRNA-mediated gene regulation. *Sci. China C. Life Sci.* **2009**, *52*, 1111–1116. [CrossRef] [PubMed]

53. Eiring, A.M.; Harb, J.G.; Neviani, P.; Garton, C.; Oaks, J.J.; Spizzo, R.; Liu, S.; Schwind, S.; Santhanam, R.; Hickey, C.J.; et al. miR-328 Functions as an RNA Decoy to Modulate hnRNP E2 Regulation of mRNA Translation in Leukemic Blasts. *Cell* **2010**, *140*, 652–665. [CrossRef] [PubMed]

54. Wilczynska, A.; Bushell, M. The complexity of mirna-mediated repression. *Cell Death Differ.* **2015**, *22*, 22–33. [CrossRef] [PubMed]

55. Neilsen, C.T.; Goodall, G.J.; Bracken, C.P. IsomiRs – the overlooked repertoire in the dynamic microRNAome. *Trends in Genetics* **2012**, *28*, 544–549. [CrossRef] [PubMed]

56. Cloonan, N.; Wani, S.; Xu, Q.; Gu, J.; Lea, K.; Heater, S.; Barbacioru, C.; Steptoe, A.L.; Martin, H.C.; Nourbakhsh, E.; et al. MicroRNAs and their isomiRs function cooperatively to target common biological pathways. *Genome Biol.* **2011**, *12*, R126. [CrossRef] [PubMed]

57. Wang, F.; Zhang, P.; Ma, Y.; Yang, J.; Moyer, M.P.; Shi, C.; Peng, J.; Qin, H. NIRF is frequently upregulated in colorectal cancer and its oncogenicity can be suppressed by let-7a microRNA. *Cancer Lett.* **2012**, *314*, 223–231. [CrossRef] [PubMed]

58. Tazawa, H.; Tsuchiya, N.; Izumiya, M.; Nakagama, H. Tumor-suppressive mir-34a induces senescence-like growth arrest through modulation of the E2F pathway in human colon cancer cells. *Proc. Natl. Acad. Sci. USA* **2007**, *104*, 15472–15477. [CrossRef] [PubMed]

59. Strillacci, A.; Griffoni, C.; Sansone, P.; Paterini, P.; Piazzi, G.; Lazzarini, G.; Spisni, E.; Pantaleo, M.A.; Biasco, G.; Tomasi, V. MiR-101 downregulation is involved in cyclooxygenase-2 overexpression in human colon cancer cells. *Exp. Cell Res.* **2009**, *315*, 1439–1447. [CrossRef] [PubMed]

60. Guo, C.; Sah, J.F.; Beard, L.; Willson, J.K.V.; Markowitz, S.D.; Guda, K. The noncoding RNA, miR-126, suppresses the growth of neoplastic cells by targeting phosphatidylinositol 3-kinase signaling and is frequently lost in colon cancers. *Genes Chromosomes Cancer* **2008**, *47*, 939–946. [CrossRef] [PubMed]

61. Pagliuca, A.; Valvo, C.; Fabrizi, E.; di Martino, S.; Biffoni, M.; Runci, D.; Forte, S.; De Maria, R.; Ricci-Vitiani, L. Analysis of the combined action of mir-143 and mir-145 on oncogenic pathways in colorectal cancer cells reveals a coordinate program of gene repression. *Oncogene* **2013**, *32*, 4806–4813. [CrossRef] [PubMed]

62. Akao, Y.; Nakagawa, Y.; Hirata, I.; Iio, A.; Itoh, T.; Kojima, K.; Nakashima, R.; Kitade, Y.; Naoe, T. Role of anti-oncomirs miR-143 and -145 in human colorectal tumors. *Cancer Gene Ther.* **2010**, *17*, 398–408. [CrossRef] [PubMed]

63. Liu, L.; Chen, L.; Xu, Y.; Li, R.; Du, X. MicroRNA-195 promotes apoptosis and suppresses tumorigenicity of human colorectal cancer cells. *Biochem. Biophys. Res. Commun.* **2010**, *400*, 236–240. [CrossRef] [PubMed]

64. Paterson, E.L.; Kazenwadel, J.; Bert, A.G.; Khew-Goodall, Y.; Ruszkiewicz, A.; Goodall, G.J. Down-Regulation of the miRNA-200 Family at the Invasive Front of Colorectal Cancers with Degraded Basement Membrane Indicates EMT is Involved in Cancer Progression. *Neoplasia* **2013**, *15*, 180–191. [CrossRef]

65. Wang, W.H.; Chen, J.; Zhao, F.; Zhang, B.R.; Yu, H.S.; Jin, H.Y.; Dai, J.H. Mir-150-5p Suppresses Colorectal Cancer Cell Migration and Invasion through Targeting MUC4. *Asian Pac. J. Cancer Prev.* **2014**, *15*, 6269–6273. [CrossRef] [PubMed]

66. Feng, J.; Yang, Y.; Zhang, P.; Wang, F.; Ma, Y.; Qin, H.; Wang, Y. miR-150 functions as a tumour suppressor in human colorectal cancer by targeting c-Myb. *J. Cell Mol. Med.* **2014**, *18*, 2125–2134. [CrossRef] [PubMed]

67. Bitarte, N.; Bandres, E.; Boni, V.; Zarate, R.; Rodriguez, J.; Gonzalez-Huarriz, M.; Lopez, I.; Javier Sola, J.; Alonso, M.M.; Fortes, P.; et al. MicroRNA-451 is Involved in the Self-renewal, Tumorigenicity, and Chemoresistance of Colorectal Cancer Stem Cells. *Stem Cells* **2011**, *29*, 1661–1671. [CrossRef] [PubMed]

68. Ma, Y.; Zhang, P.; Wang, F.; Zhang, H.; Yang, Y.; Shi, C.; Xia, Y.; Peng, J.; Liu, W.; Yang, Z.; et al. Elevated oncofoetal miR-17-5p expression regulates colorectal cancer progression by repressing its target gene P130. *Nat. Commun.* **2012**, *3*, 1291. [CrossRef] [PubMed]

69. Tsuchida, A.; Ohno, S.; Wu, W.; Borjigin, N.; Fujita, K.; Aoki, T.; Ueda, S.; Takanashi, M.; Kuroda, M. Mir-92 is a key oncogenic component of the mir-17–92 cluster in colon cancer. *Cancer Sci.* **2011**, *102*, 2264–2271. [CrossRef] [PubMed]

70. Jahid, S.; Sun, J.; Edwards, R.A.; Dizon, D.; Panarelli, N.C.; Milsom, J.W.; Sikandar, S.S.; Gümüş, Z.H.; Lipkin, S.M. *miR-23a* Promotes the Transition from Indolent to Invasive Colorectal Cancer. *Cancer Discov.* **2012**, *2*, 540–553. [CrossRef] [PubMed]

71. Chintharlapalli, S.; Papineni, S.; Abdelrahim, M.; Abudayyeh, A.; Jutooru, I.; Chadalapaka, G.; Wu, F.; Mertens-Talcott, S.; Vanderlaag, K.; Cho, S.D.; et al. Oncogenic MicroRNA-27a is a Target For Anticancer Agent Methyl 2-Cyano-3,11-dioxo-18β-olean-1,12-dien-30-oate in Colon Cancer Cells. *Int. J. Cancer* **2009**, *125*, 1965–1974. [CrossRef] [PubMed]

72. Valeri, N.; Braconi, C.; Gasparini, P.; Murgia, C.; Lampis, A.; Paulus-Hock, V.; Hart, J.R.; Ueno, L.; Grivennikov, S.I.; Lovat, F.; et al. Microrna-135b Promotes Cancer Progression by Acting as a Downstream Effector of Oncogenic Pathways in Colon Cancer. *Cancer Cell* **2014**, *25*, 469–483. [PubMed]

73. Lu, M.H.; Huang, C.C.; Pan, M.R.; Chen, H.H.; Hung, W.C. Prospero Homeobox 1 Promotes Epithelial–Mesenchymal Transition in Colon Cancer Cells by Inhibiting E-cadherin via miR-9. *Clin. Cancer Res.* **2012**, *18*, 6416–6425. [CrossRef] [PubMed]

74. Park, Y.R.; Lee, S.T.; Kim, S.L.; Liu, Y.C.; Lee, M.R.; Shin, J.H.; Seo, S.Y.; Kim, S.H.; Kim, I.H.; Lee, S.O.; et al. MicroRNA-9 suppresses cell migration and invasion through downregulation of TM4SF1 in colorectal cancer. *Int. J. Oncol.* **2016**, *48*, 2135–2143. [PubMed]

75. Fassan, M.; Pizzi, M.; Giacomelli, L.; Mescoli, C.; Ludwig, K.; Pucciarelli, S.; Rugge, M. PDCD4 nuclear loss inversely correlates with miR-21 levels in colon carcinogenesis. *Virchows Arch.* **2011**, *458*, 413–419. [CrossRef] [PubMed]

76. Xiong, B.; Cheng, Y.; Ma, L.; Zhang, C. Mir-21 Regulates Biological Behavior Through the PTEN/PI-3 K/Akt Signaling Pathway in Human Colorectal Cancer Cells. *Int. J. Oncol.* **2013**, *42*, 219–228. [CrossRef] [PubMed]

77. Yu, Y.; Kanwar, S.S.; Patel, B.B.; Oh, P.S.; Nautiyal, J.; Sarkar, F.H.; Majumdar, A.P. MicroRNA-21 induces stemness by downregulating transforming growth factor beta receptor 2 (TGFβR2) in colon cancer cells. *Carcinogenesis* **2012**, *33*, 68–76. [CrossRef] [PubMed]

78. Pan, J.H.; Abernathy, B.; Kim, Y.J.; Lee, J.H.; Kim, J.H.; Shin, E.C.; Kim, J.K. Cruciferous vegetables and colorectal cancer prevention through microRNA regulation: A review. *Crit. Rev. Food Sci. Nutr.* **2017**, 1–13. [CrossRef] [PubMed]

79. Zhang, X.; Chen, X.; Lin, J.; Lwin, T.; Wright, G.; Moscinski, L.C.; Dalton, W.S.; Seto, E.; Wright, K.; Sotomayor, E.; et al. Myc represses miR-15a/miR-16-1 expression through recruitment of HDAC3 in mantle cell and other non-Hodgkin B-cell lymphomas. *Oncogene* **2012**, *31*, 3002–3008. [CrossRef] [PubMed]

80. Wiesen, J.L.; Tomasi, T.B. Dicer is regulated by cellular stresses and interferons. *Mol. Immunol.* **2009**, *46*, 1222–1228. [CrossRef] [PubMed]

81. Cheng, X.; Ku, C.H.; Siow, R.C.M. Regulation of the Nrf2 antioxidant pathway by microRNAs: New players in micromanaging redox homeostasis. *Free Radic. Biol. Med.* **2013**, *64*, 4–11. [CrossRef] [PubMed]

82. Leung, A.K.; Vyas, S.; Rood, J.E.; Bhutkar, A.; Sharp, P.A.; Chang, P. Poly(ADP-Ribose) Regulates Stress Responses and MicroRNA Activity in the Cytoplasm. *Mol. Cell* **2011**, *42*, 489–499. [CrossRef] [PubMed]

83. Shah, N.M.; Rushworth, S.A.; Murray, M.Y.; Bowles, K.M.; MacEwan, D.J. Understanding the role of NRF2-regulated miRNAs in human malignancies. *Oncotarget* **2013**, *4*, 1130–1142. [CrossRef] [PubMed]

84. Kurinna, S.; Werner, S. Nrf2 and microRNAs: new but awaited relations. *Biochem. Soc. Trans.* **2015**, *43*, 595–601. [CrossRef] [PubMed]

85. Tili, E.; Michaille, J.J.; Wernicke, D.; Alder, H.; Costinean, S.; Volinia, S.; Croce, C.M. Mutator activity induced by microRNA-155 (*miR-155*) links inflammation and cancer. *Proc. Natl. Acad. Sci. USA* **2011**, *108*, 4908–4913. [CrossRef] [PubMed]

86. Rahman, I.; Marwick, J.; Kirkham, P. Redox modulation of chromatin remodeling: impact on histone acetylation and deacetylation, NF-κB and pro-inflammatory gene expression. *Biochem. Pharmacol.* **2004**, *68*, 1255–1267. [CrossRef] [PubMed]

87. Luxen, S.; Belinsky, S.A.; Knaus, U.G. Silencing of *DUOX* NADPH Oxidases by Promoter Hypermethylation in Lung Cancer. *Cancer Res.* **2008**, *68*, 1037–1045. [CrossRef] [PubMed]

88. Bäckdahl, L.; Bushell, A.; Beck, S. Inflammatory signalling as mediator of epigenetic modulation in tissue-specific chronic inflammation. *Int. J. Biochem. Cell Biol.* **2009**, *41*, 176–184. [CrossRef] [PubMed]

89. McBrian, M.A.; Behbahan, I.S.; Ferrari, R.; Su, T.; Huang, T.W.; Li, K. Global Histone Acetylation is Linked to pH. *Cancer Discov.* **2013**, *3*, 12.

90. Leoni, F.; Fossati, G.; Lewis, E.C.; Lee, J.K.; Porro, G.; Pagani, P.; Modena, D.; Moras, M.L.; Pozzi, P.; Reznikov, L.L.; et al. The Histone Deacetylase Inhibitor ITF2357 Reduces Production of Pro-Inflammatory Cytokines In Vitro and Systemic Inflammation In Vivo. *Mol. Med.* **2005**, *11*, 1–15. [CrossRef] [PubMed]

91. Fearon, E.R.; Vogelstein, B. A genetic model for colorectal tumorigenesis. *Cell* **1990**, *61*, 759–767. [CrossRef]

92. Slaby, O.; Svoboda, M.; Michalek, J.; Vyzula, R. MicroRNAs in colorectal cancer: Translation of molecular biology into clinical application. *Mol. Cancer* **2009**, *8*, 102. [CrossRef] [PubMed]

93. Amirkhah, R.; Schmitz, U.; Linnebacher, M.; Wolkenhauer, O.; Farazmand, A. MicroRNA-mRNA interactions in colorectal cancer and their role in tumor progression. *Genes Chromosomes Cancer* **2015**, *54*, 129–141. [CrossRef] [PubMed]

94. Chen, B.; Liu, Y.; Jin, X.; Lu, W.; Liu, J.; Xia, Z.; Yuan, Q.; Zhao, X.; Xu, N.; Liang, S. MicroRNA-26a regulates glucose metabolism by direct targeting PDHX in colorectal cancer cells. *BMC Cancer* **2014**, *14*, 443. [CrossRef] [PubMed]
95. Vander Heiden, M.G.; Cantley, L.C.; Thompson, C.B. Understanding the Warburg Effect: The Metabolic Requirements of Cell Proliferation. *Science* **2009**, *324*, 1029–1033. [CrossRef] [PubMed]
96. Thomson, J.M.; Parker, J.; Perou, C.M.; Hammond, S.M. A custom microarray platform for analysis of microRNA gene expression. *Nat. Meth.* **2004**, *1*, 47–53. [CrossRef] [PubMed]
97. miRBase. miRBase blog. Available online: http://www.mirbase.org/blog/ (accessed on 8 September 2016).
98. Sorefan, K.; Pais, H.; Hall, A.E.; Kozomara, A.; Griffiths-Jones, S.; Moulton, V.; Dalmay, T. Reducing ligation bias of small RNAs in libraries for next generation sequencing. *Silence* **2012**, *3*, 1–11. [CrossRef] [PubMed]
99. Xu, P.; Billmeier, M.; Mohorianu, I.; Green, D.; Fraser, W.D.; Dalmay, T. An improved protocol for small RNA library construction using High Definition adapters. *Methods Next Gener. Seq.* **2015**, *2*. [CrossRef]
100. Caporaso, J.G.; Lauber, C.L.; Walters, W.A.; Berg-Lyons, D.; Huntley, J.; Fierer, N.; Owens, S.M.; Betley, J.; Fraser, L.; Bauer, M.; et al. Ultra-high-throughput microbial community analysis on the Illumina HiSeq and MiSeq platforms. *ISME J.* **2012**, *6*, 1621–1624. [CrossRef] [PubMed]

Review

Multi-Targeted Molecular Effects of *Hibiscus sabdariffa* Polyphenols: An Opportunity for a Global Approach to Obesity

María Herranz-López [1], Mariló Olivares-Vicente [1], José Antonio Encinar [1], Enrique Barrajón-Catalán [1], Antonio Segura-Carretero [2,3], Jorge Joven [4] and Vicente Micol [1,5,*]

[1] Instituto de Biología Molecular y Celular (IBMC), Universidad Miguel Hernández (UMH), Edificio Torregaitán, Elche 03202, Spain; mherranz@umh.es (M.H.-L.); maria.olivares@umh.es (M.O.-V.); jant.encinar@umh.es (J.A.E.); e.barrajon@umh.es (E.B.-C.)

[2] Department of Analytical Chemistry, University of Granada, Avda. Fuentenueva s/n, Granada 18071, Spain; ansegura@ugr.es

[3] Research and Development of Functional Food Centre (CIDAF), PTS Granada, Avda. del Conocimiento s/n., Edificio BioRegión, Granada 18016, Spain

[4] Unitat de Recerca Biomèdica, Hospital Universitari Sant Joan, Institut d'Investigació Sanitària Pere Virgili, Universitat Rovira i Virgili, Reus 43201, Spain; jjoven@grupsagessa.com

[5] CIBER: CB12/03/30038, Fisiopatología de la Obesidad y la Nutrición, CIBERobn, Instituto de Salud Carlos III (ISCIII), Palma de Mallorca 07122, Spain

*   Correspondence: vmicol@umh.es; Tel.: +34-96-665-8430

Received: 31 July 2017; Accepted: 14 August 2017; Published: 20 August 2017

Abstract: Improper diet can alter gene expression by breaking the energy balance equation and changing metabolic and oxidative stress biomarkers, which can result in the development of obesity-related metabolic disorders. The pleiotropic effects of dietary plant polyphenols are capable of counteracting by modulating different key molecular targets at the cell, as well as through epigenetic modifications. *Hibiscus sabdariffa* (HS)-derived polyphenols are known to ameliorate various obesity-related conditions. Recent evidence leads to propose the complex nature of the underlying mechanism of action. This multi-targeted mechanism includes the regulation of energy metabolism, oxidative stress and inflammatory pathways, transcription factors, hormones and peptides, digestive enzymes, as well as epigenetic modifications. This article reviews the accumulated evidence on the multiple anti-obesity effects of HS polyphenols in cell and animal models, as well as in humans, and its putative molecular targets. In silico studies reveal the capacity of several HS polyphenols to act as putative ligands for different digestive and metabolic enzymes, which may also deserve further attention. Therefore, a global approach including integrated and networked omics techniques, virtual screening and epigenetic analysis is necessary to fully understand the molecular mechanisms of HS polyphenols and metabolites involved, as well as their possible implications in the design of safe and effective polyphenolic formulations for obesity.

Keywords: antioxidants; dietary supplementation; obesity; epigenetics; metabolic stress; polyphenols; metabolites; virtual screening; *Hibiscus sabdariffa*

---

1. Introduction

Diet-induced obesity and a sedentary lifestyle are consistent with the appearance of several metabolic dysfunctions leading to metabolic syndrome (MetS). MetS is a clustering of cardio-metabolic risk factors, including abdominal obesity, insulin resistance, dyslipidemia and hypertension [1,2]. Although considerable progress has been made in understanding the molecular mechanisms underlying obesity, in many cases its treatment results in failure [3]. In this respect, caloric restriction and physical activity are the primary approaches prescribed to improve an individual´s metabolism. However, in the majority of cases, this is very difficult to maintain for a prolonged period, as it implicates important changes in the individual´s lifestyle [4,5]. Likewise, current anti-obesity pharmacological treatments are in many cases inefficient, and often present important side effects when taken for a prolonged period [6–8]. For this reason, many researchers and clinicians have drawn their attention towards more traditional methods to correct the obesity-induced energy imbalance. Accordingly, the use of metabolic drugs or natural dietary products, including polyphenolic xenohormetins, are some of the most interesting and promising strategies in forthcoming studies addressing obesity [9,10]. Exploring the effects of polyphenols on specific cellular pathways and diseases may turn the preventive use of natural agents into a dietary intervention in the treatment of chronic metabolic diseases. The multifactorial nature of polyphenols, a plausible consequence of their molecular diversity, offers a new opportunity to address obesity [11–16]. To this end, the main objective of this review is to elucidate the putative molecular targets of *Hibiscus sabdariffa* (HS) L. (Malvaceae) polyphenols and provide an overview of their role in the prevention of chronic diseases such as obesity.

In this review, we have collected all the relevant evidence regarding the possible benefits of polyphenolic extracts derived from HS calyces on the management of obesity-related pathologies [17–22]. These studies include the use of cell and animal obesity models, as well as human clinical trials. Furthermore, the complete characterization of this plant extract, its effective dosage, synergistic effects, bioavailability, in silico approach on selected molecular targets and localization of bioactive metabolites in tissues have been reviewed. The aim of this review is to provide a comprehensive examination clarifying whether the proper use of HS polyphenols may present an opportunity to improve the control or prevention of obesity-related diseases through the modulation of key proteins involved in oxidative stress and inflammation and, consequently, in the regulation of metabolic and bioenergetics pathways.

In this respect, recent data have shown that AMP-activated protein kinase (AMPK) and peroxisome proliferator-activated receptors (PPAR), among others, may be possible molecular targets for HS polyphenols, as evidenced in cellular and animal models [21,23,24]. This hypothesis suggests that the active manipulation of energy sensors and effectors in obesity might be a feasible preventive therapy. Nevertheless, further research is necessary in order to delimit all the protein targets that are modulated by HS polyphenols and to elucidate their molecular mechanisms. For this purpose, metabolic profiling through "Omics" science must be used to identify the final intracellular metabolites derived from HS polyphenols that reach intracellular targets, as well as to identify endogenous metabolite biomarkers. From this perspective, validated obesity-related biomarkers could provide new diagnostic tools and establish key insights concerning the effects of natural compounds on the pathogenesis of energy-related complications.

2. *Hibiscus sabdariffa* Polyphenols as Xenohormetic Agents

"Let food be thy medicine and medicine be thy food" quoted by Hippocrates, and its more modern analog "an apple a day keeps the doctor away" both reflect the implicitness of food on health. Plants that traditionally served as food, fuel, water and fiber, are now being engineered as a source of natural bioactive compounds with particular attention focused on nutraceutical products and functional foods. This has in turn given rise to the concept of xenohormesis, or molecular networking between species, which tries to explain the multiple positive effects of plant-derived polyphenols on human health [25,26]. This consists in the idea that our body reacts to the signals that plants generate in periods of stress. In this sense, these biochemical signals, called xenohormetins, fulfill important metabolic and defensive functions in plants, such as protection against UV radiation and pathogen infection, nodulation, hormone transportation, as well as several other functions such as defense against herbivorism or pollination [27]. The induction of protective secondary metabolites, especially phenolics, allows plants to withstand the effect of environmental stressors, as a self-defense mechanism against external conditions [28]. Thus, this complex mechanism to protect themselves and improved throughout evolution can be in part extrapolated to humans through plant food consumption [9,29], challenging their own genetic inheritance by modifying innate responses [10].

Several thousand of these safe xenohormetic molecules have already been identified from a multitude of plant species [30]. Plants increase the synthesis of secondary metabolites upon aggressive conditions [31,32]. However, the expected amount of polyphenols in vegetables and fruits for human consumption is low and its bioavailability is limited [33]. A successful strategy to increase the intake of polyphenols would be using products derived from exotic or medicinal plants that may represent an important source of polyphenols, as well as through the proper extraction procedure and enrichment techniques [34–36]. Among various medicinal plants studied, HS represents a potentially optimal source of bioactive molecules for the treatment of various diseases [17,20,21,37–39].

In this respect, anthocyanins are one of the major polyphenolic compounds in HS aqueous extracts, and are responsible for the bright red color of the flowers. Anthocyanins in HS such as delphinidin-sambubioside (red pigment) and cyanidin-sambubioside (pink pigment) comprise the predominant anthocyanin compounds of the extract (Figure 1) and can be used as an alternative to the synthetic dyes used in the food industry [40,41]. Due to their remarkable in vitro antioxidant capability, anthocyanins have received increasing attention in the past two decades [42–44], and is generally accepted that anthocyanins may substantially contribute to the protective effects of HS. In fact, a correlation has been established between the antioxidant activity of HS materials and anthocyanin content, suggesting that these compounds may significantly contribute to HS′s antioxidant effect [45]. Nevertheless, extrapolation of the in vitro antioxidant properties of anthocyanins to an in vivo situation may be difficult, since these compounds are highly soluble in water and have a very short half-life, i.e., are poorly bioavailable and quite possibly not capable of reaching their molecular targets at a sufficient concentration to exert a notable effect [46]. Therefore, this suggests that the biological effect of HS extracts is not exclusively due to these antioxidant compounds. On the other hand, many studies have indicated that polyphenols present a wide variety of effects, supporting the fact that these compounds reach multiple molecular targets and exhibit molecular promiscuity [47,48], which might also endorse the xenohormesis hypothesis. In fact, HS polyphenols have recently been proposed as modulators of gene expression [21] and may be involved in several pathways related to chronic inflammation and energy metabolism [17,23,49]. Consequently, it seems necessary to discern whether the health benefits attributed to HS polyphenols are mainly due to their antioxidant effect or to their ability to modulate the activity of different target proteins, possibly through a synergic mechanism.

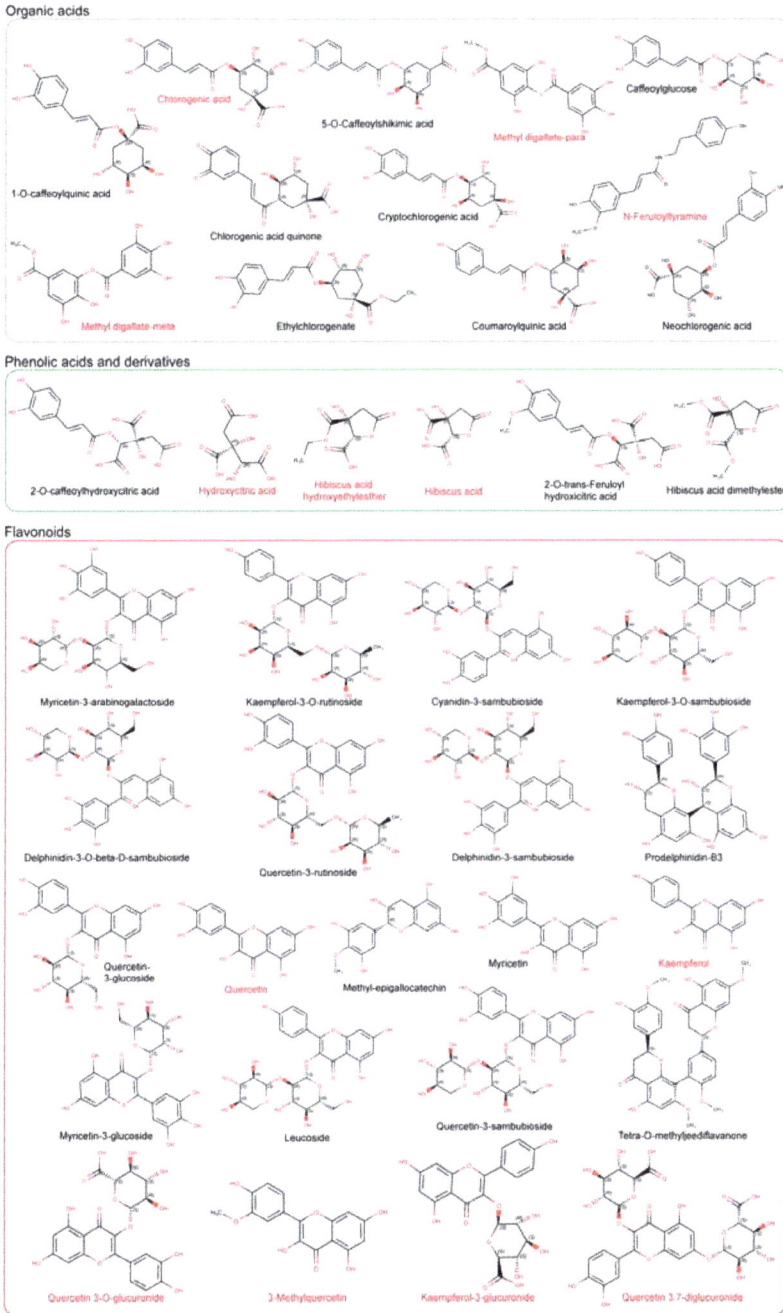

Figure 1. Chemical structures (explicit hydrogens have been eliminated) of the major compounds identified in *Hibiscus sabdariffa* extracts by liquid chromatography coupled to high resolution mass spectrometry. Compounds are grouped in families according to their chemical structure. The complete characterization of the extracts has been previously reported [18,20,37]. The common name of each compound is included under its structure. Compounds with their names in red indicate that these compounds have been identified as plasma metabolites in the bibliography.

3. Characterization and Synergy of *Hibiscus sabdariffa* Bioactive Compounds

HS herbal tea has been traditionally consumed by various cultures. In the abovementioned research studies, HS was generally prepared as a standardized aqueous extract from their dried calyces, although other leaf extractions have also been used [39]. Complex plant-derived mixtures such as herbal teas suppose an important challenge for analytical procedures, although the chromatographic profile of HS aqueous extracts have been studied in detail [50]. In an attempt to identify the major bioactive compounds, several studies have focused on the chemical characterization and quantitation of extracts from HS calyces by high-performance liquid chromatography coupled to high resolution mass spectrometry (HPLC-MS) [18,20,37]. These studies revealed that anthocyanins together with organic acids are two of the most abundant groups identified in the aqueous extract [18] (Figure 1). However, phenolic acid derivatives, flavonol derivatives and phenylpropanoids have also been identified and quantified in the extract (Figure 1). HS calyces are rich in polysaccharides and soluble fiber, mainly pectins, arabinans and arabinogalactans of low molecular mass [51], but it is proposed that these compounds do not contribute to HS bioactivity [20]. HS calyces also contain a high concentration of ascorbic, arachidic, citric, stearic, and malic acids [52], making its aqueous extract quite acidic (pH = 2.8), which is ideal for maintaining polyphenolic stability. In addition, other biomolecules are present in the aqueous extracts at a low percentage, mainly unidentified proteins and peptides [53]. Therefore, such a complex plant-derived mixture may represent an opportunity for the study of potential synergistic effects on different pharmacological targets.

Plants have developed various strategies to modulate biological processes through molecular promiscuity or polypharmacologic effects, which act as multi-target drugs [54]. Due to their nature, plant bioactive compounds could have modulated their diversity throughout evolution to act as ligands of different molecular targets in animal cells. This would be responsible for the enhancement of therapies through possible synergistic interactions with multiple targets. This particular feature of plant compounds could become an opportunity to design suitable and novel approaches for diseases with complex pathogenic mechanisms [36], such as obesity-related complications.

Several studies have postulated on the putative synergistic effect of mixed botanical compounds or extracts to explain their beneficial effects [41,55,56]. Nevertheless, only a few studies have demonstrated the increased efficiency of HS extracts with respect to its isolated compounds, suggesting a therapeutic advantage for the whole extract in the treatment of complex disorders [20]. Nevertheless, it must be noted that the synergistic effect of botanical products is difficult to prove. Synergistic pharmacological interactions between compounds or drugs must be demonstrated by using specific experimental and mathematical approaches, such as the calculation of the combination index, the fractional inhibitory concentration index or isobologram construction [10]. However, this is not always easy to approach in the case of very complex matrices [57], since multiple combinations are implicated, and in many cases the polyphenolic mixtures are not fully characterized [10,58].

Plant mixtures may also exhibit an inverted U-shaped dose-effect curve of purification level vs. bioactivity, i.e., bioactivity increases up to a certain point in which the increase of the purity leads to the loss of some important compound, with the concomitant drop in bioactivity. Therefore, bioassay-guided isolation processes usually fail to reveal the major active compounds of a complex mixture. An alternative to this problem could consist in testing the whole extract and its fractions via metabolomics-guided isolation [59]. The integration of high content screening with non-targeted metabolomics could provide an additional strategy to recognize the metabolites responsible for the biological effects and facilitate the demonstration of the putative synergy in complex herbal mixtures.

4. Bioavailability, Tissue Distribution and Cellular Metabolites Derived from *Hibiscus sabdariffa* Bioactive Compounds

Despite the considerable efforts to characterize the HS extract and the abundance of known potential bioactive compounds, the identification of the key components and their potential targets is still largely unknown. Understanding the pharmacokinetics of compounds derived from HS extracts is essential, since their absorption, metabolism and distribution might determine the mode of action and molecular targets involved in its bioactivity. Regardless, despite the abundance of anthocyanins in HS and their attributed beneficial effects on health, several studies have revealed that these compounds present low absorption and bioavailability in vivo [60]. Therefore, it is possible that other, less abundant compounds present in HS extracts or other not yet unidentified metabolites may be responsible for the observed beneficial effects. In this respect, a bioavailability study carried out in rats revealed a total of seventeen compounds in rat plasma samples, of which quercetin and kaempferol glucuronides were the most predominant one [19]. In fact, recent studies have revealed a higher bioavailability of flavonol derivatives derived from HS compared with other alternative sources [61], with an approximate plasma concentration of 5 µM after administration in rats [19]. In this vein, the metabolite quercetin-3-glucuronide (Figure 1) was detected in hepatic and immune cells in the liver and in the intestinal mucosa of hyperlipidemic mice fed with a polyphenol-enriched extract of HS, concomitantly with an improved steatohepatitis [21]. The permeability of the polyphenolic extracts in Caco-2 intestinal human cells, a model of human intestinal absorption, has also been reported [50,62]. The data from these studies revealed a low level of permeability of the complex polyphenolic mixture through the gut barrier, suggesting that the presence of high concentrations of several polyphenols in the extract could lead to a saturation of the specific transport mechanism. In these studies, only a few compounds (*N*-feruloyltyramine and quercetin (Figure 1), among others, were able to pass through the gut barrier model. Finally, a recent study in hypertrophied 3T3-L1 adipocytes revealed the cellular absorption and metabolism of both quercetin and quercetin-3-glucuronide in adipocytes, which suggests that these flavonols might also reach other intracellular targets that contribute to their bioactivity [63]. Given all this, it could be hypothesized that the compounds responsible for the multiple effects observed with HS extracts may not be the most abundant ones, being the flavonol derivatives the possible candidates for the observed effects in animal models.

5. Effect of *Hibiscus sabdariffa* Compounds on Selected Digestive Enzymes

Very few studies have focused on the interaction of HS polyphenols with digestive enzymes as their putative molecular targets to explain their effect on obesity. Polyphenols are capable of interacting with proteins through hydrophobic or hydrophilic interactions, leading to the formation of aggregates that can alter and affect its biological activity [64]. The inhibitory effect of HS polyphenols against enzymes implicated in carbohydrate digestion, such as α-amylase activity, and the ability of these compounds to block sugars and starch absorption in an α-amylase-added Caco-2 system, has been reported, which deserves further attention due to is potential implication in weight loss [65,66]. The potential inhibitory activity against pancreatic lipase was also reported by examining the effect of HS extracts on fat absorption-excretion and body weight in rats [67]. This study concluded that animals supplemented with HS polyphenols excreted significant amounts of fat in the feces, mainly palmitic and oleic acids, compared to controls. In addition, the lower weight gain observed in HS-fed animals leads to postulate on an inhibitory effect of pancreatic amylase, in agreement with that proposed by Hansawasdi et al. Thus, continuous administration of HS polyphenols might improve obesity-related metabolic disorders in a similar manner to current digestion inhibitory drugs such as orlistat, which is also associated with triglyceride reductions and adiponectin activation [68]. Although these studies considered hibiscus acid as the main component responsible for the enzymatic inhibition (Figure 1), other recently identified HS compounds could also be responsible by inhibiting pancreatic lipase and/or α-glucosidase. Unfortunately, little is known regarding the molecular mechanisms implicated in the inhibition of these or other digestive enzymes by phenolic compounds. A plausible hypothesis could be that these compounds interact directly with the catalytic site of the enzymes, especially if the site is hydrophobic. In our laboratory, we performed molecular docking experiments of a library containing all the compounds identified in HS extracts (Figure 1) against the catalytic binding sites of different enzymes present in the digestive tract. To this end, the crystal structures of pancreatic lipase and glucosidase enzymes were selected to determine whether HS polyphenols can bind these enzymes with sufficient affinity to be considered as inhibitors. Molecular docking can predict the structure of a receptor-ligand complex and calculates a theoretical Gibbs free-energy variation (ΔG, kcal/mol) for different poses of each ligand. This parameter reflects the intensity and number of atomic interactions between the amino acids of the binding site in the protein and the ligand [69,70]. Figure 2 depicts the Gibbs free-energy variation values calculated through molecular coupling experiments obtained using a library of compounds present in the HS (Figure 1) against the catalytic site of various intestinal glucosidase enzymes (Figure 2A) as well as against two ligand binding sites of the enzymatic complex of the pancreatic triacylglycerol lipase/colipase (Figure 2B). As it can be observed in Figure 2A, the majority of the compounds exhibited lower or similar ΔG values to compounds such as agarbose or naringenin [71], which are known inhibitors of intestinal glycosidases. Therefore, flavonols such as quercetin or kaempferol and their metabolites (Figure 1), which have been identified in plasma of treated animal models, could behave as competitive inhibitors of these glucosidases. Likewise, some phenolic acids or anthocyanins may act as putative ligands of intestinal glucosidase enzymes due to their reasonably low ΔG values. Similarly, the free energy variations for the binding of the HS compounds to the two sites in the lipase/colipase complex are shown in Figure 2B and compared to the drug inhibitor orlistat. As mentioned in the case of intestinal glucosidases, the majority of the HS compounds exhibited lower ΔG values than this drug, and therefore it is plausible to consider these compounds as inhibitors of this enzymatic complex, which could account for the decreased absorption of fatty acids observed in animal models using HS extract, especially for flavonols derivatives. Our computational results based on free-energy variations suggest that these molecules have the potential to inhibit digestive enzymes. Nevertheless, further in vitro and in vivo studies will be required to substantiate whether this in silico approach is an appropriate tool for the identification of new potential inhibitors of digestive enzymes.

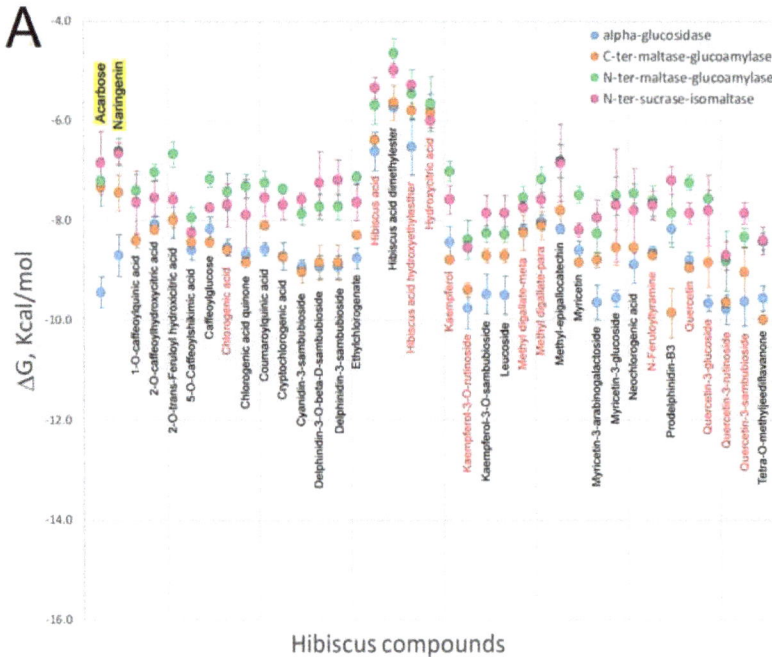

Figure 2. *Cont.*

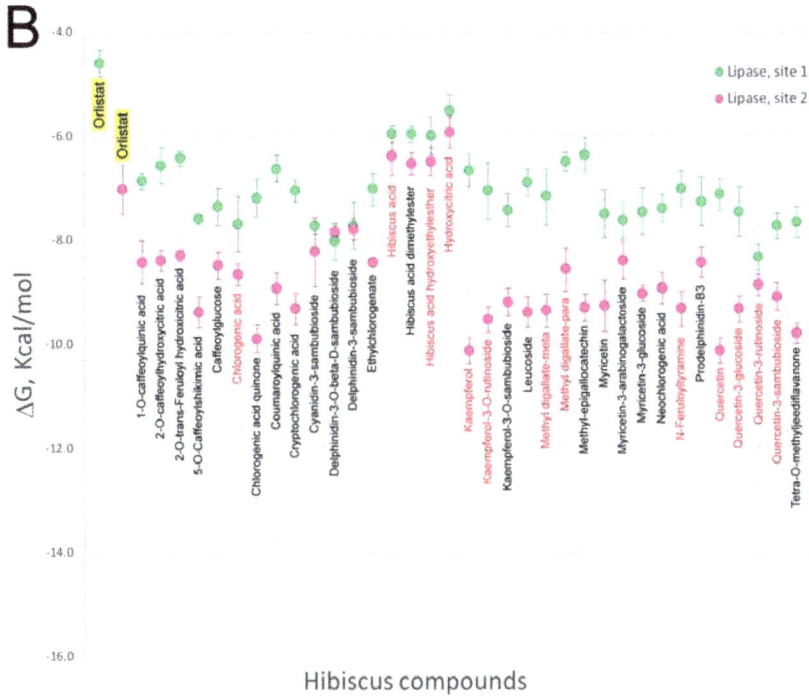

Figure 2. Comparison of the free energy variation (ΔG, kcal/mol) of the *Hibiscus sabdariffa* bioactive compounds and X-ray co-crystalized ligands or known modulators (first name of each panel and yellow background) against the main binding sites of the catalytic binding site of some glucosidase enzymes (panel A) and the two crystal binding sites of the triacylglycerol lipase/colipase complex (panel B) by molecular docking. The common name of each compound is included under mean and standard deviation plotted in each panel. Red colour names indicate that these compounds have been identified as plasma metabolites in the bibliography. For the virtual screening process, a chemical library containing the chemical structures to be screened was initially built, following ADMET criteria. In addition, the molecular target high resolution structures were obtained. Then, computer clusters with a high computation capability were used to compare target structures with the chemical library to find the best docking results according to DG binding values. Finally, all tested compounds were prioritized as a function of minor DG and proposed as candidates for subsequent in vitro and in vivo tests.

6. Molecular Effects of *Hibiscus sabdariffa* Polyphenols

6.1. Effect on Redox Homeostasis

The antioxidant capabilities of plant polyphenols have been well established by many in vitro and in vivo studies, demonstrating a clear correlation with health [16]. To this end, botanical polyphenols could be used as possible therapeutic sources to treat obesity. Oxidative stress is implicated in the development of many chronic conditions, including obesity. This is due to an imbalance between excess reactive oxygen species (ROS) and the inability of the intracellular defense system to efficiently eliminate these oxidative agents. This dysregulation leads to the oxidation and damage of macromolecules such as carbohydrates, lipids, proteins and nucleic acids, which finally cause organelle and cell dysfunction and contributes to the progression of the pathology [72]. ROS are a set of unstable molecules and free radicals derived from molecular oxygen (O_2) and are mainly generated in the oxidative respiratory chain of the mitochondria. Superoxide anion ($\cdot O_2^-$) is generally the precursor of the majority of the ROS produced, and can lead to the formation of hydrogen peroxide (H_2O_2), and consequently hydroxyl radical ($OH\cdot$) by Fenton's reaction. Oxidative damage is generally prevented through the release of intercellular and intracellular antioxidant enzymes such as superoxide dismutase (SOD), catalase (CAT), glutathione peroxidase (GPx) and glutathione reductase (GR), which act as scavengers for the different ROS (Figure 3) [73]. Additionally, $\cdot O_2^-$ and H_2O_2 can also be generated by NADPH oxidase, a membrane-bound enzymatic complex, which plays an important role in cellular proliferation, serotonin biosynthesis, endothelial signaling, regulation of renal functions, and the immune response against microorganisms, although its overexpression is associated with various neurological diseases and cancers [74].

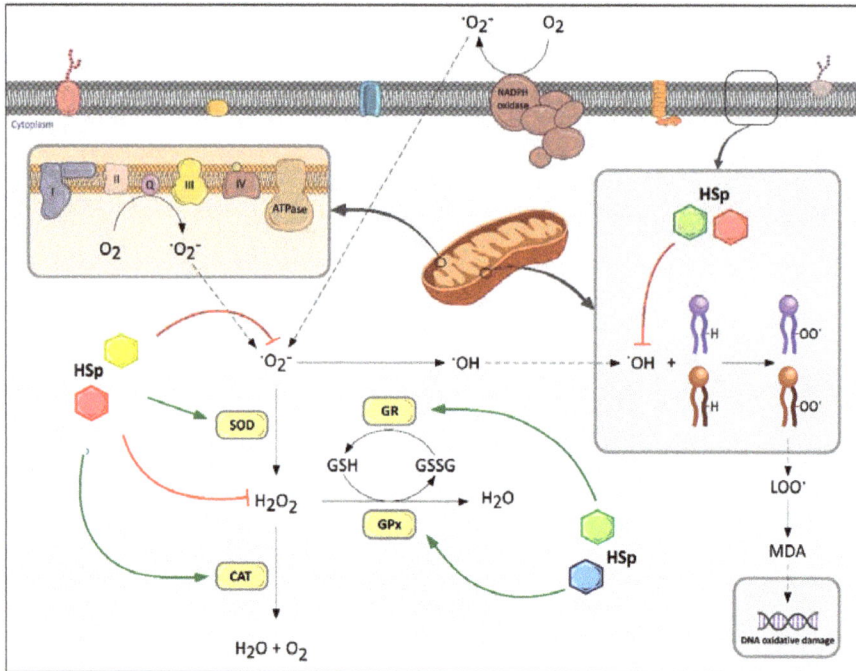

Figure 3. Multiple effects of HS polyphenols on intracellular redox homeostasis. The antioxidant abilities of HS polyphenols (HSp), represented by colored hexagons, act directly or indirectly upon intracellular ROS generation in oxidative status. HSp may directly react with ROS and free radical intermediates by halting the chain reaction, thereby stopping the ROS-induced damage. HSp antioxidant activity blocks intracellular ROS generation in hydrophilic environments and also the generation of lipoperoxy radicals, which are major responsible for DNA oxidative damage. Alternatively, HSp may act indirectly by up-regulating antioxidant enzymes expression or protein activation. SOD, superoxide dismutase; CAT, catalase; GPx, glutathione peroxidase; GR, glutathione reductase. Different colors of HSp symbols indicate different families or structural moieties of polyphenols' metabolites.

A correlation exists between obesity and oxidative stress, as a chronic inflammatory process has been detected in the adipose tissue of obese experimental animal models as well as in humans, along with an increased NADPH oxidase expression and decrease in antioxidant enzymes [75]. Other studies have reported that high levels of fatty acids and glucose increase intracellular ROS generation in adipocyte cell cultures [20,75,76]. According to these reports, adipocytes in obese individuals undergo hypertrophy as a result of an excess caloric intake and a low metabolic rate. Consequently, the increased expression of NADPH oxidase, exacerbated fatty acid oxidation in the mitochondria, and decreased expression of SOD, CAT and GPx lead to an excessive production of ROS. ROS also function as mediators for the activation of nuclear factor-κB (NF-κB) and mitogen-activated protein kinase (MAPK), contributing to the dysregulation of the expression of inflammatory adipokines and a low-grade but chronic state of inflammation (Figure 4) [77].

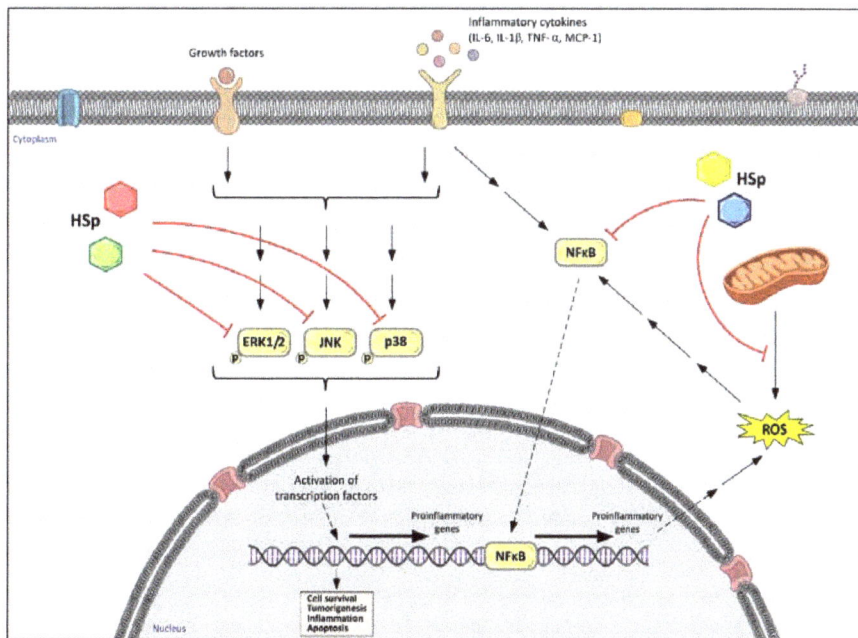

Figure 4. Effects of HS polyphenols on inflammation-related and MAPK pathways. Cellular inhibitory effect of HS polyphenols (HSp), represented by colored hexagons, is associated with a decreased phosphorylation of extracellular signal-regulated protein kinases 1 and 2 (ERK1/2) and c-Jun N-terminal kinases (JNK), as well as p38 kinase. HS polyphenols also inhibit inflammation by down-regulating NF-κB pathway and ROS generation, revealing that inflammation, oxidative stress-related pathways and mitochondrial function are closely related and interdependent processes in the HSp effects. Different colors of HSp symbols indicate different families or structural moieties of polyphenols' metabolites.

Various studies have indicated a possible role for HS polyphenols in regulating excess ROS in obesity-related disorders. For example, HS aqueous extracts have shown a higher capability to scavenge peroxyl radicals in water environments than in lipophilic systems, as well as a stronger metal-reducing effect than olive leaf extract [18]. The antioxidant properties of HS polyphenols have also been measured in the plasma of rats after an acute ingestion of a polyphenol-enriched HS extract [19]. In a later study, a correlation between the presence of phenolic acids in plasma at shorter times and its antioxidant effect through ferric ion reduction and superoxide scavenging was reported. Furthermore, the presence of flavonol glucuronides (quercetin and kaempferol) in plasma samples at longer times correlated with an inhibitory effect on lipid peroxidation. Furthermore, a polyphenol-enriched HS extract exerted a higher effect on inhibiting intracellular ROS formation than an aqueous HS extract in a culture of hypertrophied adipocytes [20]. All these data suggest that finding a suitable combination of bioactive HS polyphenols could represent an opportunity to ameliorate obesity-associated oxidative stress.

A correlation between the total phenolic content of HS extracts and their antioxidant properties has been reported [20,78]. Therefore, it must be presumed that the antioxidant properties of the extracts lie on the polyphenol's abilities (phenolic acids, flavonols and anthocyanins) to scavenge free radicals. This effect has been observed in several oxidative damage models, in both cells and experimental animals. The organic extracts from HS exhibited strong protective properties against malondialdehyde formation and cellular lysis in tert-butyl hydroperoxide-induced oxidative damage in rat primary hepatocytes, suggesting a protective effect against cytotoxicity and genotoxicity [79]. HS anthocyanin extracts have been shown in vitro to scavenge O_2^- radicals and H_2O_2, completely attenuate the CCl(4)-mediated decrease in antioxidant enzymes, as well as activate phase II drug detoxification enzymes in CCl(4)-induced oxidative damage of rat livers [80]. Furthermore, the antimutagenic activity and free radical scavenging effects on active oxygen species and lipid peroxidation of HS organic extracts (chloroform and ethyl acetate) has been reported using an oxidative stress-induced rat model, besides revealing a strong ability to scavenge $\cdot O_2^-$ and OH· radicals and H_2O_2 in vitro [81]. These results point to a strong scavenging ability of HS polyphenols at different stages by eliminating $\cdot O_2^-$, H_2O_2 and OH·, which may take place either by simple hydrogen donation or by preventing antioxidant enzyme degradation. Furthermore, the capacity of HS polyphenols to inhibit lipid peroxidation evidenced in different models may be related to their ability to scavenge OH· radicals, which are the major culprits for the generation of lipoperoxyl radicals (ROO·) (Figure 3).

As mentioned above, polyphenols from HS may not only exert their antioxidant effect by directly scavenging ROS, but also by modulating their effect and induce the expression of other antioxidant enzymes. Essa et al. [82] reported that HS extracts increased the levels of SOD, CAT and GPx, and reduce glutathione (GSH) in brain tissues of hyperammonemic rats. In addition, several studies in animal models of metabolic syndrome, diabetes or hyperlipidemia have reported an increase in the expression of SOD, CAT, GPx and GR (Figure 3) in several tissues, such as the kidney, liver and heart [83–87]. These studies support the hypothesis that HS extracts are capable of ameliorating oxidative stress in metabolic diseases.

6.2. Hibiscus sabdariffa Effects on Inflammatory and Immune Response

Adipose tissue represents not only a metabolic but also an important endocrine organ. Adipocytes are the main constituent cells in this tissue, and their primary function is to store energy as fat. Nevertheless, adipocytes also play an important role in the secretion of a large number of hormones and cytokines (adipokines) that regulate processes such as lipid metabolism, glucose homeostasis, insulin sensitivity, inflammation, blood pressure or angiogenesis. In addition to adipocytes, other cells such as fibroblasts, pre-adipocytes, vascular endothelial cells and immune cells are also present in the adipose tissue.

In an obese individual, the adipocytes are hypertrophic, leading to molecular and cellular disturbances that can affect their functionality [88]. Accordingly, the secretion of pro-inflammatory adipokines increases in this tissue, including interleukin-6 (IL-6), tumor necrosis factor-alpha (TNF-α), monocyte chemoattractant protein-1 (MCP-1) and vascular cell adhesion molecule-1 (VCAM-1) (Figure 4). This pro-inflammatory phenotype leads to a low-grade inflammation systemic condition including macrophage infiltration in adipose tissue [88,89]. Furthermore, lipid accumulation in other organs such as the muscle, pancreas or liver can induce inflammation and insulin resistance, as well as develop metabolic diseases such as liver steatosis, atherosclerosis or type 2 diabetes [1,90,91]. On the other hand, chronic inflammation and oxidative stress are closely related and interdependent processes that contribute to the pathogenesis of obesity-associated diseases [92].

The anti-inflammatory activity of HS polyphenols has been previously reported in cell and animal models and humans. For example, in the cell model of hypertrophied 3T3-L1 adipocytes, both aqueous and polyphenol-enriched HS extracts inhibited the secretion of eight pro-inflammatory adipokines [20].

Ameliorating the oxidative status might be one of the strategies by which HS polyphenols exert their anti-inflammatory effects, since oxidation and inflammation are closely associated. In this regard, the HS polyphenolic extract has been found to inhibit xanthine oxidase activity in vitro, and decrease nitrite and prostaglandin E2 secretions in LPS-induced cells. Moreover, HS polyphenols also inhibited inflammation by down-regulating cyclooxygenase-2 (COX-2) and inhibiting the activation of c-Jun N-terminal kinase (JNK) and p38 kinase, postulating a relationship between HS polyphenols, oxidative stress and suppression of nuclear factor-κB (NF-kB) translocation in a lipopolysaccharide-induced inflammation rat model (Figure 4) [93].

The production of MCP-1 in monocytes/macrophages can be mediated through the activation of extracellular signal-regulated kinases 1/2 (ERK1/2) and JNK as well as the NF-κB pathway [94]. NF-κB is a complex protein capable of regulating the transcription of several genes related to the inflammatory process. HS extracts have been demonstrated to inhibit the expression of NF-κB, as well as decrease the levels of TNF-α, IL-6 and IFN-γ, indicating a hepatoprotective effect in rats with thioacetamide (TAA)-induced hepatotoxicity [49] (Figure 4). Furthermore, oral consumption of an aqueous HS extract in healthy humans has been shown to decrease the plasma levels of MCP-1, suggesting that such an effect is not due to the antioxidant activity itself, but rather the inhibition of inflammatory and metabolic pathways [17].

Since oxidative stress, inflammation and hypertension are also closely linked, the relationship between HS consumption and blood pressure has been investigated by several authors. Joven et al. demonstrated that HS extracts lowered blood pressure in human patients with metabolic syndrome and improved the endothelial function in a rat model. In this study, the consumption of 125 mg/kg/day of HS polyphenols by human patients for four weeks decreased the majority of the inflammatory and oxidative stress biomarkers analyzed, as well as increased the anti-inflammatory hormone adiponectin [22]. The same study revealed that HS extracts reduced the blood pressure in rats, while also inhibiting TNF-α-induced cytokine secretion in rat endothelial cells, suggesting a possible transcriptional down-regulation of glyceraldehyde 3-phosphate dehydrogenase (GAPDH). Furthermore, a reduction of NF-κB and ROS generation and increased nitric oxide synthase (eNOS) and nitric oxide (NO) was observed in the same cells. The results of this study suggest that the hypotensive effect of HS extracts observed in both humans and rats might be mediated by antioxidant, anti-inflammatory and endothelium-dependent mechanisms, in agreement with Hopkins et al. [95]. However, this is in contrast with other studies that have attempted to explain the antihypertensive effectiveness of this extract only through the inhibition of the angiotensin I-converting enzyme [96,97]. In this regard, a recent meta-analysis study on randomized controlled trials using HS confirmed the potential effectiveness, reducing both systolic and diastolic pressure and considered the combination of HS supplements and antihypertensive medications to optimize the antihypertensive therapy [38].

All this evidence leads to the postulation that the anti-inflammatory effects of HS polyphenols are closely related to the modulation of oxidative stress-related pathways. Nevertheless, it must be assumed that the complex mixture of compounds contained in the polyphenolic HS extract may interact with a wide variety of molecular targets directly related to inflammatory processes; therefore, the modulation of gene expression and/or protein activity in these pathways should also be considered. Anyway, the inhibition of MAPK and NF-kB pathways seem to be the major mechanisms of action of HS polyphenols (Figure 4). Thus, regulation of the inflammatory processes in obesity through HS extract consumption could be an interesting approach to ameliorate and/or prevent other chronic inflammatory diseases related to metabolic syndrome such as atherosclerosis, steatohepatitis or cardiovascular diseases.

6.3. Modulation of Energy Metabolism and Lipid Management by Hibiscus sabdariffa Polyphenols

Polyphenols comprise of a number of molecular scaffolds with an enormous variety of substitutions with different moieties, either through intermolecular interactions or by polymerization. Therefore, it seems reasonable to consider that such a structural diversity confers the possibility of generating a limitless number of pharmacological compounds for various targets. In this sense, complex polyphenolic mixtures may have a multi-targeted mechanism of action, as reported [10].

As previously commented the different chemical structures contained in the HS extract (Figure 1) have a strong antioxidant capability and protect against chronic inflammation, which have been positively correlated to obesity-related metabolic disorders [48,98]. Nevertheless, these pathologies are complex conditions that require a multifaceted approach, including effects on the inflammatory network, antioxidant status and energy metabolism. Recently, the relationship between polyphenols interactions and energy metabolism modulation has been reported, suggesting that the consumption of plant-derived polyphenols can change lipid and energy metabolism and may facilitate weight loss and prevent weight gain [15].

The ability of HS polyphenols to modulate energy metabolism in order to explain the potential beneficial effects on lipid management and weight loss has been studied by several authors in cell and animal models. The polyphenol-enriched HS extract presented a higher efficiency in inhibiting intracellular triglyceride accumulation than aqueous HS extract that contained higher polysaccharide content in a culture of hypertrophied adipocytes [20]. This observation points to the polyphenols as the candidate molecules responsible for the lower lipid accumulation, indicating that it may be a suitable strategy to improve obesity-associated disturbances. Several studies have proven the ability of HS polyphenols to help reduce body weight through inhibition of fat accumulation, while also improving glucose tolerance and normalize the glycemic index in obese mice models [23,99].

Its effect on weight loss has prompted hibiscus extract to be examined for its potential effect on lipid profiles. Several studies have indicated that HS extracts have a lipid lowering activity, which could prevent cardiovascular disease through lipid modification [100]. Fernández-Arroyo et al. [18] tested HS polyphenols on low density lipoprotein receptor deficient mice (LDLr$^{-/-}$) under a hypercaloric diet (20% fat and 0.25% cholesterol, *w/w*). A strong correlation between HS consumption and the ability to decrease serum triglyceride concentration was reported in this study. Several studies using hyperlipidemic rat models together with continuous cholesterol feeding for four weeks resulted in a significant reduction of serum cholesterol, triglycerides, LDL and VLDL levels [101,102]. Since the majority of the plasma apolipoproteins, endogenous lipids and lipoproteins are synthetized in the liver, several studies have attempted to establish a link between the hypolipidemic ability of HS and its putative mechanism on the liver. Yang et al. [24] examined the effect of HS on liver fat metabolism and revealed that HS polyphenols reduced liver damage by promoting lipid clearance. A decrease in the plasma lipid level and hepatocyte lipid content concomitantly with the activation of AMPK, along with an inhibition of fatty acid synthase (FASN) expression, was also detected. A similar effect has been observed in hyperlipidemic mice model, in which the administration of the HS polyphenolic extract prevented fatty liver disease associated with changes in lipid and glucose metabolism, as well as AMPK activation and decreased expression of lipogenic genes such as FAS and SREBP-1c [21] (Figure 5).

It is known that the activation of AMPK, a master regulator of energy metabolism, and the subsequent inhibition of acetyl-CoA carboxylase play a crucial role in fatty acid oxidation by regulating mitochondrial availability of fatty acids [103] (Figure 5). Both AMPK and PPARs have been shown to play an important role in the pathogenesis of both alcoholic liver disease and non-alcoholic fatty liver disease, in which fatty acid oxidation is impaired. Administration of AMPK or PPAR-α activators have shown to be effective in ameliorating both diseases [104]. Polyphenols have also been shown to upregulate PPAR-γ-mediated adiponectin expression, along with AMPK activation [12]. Therefore, regulating the complex network of energy metabolism and mitochondrial function may require a multifaceted approach, including AMPK and PPARs, which can be addressed by the pleiotropic character of plant-derived polyphenols.

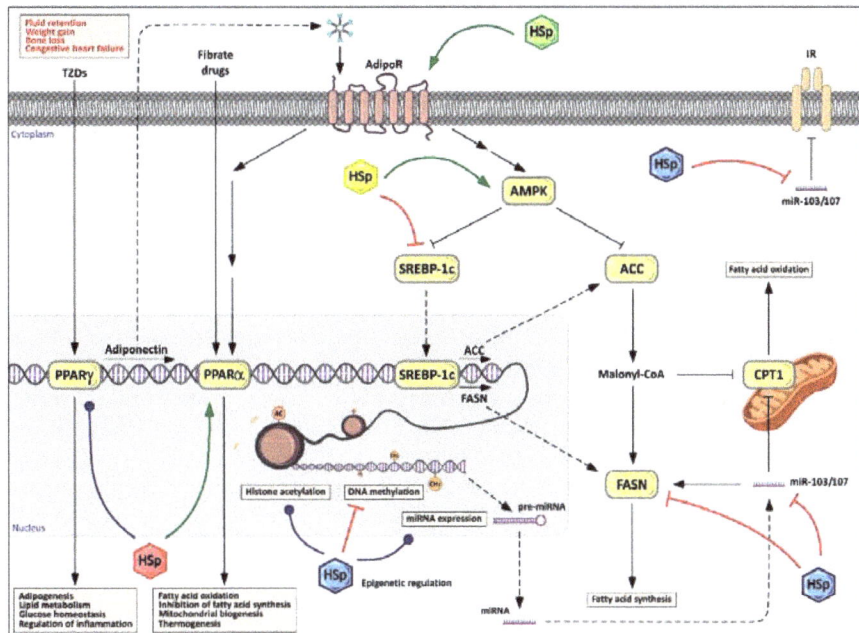

Figure 5. Pleiotropic effects of HS polyphenols on cellular energy metabolism and lipid and glucose homeostasis. HS polyphenols (HSp), represented by colored hexagons, stimulate AMP-activated protein kinase and, consequently, inhibit downstream targets, such as acetyl-CoA carboxylase (ACC) and fatty acid synthase (FASN), decreasing the fatty acids synthesis and stimulating lipolysis. Alternatively, HSp action, either directly or through AMPK, promotes downregulation of FASN by the inhibition of the sterol regulatory element-binding protein-1c (SREBP-1c). The modulation of peroxisome proliferator-activated receptors (PPAR) expression and/or activity, and the subsequent adiponectin expression increase, in response to HSp improves lipid and glucose homeostasis and mitochondrial biogenesis. The pleiotropic effects of HSp also include primary control mechanisms such as epigenetic modifications through histone acetylation, DNA methylation and miRNA expression. Different colors of HSp symbols indicate different families or structural moieties of polyphenols' metabolites.

HS polyphenols have been proven to regulate AMPK as well as several transcription factors related to lipid and glucose homeostasis, such as PPARs and SREBP-1c in hyperlipidemic mice model [21,23] (Figure 5). Growing evidence indicates that these transcription factors are critical regulators of hepatic lipid metabolism, stimulating the expression of several enzymes implicated in liver fatty-acid synthesis, glucose transport and gluconeogenesis [23]. Consistent with this hypothesis, a recently discovered low molecular weight compound acting as an adiponectin receptor (AdipoR) agonist, AdipoRon, has been proposed as a possible molecular tool to improve insulin resistance, by mediating the activation of AMPK and PPAR-pathways, mimicking the effect of adiponectin [3]. Thiazolidinediones are oral medications for type 2 diabetes that function as synthetic ligands and potent agonists of PPAR-γ, being highly effective in reducing glucose levels and improving insulin sensitivity. However, the administration of thiazolidineones is associated with the occurrence of severe side effects such as fluid retention, weight gain, cardiac hypertrophy, bone fractures and hepatotoxicity (Figure 5) [6]. The use of polyphenols as mild PPAR-γ agonists based on selective cofactor-receptor interactions, while avoiding the side effects observed in synthetic agonists, has been postulated as an opportunity for the management of obesity [69]. In fact, some polyphenols such as resveratrol [105] or scutellarin [106] have shown experimentally the capability to modulate PPAR-γ. Moreover, extensive effort has been recently made on the design of alternative synthetic PPAR modulators, such as metaglidasen, that are devoid of the typical side effects observed with thiazolidinedione anti-diabetic agents [107]. The ability of HS polyphenolic extracts to activate AMPK has also been associated with the activation of PPAR-α, which triggers FASN inhibition, stimulating fatty acid oxidation and antioxidant enzyme expression, thereby reducing lipid content and oxidative stress, which enhance mitochondrial metabolism and lipid management [21] (Figure 5). HS polyphenols has also been correlated with a significant increase in serum adiponectin and decrease in serum leptin, which can revert overweight-induced metabolic alterations [22].

6.4. Epigenetic Effects of Hibiscus sabdariffa Polyphenols

The majority of the studies on the effects of HS and other plant polyphenols have focused on specific molecular targets in order to explain their molecular mechanism in the amelioration of obesity-related pathologies. As such, several cellular pathways, transcription factors and enzymatic activities seem to be modulated by these compounds. This mode of action suggests the existence of a primary control mechanism, which appears to be in line with the pleiotropic character and structural diversity of these compounds. Recent studies have shown that plant polyphenols are capable of exerting an epigenetic control. Epigenetics include a number of extra-genetic processes such as DNA methylation, post-translational histone modifications and non-coding RNA mediated gene silencing, all of which can alter gene expression, but do not involve DNA sequence changes and can be modified by environmental stimuli [108]. Plant derived polyphenols, such as curcumin, catechins, resveratrol or some flavonols, can interact with various enzymes and important epigenetic modifiers, such as histone acetyltransferases, histone deacetylases, DNA methyltransferases, kinases and miRNA [109,110].

Several authors have reported that polyphenols can reverse the altered epigenetic modifications observed in metabolic disorders by changing DNA methylation and histone modifications [111]. For example, polyphenols are effective histone deacetylases inhibitors and thus can be used to reverse the reduced histone acetylation associated with neurodegeneration [112,113]. Quercetin, a fairly abundant flavonol in the polyphenolic fraction of HS extracts, inhibited the histone acetyltransferase activity in the promoter region of genes associated with inflammation [114]. Currently, the epigenetic regulation of energy balance and adipose tissue biology has been proposed by modulating the expression of certain microRNAs [115]. HS polyphenols have been shown to be capable of regulating microRNA expression in hyperlipidemic mice with LDL receptor deficiency [21]. The chronic oral administration of HS polyphenols in mice under a fat-enriched diet reverted the changes observed in non-specific microRNAs miR103/107 concomitantly with the prevention of diet-induced fatty liver disease and changes in lipid and glucose metabolism (Figure 5). Recently, the ability of HS polyphenols to regulate gene expression through the epigenome has been confirmed [116]. Research is currently focused on elucidating whether HS polyphenols and their metabolites are capable of regulating epigenetic modifications through DNA methylation, histone post-translational modifications of miRNA expression modulation.

7. Virtual Screening of Hibiscus sabdariffa Polyphenols on Selected Protein Targets

Many studies have indicated that HS polyphenols are capable of targeting several proteins that are implicated in obesity-associated metabolic disorders, i.e. AMPK, PPAR and FASN. In order to elucidate on the possibility that HS polyphenols may exert a direct effect on these proteins, virtual screening using a library of chemical structures was assessed (Figure 1). The abundant high resolution structural information of these proteins, especially for PPAR and AMPK, allowed us to perform molecular docking approaches in order to try to understand if HS compounds, especially those detected in blood plasma as metabolites, could interact with these proteins at their catalytic or regulatory sites. The results are shown in Figure 6A (AMPK kinase) and Figure 6B (fatty acid synthase and PPARgamma). AMPK is a cellular energy state-sensitive kinase that is activated under deficient energy states due to lack of nutrients or hypoxia [117]. In addition, it can be activated when certain proteins are phosphorylated, promoting the inactivation of energy-consuming pathways and activating the catabolism of fatty acids and other fuels [118]. The AMPK gamma subunit has three binding sites for AMP/ADP or other modulators, which activate and maintain the phosphorylation of threonine 172 within the activation loop of the catalytic alpha subunit kinase. Here, we have performed molecular docking with the x-ray co-crystallographic ligands and AMPK subunits on its binding sites as a control of the most adequate ΔG in each binding site. X-ray value for AICAR as ligand on the three sites of the AMPK gamma subunit presented a $\Delta G \approx -6.5 \pm 0.5$ kcal/mol when docked over its respective binding sites for all the available PDBs: 2UV4, 2UV5, 2UV6, 2UV7, 4CFE, 4CFF, 4RER, 4REW, and 4ZHX [118]. As can be observed in Figure 6A, with the exception of hibiscus acid and its derivatives, all HS compounds, especially those present in rat blood plasma as metabolites (its most commonly-used name is indicated in red text, Figure 1), show ΔGs decrease up to 2 kcal/mol compared to the controls included in the graph, AICAR in this case. Therefore, these results seem to indicate that most HS compounds excluding organic acids might act as potential activators for the AMPK gamma subunit. At the interface of the interaction between the alpha catalytic and beta regulatory subunits of AMPK, crystallographic data show an additional regulatory site for this enzyme. Molecular docking of the A-769662 activator against this regulatory site presents a $\Delta G \approx -10.72 \pm 0.30$ kcal/mol [118]. Our docking data against this regulatory site shows that HS compounds exhibit ΔG values 2 kcal/mol greater than A-769662 inhibitor; therefore, we would not expect them to exert any regulatory role on this binding site. When the ΔG values of the HS compounds against the catalytic site of the AMPK alpha subunit were compared with the ΔG value (-13.5 ± 1.3 kcal/mol) of staurosporine, which is an ATP-competitive kinase inhibitor with a high affinity but little selectivity, it was observed that none of the HS compounds presented such low ΔGs values. As such, they would not behave as competitive AMPK inhibitors. Finally, Figure 6B compares the ΔGs of HS compounds with those of known PPARgamma and FASN modulators or inhibitors. Only certain HS compounds (Kaempferol-3-O-rutinoside, quercetin-3-rutinoside or tetra-O-methyljeediflavanone) presented a ΔG comparable to that of scutellarine, a flavone for which there is experimental evidence [106] of its ability to bind to PPARgamma and thus could activate this nuclear receptor. Interestingly, these same three compounds, and also quercetin 3,7-diglucuronide showed a ΔG lower than -10 kcal/mol for FASN docking, which is comparable to that detected in the GSK2194069 FASN inhibitor [119], making us to think that they could also behave as inhibitors of this enzyme implicated in fatty acid biosynthesis. Taken together, the data derived from the molecular docking experiments of HS compounds *versus* several binding sites to different proteins involved in energy metabolism, seem to indicate that several compounds extracted from this plant might exert a direct effect on these proteins. Of course, further studies, both in vitro and in vivo, will be required to corroborate this hypothesis and understand the mechanisms by which these compounds exert their effect on the corresponding proteins.

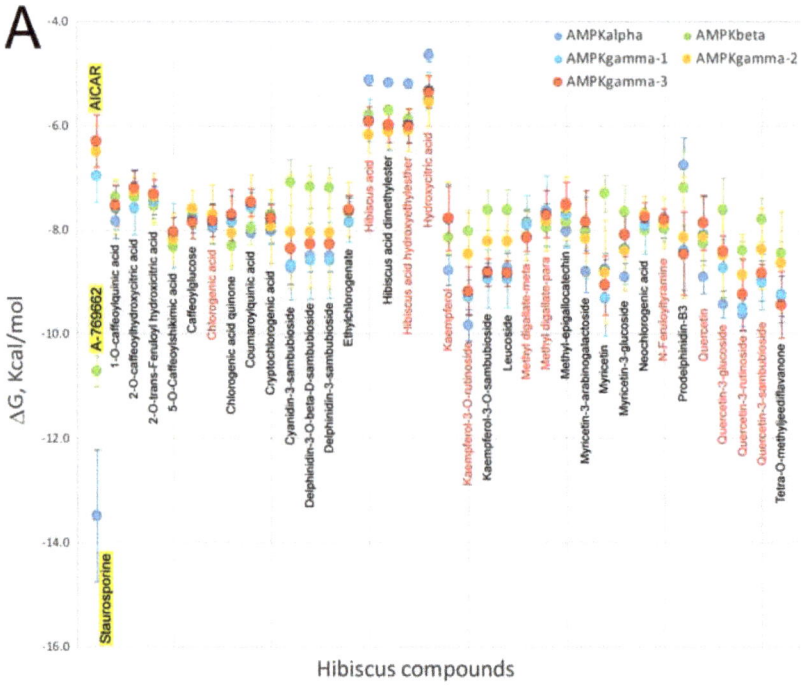

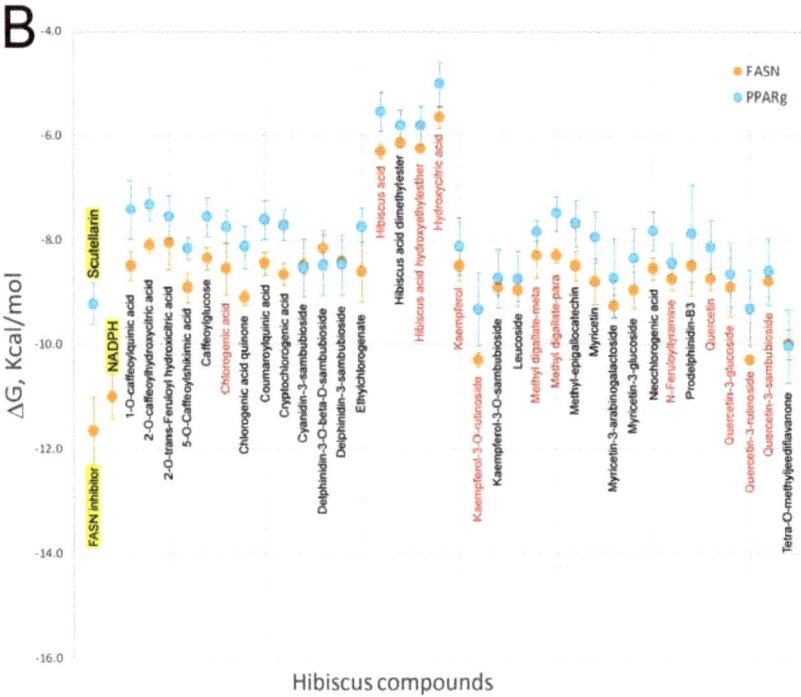

Figure 6. Comparison of the free energy variation (ΔG, kcal/mol) of the *Hibiscus sabdariffa* bioactive compounds and X-ray co-crystalized ligands or known modulators (first name of each panel and yellow background) against the main binding sites of: the catalytic binding site of the three AMPK subunits obtained (**A**); and the catalytic binding site for fatty acid synthase and binding cavity of PPAR-gamma nuclear receptor (**B**), by molecular docking. The common name of each compound is included under mean and standard deviation plotted in each all four panel. Red color names indicate that these compounds have been identified as plasma metabolites in the bibliography. Virtual screening process was performed as mentioned in Figure 2.

8. Potential Molecular Mechanism of *Hibiscus sabdariffa* in Obesity: A Global Analysis

Obesity-related metabolic disorders are characterized by early cellular events and dysregulation of normal cellular homeostasis. In this sense, obesity must be understood as a multidimensional disease, ranging from adipocyte hypertrophy to the appearance of symptoms related to metabolic alterations. HS derived polyphenols could approach the therapy of this pathology from a global perspective, taking advantage of the pleiotropic character of these compounds. Several altered metabolic biomarkers and proteins have been validated as a prequel in the development of bioenergy alterations. Thus, crucial proteins such as PPAR, FASN, lipase, adiponectin, leptin, MCP-1, AMPK, NF-kB and SOD, among others, have been proposed by several authors as specific targets in the treatment of obesity-associated metabolic disorders. In this scenario, HS polyphenols seem to interact with all the above-mentioned targets. In addition, their antioxidant capacity may explain many of the benefits granted to this plant.

Among the putative processes involved in the molecular effects of HS polyphenols on obesity are: modulation of the signaling and energy metabolism pathways, regulation of redox homeostasis and inflammation, restoration of mitochondrial functionality and epigenetic machinery regulation. The last process may probably play a key role on the other aspects.

The extensive literature on the multiple beneficial effects that HS extract exert on obesity-related diseases, as well as its putative mechanism of action, provides sufficient evidence to conclude that the use of this extract could become an adjuvant in these pathologies. Despite the fact that the majority of the cellular and animal effects observed with HS polyphenols have been corroborated in various human trials, further studies are required to define the adequate dosage for therapy as well as validate its use in the management of obesity [116]. Therefore, future research should pay attention to identify the cellular metabolites of HS acting as final effectors through in vivo experiments and powerful analytical platforms.

In this regard, a recent study on the acute multifunctional effects of HS polyphenols in humans have used a combined transcriptomics and metabolomics approach to determine the modifications in the expression of relevant genes through network-based methods [53]. Their results suggest that these polyphenols exert simultaneous effects in mitochondrial function, energy homeostasis, oxidative stress and inflammation, as well as cardiovascular-protective effects, acting upon multiple targets: proteins, hormones and physiological peptides. These findings support the hypothesis that polyphenols are extremely bioactive in humans, and the observed effects are the result of numerous beneficial and synergistic interactions, through the modulation of multiple metabolic pathways and epigenetic regulation [53].

Although the potential pharmacological and therapeutic superiority of the combination of polyphenols in HS extract with respect to their individual components for obesity-related pathologies seems to be demonstrated, several issues related to bioavailability, dosage and safety need to be addressed before its use in human nutrition. First, human equivalent dose must be established from animal trials to assure its efficacy in metabolic disorders. Second, the stability and absorption of these formulations has to be explored to maximize the efficacy of these formulations in human nutrition [10,20]. Finally, although the combination of polyphenols from HS appears to be safe from its traditional use in several cultures, the subchronic administration of these mixtures has to be assayed for toxicity due to the potential polyphenol toxicity itself on a long-term basis or to the presence of some contaminants in plants (mycotoxins, pesticides or metals) [120].

When explaining the multi-targeted effects that plant-derived polyphenols have on proteins targets, an intriguing question arises: are we facing a large variety of molecules bearing a few moieties able to interact with many protein targets, or are the responsible compounds just a few simple and yet to be identified metabolites derived from many polyphenolic structures and acting at primary control epigenetic mechanisms?

Nevertheless, the use of partial or biased approaches focused on specific targets will probably fail to elucidate the precise mechanism of the polyphenols. On the contrary, to fully understand the health and multi-target effects that polyphenols have on obesity and human health, it is necessary to integrate transcriptomic, proteomic and targeted metabolomic approaches, with the support of virtual screening techniques of metabolites on selected protein targets. Lastly, the use of epigenetic approaches will be required to understand how these effects are regulated, as well as determine the most adequate biomarkers for disease diagnostics and prognostics, as well as for treatment. Only with this global approach we will be able to succeed in the design of polyphenolic mixtures sufficiently effective to ameliorate obesity-related pathologies through nutritional intervention or pharmacological treatment.

Acknowledgments: Some of the investigations described in this review have been partially or fully supported by competitive public grants from the following institutions: AGL2011-29857-C03-03 and IDI-20120751 grants (Spanish Ministry of Science and Innovation); projects AGL2015-67995-C3-1-R, AGL2015-67995-C3-2-R and AGL2015-67995-C3-3-R from the Spanish Ministry of Economy and Competitiveness (MINECO); and PROMETEO/2012/007, PROMETEO/2016/006, ACOMP/2013/093, ACIF/2010/162, ACIF/2015/158 and ACIF/2016/230 grants from *Generalitat Valenciana* and CIBER (CB12/03/30038, Fisiopatologia de la Obesidad y la Nutricion, CIBERobn, Instituto de Salud Carlos III, Spain). We are grateful to Research, Technological Innovation and Supercomputing Center of Extremadura (CenitS) for allowing us to use their supercomputing facilities (LUSITANIA II).

Author Contributions: M.O.-V. and E.B.-C. reviewed the bibliography and built the figures. J.A.E. designed the virtual screening research and analyzed the data. J.J., A.S.-C. and V.M. designed most of the research. M.H.-L. and V.M. wrote and reviewed the manuscript.

Conflicts of Interest: The authors declare no conflict of interest.

Abbreviations

Hibiscus sabdariffa	HS
low-density lipoprotein	LDL
LDL receptor	LDLr
reactive oxygen species	ROS
Sterol regulatory element-binding proteins	SREBP-1c
AMP-activated protein kinase	AMPK
fatty acid synthase	FASN
peroxisome proliferator-activated receptor	PPAR
superoxide dismutase	SOD
catalase	CAT
glutathione peroxidase	GPx
glutathione reductase	GR
diode array detection	DAD
electrospray	ESI
high-performance liquid chromatography	HPLC

References

1. Luna-Luna, M.; Medina-Urrutia, A.; Vargas-Alarcon, G.; Coss-Rovirosa, F.; Vargas-Barron, J.; Perez-Mendez, O. Adipose tissue in metabolic syndrome: Onset and progression of atherosclerosis. *Arch. Med. Res.* **2015**, *46*. [CrossRef] [PubMed]

2. Grundy, S.M. Metabolic syndrome: Connecting and reconciling cardiovascular and diabetes worlds. *J. Am. Coll. Cardiol.* **2006**, *47*, 1093–1100. [CrossRef] [PubMed]

3. Okada-Iwabu, M.; Yamauchi, T.; Iwabu, M.; Honma, T.; Hamagami, K.; Matsuda, K.; Yamaguchi, M.; Tanabe, H.; Kimura-Someya, T.; Shirouzu, M.; et al. A small-molecule adipor agonist for type 2 diabetes and short life in obesity. *Nature* **2013**, *503*, 493–499. [CrossRef] [PubMed]

4. Rise, M.B.; Pellerud, A.; Rygg, L.Ø.; Steinsbekk, A. Making and maintaining lifestyle changes after participating in group based type 2 diabetes self-management educations: A qualitative study. *PLoS ONE* **2013**, *8*, e64009. [CrossRef] [PubMed]

5. Eckel, R.H.; Alberti, K.; Grundy, S.M.; Zimmet, P.Z. The metabolic syndrome. *Lancet* **2005**, *365*, 1415–1428. [CrossRef]

6. Ahmadian, M.; Suh, J.M.; Hah, N.; Liddle, C.; Atkins, A.R.; Downes, M.; Evans, R.M. Ppargamma signaling and metabolism: The good, the bad and the future. *Nature* **2013**, *19*, 557–566.

7. Leite, C.E.; Mocelin, C.A.; Petersen, G.O.; Leal, M.B.; Thiesen, F.V. Rimonabant: An antagonist drug of the endocannabinoid system for the treatment of obesity. *Pharmacol Rep.* **2009**, *61*, 217–224. [CrossRef]

8. Heck, A.M.; Yanovski, J.A.; Calis, K.A. Orlistat, a new lipase inhibitor for the management of obesity. *Pharmacotherapy* **2000**, *20*, 270–279. [CrossRef] [PubMed]

9. Menendez, J.A.; Joven, J.; Aragones, G.; Barrajon-Catalan, E.; Beltran-Debon, R.; Borras-Linares, I.; Camps, J.; Corominas-Faja, B.; Cufi, S.; Fernandez-Arroyo, S.; et al. Xenohormetic and anti-aging activity of secoiridoid polyphenols present in extra virgin olive oil: A new family of gerosuppressant agents. *Cell Cycle* **2013**, *12*, 555–578. [CrossRef] [PubMed]

10. Barrajon-Catalan, E.; Herranz-Lopez, M.; Joven, J.; Segura-Carretero, A.; Alonso-Villaverde, C.; Menendez, J.A.; Micol, V. Molecular promiscuity of plant polyphenols in the management of age-related diseases: Far beyond their antioxidant properties. *Adv. Exp. Med. Biol.* **2014**, *824*, 141–159. [PubMed]

11. Beltran-Debon, R.; Rull, A.; Rodriguez-Sanabria, F.; Iswaldi, I.; Herranz-Lopez, M.; Aragones, G.; Camps, J.; Alonso-Villaverde, C.; Menendez, J.A.; Micol, V.; et al. Continuous administration of polyphenols from aqueous rooibos (*Aspalathus linearis*) extract ameliorates dietary-induced metabolic disturbances in hyperlipidemic mice. *Phytomedicine* **2011**, *18*, 414–424. [CrossRef] [PubMed]

12. Herranz-López, M.; Barrajón-Catalán, E.; Segura-Carretero, A.; Menéndez, J.A.; Joven, J.; Micol, V. Lemon verbena (*Lippia citriodora*) polyphenols alleviate obesity-related disturbances in hypertrophic adipocytes through ampk-dependent mechanisms. *Phytomedicine* **2015**, *22*, 605–614. [CrossRef] [PubMed]

13. Ali, F.; Ismail, A.; Kersten, S. Molecular mechanisms underlying the potential antiobesity-related diseases effect of cocoa polyphenols. *Mol. Nutr. Food Res.* **2014**, *58*, 33–48. [CrossRef] [PubMed]

14. Amiot, M.J.; Riva, C.; Vinet, A. Effects of dietary polyphenols on metabolic syndrome features in humans: A systematic review. *Obes. Rev.* **2016**, *17*, 573–586. [CrossRef] [PubMed]

15. Wang, S.; Moustaid-Moussa, N.; Chen, L.; Mo, H.; Shastri, A.; Su, R.; Bapat, P.; Kwun, I.; Shen, C.L. Novel insights of dietary polyphenols and obesity. *J. Nutr. Biochem.* **2014**, *25*, 1–18. [CrossRef] [PubMed]

16. Halliwell, B.; Rafter, J.; Jenner, A. Health promotion by flavonoids, tocopherols, tocotrienols, and other phenols: Direct or indirect effects? Antioxidant or not? *Am. J. Clin. Nutr.* **2005**, *81*, 268S–276S. [PubMed]

17. Beltran-Debon, R.; Alonso-Villaverde, C.; Aragones, G.; Rodriguez-Medina, I.; Rull, A.; Micol, V.; Segura-Carretero, A.; Fernandez-Gutierrez, A.; Camps, J.; Joven, J. The aqueous extract of *hibiscus sabdariffa* calices modulates the production of monocyte chemoattractant protein-1 in humans. *Phytomedicine* **2010**, *17*, 186–191. [CrossRef] [PubMed]

18. Fernández-Arroyo, S.; Rodríguez-Medina, I.C.; Beltrán-Debón, R.; Pasini, F.; Joven, J.; Micol, V.; Segura-Carretero, A.; Fernández-Gutiérrez, A. Quantification of the polyphenolic fraction and in vitro antioxidant and in vivo anti-hyperlipemic activities of hibiscus sabdariffa aqueous extract. *Food Res. Int.* **2011**, *44*, 1490–1495. [CrossRef]

19. Fernandez-Arroyo, S.; Herranz-Lopez, M.; Beltran-Debon, R.; Borras-Linares, I.; Barrajon-Catalan, E.; Joven, J.; Fernandez-Gutierrez, A.; Segura-Carretero, A.; Micol, V. Bioavailability study of a polyphenol-enriched extract from hibiscus sabdariffa in rats and associated antioxidant status. *Mol. Nutr. Food Res.* **2012**, *56*, 1590–1595. [CrossRef] [PubMed]

20. Herranz-Lopez, M.; Fernandez-Arroyo, S.; Perez-Sanchez, A.; Barrajon-Catalan, E.; Beltran-Debon, R.; Menendez, J.A.; Alonso-Villaverde, C.; Segura-Carretero, A.; Joven, J.; Micol, V. Synergism of plant-derived polyphenols in adipogenesis: Perspectives and implications. *Phytomedicine* **2012**, *19*, 253–261. [CrossRef] [PubMed]

21. Joven, J.; Espinel, E.; Rull, A.; Aragones, G.; Rodriguez-Gallego, E.; Camps, J.; Micol, V.; Herranz-Lopez, M.; Menendez, J.A.; Borras, I.; et al. Plant-derived polyphenols regulate expression of mirna paralogs mir-103/107 and mir-122 and prevent diet-induced fatty liver disease in hyperlipidemic mice. *Biochim. Biophys. Acta.* **2012**, *1820*, 894–899. [CrossRef] [PubMed]

22. Joven, J.; March, I.; Espinel, E.; Fernandez-Arroyo, S.; Rodriguez-Gallego, E.; Aragones, G.; Beltran-Debon, R.; Alonso-Villaverde, C.; Rios, L.; Martin-Paredero, V.; et al. Hibiscus sabdariffa extract lowers blood pressure and improves endothelial function. *Mol. Nutr. Food Res.* **2014**, *58*, 1374–1378. [CrossRef] [PubMed]

23. Villalpando-Arteaga, E.V.; Mendieta-Condado, E.; Esquivel-Solis, H.; Canales-Aguirre, A.A.; Galvez-Gastelum, F.J.; Mateos-Diaz, J.C.; Rodriguez-Gonzalez, J.A.; Marquez-Aguirre, A.L. Hibiscus sabdariffa l. Aqueous extract attenuates hepatic steatosis through down-regulation of ppar-gamma and srebp-1c in diet-induced obese mice. *Food Funct.* **2013**, *4*, 618–626. [CrossRef] [PubMed]

24. Yang, M.Y.; Peng, C.H.; Chan, K.C.; Yang, Y.S.; Huang, C.N.; Wang, C.J. The hypolipidemic effect of hibiscus sabdariffa polyphenols via inhibiting lipogenesis and promoting hepatic lipid clearance. *J. Agric. Food Chem.* **2010**, *58*, 850–859. [CrossRef] [PubMed]

25. Howitz, K.T.; Sinclair, D.A. Xenohormesis: Sensing the chemical cues of other species. *Cell* **2008**, *133*, 387–391. [CrossRef] [PubMed]

26. Lamming, D.W.; Wood, J.G.; Sinclair, D.A. Small molecules that regulate lifespan: Evidence for xenohormesis. *Mol. Microbiol.* **2004**, *53*, 1003–1009. [CrossRef] [PubMed]

27. Falcone Ferreyra, M.L.; Rius, S.P.; Casati, P. Flavonoids: Biosynthesis, biological functions, and biotechnological applications. *Front. Plant Sci.* **2012**, *3*, 222. [CrossRef] [PubMed]

28. Cramer, G.R.; Urano, K.; Delrot, S.; Pezzotti, M.; Shinozaki, K. Effects of abiotic stress on plants: A systems biology perspective. *BMC Plant Biol.* **2011**, *11*, 163. [CrossRef] [PubMed]

29. Hooper, P.L.; Hooper, P.L.; Tytell, M.; Vígh, L. Xenohormesis: Health benefits from an eon of plant stress response evolution. *Cell Stress Chaperones* **2010**, *15*, 761–770. [CrossRef] [PubMed]

30. Pandey, K.B.; Rizvi, S.I. Plant polyphenols as dietary antioxidants in human health and disease. *Oxid. Med. Cell. Longev.* **2009**, *2*, 270–278. [CrossRef] [PubMed]

31. Lima, G.; Vianello, F.; Corrêa, C.; Campos, R.; Borguini, M. Polyphenols in fruits and vegetables and its effect on human health. *Food Nutr. Sci.* **2014**, *5*, 1065–1082. [CrossRef]

32. Manach, C.; Scalbert, A.; Morand, C.; Rémésy, C.; Jiménez, L. Polyphenols: Food sources and bioavailability. *Am. J. Clin. Nutr.* **2004**, *79*, 727–747. [PubMed]

33. Scalbert, A.; Williamson, G. Dietary intake and bioavailability of polyphenols. *J. Nutr.* **2000**, *130*, 2073s–2085s. [PubMed]

34. Devalaraja, S.; Jain, S.; Yadav, H. Exotic fruits as therapeutic complements for diabetes, obesity and metabolic syndrome. *Food Res. Int.* **2011**, *44*, 1856–1865. [CrossRef] [PubMed]

35. Tufts, H.R.; Harris, C.S.; Bukania, Z.N.; Johns, T. Antioxidant and anti-inflammatory activities of kenyan leafy green vegetables, wild fruits, and medicinal plants with potential relevance for kwashiorkor. *Evid. Based Complement Alternat. Med.* **2015**, *2015*, 807158. [CrossRef] [PubMed]

36. Brglez Mojzer, E.; Knez Hrncic, M.; Skerget, M.; Knez, Z.; Bren, U. Polyphenols: Extraction methods, antioxidative action, bioavailability and anticarcinogenic effects. *Molecules* **2016**, *21*, 901. [CrossRef] [PubMed]

37. Rodriguez-Medina, I.C.; Beltran-Debon, R.; Molina, V.M.; Alonso-Villaverde, C.; Joven, J.; Menendez, J.A.; Segura-Carretero, A.; Fernandez-Gutierrez, A. Direct characterization of aqueous extract of hibiscus sabdariffa using HPLC with diode array detection coupled to ESI and ion trap MS. *J. Sep. Sci.* **2009**, *32*, 3441–3448. [CrossRef] [PubMed]

38. Serban, C.; Sahebkar, A.; Ursoniu, S.; Andrica, F.; Banach, M. Effect of sour tea (Hibiscus sabdariffa L.) on arterial hypertension: A systematic review and meta-analysis of randomized controlled trials. *J. Hypertens.* **2015**, *33*, 1119–1127. [CrossRef] [PubMed]

39. Chin, K.L.; Zhen, J.; Qi, Y.; Chin, S.L.; Breithaupt, M.; Wu, Q.L.; Simon, J.; Henson, J.; Ferchaud, V. A comparative evaluation: Phytochemical composition and antioxidant capacity of three roselle (Hibiscus sabdariffa L.) accessions. *Acta. Hortic.* **2016**, *1125*, 99–107. [CrossRef]

40. Sindi, H.A.; Marshall, L.J.; Morgan, M.R.A. Comparative chemical and biochemical analysis of extracts of hibiscus sabdariffa. *Food Chem.* **2014**, *164*, 23–29. [CrossRef] [PubMed]

41. Villani, T.; Juliani, H.R.; Simon, J.E.; Wu, Q.L. Hibiscus sabdariffa: Phytochemistry, quality control and health properties. *ACS Symp. Ser.* **2013**, *1127*, 209–230.

42. Huang, W.Y.; Zhang, H.C.; Liu, W.X.; Li, C.Y. Survey of antioxidant capacity and phenolic composition of blueberry, blackberry, and strawberry in Nanjing. *J. Zhejiang Univ. Sci. B* **2012**, *13*, 94–102. [CrossRef] [PubMed]

43. Bowen-Forbes, C.S.; Zhang, Y.; Nair, M.G. Anthocyanin content, antioxidant, anti-inflammatory and anticancer properties of blackberry and raspberry fruits. *Food Comp. Anal.* **2010**, *23*, 554–560. [CrossRef]

44. Chang, Y.C.; Huang, K.X.; Huang, A.C.; Ho, Y.C.; Wang, C.J. Hibiscus anthocyanins-rich extract inhibited LDL oxidation and oxLDL-mediated macrophages apoptosis. *Food Chem. Toxicol.* **2006**, *44*, 1015–1023. [CrossRef] [PubMed]

45. Tsai, P.J.; McIntosh, J.; Pearce, P.; Camden, B.; Jordan, B.R. Anthocyanin and antioxidant capacity in roselle (Hhibiscus sabdariffa L.) extract. *Food Res. Int.* **2002**, *35*, 351–356. [CrossRef]

46. Mahdavi, S.A.; Jafari, S.M.; Ghorbani, M.; Assadpoor, E. Spray-drying microencapsulation of anthocyanins by natural biopolymers: A review. *Drying Technol.* **2014**, *32*, 509–518. [CrossRef]

47. Avior, Y.; Bomze, D.; Ramon, O.; Nahmias, Y. Flavonoids as dietary regulators of nuclear receptor activity. *Food Funct.* **2013**, *4*, 831–844. [CrossRef] [PubMed]

48. Joven, J.; Micol, V.; Segura-Carretero, A.; Alonso-Villaverde, C.; Menéndez, J.A.; Aragonès, G.; Barrajón-Catalán, E.; Beltrán-Debón, R.; Camps, J.; Cufí, S.; et al. Polyphenols and the modulation of gene expression pathways: Can we eat our way out of the danger of chronic disease? *Crit. Rev. Food Sci. Nutr.* **2014**, *54*, 985–1001. [CrossRef] [PubMed]

49. Ezzat, S.M.; Salama, M.M.; Seif El-Din, S.H.; Saleh, S.; El-Lakkany, N.M.; Hammam, O.A.; Salem, M.B.; Botros, S.S. Metabolic profile and hepatoprotective activity of the anthocyanin-rich extract of Hibiscus sabdariffa calyces. *Pharm. Biol.* **2016**, *54*, 3172–3181. [CrossRef] [PubMed]

50. Borrás-Linares, I.; Herranz-López, M.; Barrajón-Catalán, E.; Arráez-Román, D.; González-Álvarez, I.; Bermejo, M.; Gutiérrez, A.F.; Micol, V.; Segura-Carretero, A. Permeability study of polyphenols derived from a phenolic-enriched hibiscus sabdariffa extract by UHPLC-ESI-UHR-QQ-TOF-MS. *Int. J. Mol. Sci.* **2015**, *16*, 18396–18411. [CrossRef] [PubMed]

51. Muller, B.M.; Franz, G. Chemical structure and biological activity of polysaccharides from Hibiscus sabdariffa. *Planta Med.* **1992**, *58*, 60–67. [CrossRef] [PubMed]

52. Sayago-Ayerdi, S.G.; Arranz, S.; Serrano, J.; Goni, I. Dietary fiber content and associated antioxidant compounds in roselle flower (Hibiscus sabdariffa L.) beverage. *J. Agric. Food Chem.* **2007**, *55*, 7886–7890. [CrossRef] [PubMed]

53. Beltran-Debon, R.; Rodriguez-Gallego, E.; Fernandez-Arroyo, S.; Senan-Campos, O.; Massucci, F.A.; Hernandez-Aguilera, A.; Sales-Pardo, M.; Guimera, R.; Camps, J.; Menendez, J.A.; et al. The acute impact of polyphenols from hibiscus sabdariffa in metabolic homeostasis: An approach combining metabolomics and gene-expression analyses. *Food Funct.* **2015**, *6*, 2957–2966. [CrossRef] [PubMed]

54. Efferth, T.; Koch, E. Complex interactions between phytochemicals. The multi-target therapeutic concept of phytotherapy. *Curr. Drug Targets* **2011**, *12*, 122–132. [CrossRef] [PubMed]

55. Yang, Y.; Zhang, Z.; Li, S.; Ye, X.; Li, X.; He, K. Synergy effects of herb extracts: Pharmacokinetics and pharmacodynamic basis. *Fitoterapia* **2014**, *92*, 133–147. [CrossRef] [PubMed]

56. Gertsch, J. Botanical drugs, synergy, and network pharmacology: Forth and back to intelligent mixtures. *Planta Med.* **2011**, *77*, 1086–1098. [CrossRef] [PubMed]

57. Keith, C.T.; Borisy, A.A.; Stockwell, B.R. Multicomponent therapeutics for networked systems. *Nat. Rev. Drug Discov.* **2005**, *4*, 71–78. [CrossRef] [PubMed]

58. Wagner, H.; Ulrich-Merzenich, G. Synergy research: Approaching a new generation of phytopharmaceuticals. *Phytomedicine* **2009**, *16*, 97–110. [CrossRef] [PubMed]

59. Olsen, E.K.; Søderholm, K.L.; Isaksson, J.; Andersen, J.H.; Hansen, E. Metabolomic profiling reveals the n-acyl-taurine geodiataurine in extracts from the marine sponge Geodia macandrewii (Bowerbank). *J. Nat. Prod.* **2016**, *79*, 1285–1291. [CrossRef] [PubMed]

60. Frank, T.; Janssen, M.; Netzel, M.; Strass, G.; Kler, A.; Kriesl, E.; Bitsch, I. Pharmacokinetics of anthocyanidin-3-glycosides following consumption of Hibiscus sabdariffa L. Extract. *J. Clin. Pharmacol.* **2005**, *45*, 203–210. [CrossRef] [PubMed]

61. Nemeth, K.; Piskula, M.K. Food content, processing, absorption and metabolism of onion flavonoids. *Crit. Rev. Food Sci. Nutr.* **2007**, *47*, 397–409. [CrossRef] [PubMed]

62. Del Mar Contreras, M.; Borras-Linares, I.; Herranz-Lopez, M.; Micol, V.; Segura-Carretero, A. Further exploring the absorption and enterocyte metabolism of quercetin forms in the caco-2 model using NANO-LC-TOF-MS. *Electrophoresis* **2016**, *37*, 998–1006. [CrossRef] [PubMed]

63. Herranz-López, M.; Borrás-Linares, I.; Olivares-Vicente, M.; Gálvez, J.; Segura-Carretero, A.; Micol, V. Correlation between the cellular metabolism of quercetin and its glucuronide metabolite and oxidative stress in hypertrophied 3t3-l1 adipocytes. *Phytomedicine* **2017**, *25*, 25–28. [CrossRef] [PubMed]

64. Bandyopadhyay, P.; Ghosh, A.K.; Ghosh, C. Recent developments on polyphenol-protein interactions: Effects on tea and coffee taste, antioxidant properties and the digestive system. *Food Funct.* **2012**, *3*, 592–605. [CrossRef] [PubMed]

65. Hansawasdi, C.; Kawabata, J.; Kasai, T. A-amylase inhibitors from roselle (Hibiscus sabdariffa L.) tea. *Biosci. Biotechnol. Biochem.* **2000**, *64*, 1041–1043. [CrossRef] [PubMed]

66. Hansawasdi, C.; Kawabata, J.; Kasai, T. Hibiscus acid as an inhibitor of starch digestion in the caco-2 cell model system. *Biosci. Biotechnol. Biochem.* **2001**, *65*, 2087–2089. [CrossRef] [PubMed]

67. Carvajal-Zarrabal, O.; Hayward-Jones, P.M.; Orta-Flores, Z.; Nolasco-Hipolito, C.; Barradas-Dermitz, D.M.; Aguilar-Uscanga, M.G.; Pedroza-Hernandez, M.F.; et al. Effect of hibiscus sabdariffa l. Dried calyx ethanol extract on fat absorption-excretion, and body weight implication in rats. *J. Biomed. Biotechnol.* **2009**, *2009*, 5. [CrossRef] [PubMed]

68. Rodina, A.V.; Severin, S.E. The role of adiponectin in the pathogenesis of the metabolic syndrome and approach to therapy. *Patol. Fiziol. Eksp. Ter.* **2013**, 15–26.

69. Encinar, J.A.; Fernandez-Ballester, G.; Galiano-Ibarra, V.; Micol, V. In silico approach for the discovery of new ppargamma modulators among plant-derived polyphenols. *Drug Des. Devel. Ther.* **2015**, *9*, 5877–5895. [CrossRef] [PubMed]

70. Galiano, V.; Garcia-Valtanen, P.; Micol, V.; Encinar, J.A. Looking for inhibitors of the dengue virus ns5 rna-dependent rna-polymerase using a molecular docking approach. *Drug Des. Devel. Ther.* **2016**, *10*, 3163–3181. [CrossRef] [PubMed]

71. Priscilla, D.H.; Roy, D.; Suresh, A.; Kumar, V.; Thirumurugan, K. Naringenin inhibits alpha-glucosidase activity: A promising strategy for the regulation of postprandial hyperglycemia in high fat diet fed streptozotocin induced diabetic rats. *Chem. Biol. Interact.* **2014**, *210*, 77–85. [CrossRef] [PubMed]

72. Murphy, M.P. How mitochondria produce reactive oxygen species. *Biochem. J.* **2009**, *417*, 1–13. [CrossRef] [PubMed]

73. Ray, P.D.; Huang, B.W.; Tsuji, Y. Reactive oxygen species (ROS) homeostasis and redox regulation in cellular signaling. *Cell. Signal.* **2012**, *24*, 981–990. [CrossRef] [PubMed]

74. Skonieczna, M.; Hejmo, T.; Poterala-Hejmo, A.; Cieslar-Pobuda, A.; Buldak, R.J. Nadph oxidases: Insights into selected functions and mechanisms of action in cancer and stem cells. *Oxid. Med. Cell. Longev.* **2017**, *2017*, 9420539. [CrossRef] [PubMed]

75. Furukawa, S.; Fujita, T.; Shimabukuro, M.; Iwaki, M.; Yamada, Y.; Nakajima, Y.; Nakayama, O.; Makishima, M.; Matsuda, M.; Shimomura, I. Increased oxidative stress in obesity and its impact on metabolic syndrome. *J. Clin. Invest.* **2004**, *114*, 1752–1761. [CrossRef] [PubMed]

76. Yeop Han, C.; Kargi, A.Y.; Omer, M.; Chan, C.K.; Wabitsch, M.; O'Brien, K.D.; Wight, T.N.; Chait, A. Differential effect of saturated and unsaturated free fatty acids on the generation of monocyte adhesion and chemotactic factors by adipocytes: Dissociation of adipocyte hypertrophy from inflammation. *Diabetes* **2010**, *59*, 386–396. [PubMed]

77. Sakurai, T.; Ogasawara, J.; Shirato, K.; Izawa, T.; Oh-Ishi, S.; Ishibashi, Y.; Radak, Z.; Ohno, H.; Kizaki, T. Exercise training attenuates the dysregulated expression of adipokines and oxidative stress in white adipose tissue. *Oxid. Med. Cell. Longev.* **2017**, *2017*, 9410954. [CrossRef] [PubMed]

78. Yang, L.; Gou, Y.; Zhao, T.; Zhao, J.; Li, F.; Zhang, B.; Wu, X. Antioxidant capacity of extracts from calyx fruits of roselle (Hibiscus sabdariffa l.). *Afr. J. Biotechnol.* **2012**, *11*, 4063–4068.
79. Tseng, T.H.; Kao, E.S.; Chu, C.Y.; Chou, F.P.; Lin Wu, H.W.; Wang, C.J. Protective effects of dried flower extracts of hibiscus sabdariffa l. Against oxidative stress in rat primary hepatocytes. *Food Chem. Toxicol.* **1997**, *35*, 1159–1164. [CrossRef]
80. Ajiboye, T.O.; Salawu, N.A.; Yakubu, M.T.; Oladiji, A.T.; Akanji, M.A.; Okogun, J.I. Antioxidant and drug detoxification potentials of hibiscus sabdariffa anthocyanin extract. *Drug Chem. Toxicol.* **2011**, *34*, 109–115. [CrossRef] [PubMed]
81. Farombi, E.O.; Fakoya, A. Free radical scavenging and antigenotoxic activities of natural phenolic compounds in dried flowers of Hibiscus sabdariffa l. *Mol. Nutr. Food Res.* **2005**, *49*, 1120–1128. [CrossRef] [PubMed]
82. Essa, M.M.; Subramanian, P. Hibiscus sabdariffa affects ammonium chloride-induced hyperammonemic rats. *Evid. Based Complement Alternat. Med.* **2007**, *4*, 321–325. [CrossRef] [PubMed]
83. Farombi, E.O.; Ige, O.O. Hypolipidemic and antioxidant effects of ethanolic extract from dried calyx of Hibiscus sabdariffa in alloxan-induced diabetic rats. *Fundam. Clin. Pharmacol.* **2007**, *21*, 601–609. [CrossRef] [PubMed]
84. Ekor, M.; Adesanoye, O.A.; Udo, I.E.; Adegoke, O.A.; Raji, J.; Farombi, E.O. Hibiscus sabdariffa ethanolic extract protects against dyslipidemia and oxidative stress induced by chronic cholesterol administration in rabbits. *Afr. J. Med. Med. Sci.* **2010**, *39*, 161–170. [PubMed]
85. Adeyemi, D.O.; Ukwenya, V.O.; Obuotor, E.M.; Adewole, S.O. Anti-hepatotoxic activities of hibiscus sabdariffa l. In animal model of streptozotocin diabetes-induced liver damage. *BMC Complement Altern. Med.* **2014**, *14*, 277. [CrossRef] [PubMed]
86. Ajiboye, T.O.; Raji, H.O.; Adeleye, A.O.; Adigun, N.S.; Giwa, O.B.; Ojewuyi, O.B.; Oladiji, A.T. Hibiscus sabdariffa calyx palliates insulin resistance, hyperglycemia, dyslipidemia and oxidative rout in fructose-induced metabolic syndrome rats. *J. Sci. Food Agric.* **2016**, *96*, 1522–1531. [CrossRef] [PubMed]
87. Zuniga-Munoz, A.; Guarner, V.; Diaz-Cruz, A.; Diaz-Diaz, E.; Beltran-Rodriguez, U.; Perez-Torres, I. Modulation of oxidative stress in fatty liver of rat with metabolic syndrome by Hibiscus sabdariffa. *Immunol. Endocr. Metab. Agents Med. Chem* **2013**, *13*, 196–205.
88. Greenberg, A.S.; Obin, M.S. Obesity and the role of adipose tissue in inflammation and metabolism. *Am. J. Clin. Nutr.* **2006**, *83*, 461s–465s. [PubMed]
89. Kang, Y.E.; Kim, J.M.; Joung, K.H.; Lee, J.H.; You, B.R.; Choi, M.J.; Ryu, M.J.; Ko, Y.B.; Lee, M.A.; Lee, J.; et al. The roles of adipokines, proinflammatory cytokines, and adipose tissue macrophages in obesity-associated insulin resistance in modest obesity and early metabolic dysfunction. *PLoS ONE* **2016**, *11*, e0154003. [CrossRef] [PubMed]
90. O'Neill, S.; O'Driscoll, L. Metabolic syndrome: A closer look at the growing epidemic and its associated pathologies. *Obes. Rev.* **2015**, *16*, 1–12. [CrossRef] [PubMed]
91. Cancello, R.; Clement, K. Is obesity an inflammatory illness? Role of low-grade inflammation and macrophage infiltration in human white adipose tissue. *Bjog* **2006**, *113*, 1141–1147. [CrossRef] [PubMed]
92. Marseglia, L.; Manti, S.; D'Angelo, G.; Nicotera, A.; Parisi, E.; Di Rosa, G.; Gitto, E.; Arrigo, T. Oxidative stress in obesity: A critical component in human diseases. *Int. J. Mol. Sci.* **2014**, *16*, 378–400. [CrossRef] [PubMed]
93. Kao, E.S.; Hsu, J.D.; Wang, C.J.; Yang, S.H.; Cheng, S.Y.; Lee, H.J. Polyphenols extracted from Hibiscus sabdariffa l. Inhibited lipopolysaccharide-induced inflammation by improving antioxidative conditions and regulating cyclooxygenase-2 expression. *Biosci. Biotechnol. Biochem.* **2009**, *73*, 385–390. [CrossRef] [PubMed]
94. Hashimoto, K.; Ichiyama, T.; Hasegawa, M.; Hasegawa, S.; Matsubara, T.; Furukawa, S. Cysteinyl leukotrienes induce monocyte chemoattractant protein-1 in human monocyte/macrophages via mitogen-activated protein kinase and nuclear factor-kappab pathways. *Int. Arch. Allergy. Immunol.* **2009**, *149*, 275–282. [CrossRef] [PubMed]
95. Hopkins, A.L.; Lamm, M.G.; Funk, J.L.; Ritenbaugh, C. Hibiscus sabdariffa l. In the treatment of hypertension and hyperlipidemia: A comprehensive review of animal and human studies. *Fitoterapia* **2013**, *85*, 84–94. [CrossRef] [PubMed]

96. Herrera-Arellano, A.; Miranda-Sanchez, J.; Avila-Castro, P.; Herrera-Alvarez, S.; Jimenez-Ferrer, J.E.; Zamilpa, A.; Roman-Ramos, R.; Ponce-Monter, H.; Tortoriello, J. Clinical effects produced by a standardized herbal medicinal product of hibiscus sabdariffa on patients with hypertension. A randomized, double-blind, lisinopril-controlled clinical trial. *Planta Med.* **2007**, *73*, 6–12. [CrossRef] [PubMed]

97. Nwachukwu, D.C.; Aneke, E.; Nwachukwu, N.Z.; Obika, L.F.; Nwagha, U.I.; Eze, A.A. Effect of hibiscus sabdariffaon blood pressure and electrolyte profile of mild to moderate hypertensive nigerians: A comparative study with hydrochlorothiazide. *Niger. J. Clin. Pract.* **2015**, *18*, 762–770. [CrossRef] [PubMed]

98. Perez-Torres, I.; Ruiz-Ramirez, A.; Banos, G.; El-Hafidi, M. Hibiscus sabdariffa Linnaeus (Malvaceae), curcumin and resveratrol as alternative medicinal agents against metabolic syndrome. *Cardiovasc. Hematol. Agents Med. Chem.* **2013**, *11*, 25–37. [CrossRef] [PubMed]

99. Alarcon-Aguilar, F.J.; Zamilpa, A.; Perez-Garcia, M.D.; Almanza-Perez, J.C.; Romero-Nunez, E.; Campos-Sepulveda, E.A.; Vazquez-Carrillo, L.I.; Roman-Ramos, R. Effect of Hibiscus sabdariffa on obesity in msg mice. *J. Ethnopharmacol.* **2007**, *114*, 66–71. [CrossRef] [PubMed]

100. Gurrola-Díaz, C.M.; García-López, P.M.; Sánchez-Enríquez, S.; Troyo-Sanromán, R.; Andrade-González, I.; Gómez-Leyva, J.F. Effects of hibiscus sabdariffa extract powder and preventive treatment (diet) on the lipid profiles of patients with metabolic syndrome (mesy). *Phytomedicine* **2010**, *17*, 500–505. [CrossRef] [PubMed]

101. Gosain, S.; Ircchiaya, R.; Sharma, P.C.; Thareja, S.; Kalra, A.; Deep, A.; Bhardwaj, T.R. Hypolipidemic effect of ethanolic extract from the leaves of Hibiscus sabdariffa l. In hyperlipidemic rats. *Acta. Pol. Pharm.* **2010**, *67*, 179–184. [PubMed]

102. Ochani, P.C.; D'Mello, P. Antioxidant and antihyperlipidemic activity of Hibiscus sabdariffa linn. Leaves and calyces extracts in rats. *Indian. J. Exp. Biol.* **2009**, *47*, 276–282. [PubMed]

103. Srivastava, R.A.K.; Pinkosky, S.L.; Filippov, S.; Hanselman, J.C.; Cramer, C.T.; Newton, R.S. Amp-activated protein kinase: An emerging drug target to regulate imbalances in lipid and carbohydrate metabolism to treat cardio-metabolic diseases: Thematic review series: New lipid and lipoprotein targets for the treatment of cardiometabolic diseases. *J. Lipid. Res.* **2012**, *53*, 2490–2514. [CrossRef] [PubMed]

104. Yamauchi, T.; Kamon, J.; Minokoshi, Y.; Ito, Y.; Waki, H.; Uchida, S.; Yamashita, S.; Noda, M.; Kita, S.; Ueki, K.; et al. Adiponectin stimulates glucose utilization and fatty-acid oxidation by activating amp-activated protein kinase. *Nat. Med.* **2002**, *8*, 1288–1295. [CrossRef] [PubMed]

105. Calleri, E.; Pochetti, G.; Dossou, K.S.S.; Laghezza, A.; Montanari, R.; Capelli, D.; Prada, E.; Loiodice, F.; Massolini, G.; Bernier, M.; et al. Resveratrol and its metabolites bind to ppars. *Chembiochem* **2014**, *15*, 1154–1160. [CrossRef] [PubMed]

106. Lu, K.; Han, M.; Ting, H.L.; Liu, Z.; Zhang, D. Scutellarin from scutellaria baicalensis suppresses adipogenesis by upregulating pparalpha in 3T3-L1 cells. *J. Nat. Prod.* **2013**, *76*, 672–678. [CrossRef] [PubMed]

107. Laghezza, A.; Montanari, R.; Lavecchia, A.; Piemontese, L.; Pochetti, G.; Iacobazzi, V.; Infantino, V.; Capelli, D.; De Bellis, M.; Liantonio, A.; et al. On the metabolically active form of metaglidasen: Improved synthesis and investigation of its peculiar activity on peroxisome proliferator-activated receptors and skeletal muscles. *ChemMedChem* **2015**, *10*, 555–565. [CrossRef] [PubMed]

108. Vanden Berghe, W. Epigenetic impact of dietary polyphenols in cancer chemoprevention: Lifelong remodeling of our epigenomes. *Pharmacol. Res.* **2012**, *65*, 565–576. [CrossRef] [PubMed]

109. Li, W.; Guo, Y.; Zhang, C.; Wu, R.; Yang, A.Y.; Gaspar, J.; Kong, A.N.T. Dietary phytochemicals and cancer chemoprevention: A perspective on oxidative stress, inflammation, and epigenetics. *Chem. Res. Toxicol.* **2016**, *29*, 2071–2095. [CrossRef] [PubMed]

110. Thakur, V.S.; Deb, G.; Babcook, M.A.; Gupta, S. Plant phytochemicals as epigenetic modulators: Role in cancer chemoprevention. *AAPS J.* **2014**, *16*, 151–163. [CrossRef] [PubMed]

111. Pan, M.H.; Lai, C.S.; Wu, J.C.; Ho, C.T. Epigenetic and disease targets by polyphenols. *Curr. Pharm. Des.* **2013**, *19*, 6156–6185. [CrossRef] [PubMed]

112. Bhat, M.I.; Kapila, R. Dietary metabolites derived from gut microbiota: Critical modulators of epigenetic changes in mammals. *Nutr. Rev.* **2017**, *75*, 374–389. [CrossRef] [PubMed]

113. Stilling, R.M.; Dinan, T.G.; Cryan, J.F. Microbial genes, brain & behaviour-epigenetic regulation of the gut-brain axis. *Genes Brain Behav.* **2014**, *13*, 69–86. [PubMed]

114. Ruiz, P.A.; Braune, A.; Hölzlwimmer, G.; Quintanilla-Fend, L.; Haller, D. Quercetin inhibits tnf-induced nf-κb transcription factor recruitment to proinflammatory gene promoters in murine intestinal epithelial cells. *J. Nutr.* **2007**, *137*, 1208–1215. [PubMed]

115. Rottiers, V.; Naar, A.M. Micrornas in metabolism and metabolic disorders. *Nat. Rev. Mol. Cell Biol.* **2012**, *13*, 239–250. [CrossRef] [PubMed]

116. Guardiola, S.; Mach, N. Therapeutic potential of hibiscus sabdariffa: A review of the scientific evidence. *Endocrinol. Nutr.* **2014**, *63*, 274–295. [CrossRef] [PubMed]

117. Hardie, D.G. Amp-activated protein kinase: An energy sensor that regulates all aspects of cell function. *Genes Dev.* **2011**, *25*, 1895–1908. [CrossRef] [PubMed]

118. Jimenez-Sanchez, C.; Olivares-Vicente, M.; Rodriguez-Perez, C.; Herranz-Lopez, M.; Lozano-Sanchez, J.; Segura Carretero, A.; Fernandez-Gutierrez, A.; Micol, V. Ampk modulatory activity of olive–tree leaves phenolic compounds: Bioassay-guided isolation on adipocyte model and in silico approach. *PLoS ONE* **2017**. [CrossRef] [PubMed]

119. Hardwicke, M.A.; Rendina, A.R.; Williams, S.P.; Moore, M.L.; Wang, L.; Krueger, J.A.; Plant, R.N.; Totoritis, R.D.; Zhang, G.; Briand, J.; et al. A human fatty acid synthase inhibitor binds beta-ketoacyl reductase in the keto-substrate site. *Nat. Chem. Biol.* **2014**, *10*, 774–779. [CrossRef] [PubMed]

120. Piemontese, L. Plant food supplements with antioxidant properties for the treatment of chronic and neurodegenerative diseases: Benefits or risks? *J. Diet Suppl.* **2017**, *14*, 478–484. [CrossRef] [PubMed]

nutrients

MDPI

Review

Antioxidant Therapeutic Strategies for Cardiovascular Conditions Associated with Oxidative Stress

Jorge G. Farías [1], Víctor M. Molina [2,3], Rodrigo A. Carrasco [4,5], Andrea B. Zepeda [1], Elías Figueroa [1,6], Pablo Letelier [1,7] and Rodrigo L. Castillo [4,8,*]

[1] Departamento de Ingeniería Química, Facultad de Ingeniería y Ciencias, Universidad de La Frontera, Temuco 4780000, Chile; jorge.farias@ufrontera.cl (J.G.F.); andrea.zepeda.p@gmail.com (A.B.Z.); efigueroavillalobos@gmail.com (E.F.); pablolete@gmail.com (P.L.)
[2] Unidad de Cuidados Intensivos, Hospital de Niños Roberto del Río, Santiago 7500922, Chile; victor.molina.cancino@gmail.com
[3] Unidad de Cuidados Intensivos Pediátricos, Hospital Clínico Pontificia Universidad Católica de Chile, Santiago 7500922, Chile
[4] Laboratorio de Investigación Biomédica, Departamento de Medicina Interna, Hospital del Salvador, Santiago 7500922, Chile; r_carrasco_l@yahoo.com
[5] Departamento de Cardiología, Clínica Alemana, Santiago 7500922, Chile
[6] Núcleo de Investigación en Producción Alimentaria, BIOACUI, Escuela de Acuicultura, Universidad Católica de Temuco, Temuco 4780000, Chile
[7] School of Health Sciences, Universidad Católica de Temuco, Temuco 4780000, Chile
[8] Programa de Fisiopatología Oriente, Instituto de Ciencias Biomédicas, Facultad de Medicina, Universidad de Chile, Santiago 7500922, Chile
* Correspondence: rcastillo@med.uchile.cl or rodrigouch@gmail.com

Received: 4 July 2017; Accepted: 18 August 2017; Published: 1 September 2017

Abstract: Oxidative stress (OS) refers to the imbalance between the generation of reactive oxygen species (ROS) and the ability to scavenge these ROS by endogenous antioxidant systems, where ROS overwhelms the antioxidant capacity. Excessive presence of ROS results in irreversible damage to cell membranes, DNA, and other cellular structures by oxidizing lipids, proteins, and nucleic acids. Oxidative stress plays a crucial role in the pathogenesis of cardiovascular diseases related to hypoxia, cardiotoxicity and ischemia–reperfusion. Here, we describe the participation of OS in the pathophysiology of cardiovascular conditions such as myocardial infarction, anthracycline cardiotoxicity and congenital heart disease. This review focuses on the different clinical events where redox factors and OS are related to cardiovascular pathophysiology, giving to support for novel pharmacological therapies such as omega 3 fatty acids, non-selective betablockers and microRNAs.

Keywords: hypoxia; oxidative stress; cardiac tissue; microRNA; omega 3 fatty acids; carvedilol; congenital heart disease

1. Introduction

Hypoxia-related cardiovascular pathologies, such as myocardial infarction, stroke, peripheral vascular disease and renal ischemia, are among the most frequent causes of death and disability [1]. Hypoxia is defined as the threshold where the oxygen concentration is a limiting factor for normal cellular processes, including ATP synthesis. The integration of local responses defines hypoxia as a paradigm of reactions affecting the entire body [2]. Subsequently, an oxygen gradient arises between affected and non-affected tissues, stimulating the migration and proliferation of endothelial cells and fibroblasts, thereby reconstituting normal oxygen supply by increasing perfusion [3]. If this process fails, a prolonged inadequate vascular supply of oxygen leads to chronic hypoxia and can cause chronic diseases. Conversely, some cardiovascular diseases are related to the re-exposure to physiologic or supra-normal oxygen concentrations after a hypoxic insult, which constitute the basis for ischemia–reperfusion injury. Oxidative stress (OS) seems to be a common pathway in several morbid states in which myocardial injury is the primary determinant. In this review, the involvement of oxidative stress in cardiovascular disease is explored and redox-based strategies are reviewed in representative conditions that serve as prototypical models for antioxidant therapies development.

2. Oxidative Stress in Cardiovascular Disease

For decades, oxidative stress (OS) was defined as an imbalance between the production of reactive oxygen species (ROS) and antioxidant defenses in the cell, which leads to oxidative damage of cell structures, including lipids, membranes, proteins and DNA [4]. This process results in inactivation of essential metabolic enzymes and disruption of signal transduction pathways [5]. Now it is clear that differences in subcellular and tissue compartmentalization of ROS contribute to stress responses [6]. It is important to know that ROS is produced as a result of normal cellular metabolism processes [7], while antioxidants eliminate oxidants and repair the damage caused by ROS [8]. Intracellular oxidative stress is produced in normal conditions by the formation of ROS as the result of normal mitochondrial respiration, but also during reperfusion in hypoxic tissue and in association with infection and inflammation [9]. ROS overproduction has been implicated in endothelial injury and extracellular/intracellular OS [10]. Additionally, OS has been implicated in a wide array of diseases such as neurodegenerative disorders, autoimmune diseases, complex lifestyle diseases and cancer [5], and it is central in the pathogenesis of more than 100 inflammatory disorders like periodontitis, diabetes, rheumatoid arthritis, stroke and inflammatory lung diseases [11–14].

Sources of ROS in Cardiovascular Pathologies

Reactive oxygen species is a collective common term that includes highly oxidative radicals such as hydroxyl (OH-) and superoxide ($O_2^{\bullet-}$) radicals, and non-radical species such as hydrogen peroxide (H_2O_2). The term can also include reactive nitrogen species, and both species are normal metabolism byproducts [15,16]. Low concentrations of ROS are required for many cellular processes, whereas overproduction is controlled and/or ameliorated by antioxidants [17].

Mitochondria are a major source for intracellular ROS generation. Within the electron transport chain, a premature leak of a small percentage of electrons to oxygen results in physiological ROS production. Antioxidants in the mitochondria such as superoxide dismutase (SOD)-2 and glutathione rapidly degrade or sequester $O_2^{\bullet-}$ to reduce reactivity. Perhaps due to high concentrations of mitochondria in cardiac tissue, reduced mitochondrial antioxidant capacity results in cardiac dysfunction [18]. Accordingly, mitochondrial damage or dysfunction results in mitochondrial cellular oxidative stress [19].

In addition, ROS have been involved in a wide range of vascular diseases associated with the functional properties of the endothelial cell barrier [20]. It has been proven that oxidized low-density lipoprotein (ox-LDL) increases ROS formation in human umbilical vein endothelial cells (HUVECs) through association with a specific endothelial receptor, which may trigger nuclear factor-κB (NF-κB) activation to induce ROS formation [21]. Among other known sources that increase ROS levels are angiotensin II and uremic toxin indoxyl sulfate-induced endothelial cell dysfunction [22]. Glucose can generate ROS via various pathways including mitochondria, nicotinamide adenine dinucleotide phosphate (NADPH)-oxidase, the sorbitol pathway, activated glycation and the insulin pathway, suggesting that sugars are involved in the development of atherosclerosis, hypertension, peripheral vascular disease, coronary artery disease, cardiomyopathy, heart failure and cardiac arrhythmias, and that these effects of added sugars are mediated through ROS [23]. NADPH oxidase (Nox) signaling is essential for normal physiology, but upregulated and overactive Nox enzymes contribute to oxidative stress and cardiovascular disease [24]. Enzymes of the lipoxygenase (Lox) family catalyze the oxidation of polyunsaturated fatty acids, and increased expression of 5-lipoxygenase (5-LO) has been found in atherosclerotic plaque and abdominal aortic aneurysms [25]. Myeloperoxidase (MPO) is a heme-containing peroxidase expressed in neutrophils and monocytes, and it is believed to produce ROS that contribute to lipid oxidation in atherosclerosis [26].

3. Cardiac Diseases Related to Myocardial Hypoxia/Ischemia

Cardiac hypoxia is the result of a disproportion between oxygen supply and demand. Due to high coronary arteriovenous differences, the myocardium is unable to bring about substantial improvements in oxygen supply by increasing oxygen extraction from the blood. Thus, the only way to meet the higher oxygen demand is by increasing the blood supply. Theoretically, any of the known mechanisms leading to tissue hypoxia could be responsible for a reduction in oxygen supply in the cardiac tissue. However, the most common causes are undoubtedly: (1) ischemic hypoxia (often described as "cardiac ischemia") induced by the reduction or interruption of the coronary blood flow; and (2) systemic (hypoxic) hypoxia ("cardiac hypoxia") characterized by a fall in PO_2 levels in the arterial blood, but with adequate perfusion [27,28]. The effects of ischemia are usually more severe than hypoxia and typically include lactic acidosis due to anaerobic glycolysis, impaired mitochondrial energy production, and cell death [29].

Hypoxia is a major contributor to cardiac pathophysiology, including myocardial infarction, cyanotic congenital heart disease and chronic cor pulmonale [30]. A chronic lack of oxygen leads to an increase in pulmonary vasoconstriction that redistributes pulmonary blood flow from low to high PO_2 regions [31]. However, chronic pulmonary vasoconstriction may result in pulmonary hypertension, increasing the afterload on the right ventricle, which may eventually lead to heart failure.

3.1. Therapeutic Strategies Related with Hypoxia/Ischemia

The most effective treatment to reduce infarct size following hypoxic insult is the re-opening of the culprit-occluded coronary artery by coronary angioplasty or thrombolysis. Adjunctive treatments at reperfusion, such as β-blockers and angiotensin-converting enzyme inhibitors, can ameliorate morbidity and mortality, although not via a reduction in infarct size [32]. Despite these improvements in treatment, mortality remains elevated in high-risk patients and the prevalence of heart failure is also increasing [33], which justifies the search for therapies that would effectively reduce infarct size. These strategies that relate the control of the magnitude and time of tissue hypoxia to functional/structural outcome (e.g., left ventricular function) can be separated into pharmacological or non-pharmacological strategies.

3.1.1. Cardiac Ischemic Preconditioning

Cardiac ischemic preconditioning was first described in 1986 in dogs subjected to brief, intermittent episodes of ischemia. These episodes had a protective effect on the myocardium that was later subjected to a sustained ischemia cycle [34]. This protocol revealed that infarct size was reduced by 75% in dogs exposed to intermittent occlusion of the circumflex artery for 4–5 min followed by 40 min of total occlusion. Accordingly, brief ischemia of the myocardium initiated a cascade of biochemical events in cardiomyocytes that protect the heart muscle during subsequent ischemic insults [35]. The mechanisms underlying these endogenous cardioprotective phenomena are complex in nature and are conventionally divided into triggers, mediators and effectors. Signaling pathways involve surface receptors such as adenosine, bradykinin and opioid-signaling kinases (for example PI3K-Akt-eNOS, JAK–STAT3, PKC, among others) and mitochondrial components (mitochondrial potassium channel dependent on ATP (mKATP), mitochondrial permeability transition pore (mPTP) and protein kinase C) [36]. The cellular and paracrine effects of preconditioning in cardiomyocytes include the induction of angiogenesis and progenitors, stem cell activation and the attenuation of cell death, inflammation, and adverse remodeling [37]. Because classic preconditioning must be implemented before the onset of severe myocardial ischemia, its clinical use is largely restricted to specific situations, such as cardiac surgery, in which the ischemic injury can be anticipated. The non-genetic approach of ischemic preconditioning to enhance cell- and tissue-based therapies has received a great deal of attention in recent years due to its non-invasive, drug-free application. Therefore, the use of pharmacological preconditioning strategies to obtain cardioprotection through classic cellular targets and studies on new targets are currently in development. Clinical use of the ischemic approach has been controversial to date. However, the current design of a reperfusion intervention has provided the basis for new therapy in cardiovascular pharmacology against ischemia–reperfusion injury [38].

3.1.2. Pharmacological Preconditioning

The discovery of ischemic preconditioning has shown the existence of intrinsic systems of cytoprotection, the activation of which can stave off the progression of irreversible tissue damage. Deciphering the molecular mediators that underlie the cytoprotective effects of preconditioning can pave the way to important therapeutic possibilities. Pharmacological activation of critical mediators of ischemic preconditioning would be expected to emulate or even intensify its salubrious effects. Indeed, it is possible to avoid the detrimental effects of ischemia as a maneuver of cardioprotection [36].

Extensive experimental research carried over the past two decades to elucidate the complex signaling pathways underlying the cardioprotective effects of 'conditioning' therapies have led to the discovery of several pharmacological agents able to reproduce the benefits of these mechanical procedures [39]; hence, the term pharmacological preconditioning has been introduced. For example, anesthetic drugs have been shown to possess such properties and, therefore, allowed for anesthetic preconditioning [40]. These denominations refer to cardioprotection triggered by anesthetics administered in this setting (i.e., before prolonged ischemia). In the case of volatile anesthetic conditioning, it is considered less risky and safer in clinical application than its ischemic counterparts, particularly in the diseased myocardium because it does not require the direct administration of the therapeutic agent into the coronary artery; moreover, after inhalation, it provides systemic protection [41]. In the context of protection of the myocardium, nervous system, gut, and kidney beyond the duration of exposure to the volatile agent, anesthetic conditioning may offer additional benefits during the critical postoperative period and may also have a direct impact on long-term prognosis and clinical outcome [42]. Cardioprotective effects of anesthetic agents depend on the interaction of various factors such as administration protocols, choice of specific agents, concomitant use of other drugs, and the variables used to assess myocardial function. There are many confounding factors that limit the applicability of anesthetic conditioning in humans. Some issues to be resolved by future experimental studies are: gender differences, maximum duration of index ischemia, optimal duration and concentration of drug, and potential interactions with other drugs [38].

The mechanistic evidence that supports the cardioprotective effects of antioxidants, and the molecular mechanisms triggered at cardiac and vascular tissue are shown in Figure 1.

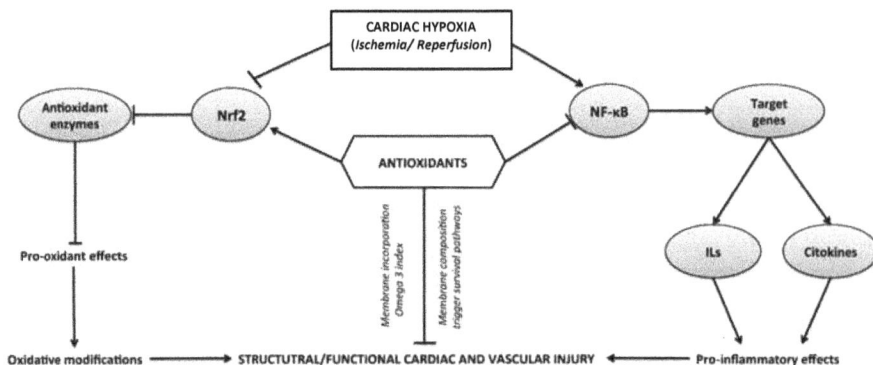

Figure 1. Representation of the cellular and molecular pathways of damage induced following a time course of hypoxia and ischemia–reperfusion cycle in different tissues. The activation of enzymatic and non-enzymatic sources of reactive oxygen species (ROS) is associated with the modulation of redox-sensitive transcriptional factors, such as the activation of nuclear factor (NF)-κB and inhibition of nuclear factor erythroid 2—related factor 2 (Nrf2). Both cellular pathways are implicated in the oxidative modifications or pro-inflammatory effects that can mediate structural or functional cardiovascular impairment.

3.2. Novel Antioxidant-Based Therapies in Ischemia–Reperfusion

3.2.1. Cardiac Preconditioning with Omega 3

In vitro studies, animal experiments, observational studies and randomized clinical trials have examined the cardiovascular effects of seafood consumption and long-chain omega-3 polyunsaturated fatty acids. These types of fatty acids are composed of eicosapentaenoic acid (EPA; 20:5 *n*-3), docosahexaenoic acid (DHA; 22:6 *n*-3) and α-linolenic acid (ALA; 18:3 *n*-3). Alpha-linolenic acid is a plant-derived omega 3 found in a relatively limited set of seeds, nuts and their oils. Alpha-linolenic acid cannot be synthesized in humans and it is an essential dietary fatty acid [43]. There are biochemical pathways to convert ALA to EPA and EPA to DHA, but such endogenous conversion is limited in humans: between 0.2% and 8% of ALA is converted to EPA (with conversion generally higher in women) and 0% to 4% of ALA to DHA (10–14). Thus, tissue and circulating EPA and DHA levels are primarily determined by their direct dietary consumption [44]. Recent studies suggest that the beneficial effects of fish oil are due, in part, to the generation of various free radical-generated non-enzymatic bioactive oxidation products from omega 3, although the specific molecular species responsible for these effects have not been identified. It is of interest to note that the beneficial effects of EPA and DHA could arise from both direct short-term or long-term effects mediated by changes in some intracellular pathways as discussed below. Direct actions of omega 3 have been confirmed by experimental studies of sudden cardiac death in a reliable dog model, showing that these compounds electrically stabilize heart cell membranes through the modulation of the fast voltage-dependent Na^+ currents and the L type Ca^{2+} channels. Derived from this effect, cardiac cells become resistant to arrhythmias [45,46]. Moreover, it has been pointed out that *n*-3 polyunsaturated fatty acids (PUFA) can exert a reversible modulation in the kinetics of several ion channels by binding to specific sites on channel proteins and by non-specifically incorporating them into lipid cell membranes [47]. These changes are consistent with the type of fatty acids incorporated into the cardiac tissue membrane.

With regard to diet supplementation, it has been noted that these rich in omega-3 polyunsaturated fatty acids are associated with decreased incidences of cardiovascular disease. The extent to which these beneficial fats are incorporated into and distributed throughout body tissues is uncertain. In some animals supplemented for more than two weeks with diets enriched with omega 3 as fish oil, the incorporation kinetics of both EPA and DHA have been measured, and this might be associated with the tissue response profile to an injury [48]. In the case of the heart and blood vessels, this would determine the type of hemodynamic response to a pro-inflammatory and pro-oxidant injury. For example, controlled ischemia in an ex vivo model may induce a greater recovery of ventricular function if the supplementation has a high incorporation of DHA in the cardiac tissue [1,49]. These kinetics would also allow a more efficient anti-oxidant and anti-inflammatory response at the heart tissue level [50,51]. However, consumption of dietary flaxseed appears to be an effective means to increase ALA content in body tissues, but the degree will depend upon the tissues examined.

In relation to chronic consumption, it has been reported that omega 3 is selectively incorporated into cardiac cell membranes in a dose-related manner after 8 weeks of supplementation [52]. Also, omega 3 can improve post-ischemic functional recovery in the ex vivo Langendorff perfusion of rat heart, also suggesting the benefit of a diet highly enriched with omega 3 content [53,54]. Regular intake can slow the heart rate, reduce myocardial oxygen consumption, and increase coronary reserve. These properties contribute to preconditioning-like effects of resistance to myocardial infarction and improved post-hypoxic recovery. These effects can be demonstrated in isolated hearts, regardless of the effects of omega 3 on neural or blood parameters. Also, the enrichment of myocardial membranes with omega 3 reduces vulnerability to cardiac arrhythmias, particularly ventricular fibrillation, and attenuates heart failure and cardiac hypertrophy [55].

3.2.2. Antioxidant Mechanism Induced by Omega 3

Experimental evidence demonstrates that antioxidant effects of omega 3 are related to the incorporation of these compounds into the cell membrane and the modulation of redox signaling pathways. In this view, omega 3 supplementation increases the expression and activity of the antioxidants enzymes and attenuates thiobarbituric acid-reactive substances (TBARS) increased in rats ([56], Erdogan et al., 2004). Oxidized omega 3 reacts directly with Keap1, a negative regulator of Nrf2, initiating Keap1 dissociation with Cullin3 and thereby inducing Nrf2-dependent target antioxidant genes such as heme oxygenase-1 ([57], Gao et al., 2007). This omega 3-antioxidant reinforcement is associated with a reduction in the susceptibility of myocytes to ROS-induced IR injury, and to an increase in SOD and GSH-Px expressions ([46], Jahangiri et al., 2006). Animal studies showed that the cardioprotective effects of PUFA can be exerted through the upregulation of heat shock protein 72, a key preconditioning protein, and higher omega 3 content of myocardial membranes, which appears to facilitate the protective response to hypoxic injury ([58], McGuinness et al., 2006). Recently, hearts supplemented with omega 3 showed lower infarct size and a higher left ventricular pressure compared with non-supplemented rats. Hearts in the supplemented group with omega 3 showed lower levels of oxidative stress markers and higher antioxidant activity, decreased activity and NF-κB and Nrf2 activation, compared with the non-supplemented group.

3.2.3. Microribonucleic Acids (miRNAs)

Hypoxia is a powerful stimulus, regulating the expression of a specific subset of microribonucleic acids (miRNAs), called hypoxia-associated miRNA (hypoxamiRs). Accordingly, several hypoxamiRs are involved in cardiac development and ischemic cardiovascular diseases [59,60]. miR-210 represents major hypoxia-inducible miRNAs, and has been studied for its effects such as promotion of cell survival and improvement of heart function [61], possibly via the upregulation of angiogenesis and inhibition of cardiomyocyte apoptosis [62]. Moreover, evidence indicates that some miRNAs may have a direct (synergistic) effect on the expression of the hypoxia-inducible factor (HIF), a key regulator of the transcriptional response to hypoxia, and thus a negative-feedback loop in cells exposed to prolonged hypoxia [63]. On the other hand, the vascular endothelial growth factor (VEGF) and HIF-VEGF pathways are related to the pathophysiology of ischemic vascular disease [64]. These pathways can also be regulated epigenetically by miRNAs. Interestingly, some of these miRNAs, such as miR-206, significantly suppress the viability and invasion and promote the apoptosis of endothelial progenitor cells in coronary artery disease patients by modulating VEGF expression [64]. Accumulating evidence suggests that miRNAs have antiangiogenic properties, for example, miR20b modulates HIF-1a, STAT3 and VEGF expression [65,66], mir-221 and miR-222 target c-kit and eNOS [65–67], miR-320 targets insulin-like growth factor (IGF)-1 in the diabetic endothelial cells [68], and miR-145 reduces microvascular cell migration in vitro [69]. On the other hand, some pro-angiogenic factors have been described, such as mir-296 [70] miR-130a [71], miR-126 [72–74], miR-210 [75,76], let-7f and miR-27b [77].

miRNA as a Therapeutic Strategy

miRNAs are new and powerful candidates for therapeutic intervention against various pathological conditions, including cardiovascular diseases [78]. miR-1 is the most abundant miRNA specific to cardiac and skeletal muscle [79,80]. It is an important regulator of cardiomyocyte growth in the adult heart as well as a pro-apoptotic factor in myocardial ischemia [81], relating to diseases such as hypertrophy, myocardial infarction and arrhythmias [82,83], and can be used as a myocardial infarction biomarker [84]. It has been demonstrated that increased miR-1 levels significantly reduce infarction size [85]. In addition, intracardiac injection of miR-21 along with miR-1 and miR-24 was reported to reduce infarct size in a rat model [81]. The transplantation of mesenchymal stem cell (MSC) delivery overexpressing miR-210 and miR-1 to the infarcted rat hearts improved cardiac function [86]. By contrast, repressed miRNAs have a protective effect, such as the inhibition of microRNA-377 function by antagomir transplantation of MSC. It was observed to reduce fibrosis and improve myocardial function [87], and anti-miR-29 antagomirs significantly reduced myocardial infarct size [88].

Some studies with natural compounds indicate that miRNA expression is modified following administration of antioxidant compounds, functioning as a protective mechanism for cardiovascular disease. For example, miR-126 expression was increased in colon-derived myofibroblast cells upon treatment with wine-derived polyphenols, and that response was associated with a reduced expression of inflammatory genes [89]. Also, therapy with luteolin-7-diglucuronide (L7DG), a naturally-occurring antioxidant found in edible plants, attenuated altered isoproterenol (ISO) expression of miRNAs associated with induced myocardium injury and fibrosis in mice [90]. In addition, H_2O_2 modifies the expression of microRNAs ([91], Wei, Gan et al., 2016). These studies show that miRNAs are modulated by antioxidant compounds and ROS, however they also can be involved in ROS production. For example miR-135a participates in the regulation of H_2O_2-mediated apoptosis in embryonic rat cardiac myoblast cell lines ([92], Liu, Shi et al., 2017). A study in an animal model revealed that miR-133a via targeting uncoupling protein 2 (UCP2) participates in inflammatory bowel disease by altering downstream inflammation, oxidative stress and markers of energy metabolism ([93], Jin, Chen et al., 2017). The silencing of miR-155 modulated stress oxidatives by decreased ROS and promoted nitric oxide (NO) generation in human brain microvessel endothelial cells (HBMECs) via regulating diverse gene expression (caspase-3, ICAM-1, EGFR/ERK/p38 MAPK and PI3K/Akt pathways) ([94], Liu, Pan et al., 2015). The overexpression of miR-103 abrogated cell activity and ROS production induced by H_2O_2, via targeting Bcl2/adenovirus E1B 19 kDa interacting protein 3 (BNIP3) in HUVEC cell lines ([95], Xu, Gao et al., 2015). These studies provide novel clues and potential future therapeutic targets for the treatment cardiovascular disease.

The possible future clinical applications include the use of different strategies based on inducing or repressing miRNA expression [96]. For repression, antagonists (antagomirs) inhibit the activity of specific miRNAs. In contrast, miRNA mimics are used to restore miRNAs that show a loss of function [96,97]. However, one of the major obstacles of this therapy is low stability and bioavailability. Small molecules (less than 50 kDa) are generally filtered by the kidney and subsequently excreted. Furthermore, macrophages and monocytes rapidly remove RNA complexes from the circulatory system. For this reason, different modifications have been generated in microRNAs, which improve the availability and the effect in vivo [97] Also, several strategies (delivery vehicles for miRNA therapeutics) have been used to improve their bioavailability, such as liposomes, polymers (the cationic polymer polyethylene imine), conjugates, exosomes and bacteriophages [98]. miR-122 was the first miRNA that underwent successful clinical trials in hepatitis C virus (HCV)-infected patients [96] and it is a unique miRNA with therapeutic potential both as an anti-mir in combating HCV infection and as a mimic against various liver diseases [99]. Other clinical studies are being carried out, mainly in oncological disease, where the partial responses observed are very promising [100]. In the cardiovascular field, promising preclinical studies suggest that miRNAs could be useful in treating these disorders, although several challenges related to specificity and targeted delivery remain to be overcome.

4. Oxidative Stress and Cardiotoxicity

As an inclusive definition, cardiotoxicity is the development of myocardial injury as a response to an endogenous or exogenous agent. Several forms of cardiotoxicity include oxidative stress among their pathophysiological mechanisms, such as cardiotoxicity in takotsubo cardiomyopathy, cocaine-mediated cardiotoxicity, sepsis-induced myocardial dysfunction, and others [101–103]. Oxidative stress also counts among the mechanisms of cardiotoxicity of some pharmacological treatments in the context of adverse effects, where chemotherapeutic drugs have been studied the most [104].

4.1. Chemotherapy-Induced Cardiotoxicity Secondary to the Collateral Damage of Oxidative Stress on Non-Target Tissues

Injury caused by chemotherapy in non-target tissues often complicates cancer treatments by limiting the use of optimal therapeutic doses of anticancer drugs and impairing the quality of life of patients during and after treatment. Oxidative stress, directly or indirectly caused by chemotherapeutics, is one of the underlying mechanisms of the toxicity of anticancer drugs in non-cancerous tissues, with the effects on the heart being the most studied for their great impact on the survival prognosis of these patients [105]. Many of the most commonly used chemotherapy drugs have been reported to induce oxidative stress, including anthracyclines, cyclophosphamide, cisplatin, busulfan, mitomycin, fluorouracil, cytarabine, and bleomycin [105]. In an exceptional way, some of these chemotherapeutic agents, such as bleomycin, could potentially use the generation of oxidative stress as a mechanism for killing cancer cells [106]. In the vast majority of chemotherapeutic agents the generation of oxidative stress has no role in antineoplastic effectiveness and the induction of oxidative stress occurs in non-target tissues and thereby leads to "normal tissue injury" [105]. In addition, some of the new molecularly targeted therapies in oncology may also induce oxidative stress, such as trastuzumab [107]. Trastuzumab, a monoclonal antibody against ErbB-2 (HER2), induces cardiac dysfunction through the alteration of NADPH oxidase and mitogen-activated protein kinase (MAPK) signaling pathways [108,109]. Furthermore, this alteration of HER2 signaling through NADPH oxidase and MAPKs has been associated with an increase in oxidative stress, leading to dilated cardiomyopathy [109,110].

Thus, although the use of classical chemotherapeutic agents with new molecularly targeted treatments has greatly improved survival rates, leading in some cases to curing the cancer, the oxidative stress-mediated impairment of normal tissues is a significant side effect and decreases patients' quality of life [105], with particular relevance to the cardiovascular effects, since these are the ones that most determine the prognosis of these patients. A better understanding of the mechanisms involved in oxidative heart injury is essential to the design of intervention strategies that will attenuate the cardiotoxicity of chemotherapeutic agents without compromising their anticancer efficacy.

4.2. Mechanisms of Anthracycline-Induced Cardiotoxicity

Anthracyclines tend to accumulate in the mitochondria, which explain their predilection for myocardial tissue that has a high mitochondrial density due to its high metabolic demand [111]. The classic anthracycline cardiotoxicity hypothesis proposed that ROS generation is the initial event that leads to redox imbalance [112]. ROS generation may be due to the anthracycline effects on complex I of the electron transport chain, after reduction of anthracycline ring C, leading to the formation of the free radical semiquinone [113,114]. This radical is relatively stable in an anoxic environment medium, but in normoxic conditions, the unpaired electron is donated to the oxygen, forming superoxide radicals. Complex I, through flavoproteins, catalyzes the formation of the reduced semiquinone radical, first accepting electrons from NADH or NADPH, and then delivering them to anthracyclines. This sequence of reactions is known as "redox cycling" and may be highly detrimental because a relatively small number of anthracyclines is sufficient for the formation of numerous superoxide radicals with the ensuing oxidative injury [115,116].

However, in recent years a new hypothesis has focused on "Top2β" as the initial event of cardiotoxicity [117]. In this sense, from a pathophysiological point of view, this new hypothesis has displaced the generation of reactive oxygen species as the first initiator event at an early stage of damage, putting ROS generation at a later stage or as a downstream event, being a consequence of the alterations produced by the interaction between anthracyclines and Top2β.

Whatever the case, either as an initiating or a downstream event, oxidative stress emerges as an attractive target to prevent cardiotoxicity with antioxidant therapies without compromising the anticancer effectiveness of anthracyclines.

4.3. Preventive Therapies for Anthracycline-Induced Cardiotoxicity with Direct or Indirect Antioxidant Effects

4.3.1. Reactive Oxygen Species Scavengers

Several compounds with antioxidant properties have been studied in vitro with some degree of success [118,119]. Also, in addition to preventing direct damage by oxidative stress, the use of antioxidants could indirectly block the induction of ROS-induced apoptosis. However, although these previous in vitro studies with antioxidants and free radical scavengers have shown an inhibition of myocardial apoptosis [120,121], the success of these interventions in in vivo studies has been less satisfactory. In fact, molecules with antioxidant characteristics such as vitamin E and selenium or nimesulide that had shown good results in vitro showed a poor preventive effect of anthracycline-induced cardiotoxicity in in vivo models [122,123].

This dissociation between the effectiveness of the interventions in vitro and in vivo could be due to the fact that the concentrations of antioxidants that should be reached in myocardial tissue to prevent damage are too high [124]. It is known that the dose indicated to obtain an effective action to eliminate free radicals with vitamin E and C at the myocardial level cannot be obtained with oral contributions; therefore, the findings of a null clinical action of therapies provided by this route are highly predictable.

4.3.2. Prevention of Reactive Oxygen Species Generation

Another antioxidant strategy focused on preventing the generation of ROS may seem more effective than the classic interventions with antioxidant free radical scavengers. In this sense, most of the cardiotoxicity preventive strategies with antioxidant effects currently under study are characterized by their potential mechanism of action being directed toward mitochondrial ROS generation.

Carvedilol

Carvedilol, a competitive blocker of β1, β2 and α1-adrenergic receptors, is widely used clinically for the treatment of heart failure, hypertension and acute myocardial infarction.

One distinctive feature of carvedilol is a potent antioxidant property, which is not shared by other β-adrenergic receptor antagonists [125]. The observation that carvedilol also acts as an inhibitor of mitochondrial complex-I is of importance, since this mitochondrial system was proposed to be involved in the mechanisms of anthracycline-induced cardiotoxicity [126]. Carvedilol is therefore superior to other beta blockers, such as atenolol, in reducing the negative impact induced by doxorubicin on the left ventricular ejection fraction, as well as increased lipoperoxidation in in vivo models [127]. This has been further confirmed by other in vivo studies in which carvedilol decreased both mitochondrial and histopathological cardiac toxicity caused by anthracyclines [128].

A clinical study evaluated the cardioprotective role of carvedilol against the cardiotoxic effect of anthracyclines, determining that the preventive use of carvedilol allowed the preservation of left ventricular systolic function at six months, based on echocardiographic observation variables [129]. Another study, the OVERCOME Trial (prevention of left ventricular dysfunction with enalapril and carvedilol in patients undergoing intensive chemotherapy for the treatment of malignant hemopathies), a randomized clinical trial evaluating a combined treatment of enalapril and carvedilol, was able to prevent reduction in the left ventricular ejection fraction in hemato-oncologic patients who had received intensive chemotherapy. This was an encouraging strategy that should be confirmed by larger clinical trials [130].

Omega-3

Omega-3 represents an attractive preventive strategy due its ability to reduce the susceptibility to oxidative stress injury in myocardial cells. This effect is explained by previously discussed mechanisms, such as increased antioxidant defenses, changes in membrane fluidity and the ability to prevent the release of intracellular calcium in response to oxidative stress [44].

The first clinical studies with omega-3 in oncology patients were aimed at improving the antineoplastic effect of chemotherapy. A study in breast cancer patients receiving anthracycline chemotherapy, which used DHA to improve sensitivity to chemotherapy, found no major adverse side effects [131]. Subsequently, other studies that used omega-3 before or during chemotherapy were able to improve the effectiveness of the chemotherapy [132]. In relation to the potential benefit of a non-ischemic preconditioning that could be offered with the use of omega-3, several animal studies have evaluated the effectiveness of preconditioning as a cardioprotection mechanism against anthracycline-induced cardiotoxicity. It has been established that ischemic preconditioning decreases the cardiotoxicity due to anthracycline, assessed with echocardiographic control of left ventricular function. This type of cardioprotection can also ameliorate the apoptosis rate in cardiomyocytes [133–135].

Specifically, there are not many studies that have evaluated the effect of omega-3 to prevent anthracycline-induced cardiotoxicity. Among the few studies available, an animal model study found that omega-3 did not increase anthracycline cardiotoxicity, which contrasted with a study by Carbone et al., where omega-3 not only failed to prevent cardiotoxicity, but exacerbated anthracycline cardiotoxicity [136,137]. This paradoxical situation has not been evidenced in clinical studies in cancer patients who have used omega-3. This could be explained by the fact that this study was carried out on a sheep model, which involves a totally different metabolic model of the fatty acids compared to human metabolism, because this is an herbivorous animal model.

Finally, a study carried out on rats, a metabolic model more similar to humans with respect to fatty acids, evaluated the cardiac effect of omega-3 on the function and histology in a model of anthracycline heart failure. In this study, the diet with omega-3 supplementation attenuated anthracycline-induced cardiac dysfunction, suggesting that this might be associated with an earlier recovery of cytokine imbalance caused by anthracyclines [138].

4.4. Chemotherapy-Induced Cardiotoxicity of Other Non-Anthracycline Agents

The non-anthracyclines agents may have multiple manifestations of cardiovascular toxicity, including left ventricular dysfunction, hypertension, ischemia and QT prolongation [139]. In addition, its cumulative incidence can be high, as for example, heart failure at 10 years is 32.5% after non-anthracycline chemotherapy regimens, therefore it also represents an important impact on surviving chemotherapy patients [139]. Although the antineoplastic mechanisms of action of these chemotherapeutics agents are multiple, they have in common the induction of oxidative stress in non-target tissues, thereby leading to "normal tissue injury" [140], which occurs for example with cyclophosphamide, cisplatin, busulfan, mitomycin, fluorouracil, cytarabine, and bleomycin [140].Among these agents, 5-fluorouracil and its prodrug capecitabine are some of the most important agents because they are the most common causes of chemotherapy-related cardiotoxicity after the anthracyclines, and depending on the study, its rates of toxicity range from 1 to 19% [141].

The influence of 5-FU treatment on the antioxidant system in myocardial tissue was studied by Durak et al. [142], They found lowered activities of superoxide dismutase and glutathione peroxidase accompanied by higher catalase activity in 5-FU-treated female guinea pigs. The antioxidant potential, defined relative to malondialdehyde (MDA) levels, declined in 5-FU-treated animals compared with controls, while MDA levels increased [142]. However, the role of oxidative stress in the pathogenesis of 5-FU cardiotoxicity is not well-established, and the source of ROS formation remains undefined. In vitro studies of free radical formation and animal studies investigating the role of iron-chelators may confirm or disprove this hypothesis [143].

More important clinical trials with antioxidant therapies in cardiovascular pathologies related with oxidative stress are shown in Table 1.

Table 1. More important clinical trials with antioxidant therapies in cardiovascular pathologies related with oxidative stress.

Trial (N)	Primary End Point	Treatment/Results (R)	Reference
Kalay et al., 2006 ($n = 50$)	Reduction in LVEF between baseline and 6 months	**Treatment:** Carvedilol 12.5 mg daily vs. placebo. The interventions were initiated prior to the start of chemotherapy and maintained for 6 months. **Results:** Placebo: LVEF 68.9%→52.3%, statistically significant reduction ($p < 0.001$); Carvedilol: LVEF 70.5%→69.7%, no statistically significant reduction ($p = 0.3$).	[129]
OVERCOME Trial ($n = 90$)	The primary efficacy endpoint was the absolute change in LVEF between baseline and 6 months	**Treatment:** Enalapril + carvedilol vs. no treatment Medications titrated as tolerated. Medications started within 1 week before the first chemotherapy cycle and continued for 6 months. **Results:** Control: LVEF 64.6%→57.9%, statistically significant reduction, resulting in a −3.1% absolute difference by echocardiography and −3.4% by cardiac magnetic resonance. Enalapril + carvedilol: LVEF 63.3%→62.9%, no statistically significant changes.	[130]
POAF, Chilean Trial ($n = 203$)	Relative risk of reduction the occurrence of electrocardiographically confirmed POAF from surgery until hospital discharge. Follow-up 14 days.	Patients were randomized to placebo or supplementation with n-3 polyunsaturated fatty acids (2 g/day) (EPA: DHA ratio 1:2), vitamin C (1 g/day), and vitamin E (400 IU/day). **Results:** Supplemented group versus placebo group (relative risk (RR): 0.28) ($p < 0.01$).	[144]
OPERA Trial ($n = 564$)	Incident POAF lasting ≥30 s, centrally adjudicated, and confirmed by rhythm strip or electrocardiography	Fish oil or placebo supplementation (10 g over 3 to 5 days, or 8 g over 2 days). **R:** neither higher habitual circulating omega 3 levels, nor achieved levels or changes following short-term fish oil supplementation are associated with risk of POAF.	[145]
The OMEGA-Study in Critical Ill Patients ($n = 272$)	Patients with acute lung injury would increase ventilator-free days to study day 28.	Twice-daily enteral supplementation of n-3 fatty acids, γ-linolenic acid, and antioxidants compared with an isocaloric control. **R:** patients receiving the omega 3 supplement had fewer ventilator-free days (14.0 vs. 17.2; $p = 0.02$) (difference, −3.2 (95% CI, −5.8 to −0.7)) and intensive care unit-free days (14.0 vs. 16.7; $p = 0.04$). The study was stopped	[146]

LVEF, left ventricular ejection fraction; MI, myocardial infarction; EPA, eicosapentaenoic acid; POAF, postoperative atrial fibrillation; DHA, docosaexaenoic acid.

5. Antioxidant-Based Strategies in Congenital Heart Disease Surgical Correction

Cardiopulmonary bypass (CPB) is known to be associated with postoperative organ dysfunction and with a systemic inflammatory response [147]. Oxidative stress is believed to participate in the pathogenesis of this response, thereby being a potential therapeutic target [148,149]. Major inflammation triggers in these patients include blood–CPB circuit contact, translocation of intestinal endotoxin and myocardial ischemia–reperfusion injury, and also surgical trauma, hypothermia and hemolysis [147]. The contact of blood with the cardiopulmonary circuit elicits an inflammatory response that includes neutrophil activation and superoxide production [150] through the well-known NADPH oxidase-mediated oxidative burst.

The patient's ability to withstand the inflammatory and oxidative insult depends on the balance between the magnitude of the pro-inflammatory and pro-oxidative insult and the anti-inflammatory and anti-oxidative response, in addition of course to the previous organ function and comorbidities. In this regard, children, and especially newborns, are a particularly vulnerable population due to distinctive characteristics of congenital heart surgery: (1) longer CPB and circulatory arrest duration; (2) greater CPB circuit surface area/patient size ratio; (3) low antioxidant reserve in patients with cyanotic heart defects that will be abruptly re-oxygenated [151,152]; and (4) reduced antioxidant defenses and higher levels of free iron in newborns and especially in pre-term infants [153]. Indeed, in children the reduction in antioxidant defenses during CPB, measured as the total blood glutathione concentration, is inversely related to the CPB duration, and the resulting lipid peroxidation does not return to normal values at 24 h postoperatively [154]. Temporal analysis of oxidative stress biomarkers in children shows that a reduction of plasma ascorbate levels, an increase in its oxidation product (dehydroascorbic acid) and an increase in plasmatic MDA concentration occur early after cross-clamp removal. This study also showed that peak concentrations of IL-6 and IL-8 occur later (3-12 h post-CPB), and that the loss of ascorbate and cytokine concentration correlates with CPB time [155].

Besides systemic oxidative stress, surgery-related myocardial injury in infants with congenital heart disease is of foremost importance, because these hearts almost never have a normal myocardial function and an absolutely normal anatomy is almost never achieved. In patients under 1 year of age undergoing surgical reparation of ventricular septal defect (VSD) or tetralogy of Fallot (TOF), an increase of TBARS, 8-isoprostane and protein carbonyl concentrations in coronary sinus blood after 1–3–5–10 min following aortic cross-clamp removal has been observed [156]. Accordingly, histopathological analysis of the myocardium in infants dying from heart failure after cardiac surgery show ischemic lesions that colocalize with the expression of 4-hydroxynonenal, a lipid peroxidation marker, which may imply a role of oxidative injury in the pathogenesis of these lesions [157].

Despite the abundant evidence showing the effect of CPB on redox balance, the implications of oxidative stress in the clinical outcome of these children is less clear. In a study that compared children after heart surgery with and without low cardiac output syndrome, no differences were found between these two groups in TBARS and carbonyl serum levels in peripheral blood [158]. This study, however, was very heterogeneous in the types of congenital heart malformations that were included. Also, the use of peripheral blood is a limitation when assessing myocardial oxidative damage. By contrast, children undergoing stage II univentricular staging surgery have increased plasma F_2-isoprostane concentration after CPB that associates with decreased lung compliance, higher PCO_2 and lower pH, which may imply a role of oxidative stress in postoperative behavior in this specific patient subset [159].

Several oxidative stress therapeutic strategies have been studied in these patients:

5.1. Glucocorticoids

Glucocorticoids have been widely used in the past as a way of controlling the inflammatory response, but no clear benefit has been proven [160].

5.2. Antioxidants

Although several antioxidant-based strategies in adults have been evaluated, such studies in children are almost non-existent. A small study evaluated the effect of allopurinol supplementation in TOF surgery, showing less ROS expression in myocardial tissue, but no difference in MDA concentration in coronary sinus blood was observed [161]. Also, the use of curcumin, a potent ROS scavenger, in TOF surgery results in decreased c-Jun N-terminal kinase activity in cardiomyocyte nuclei and less caspase-3 expression, which relates to better right and left ventricle systolic function [162].

5.3. Controlling Oxygen Supply

The use of normoxic instead of hyperoxic CPB in patients with cyanotic heart disease undergoing surgery results in lower plasma troponin I, lower F_2-isoprostane concentration, lower protein S100 release (a marker of cerebral injury) and lower alpha-glutamate transferase release (a marker of hepatic injury) [163]. In addition, controlled re-oxygenation in CPB, instead of standard/hyperoxic CPB, results in lower troponin I, F_2-isoprostanes, IL-6, IL-8, IL-10 and C3-alpha peripheral blood concentrations in single-ventricle patients [164].

5.4. Propofol Anesthesia

Propofol is a widely used anesthetic agent working as a ROS scavenger with a chemical structure that resembles vitamin E. Propofol can reduce post-CPB inflammatory markers and lipoperoxidation in adults [165]. In children undergoing CPB for atrial septal defect and VSD repair, the use of propofol resulted in less extubation time after surgery, in addition to a higher serum SOD activity and a lower serum IL-6 concentration during CPB and after cross-clamp removal. Accordingly, less inflammatory cell infiltration and a lower NF-κB expression was observed in myocardial tissue after CPB [166]. In another study that included several complex congenital heart malformations in children, the use of propofol also resulted in lower IL-6, IL-8 and MDA serum concentrations and higher serum SOD activity after CPB [167].

Definite evidence of the participation of oxidative stress in the postoperative clinical evolution of these patients is still lacking, but the available pathophysiological evidence makes it an attractive therapeutic target. Overall, antioxidant-based strategies have not still been properly explored in CPB-induced myocardial and multiorgan dysfunction in children.

6. Novel Experimental Antioxidant-Based Therapies

Many of the attempts in modulating oxidative stress in several disease models have been futile. The majority of the tested strategies have been based in antioxidant reinforcement by means of antioxidant supplementation. A general theoretical explanation for the failure of these treatments could be non-selective ROS modulation, which may interfere with physiological ROS-dependent signaling pathways [168], or might not be sufficiently effective in the required cellular type or sub-cellular compartment. Conversely, directed redox modulation could be of more success. Several alternative experimental approaches are being developed following this line of thought. Activation of the Nrf2 pathway by derivatives of fumaric acid can result in an antioxidant effect [169]. Targeting ROS-producing enzymes such as Nox ([170], and myeloperoxidase (MPO) [171], may also result in a more selective ROS modulation in pathologic conditions. An even more innovative approach could be to treat the consequences of the oxidative damage, by regaining loss of enzyme function. As an example, drugs with the potential to prevent or revert ROS-induced eNOS uncoupling might be promising in oxidative stress-related diseases [172]. As discussed, oxidative stress is believed to be a part of the pathogenesis of conditions that may require a much more selective approach, such as neurodegenerative diseases. As an example, modulation of the expression of antioxidant enzymes by using viral-delivery gene therapy may prove to be useful in conditions in which a definite cellular type is identified as target [173]. A highly promising field in experimental medicine is the development of cell therapy, and cardiovascular diseases are no exception [174]. Even in this type of approach, antioxidant-based therapies can be of relevance. Stem cells can be preconditioned to exert an antioxidant effect in target tissues, this way improving their viability and working as a vessel for directed antioxidant delivery [175]. Overall, several innovative antioxidant-based strategies are being developed, but their application in the clinical management of cardiovascular diseases still needs to be clarified.

7. Concluding Remarks

A continuously growing body of evidence shows that OS seems to be of key importance in the pathogenesis of several types of cardiovascular diseases. Accordingly, its modulation looks highly attractive from a therapeutic standpoint. Many of the myocardial injuries, such as those seen in ischemia–reperfusion, pharmacologic cardiotoxicity and congenital heart disease surgical correction, are relatively predictable, which offers a unique opportunity for the design of preventive or timely initiated antioxidant-based strategies. These interventions can be as simple as the use of controlled oxygen concentration in CPB or the administration of a specific type of anesthetic, or as complex as the design of multiple-drug protocols. Among the most relevant agents currently being evaluated, omega 3 polyunsaturated fatty acids are promising agents for ischemia preconditioning and anthracycline cardiotoxicity, and carvedilol is a unique beta-blocker with antioxidant properties, besides its role as a first-line heart failure drug. Also, miRNAs are starting to be explored in cardiovascular disease therapeutics. However, proper design clinical trials are still scarce in many of these diseases. It appears that a long road is still ahead before the clinical utility and proper treatment schemes of these agents are properly defined but, so far, the application of redox-based therapeutics in cardiovascular diseases seems most auspicious.

Acknowledgments: Authors of this manuscript are supported by the Fundação de Amparo à Pesquisa do Estado de São Paulo (FAPESP)—Processo Número (Grant No. 2012/50210-9) (J.G.F.) and the National Fund for Scientific and Technological Development (FONDECYT-Chile) (Grant Nos. Regular Competition No. 1151315.) & Santander-Universia Grant, 2015.

Conflicts of Interest: The authors declare no conflict of interest.

References

1. Nichols, M.; Townsend, N.; Scarborough, P.; Rayner, M. Trends in age-specific coronary heart disease mortality in the European Union over three decades: 1980–2009. *Eur. Heart J.* **2013**, *34*, 3017–3027. [CrossRef] [PubMed]
2. Zhao, M.; Zhu, P.; Fujino, M.; Zhuang, J.; Guo, H.; Sheikh, I.; Zhao, L.; Li, X.K. Oxidative Stress in Hypoxic-Ischemic Encephalopathy: Molecular Mechanisms and Therapeutic Strategies. *Int. J. Mol. Sci.* **2016**, *17*, 2078. [CrossRef] [PubMed]
3. Zimmermann, A.S.; Morrison, S.D.; Hu, M.S.; Li, S.; Nauta, A.L.; Sorkin, M.; Meyer, N.P.; Walmsley, G.G.; Maan, Z.N.; Chan, D.A.; et al. Epidermal or dermal specific knockout of PHD-2 enhances wound healing and minimizes ischemic injury. *PLoS ONE* **2014**, *9*, e93373. [CrossRef] [PubMed]
4. Rathore, K.I.; Kerr, B.J.; Redensek, A.; López-Vales, R.; Jeong, S.Y.; Ponka, P.; David, S. Ceruloplasmin protects injured spinal cord from iron-mediated oxidative damage. *J. Neurosci.* **2008**, *28*, 12736–12747. [CrossRef] [PubMed]
5. Bisht, S.; Dada, R. Oxidative stress: Major executioner in disease pathology, role in sperm DNA damage and preventive strategies. *Front. Biosci. (Schol. Ed.)* **2017**, *9*, 420–447. [PubMed]
6. Go, Y.M.; Park, H.; Koval, M.; Orr, M.; Reed, M.; Liang, Y.; Smith, D.; Pohl, J.; Jones, D.P. A key role for mitochondria in endothelial signaling by plasma cysteine/cystine redox potential. *Free Radic. Biol. Med.* **2010**, *48*, 275–283. [CrossRef] [PubMed]
7. Vakifahmetoglu-Norberg, H.; Ouchida, A.T.; Norberg, E. The role of mitochondria in metabolism and cell death. *Biochem. Biophys. Res. Commun.* **2017**, *482*, 426–431. [CrossRef] [PubMed]
8. Halliwell, B. Free radicals and antioxidants: Updating a personal view. *Nutr. Rev.* **2012**, *70*, 257–265. [CrossRef] [PubMed]

9. Sekhon, M.S.; Ainslie, P.N.; Griesdale, D.E. Clinical pathophysiology of hypoxic ischemic brain injury after cardiac arrest: A "two-hit" model. *Crit. Care* **2017**, *21*, 90. [CrossRef] [PubMed]

10. Incalza, M.A.; D'Oria, R.; Natalicchio, A.; Perrini, S.; Laviola, L.; Giorgino, F. Oxidative stress and reactive oxygen species in endothelial dysfunction associated with cardiovascular and metabolic diseases. *Vascul. Pharmacol.* **2017**, S1537–S1891. [CrossRef] [PubMed]

11. Tóthová, L.; Kamodyová, N.; Červenka, T.; Celec, P. Salivary markers of oxidative stress in oral diseases. *Front. Cell. Infect. Microbiol.* **2015**, *5*, 73. [CrossRef] [PubMed]

12. Baltacıoğlu, E.; Akalin, F.A.; Alver, A.; Balaban, F.; Unsal, M.; Karabulut, E. Total antioxidant capacity and superoxide dismutase activity levels in serum and gingival crevicular fluid in post-menopausal women with chronic periodontitis. *J. Clin. Periodontol.* **2006**, *33*, 385–392. [CrossRef] [PubMed]

13. Norouzirad, R.; González-Muniesa, P.; Ghasemi, A. Hypoxia in Obesity and Diabetes: Potential Therapeutic Effects of Hyperoxia and Nitrate. *Oxid. Med. Cell. Longev.* **2017**, *2017*, 5350267. [CrossRef] [PubMed]

14. Zhang, H.; Forman, H.J. 4-hydroxynonenal-mediated signaling and aging. *Free Radic. Biol. Med.* **2017**, *111*, 219–225. [CrossRef] [PubMed]

15. Doshi, S.B.; Khullar, K.; Sharma, R.K.; Agarwal, A. Role of reactive nitrogen species in male infertility. *Reprod. Biol. Endocrinol.* **2012**, *10*, 109. [CrossRef] [PubMed]

16. Valko, M.; Leibfritz, D.; Moncol, J.; Cronin, M.T.; Mazur, M.; Telser, J. Free radicals and antioxidants in normal physiological functions and human disease. *Int. J. Biochem. Cell Biol.* **2007**, *39*, 44–84. [CrossRef] [PubMed]

17. Labunskyy, V.M.; Gladyshev, V.N. Role of reactive oxygen species-mediated signaling in aging. *Antioxid. Redox Signal.* **2013**, *19*, 1362–1372. [CrossRef] [PubMed]

18. Sánchez-Villamil, J.P.; D'Annunzio, V.; Finocchietto, P.; Holod, S.; Rebagliati, I.; Pérez, H.; Peralta, J.G.; Gelpi, R.J.; Poderoso, J.J.; Carreras, M.C. Cardiac-specific overexpression of thioredoxin 1 attenuates mitochondrial and myocardial dysfunction in septic mice. *Int. J. Biochem. Cell Biol.* **2016**, *81*, 323–334. [CrossRef] [PubMed]

19. Kim, H.K.; Han, J. Mitochondria-Targeted Antioxidants for the Treatment of Cardiovascular Disorders. *Adv. Exp. Med. Biol.* **2017**, *982*, 621–646. [CrossRef] [PubMed]

20. Irwin, D.C.; McCord, J.M.; Nozik-Grayck, E.; Beckly, G.; Foreman, B.; Sullivan, T.; White, M.; Crossno, J., Jr.; Bailey, D.; Flores, S.C.; et al. A potential role for reactive oxygen species and the HIF-1alpha-VEGF pathway in hypoxia-induced pulmonary vascular leak. *Free Radic. Biol. Med.* **2009**, *47*, 55–61. [CrossRef] [PubMed]

21. Cominacini, L.; Garbin, U.; Pasini, A.F.; Davoli, A.; Campagnola, M.; Contessi, G.B.; Pastorino, A.M.; Lo Cascio, V. Antioxidants inhibit the expression of intercellular cell adhesion molecule-1 and vascular cell adhesion molecule-1 induced by oxidized LDL on human umbilical vein endothelial cells. *Free Radic. Biol. Med.* **1997**, *22*, 117–127. [CrossRef]

22. Yang, L.L.; Li, D.Y.; Zhang, Y.B.; Zhu, M.Y.; Chen, D.; Xu, T.D. Salvianolic acid A inhibits angiotensin II-induced proliferation of human umbilical vein endothelial cells by attenuating the production of ROS. *Acta Pharmacol. Sin* **2012**, *33*, 41–48. [CrossRef] [PubMed]

23. Prasad, K.; Dhar, I. Oxidative stress as a mechanism of added sugar-induced cardiovascular disease. *Int. J. Angiol.* **2014**, *23*, 217–226. [CrossRef] [PubMed]

24. Brown, D.I.; Griendling, K.K. Regulation of signal transduction by reactive oxygen species in the cardiovascular system. *Circ. Res.* **2015**, *116*, 531–549. [CrossRef] [PubMed]

25. Mehrabian, M.; Allayee, H.; Wong, J.; Shi, W.; Wang, X.P.; Shaposhnik, Z.; Funk, C.D.; Lusis, A.J. Identification of 5-lipoxygenase as a major gene contributing to atherosclerosis susceptibility in mice. *Circ. Res.* **2002**, *91*, 120–126. [CrossRef] [PubMed]

26. Sugamura, K.; Keaney, J.F., Jr. Reactive oxygen species in cardiovascular disease. *Free Radic. Biol. Med.* **2011**, *51*, 978–992. [CrossRef] [PubMed]

27. Clanton, T.L. Hypoxia-induced reactive oxygen species formation in skeletal muscle. *J. Appl. Physiol.* **2007**, *102*, 2379–2388. [CrossRef] [PubMed]

28. Sylvester, J.T.; Shimoda, L.A.; Aaronson, P.I.; Ward, J.P. Hypoxic pulmonary vasoconstriction. *Physiol. Rev.* **2012**, *92*, 367–520. [CrossRef] [PubMed]

29. Essop, M.F. Cardiac metabolic adaptations in response to chronic hypoxia. *J. Physiol.* **2007**, *584*, 715–726. [CrossRef] [PubMed]

30. Budev, M.M.; Arroliga, A.C.; Wiedemann, H.P.; Matthay, R.A. Cor pulmonale: An overview. *Semin. Respir. Crit. Care Med.* **2003**, *24*, 233–244. [CrossRef] [PubMed]

31. Hislop, A.; Reid, L. New findings in pulmonary arteries of rats with hypoxia-induced pulmonary hypertension. *Br. J. Exp. Pathol.* **1976**, *57*, 542–554. [PubMed]

32. Thibault, H.; Angoulvant, D.; Bergerot, C.; Ovize, M. Postconditioning the human heart. *Heart Metab.* **2007**, *37*, 19–22. [CrossRef]

33. Jhund, P.S.; McMurray, J.J. Heart failure after acute myocardial infarction: A lost battle in the war on heart failure? *Circulation* **2008**, *118*, 2019–2021. [CrossRef] [PubMed]

34. Murry, C.E.; Jennings, R.B.; Reimer, K.A. Preconditioning with ischemia: A delay of lethal cell injury in ischemic myocardium. *Circulation* **1986**, *74*, 1124–1136. [CrossRef] [PubMed]

35. Rezkalla, S.H.; Kloner, R.A. Preconditioning in humans. *Heart Fail. Rev.* **2007**, *12*, 201–206. [CrossRef] [PubMed]

36. Rodrigo, R.; Cereceda, M.; Castillo, R.L.; Asenjo, R.; Zamorano, J.; Araya, J.; Castillo-Koch, R.; Espinoza, J.; Larraín, E. Prevention of atrial fibrillation following cardiac surgery: Basis for a novel therapeutic strategy based on non-hypoxic myocardial preconditioning. *Pharmacol. Ther.* **2008**, *118*, 104–127. [CrossRef] [PubMed]

37. Hsiao, S.T.; Dilley, R.J.; Dusting, G.J.; Lim, S.Y. Ischemic preconditioning for cell based therapy and tissue engineering. *Pharmacol. Ther.* **2014**, *142*, 141–153. [CrossRef] [PubMed]

38. Álvarez, P.; Tapia, L.; Mardones, L.A.; Pedemonte, J.C.; Farías, J.G.; Castillo, R.L. Cellular mechanisms against ischemia reperfusion injury induced by the use of anesthetic pharmacological agents. *Chem. Biol. Interact.* **2014**, *218*, 89–98. [CrossRef] [PubMed]

39. Kilić, A.; Huang, C.X.; Rajapurohitam, V.; Madwed, J.B.; Karmazyn, M. Early and transient sodium-hydrogen exchanger isoform 1 inhibition attenuates subsequent cardiac hypertrophy and heart failure following coronary artery ligation. *J. Pharmacol. Exp. Ther.* **2014**, *351*, 492–499. [CrossRef] [PubMed]

40. Stadnicka, A.; Marinovic, J.; Ljubkovic, M.; Bienengraeber, M.W.; Bosnjak, Z.J. Volatile anesthetic-induced cardiac preconditioning. *J. Anesth.* **2007**, *21*, 212–219. [CrossRef] [PubMed]

41. Lemoine, S.; Tritapepe, L.; Hanouz, J.L.; Puddu, P.E. The mechanisms of cardio-protective effects of desflurane and sevoflurane at the time of reperfusion: Anaesthetic post-conditioning potentially translatable to humans? *Br. J. Anaesth.* **2016**, *116*, 456–475. [CrossRef] [PubMed]

42. Lorsomradee, S.; Cromheecke, S.; Lorsomradee, S.; De Hert, S.G. Cardioprotection with volatile anesthetics in cardiac surgery. *Asian Cardiovasc. Thorac. Ann.* **2008**, *16*, 256–264. [CrossRef] [PubMed]

43. Mozaffarian, D.; Aro, A.; Willett, W.C. Health effects of trans-fatty acids: Experimental and observational evidence. *Eur. J. Clin. Nutr.* **2009**, *63*, S5–S21. [CrossRef] [PubMed]

44. Mozaffarian, D.; Wu, J.H. Omega-3 fatty acids and cardiovascular disease: Effects on risk factors, molecular pathways, and clinical events. *J. Am. Coll. Cardiol.* **2011**, *58*, 2047–2067. [CrossRef] [PubMed]

45. Leaf, A.; Kang, J.X.; Xiao, Y.F.; Billman, G.E. Clinical prevention of sudden cardiac death by n-3 polyunsaturated fatty acids and mechanism of prevention of arrhythmias by n-3 fish oils. *Circulation* **2003**, *107*, 2646–2652. [CrossRef] [PubMed]

46. Jahangiri, A.; Leifert, W.R.; Kind, K.L.; McMurchie, E.J. Dietary fish oil alters cardiomyocyte Ca^{2+} dynamics and antioxidant status. *Free Radic. Biol. Med.* **2006**, *40*, 1592–1602. [CrossRef] [PubMed]

47. Xiao, Y.F.; Sigg, D.C.; Leaf, A. The antiarrhythmic effect of n-3 polyunsaturated fatty acids: Modulation of cardiac ion channels as a potential mechanism. *J. Membr. Biol.* **2005**, *206*, 141–154. [CrossRef] [PubMed]

48. Soni, N.K.; Ross, A.B.; Scheers, N.; Savolainen, O.I.; Nookaew, I.; Gabrielsson, B.G.; Sandberg, A.S. Eicosapentaenoic and Docosahexaenoic Acid-Enriched High Fat Diet Delays Skeletal Muscle Degradation in Mice. *Nutrients* **2016**, *8*, 543. [CrossRef] [PubMed]

49. Farías, J.G.; Carrasco-Pozo, C.; Carrasco-Loza, R.; Sepúlveda, N.; Álvarez, P.; Quezada, M.; Quiñones, J.; Molina, V.; Castillo, R.L. Polyunsaturated fatty acid induces cardioprotection against ischemia-reperfusion through the inhibition of NF-kappaB and induction of Nrf2. *Exp. Biol. Med.* **2017**, *242*, 1104–1114. [CrossRef] [PubMed]

50. Herrera, E.A.; Farías, J.G.; González-Candia, A.; Short, S.E.; Carrasco-Pozo, C.; Castillo, R.L. Ω3 Supplementation and intermittent hypobaric hypoxia induce cardioprotection enhancing antioxidant mechanisms in adult rats. *Mar. Drugs* **2015**, *13*, 838–860. [CrossRef] [PubMed]

51. Yamanushi, T.T.; Kabuto, H.; Hirakawa, E.; Janjua, N.; Takayama, F.; Mankura, M. Oral administration of eicosapentaenoic acid or docosahexaenoic acid modifies cardiac function and ameliorates congestive heart failure in male rats. *J. Nutr.* **2014**, *144*, 467–474. [CrossRef] [PubMed]

52. Xie, N.; Zhang, W.; Li, J.; Liang, H.; Zhou, H.; Duan, W.; Xu, X.; Yu, S.; Zhang, H.; Yi, D. α-Linolenic acid intake attenuates myocardial ischemia/reperfusion injury through anti-inflammatory and anti-oxidative stress effects in diabetic but not normal rats. *Arch. Med. Res.* **2011**, *42*, 171–181. [CrossRef] [PubMed]

53. Allahdadi, K.J.; Walker, B.R.; Kanagy, N.L. Augmented endothelin vasoconstriction in intermittent hypoxia-induced hypertension. *Hypertension* **2005**, *45*, 705–709. [CrossRef] [PubMed]

54. Ip, W.T.; McAlindon, A.; Miller, S.E.; Bell, J.R.; Curl, C.L.; Huggins, C.E.; Mellor, K.M.; Raaijmakers, A.J.; Bienvenu, L.A.; McLennan, P.L.; et al. Dietary omega-6 fatty acid replacement selectively impairs cardiac functional recovery after ischemia in female (but not male) rats. *Am. J. Physiol. Heart Circ. Physiol.* **2016**, *311*, H768–H780. [CrossRef] [PubMed]

55. McLennan, P.L.; Owen, A.J.; Slee, E.L.; Theiss, M.L. Myocardial function, ischaemia and n-3 polyunsaturated fatty acids: A membrane basis. *J. Cardiovasc. Med.* **2007**, *8* (Suppl. 1), S15–S18. [CrossRef] [PubMed]

56. Erdogan, H.; Fadillioglu, E.; Ozgocmen, S.; Sogut, S.; Ozyurt, B.; Akyol, O.; Ardicoglu, O. Effect of fish oil supplementation on plasma oxidant/antioxidant status in rats. *Prostaglandins Leukot. Essent. Fatty Acids* **2004**, *71*, 149–152. [CrossRef] [PubMed]

57. Gao, L.; Wang, J.; Sekhar, K.R.; Yin, H.; Yared, N.F.; Schneider, S.N.; Sasi, S.; Dalton, T.P.; Anderson, M.E.; Chan, J.Y.; et al. Novel *n*-3 fatty acid oxidation products activate Nrf2 by destabilizing the association between Keap1 and Cullin3. *J. Biol. Chem.* **2007**, *282*, 2529–2537. [CrossRef] [PubMed]

58. McGuinness, J.; Neilan, T.G.; Sharkasi, A.; Bouchier-Hayes, D.; Redmond, J.M. Myocardial protection using an omega-3 fatty acid infusion: Quantification and mechanism of action. *J. Thorac. Cardiovasc. Surg.* **2006**, *132*, 72–79. [CrossRef] [PubMed]

59. Azzouzi, H.E.; Leptidis, S.; Doevendans, P.A.; De Windt, L.J. HypoxamiRs: Regulators of cardiac hypoxia and energy metabolism. *Trends Endocrinol. Metab.* **2015**, *26*, 502–508. [CrossRef] [PubMed]

60. Greco, S.; Gaetano, C.; Martelli, F. HypoxamiR regulation and function in ischemic cardiovascular diseases. *Antioxid. Redox Signal.* **2014**, *21*, 1202–1219. [CrossRef] [PubMed]

61. Kim, H.W.; Jiang, S.; Ashraf, M.; Haider, K.H. Stem cell-based delivery of Hypoxamir-210 to the infarcted heart: Implications on stem cell survival and preservation of infarcted heart function. *J. Mol. Med.* **2012**, *90*, 997–1010. [CrossRef] [PubMed]

62. Hu, S.; Huang, M.; Li, Z.; Jia, F.; Ghosh, Z.; Lijkwan, M.A.; Fasanaro, P.; Sun, N.; Wang, X.; Martelli, F.; et al. MicroRNA-210 as a novel therapy for treatment of ischemic heart disease. *Circulation* **2010**, *122*, S124–S131. [CrossRef] [PubMed]

63. Bruning, U.; Cerone, L.; Neufeld, Z.; Fitzpatrick, S.F.; Cheong, A.; Scholz, C.C.; Simpson, D.A.; Leonard, M.O.; Tambuwala, M.M.; Cummins, E.P.; et al. MicroRNA-155 promotes resolution of hypoxia-inducible factor 1alpha activity during prolonged hypoxia. *Mol. Cell. Biol.* **2011**, *31*, 4087–4096. [CrossRef] [PubMed]

64. Wang, Y.; Huang, Q.; Liu, J.; Zheng, G.; Lin, L.; Yu, H.; Tang, W.; Huang, Z. Vascular endothelial growth factor A polymorphisms are associated with increased risk of coronary heart disease: A meta-analysis. *Oncotarget* **2017**, *8*, 30539–30551. [CrossRef] [PubMed]

65. Suarez, Y.; Fernandez-Hernando, C.; Pober, J.S.; Sessa, W.C. Dicer dependent microRNAs regulate gene expression and functions in human endothelial cells. *Circ. Res.* **2007**, *100*, 1164–1173. [CrossRef] [PubMed]

66. Poliseno, L.; Tuccoli, A.; Mariani, L.; Evangelista, M.; Citti, L.; Woods, K.; Mercatanti, A.; Hammond, S.; Rainaldi, G. MicroRNAs modulate the angiogenic properties of HUVECs. *Blood* **2006**, *108*, 3068–3071. [CrossRef] [PubMed]

67. Le Sage, C.; Nagel, R.; Egan, D.A.; Schrier, M.; Mesman, E.; Mangiola, A.; Anile, C.; Maira, G.; Mercatelli, N.; Ciafre, S.A.; et al. Regulation of the p27(Kip1) tumor suppressor by miR-221 and miR-222 promotes cancer cell proliferation. *EMBO J.* **2007**, *26*, 3699–3708. [CrossRef] [PubMed]

68. Wang, X.H.; Qian, R.Z.; Zhang, W.; Chen, S.F.; Jin, H.M.; Hu, R.M. MicroRNA-320 expression in myocardial microvascular endothelial cells and its relationship with insulin-like growth factor-1 in type 2 diabetic rats. *Clin. Exp. Pharmacol. Physiol.* **2009**, *36*, 181–188. [CrossRef] [PubMed]

69. Larsson, E.; Fredlund, F.P.; Heldin, J.; Barkefors, I.; Bondjers, C.; Genove, G.; Arrondel, C.; Gerwins, P.; Kurschat, C.; Schermer, B.; et al. Discovery of microvascular miRNAs using public gene expression data: miR-145 is expressed in pericytes and is a regulator of Fli1. *Genome Med.* **2009**, *1*, 108. [CrossRef] [PubMed]

70. Wurdinger, T.; Tannous, B.A.; Saydam, O.; Skog, J.; Grau, S.; Soutschek, J.; Weissleder, R.; Breakefield, X.O.; Krichevsky, A.M. miR-296 regulates growth factor receptor overexpression in angiogenic endothelial cells. *Cancer Cell* **2008**, *14*, 382–393. [CrossRef] [PubMed]

71. Chen, Y.; Gorski, D.H. Regulation of angiogenesis through a microRNA (miR-130a) that down-regulates antiangiogenic homeobox genes GAX and HOXA5. *Blood* **2008**, *111*, 1217–1226. [CrossRef] [PubMed]

72. Fish, J.E.; Santoro, M.M.; Morton, S.U.; Yu, S.; Yeh, R.F.; Wythe, J.D.; Ivey, K.N.; Bruneau, B.G.; Stainier, D.Y.; Srivastava, D. miR-126 regulates angiogenic signaling and vascular integrity. *Dev. Cell* **2008**, *15*, 272–284. [CrossRef] [PubMed]

73. Wang, S.; Aurora, A.B.; Johnson, B.A.; Qi, X.; McAnally, J.; Hill, J.A.; Richardson, J.A.; Bassel-Duby, R.; Olson, E.N. The endothelial-specific microRNA miR-126 governs vascular integrity and angiogenesis. *Dev. Cell* **2008**, *15*, 261–271. [CrossRef] [PubMed]

74. Kuhnert, F.; Mancuso, M.R.; Hampton, J.; Stankunas, K.; Asano, T.; Chen, C.Z.; Kuo, C.J. Attribution of vascular phenotypes of the murine Egfl7 locus to the microRNA miR-126. *Development* **2008**, *135*, 3989–3993. [CrossRef] [PubMed]

75. Pulkkinen, K.; Malm, T.; Turunen, M.; Koistinaho, J.; Yla-Herttuala, S. Hypoxia induces microRNA miR-210 in vitro and in vivo ephrin-A3 and neuronal pentraxin 1 are potentially regulated by miR-210. *FEBS Lett.* **2008**, *582*, 2397–2401. [CrossRef] [PubMed]

76. Fasanaro, P.; D'Alessandra, Y.; Di Stefano, V.; Melchionna, R.; Romani, S.; Pompilio, G.; Capogrossi, M.C.; Martelli, F. MicroRNA-210 modulates endothelial cell response to hypoxia and inhibits the receptor tyrosine kinase ligand Ephrin-A3. *J. Biol. Chem.* **2008**, *283*, 15878–15883. [CrossRef] [PubMed]

77. Kuehbacher, A.; Urbich, C.; Zeiher, A.M.; Dimmeler, S. Role of Dicer and Drosha for endothelial microRNA expression and angiogenesis. *Circ. Res.* **2007**, *101*, 59–68. [CrossRef] [PubMed]

78. Ha, T.Y. MicroRNAs in Human Diseases: From Cancer to Cardiovascular Disease. *Immune Netw.* **2011**, *11*, 135–154. [CrossRef] [PubMed]

79. Mishima, Y.; Stahlhut, C.; Giraldez, A.J. miR-1-2 gets to the heart of the matter. *Cell* **2007**, *129*, 247–249. [CrossRef] [PubMed]

80. Zhao, Y.; Ransom, J.F.; Li, A.; Vedantham, V.; von Drehle, M.; Muth, A.N.; Tsuchihashi, T.; McManus, M.T.; Schwartz, R.J.; Srivastava, D. Dysregulation of cardiogenesis, cardiac conduction, and cell cycle in mice lacking miRNA-1-2. *Cell* **2007**, *129*, 303–317. [CrossRef] [PubMed]

81. Schulte, C.; Zeller, T. microRNA-based diagnostics and therapy in cardiovascular disease—Summing up the facts. *Cardiovasc. Diagn. Ther.* **2015**, *5*, 17–36. [CrossRef] [PubMed]

82. Cai, B.; Pan, Z.; Lu, Y. The roles of microRNAs in heart diseases: A novel important regulator. *Curr. Med. Chem.* **2010**, *17*, 407–411. [CrossRef] [PubMed]

83. Silvestri, P.; Di Russo, C.; Rigattieri, S.; Fedele, S.; Todaro, D.; Ferraiuolo, G.; Altamura, G.; Loschiavo, P. MicroRNAs and ischemic heart disease: Towards a better comprehension of pathogenesis, new diagnostic tools and new therapeutic targets. *Recent Pat. Cardiovasc. Drug Discov.* **2009**, *4*, 109–118. [CrossRef] [PubMed]

84. D'Alessandra, Y.; Devanna, P.; Limana, F.; Straino, S.; Di Carlo, A.; Brambilla, P.G.; Rubino, M.; Carena, M.C.; Spazzafumo, L.; De Simone, M.; et al. Circulating microRNAs are new and sensitive biomarkers of myocardial infarction. *Eur. Heart J.* **2010**, *31*, 2765–2773. [CrossRef] [PubMed]

85. Cheng, Y.; Tan, N.; Yang, J.; Liu, X.; Cao, X.; He, P.; Dong, X.; Qin, S.; Zhang, C. A translational study of circulating cell-free microRNA-1 in acute myocardial infarction. *Clin. Sci.* **2010**, *119*, 87–95. [CrossRef] [PubMed]

86. Huang, F.; Li, M.L.; Fang, Z.F.; Hu, X.Q.; Liu, Q.M.; Liu, Z.J.; Tang, L.; Zhao, Y.S.; Zhou, S.H. Overexpression of MicroRNA-1 improves the efficacy of mesenchymal stem cell transplantation after myocardial infarction. *Cardiology* **2013**, *125*, 18–30. [CrossRef] [PubMed]

87. Wen, Z.; Huang, W.; Feng, Y.; Cai, W.; Wang, Y.; Wang, X.; Liang, J.; Wani, M.; Chen, J.; Zhu, P.; et al. MicroRNA-377 regulates mesenchymal stem cell-induced angiogenesis in ischemic hearts by targeting VEGF. *PLoS ONE* **2014**, *9*, e104666. [CrossRef] [PubMed]

88. Ye, Y.; Hu, Z.; Lin, Y.; Zhang, C.; Perez-Polo, J.R. Downregulation of microRNA-29 by antisense inhibitors and a PPAR-gamma agonist protects against myocardial ischaemia-reperfusion injury. *Cardiovasc. Res.* **2010**, *87*, 535–544. [CrossRef] [PubMed]

89. Angel-Morales, G.; Noratto, G.; Mertens-Talcott, S. Red wine polyphenolics reduce the expression of inflammation markers in human colon-derived CCD-18Co myofibroblast cells: Potential role of microRNA-126. *Food Funct.* **2012**, *3*, 745–752. [CrossRef] [PubMed]

90. Ning, B.B.; Zhang, Y.; Wu, D.D.; Cui, J.G.; Liu, L.; Wang, P.W.; Wang, W.J.; Zhu, W.L.; Chen, Y.; Zhang, T. Luteolin-7-diglucuronide attenuates isoproterenol-induced myocardial injury and fibrosis in mice. *Acta Pharmacol. Sin.* **2017**, *38*, 331–341. [CrossRef] [PubMed]

91. Wei, M.; Gan, L.; Yang, X.; Jiang, X.; Liu, J.; Chen, L. The down-regulation of miR-125b-5p and up-regulation of Smad4 expression in human umbilical vein endothelial cells treated with hydrogen peroxide. *Xi Bao Yu Fen Zi Mian Yi Xue Za Zhi* **2016**, *32*, 1088–1093. [PubMed]

92. Liu, N.; Shi, Y.F.; Diao, H.Y.; Li, Y.X.; Cui, Y.; Song, X.J.; Tian, X.; Li, T.Y.; Liu, B. MicroRNA-135a Regulates Apoptosis Induced by Hydrogen Peroxide in Rat Cardiomyoblast Cells. *Int. J. Biol. Sci.* **2017**, *13*, 13–21. [CrossRef] [PubMed]

93. Jin, X.; Chen, D.; Zheng, R.H.; Zhang, H.; Chen, Y.P.; Xiang, Z. miRNA-133a-UCP2 pathway regulates inflammatory bowel disease progress by influencing inflammation, oxidative stress and energy metabolism. *World J. Gastroenterol.* **2017**, *23*, 76–86. [CrossRef] [PubMed]

94. Liu, Y.; Pan, Q.; Zhao, Y.; He, C.; Bi, K.; Chen, Y.; Zhao, B.; Chen, Y.; Ma, X. MicroRNA-155 Regulates ROS Production, NO Generation, Apoptosis and Multiple Functions of Human Brain Microvessel Endothelial Cells Under Physiological and Pathological Conditions. *J. Cell. Biochem.* **2015**, *116*, 2870–2881. [CrossRef] [PubMed]

95. Xu, M.C.; Gao, X.F.; Ruan, C.; Ge, Z.R.; Lu, J.D.; Zhang, J.J.; Zhang, Y.; Wang, L.; Shi, H.M. miR-103 Regulates Oxidative Stress by Targeting the BCL2/Adenovirus E1B 19 kDa Interacting Protein 3 in HUVECs. *Oxid. Med. Cell. Longev.* **2015**, *2015*, 489647. [CrossRef] [PubMed]

96. Broderick, J.A.; Zamore, P.D. MicroRNA therapeutics. *Gene Ther.* **2011**, *18*, 1104–1110. [CrossRef] [PubMed]

97. Iorio, M.V.; Croce, C.M. MicroRNA dysregulation in cancer: Diagnostics, monitoring and therapeutics. A comprehensive review. *EMBO Mol. Med.* **2012**, *4*, 143–159. [CrossRef] [PubMed]

98. Baumann, V.; Winkler, J. miRNA-based therapies: Strategies and delivery platforms for oligonucleotide and non-oligonucleotide agents. *Future Med. Chem.* **2015**, *6*, 1967–1984. [CrossRef] [PubMed]

99. Thakral, S.; Ghoshal, K. miR-122 is a unique molecule with great potential in diagnosis, prognosis of liver disease, and therapy both as miRNA mimic and antimir. *Curr. Gene Ther.* **2015**, *15*, 142–150. [CrossRef] [PubMed]

100. Reid, G.; Kao, S.C.; Pavlakis, N.; Brahmbhatt, H.; MacDiarmid, J.; Clarke, S.; Boyer, M.; van Zandwijk, N. Clinical development of TargomiRs, a miRNA mimic-based treatment for patients with recurrent thoracic cancer. *Epigenomics* **2016**, *8*, 1079–1085. [CrossRef] [PubMed]

101. Komamura, K.; Fukui, M.; Iwasaku, T.; Hirotani, S.; Masuyama, T. Takotsubo cardiomyopathy: Pathophysiology, diagnosis and treatment. *World J. Cardiol.* **2014**, *6*, 602–609. [CrossRef] [PubMed]

102. Liaudet, L.; Calderari, B.; Pacher, P. Pathophysiological mechanisms of catecholamine and cocaine-mediated cardiotoxicity. *Heart Fail. Rev.* **2014**, *19*, 815–824. [CrossRef] [PubMed]

103. Kakihana, Y.; Ito, T.; Nakahara, M.; Yamaguchi, K.; Yasuda, T. Sepsis-induced myocardial dysfunction: Pathophysiology and management. *J. Intensive Care* **2016**, *4*, 22. [CrossRef] [PubMed]

104. Angsutararux, P.; Luanpitpong, S.; Issaragrisil, S. Chemotherapy-Induced Cardiotoxicity: Overview of the Roles of Oxidative Stress. *Oxid. Med. Cell. Longev.* **2015**, *2015*, 795602. [CrossRef] [PubMed]

105. Chen, Y.; Jungsuwadee, P.; Vore, M.; Butterfield, D.A.; St Clair, D.K. Collateral damage in cancer chemotherapy: Oxidative stress in nontargeted tissues. *Mol. Interv.* **2007**, *7*, 147–156. [CrossRef] [PubMed]

106. Hug, H.; Strand, S.; Grambihler, A.; Galle, J.; Hack, V.; Stremmel, W.; Krammer, P.H.; Galle, P.R. Reactive oxygen intermediates are involved in the induction of CD95 ligand mRNA expression by cytostatic drugs in hepatoma cells. *J. Biol. Chem.* **1997**, *272*, 28191–28193. [CrossRef] [PubMed]

107. Onitilo, A.A.; Engel, J.M.; Stankowski, R.V. Cardiovascular toxicity associated with adjuvant trastuzumab therapy: Prevalence, patient characteristics, and risk factors. *Ther. Adv. Drug Saf.* **2014**, *5*, 154–166. [CrossRef] [PubMed]

108. Nakagami, H.; Takemoto, M.; Liao, J.K. NADPH oxidase-derived superoxide anion mediates angiotensin II-induced cardiac hypertrophy. *J. Mol. Cell. Cardiol.* **2003**, *35*, 851–859. [CrossRef]

109. Cardinale, D.; Colombo, A.; Sandri, M.T.; Lamantia, G.; Colombo, N.; Civelli, M.; Martinelli, G.; Veglia, F.; Fiorentini, C.; Cipolla, C.M. Prevention of high-dose chemotherapy-induced cardiotoxicity in high-risk patients by angiotensin-converting enzyme inhibition. *Circulation* **2006**, *114*, 2474–2781. [CrossRef] [PubMed]

110. Zeglinski, M.; Ludke, A.; Jassal, D.S.; Singal, P.K. Trastuzumab-induced cardiac dysfunction: A "dual-hit". *Exp. Clin. Cardiol.* **2011**, *16*, 70–74. [PubMed]

111. Anderson, A.B.; Arriaga, E.A. Subcellular metabolite profiles of the parent CCRF-CEM and the derived CEM/C2 cell lines after treatment with doxorubicin. *J. Chromatogr. B Anal. Technol. Biomed. Life Sci.* **2004**, *808*, 295–302. [CrossRef] [PubMed]

112. Octavia, Y.; Tocchetti, C.G.; Gabrielson, K.L.; Janssens, S.; Crijns, H.J.; Moens, A.L. Doxorubicin-induced cardiomyopathy: From molecular mechanisms to therapeutic strategies. *J. Mol. Cell. Cardiol.* **2012**, *52*, 1213–1225. [CrossRef] [PubMed]

113. Lebrecht, D.; Kokkori, A.; Ketelsen, U.P.; Setzer, B.; Walker, U.A. Tissue-specific mtDNA lesions and radical-associated mitochondrial dysfunction in human hearts exposed to doxorubicin. *J. Pathol.* **2005**, *207*, 436–444. [CrossRef] [PubMed]

114. Simunek, T.; Stérba, M.; Popelová, O.; Adamcová, M.; Hrdina, R.; Gersl, V. Anthracycline-induced cardiotoxicity: Overview of studies examining the roles of oxidative stress and free cellular iron. *Pharmacol. Rep.* **2009**, *61*, 154–171. [CrossRef]

115. Keizer, H.G.; Pinedo, H.M.; Schuurhuis, G.J.; Joenje, H. Doxorubicin (adriamycin): A critical review of free radical-dependent mechanisms of cytotoxicity. *Pharmacol. Ther.* **1990**, *47*, 219–231. [CrossRef]

116. Doroshow, J. Anthracycline antibiotic-stimulated superoxide; hydrogen peroxide; and hydroxyl radical production by NADH dehydrogenase. *Cancer Res.* **1983**, *43*, 4543–4551. [PubMed]

117. Zhang, S.; Liu, X.; Bawa-Khalfe, T.; Lu, L.-S.; Lyu, Y.L.; Liu, L.F.; Yeh, E.T.H. Identification of the molecular basis of doxorubicin-induced cardiotoxicity. *Nat. Med.* **2012**, *18*, 1639–1642. [CrossRef] [PubMed]

118. DeAtley, S.; Aksenov, M.; Aksenova, M.; Harris, B.; Hadley, R.; Harper, P.; Carney, J.; Butterfield, D. Antioxidants protect against reactive oxygen species associated with adriamycin-treated cardiomyocytes. *Cancer Lett.* **1999**, *136*, 41–46. [CrossRef]

119. Monti, E.; Cova, D.; Guido, E.; Morelli, R.; Oliva, C. Protective effect of the nitroxide tempol against the cardiotoxicity of adriamycin. *Free Radic. Biol. Med.* **1996**, *21*, 463–470. [CrossRef]

120. Galang, N.; Sasaki, H.; Maulik, N. Apoptotic cell death during ischemia/reperfusion and its attenuation by antioxidant therapy. *Toxicology* **2000**, *148*, 111–118. [CrossRef]

121. Ambrosio, G.; Zweier, J.L.; Becker, L.C. Apoptosis is prevented by administration of superoxide dismutase in dogs with reperfused myocardial infarction. *Basic Res. Cardiol.* **1998**, *93*, 94–96. [CrossRef] [PubMed]

122. Van Vleet, J.F.; Ferrans, V.J.; Weirich, W.E. Cardiac disease induced by chronic adriamycin administration in dogs and evaluation of vitamin E and selenium as cardioprotectants. *Am. J. Pathol.* **1980**, *99*, 13–42. [PubMed]

123. Kotsinas, A.; Gorgoulis, V.; Zacharato, P.; Zioris, H.; Triposkiadis, F.; Donta, I.; Kyriakidis, M.; Karayannacos, P.; Kittas, C. Antioxidant agent nimesulid and beta-blocker metoprolol do not exert protective effects against rat mitochondrial DNA alterations in adriamycin-induced cardiotoxicity. *Biochem. Biophys. Res. Commun.* **1999**, *254*, 651–656. [CrossRef] [PubMed]

124. Rodrigo, R.; Prieto, J.C.; Castillo, R. Cardioprotection against ischaemia/reperfusion by vitamins C and E plus *n*-3 fatty acids: Molecular mechanisms and potential clinical applications. *Clin. Sci.* **2013**, *124*, 1–15. [CrossRef] [PubMed]

125. Yue, T.L.; McKenna, P.J.; Ruffolo, R.R.; Feuerstein, G. Carvedilol: A new alfa-adrenoceptor antagonist and vasodilator antihypertensive drug; inhibits superoxide release from human neutrophils. *Eur. J. Pharmacol.* **1992**, *214*, 277–280. [PubMed]

126. Oliveira, P.J.; Gonçalves, L.; Monteiro, P.; Providencia, L.A.; Moreno, A.J. Are the antioxidant properties of carvedilol important for the protection of cardiac mitochondria? *Curr. Vasc. Pharmacol.* **2005**, *3*, 147–158. [CrossRef] [PubMed]

127. Matsui, H.; Morishima, I.; Numaguchi, Y.; Toki, Y.; Okumura, K.; Hayakawa, T. Protective effects of carvedilol against doxorubicin-induced cardiomyopathy in rats. *Life Sci.* **1999**, *65*, 1265–1274. [CrossRef]

128. Santos, D.L.; Moreno, A.J.M.; Leino, R.L.; Froberg, M.K.; Wallace, K.B. Carvedilol protects against doxorubicin-induced mitochondrial cardiomyopathy. *Toxicol. Appl. Pharmacol.* **2002**, *185*, 218–227. [CrossRef] [PubMed]

129. Kalay, N.; Basar, E.; Ozdogru, I.; Er, O.; Cetinkaya, Y.; Dogan, A.; Inanc, T.; Oguzhan, A.; Eryol, N.K.; Topsakal, R.; et al. Protective effects of carvedilol against anthracycline-induced cardiomyopathy. *J. Am. Coll. Cardiol.* **2006**, *48*, 2258–2262. [CrossRef] [PubMed]

130. Bosch, X.; Rovira, M.; Sitges, M.; Domènech, A.; Ortiz-Pérez, J.T.; de Caralt, T.M.; Morales-Ruiz, M.; Perea, R.J.; Monzó, M.; Esteve, J. Enalapril and carvedilol for preventing chemotherapy-induced left ventricular systolic dysfunction in patients with malignant hemopathies: The OVERCOME trial (preventiOn of left Ventricular dysfunction with Enalapril and carvedilol in patients submitted to intensive ChemOtherapy for the treatment of Malignant hemopathies). *J. Am. Coll. Cardiol.* **2013**, *61*, 2355–2362. [CrossRef] [PubMed]

131. Bougnoux, P.; Hajjaji, N.; Ferrasson, M.N.; Giraudeau, B.; Couet, C.; Le Floch, O. Improving outcome of chemotherapy of metastatic breast cancer by docosahexaenoic acid: A phase II trial. *Br. J. Cancer* **2009**, *101*, 1978–1985. [CrossRef] [PubMed]

132. Biondo, P.D.; Brindley, D.; Sawyer, M.; Fielda, C. The potential for treatment with dietary long-chain polyunsaturated *n*-3 fatty acids during chemotherapy. *J. Nutr. Biochem.* **2008**, *19*, 787–796. [CrossRef] [PubMed]

133. Hydock, D.S.; Lien, C.Y.; Schneider, C.M.; Hayward, R. Exercise preconditioning protects against doxorubicin-induced cardiac dysfunction. *Med. Sci. Sports Exerc.* **2008**, *40*, 808–817. [CrossRef] [PubMed]

134. Schjøtt, J.; Olsen, H.; Berg, K.; Jynge, P. Pretreatment with ischaemia attenuates acute epirubicin-induced cardiotoxicity in isolated rat hearts. *Pharmacol. Toxicol.* **1996**, *78*, 381–386. [CrossRef] [PubMed]

135. Wonders, K.Y.; Hydock, D.S.; Schneider, C.M.; Hayward, R. Acute exercise protects against doxorubicin cardiotoxicity. *Integr. Cancer Ther.* **2008**, *7*, 147–154. [CrossRef] [PubMed]

136. Germain, E.; Bonnet, P.; Aubourg, L.; Grangeponte, M.C.; Chajès, V.; Bougnoux, P. Anthracycline-induced cardiac toxicity is not increased by dietary omega-3 fatty acids. *Pharmacol. Res.* **2003**, *47*, 111–117. [CrossRef]

137. Carbone, A.; Psaltis, P.J.; Nelson, A.J.; Metcalf, R.; Richardson, J.D.; Weightman, M.; Thomas, A.; Finnie, J.W.; Young, G.D.; Worthley, S.G. Dietary omega-3 supplementation exacerbates left ventricular dysfunction in an ovine model of anthracycline-induced cardiotoxicity. *J. Card. Fail.* **2012**, *18*, 502–511. [CrossRef] [PubMed]

138. Teng, L.L.; Shao, L.; Zhao, Y.T.; Yu, X.; Zhang, D.F.; Zhang, H. The beneficial effect of *n*-3 polyunsaturated fatty acids on doxorubicin-induced chronic heart failure in rats. *J. Int. Med. Res.* **2010**, *38*, 940–948. [CrossRef] [PubMed]

139. Curigliano, G.; Cardinale, D.; Dent, S.; Criscitiello, C.; Aseyev, O.; Lenihan, D.; Cipolla, C.M. Cardiotoxicity of anticancer treatments: Epidemiology, detection, and management. *CA Cancer J. Clin.* **2016**, *66*, 309–325. [CrossRef] [PubMed]

140. Moding, E.J.; Kastan, M.B.; Kirsch, D.G. Strategies for optimizing the response of cancer and normal tissues to radiation. *Nat. Rev. Drug Discov.* **2013**, *12*, 526–542. [CrossRef] [PubMed]

141. Broder, H.; Gottlieb, R.A.; Lepor, N.E. Chemotherapy and cardiotoxicity. *Rev. Cardiovasc. Med.* **2008**, *9*, 75–83. [PubMed]

142. Durak, I.; Karaayvaz, M.; Kavutcu, M.; Cimen, M.Y.; Kaçmaz, M.; Büyükkoçak, S.; Oztürk, H.S. Reduced antioxidant defense capacity in myocardial tissue from guinea pigs treated with 5-fluorouracil. *J. Toxicol. Environ. Health A* **2000**, *59*, 585–589. [PubMed]

143. Polk, A.; Vistisen, K.; Vaage-Nilsen, M.; Nielsen, D.L. A systematic review of the pathophysiology of 5-fluorouracil-induced cardiotoxicity. *BMC Pharmacol. Toxicol.* **2014**, *15*, 47. [CrossRef] [PubMed]

144. Rodrigo, R.; Korantzopoulos, P.; Cereceda, M.; Asenjo, R.; Zamorano, J.; Villalabeitia, E.; Baeza, C.; Aguayo, R.; Castillo, R.; Carrasco, R.; et al. A randomized controlled trial to prevent post-operative atrial fibrillation by antioxidant reinforcement. *J. Am. Coll. Cardiol.* **2013**, *62*, 1457–1465. [CrossRef] [PubMed]

145. Wu, J.H.; Marchioli, R.; Silletta, M.G.; Macchia, A.; Song, X.; Siscovick, D.S.; Harris, W.S.; Masson, S.; Latini, R.; Albert, C.; et al. Plasma phospholipid omega-3 fatty acids and incidence of postoperative atrial fibrillation in the OPERA trial. *J. Am. Heart Assoc.* **2013**, *2*, e000397. [CrossRef] [PubMed]

146. Schott, C.K.; Huang, D.T. Omega-3 fatty acids, γ-linolenic acid, and antioxidants: Immunomodulators or inert dietary supplements? *Crit. Care* **2012**, *16*, 325. [CrossRef] [PubMed]

147. Laffey, J.; Boylan, J.; Cheng, D. The systemic inflammatory response to cardiac surgery: Implications for the anesthesiologist. *Anesthesiology* **2002**, *97*, 215–252. [PubMed]

148. McDonald, C.; Fraser, J.; Coombes, J.; Fung, Y. Oxidative stress during extracorporeal circulation. *Eur. J. Cardiothorac. Surg.* **2014**, *46*, 937–943. [CrossRef] [PubMed]

149. Zakkar, M.; Guida, G.; Suleiman, M.; Angelini, G. Cardiopulmonary bypass and oxidative stress. *Oxid. Med. Cell. Longev.* **2015**, *2015*, 189863. [CrossRef] [PubMed]

150. Kawahito, K.; Kobayashi, E.; Ohmori, M.; Harada, K.; Kitoh, Y.; Fujimura, A.; Fuse, K. Enhanced responsiveness of circulatory neutrophils after cardiopulmonary bypass: Increased aggregability and superoxide producing capacity. *Artif. Organs* **2000**, *24*, 37–42. [CrossRef] [PubMed]

151. Pirinccioglu, A.; Alyan, O.; Kizil, G.; Kangin, M.; Beyazit, N. Evaluation of oxidative stress in children with congenital heart defects. *Pediatr. Int.* **2012**, *54*, 94–98. [CrossRef] [PubMed]

152. Ercan, S.; Cakmak, A.; Kösecik, M.; Erel, O. The oxidative state of children with cyanotic and acyanotic congenital heart disease. *Anadolu Kardiyol. Derg.* **2009**, *9*, 486–490. [PubMed]

153. Gitto, E.; Pellegrino, S.; D'Arrigo, S.; Barberi, I.; Reiter, R. Oxidative stress in resuscitation and in ventilation of newborns. *Eur. Respir. J.* **2009**, *34*, 1461–1469. [CrossRef] [PubMed]

154. Gil-Gómez, R.; Blasco-Alonso, J.; Castillo Martín, R.; González-Correa, J.A.; de la Cruz-Cortés, J.P.; Milano-Manso, G. Oxidative stress response after cardiac surgery in children. *Rev. Esp. Cardiol.* **2015**, *68*, 256–257. [CrossRef] [PubMed]

155. Christen, S.; Finckh, B.; Lykkesfeldt, J.; Gessler, P.; Frese-Schaper, M.; Nielsen, P.; Schmid, E.R.; Schmitt, B. Oxidative stress precedes peak systemic inflammatory response in pediatric patients undergoing cardiopulmonary bypass operation. *Free Radic. Biol. Med.* **2005**, *38*, 1323–1332. [CrossRef] [PubMed]

156. Sznycer-Taub, N.; Mackie, S.; Peng, Y. Myocardial Oxidative Stress in Infants Undergoing Cardiac Surgery. *Pediatr. Cardiol.* **2016**, *37*, 746–750. [CrossRef] [PubMed]

157. Oliveira, M.; Floriano, E.; Mazin, S.; Martinez, E.Z.; Vicente, W.V.; Peres, L.C.; Rossi, M.A.; Ramos, S.G. Ischemic myocardial injuries after cardiac malformation repair in infants may be associated with oxidative stress mechanisms. *Cardiovasc. Pathol.* **2011**, *20*, e43–e52. [CrossRef] [PubMed]

158. Manso, P.; Carmona, F.; Dal-Pizzol, F.; Petronilho, F.; Cardoso, F.; Castro, M.; Carlotti, A.P. Oxidative stress markers are not associated with outcomes after pediatric heart surgery. *Paediatr. Anaesth.* **2013**, *23*, 188–194. [CrossRef] [PubMed]

159. Albers, E.; Donahue, B.; Milne, G.; Saville, B.R.; Wang, W.; Bichell, D.; McLaughlin, B. Perioperative plasma F(2)-Isoprostane levels correlate with markers of impaired ventilation in infants with single-ventricle physiology undergoing stage 2 surgical palliation on the cardiopulmonary bypass. *Pediatr. Cardiol.* **2012**, *33*, 562–568. [CrossRef] [PubMed]

160. Whitlock, R.; Devereaux, P.; Teoh, K.; Lamy, A.; Vincent, J.; Pogue, J.; Paparella, D.; Sessler, D.I.; Karthikeyan, G.; Villar, J.C.; et al. Methylprednisolone in patients undergoing cardiopulmonary bypass (SIRS): A randomised, double-blind, placebo-controlled trial. *Lancet* **2015**, *386*, 1243–1253. [CrossRef]

161. Rachmat, F.; Rachmat, J.; Sastroasmoro, S.; Wanandi, S. Effect of allopurinol on oxidative stress and hypoxic adaptation response during surgical correction of tetralogy of fallot. *Acta Med. Indones.* **2013**, *45*, 94–100. [PubMed]

162. Sukardi, R.; Sastroasmoro, S.; Siregar, N.; Djer, M.M.; Suyatna, F.D.; Sadikin, M.; Ibrahim, N.; Rahayuningsih, S.E.; Witarto, A.B. The role of curcumin as an inhibitor of oxidative stress caused by ischaemia re-perfusion injury in tetralogy of Fallot patients undergoing corrective surgery. *Cardiol. Young* **2016**, *26*, 431–438. [CrossRef] [PubMed]

163. Caputo, M.; Mokhtari, A.; Rogers, C.; Panayiotou, N.; Chen, Q.; Ghorbel, M.T.; Angelini, G.D.; Parry, A.J. The effects of normoxic versus hyperoxic cardiopulmonary bypass on oxidative stress and inflammatory response in cyanotic pediatric patients undergoing open cardiac surgery: A randomized controlled trial. *J. Thorac. Cardiovasc. Surg.* **2009**, *138*, 206–214. [CrossRef] [PubMed]

164. Caputo, M.; Mokhtari, A.; Miceli, A. Controlled reoxygenation during cardiopulmonary bypass decreases markers of organ damage, inflammation, and oxidative stress in single-ventricle patients undergoing pediatric heart surgery. *J. Thorac. Cardiovasc. Surg.* **2014**, *148*, 792–801. [CrossRef] [PubMed]

165. Zhang, S.H.; Wang, S.Y.; Yao, S.L. Antioxidative effect of propofol during cardiopulmonary bypass in adults. *Acta Pharmacol. Sin.* **2004**, *25*, 334–340. [PubMed]

166. Xia, W.; Liu, Y.; Zhou, Q.; Tang, Q.; Zou, H. Protective effect of propofol and its relation to postoperation recovery in children undergoing cardiac surgery with cardiopulmonary bypass. *Pediatr. Cardiol.* **2011**, *32*, 940–946. [CrossRef] [PubMed]

167. Xia, W.; Liu, Y.; Zhou, Q.; Tang, Q.; Zou, H. Comparison of the effects of propofol and midazolam on inflammation and oxidase stress in children with congenital heart disease undergoing cardiac surgery. *Yonsei Med. J.* **2011**, *52*, 326–332. [CrossRef] [PubMed]

168. Dao, V.; Casas, A.; Maghzal, G.; Seredenina, T.; Kaludercic, N.; Robledinos-Anton, N.; Di Lisa, F.; Stocker, R.; Ghezzi, P.; Jaquet, V.; et al. Pharmacology and Clinical Drug Candidates in Redox Medicine. *Antioxid. Redox Signal.* **2015**, *23*, 1113–1129. [CrossRef]

169. Linker, R.; Lee, D.; Ryan, S.; van Dam, A.; Conrad, R.; Bista, P.; Zeng, W.; Hronowsky, X.; Buko, A.; Chollate, S.; et al. Fumaric acid esters exert neuroprotective effects in neuroinflammation via activation of the Nrf2 antioxidant pathway. *Brain* **2011**, *134*, 678–692. [CrossRef] [PubMed]

170. Stielow, C.; Catar, R.; Muller, G.; Wingler, K.; Scheurer, P.; Schmidt, H.; Morawietz, H. Novel Nox inhibitor of oxLDL-induced reactive oxygen species formation in human endothelial cells. *Biochem. Biophys. Res. Commun.* **2006**, *344*, 200–205. [CrossRef] [PubMed]

171. Jucaite, A.; Svenningsson, P.; Rinne, J.; Cselényi, Z.; Varnäs, K.; Johnström, P.; Amini, N.; Kirjavainen, A.; Helin, S.; Minkwitz, M.; et al. Effect of the myeloperoxidase inhibitor AZD3241 on microglia: A PET study in Parkinson's disease. *Brain* **2015**, *138*, 2687–2700. [CrossRef] [PubMed]

172. Wenzel, P.; Schulz, E.; Oelze, M.; Müller, J.; Schuhmacher, S.; Alhamdani, M.; Debrezion, J.; Hortmann, M.; Reifenberg, K.; Fleming, I.; et al. AT1-receptor blockade by telmisartan upregulates GTP-cyclohydrolase I and protects eNOS in diabetic rats. *Free Radic. Biol. Med.* **2008**, *45*, 619–626. [CrossRef] [PubMed]

173. Nanou, A.; Higginbottom, A.; Valori, C.; Wyles, M.; Ning, K.; Shaw, P.; Azzouz, M. Viral delivery of antioxidant genes as a therapeutic strategy in experimental models of amyotrophic lateral sclerosis. *Mol. Ther.* **2013**, *21*, 1486–1496. [CrossRef] [PubMed]

174. Fischbach, M.; Bluestone, J.; Lim, W. Cell-based therapeutics: The next pillar of medicine. *Sci. Transl. Med.* **2013**, *5*, 179ps7. [CrossRef] [PubMed]

175. Kim, H.; Yun, J.; Kwon, S. Therapeutic Strategies for Oxidative Stress-Related Cardiovascular Diseases: Removal of Excess Reactive Oxygen Species in Adult Stem Cells. *Oxid. Med. Cell. Longev.* **2016**, *2016*, 2483163. [CrossRef] [PubMed]

MDPI

Article

The Effect of Consumption of Citrus Fruit and Olive Leaf Extract on Lipid Metabolism

Nicola Merola [1], Julián Castillo [2], Obdulio Benavente-García [2], Gaspar Ros [1] and Gema Nieto [1,*]

[1]   Department of Food Technology, Nutrition and Food Science, Veterinary Faculty University of Murcia, Campus de Espinardo, 30100 Espinardo, Murcia, Spain; nicola_merola@um.es (N.M.); gros@um.es (G.R.)
[2]   Research and Development Department of Nutrafur-Frutarom Group, Camino Viejo de Pliego s/n, 80320 Alcantarilla, Murcia, Spain; j.castillo@Nutrafur.com (J.C.); o.benavente@Nutrafur.com (O.B.-G.)
*   Correspondence: gnieto@um.es; Tel./Fax: +34-868-889-624

Received: 5 July 2017; Accepted: 19 September 2017; Published: 26 September 2017

Abstract: Citrus fruit and olive leaves are a source of bioactive compounds such as biophenols which have been shown to ameliorate obesity-related conditions through their anti-hyperlipidemic and anti-inflammatory effect, and by regulating lipoproteins and cholesterol body levels. Citrolive™ is a commercial extract which is obtained from the combination of both citrus fruit and olive leaf extracts; hence, it is hypothesised that Citrolive™ may moderate metabolic disorders that are related to obesity and their complications. Initially, an in vitro study of the inhibition of pancreatic lipase activity was made, however, no effect was found. Both preliminary and long-term evaluations of Citrolive™ on lipid metabolism were conducted in an animal model using Wistar rats. In the preliminary in vivo screening, Citrolive™ was tested on postprandial plasma triglyceride level after the administration of an oil emulsion, and a significant reduction in postprandial triacylglycerol (TAG) levels was observed. In the long-term study, Citrolive™ was administered for 60 days on Wistar rats that were fed a high-fat diet. During the study, several associated lipid metabolism indicators were analysed in blood and faeces. At the end of the experiment, the livers were removed and weighed for group comparison. Citrolive™ treatment significantly reduced the liver-to-body-weight ratio, as supported by reduced plasma transaminases compared with control, but insignificantly reduced plasma low density lipoprotein (LDL) and postprandial TAG plasma levels. In addition, faecal analysis showed that the treatment significantly increased total cholesterol excretion. On the other hand, no effect was found on faecal TAG and pancreatic lipase in vitro. In conclusion, treatment ameliorates liver inflammation symptoms that are worsened by the effects of high fat diet.

Keywords: olive and citrus extract; cholesterol; flavonoids; phenolics; oleuropein; FOT—Fat Oral Test-; weight control

---

1. Introduction

Obesity is a worldwide metabolic dysfunction that is characterised by an accumulation of excessive amounts of body fat and it is associated with the onset of several pathological conditions such as type 2 diabetes, coronary heart disease, steatosis, and dyslipidemia [1,2]. In today's increasingly overweight society, the problems that are associated with excess caloric intake are well recognised. In animal models, excessive consumption of dietary fat causes a strong inhibition of lipogenesis by altering both blood hypertriglyceridemia and hepatic lipid levels [3]. These physiological effects are further exacerbated by uncontrolled diabetes mellitus, obesity, and a sedentary lifestyle [4]. In order to treat obesity, several therapeutic strategies have been developed to fight the worldwide epidemic. One of these is focused on the inhibition of the pancreatic lipase (PL) enzyme [5]; another, by decreasing blood cholesterol and triacylglycerol (TAG) levels; and finally, another strategy focuses on blood lipoprotein level management [6].

In this sense, there is a tendency towards searching for new extracts with biologically active components and potential health-promoting properties, including the potential to prevent obesity [6,7]. Phenolics and biophenols are compounds which are formed during plant secondary metabolism and are widely present in the plant kingdom. They are distributed into several classes, i.e., flavonoids (flavanones, flavones, flavonols, isoflavones, flavan-3-ols, anthocyanins, etc.), lignans, stilbenes, terpenoids, iridoids, caffeoyl compounds, and other phenolic derivatives, which are distributed in plants and food of plant origin [8].

The extract presented in this study is called 'Citrolive™' and it is obtained from the combination of olive iridoids (oleuropein family) and citrus flavonoids. These chemical classes are being increasingly studied for their influence on lipid metabolism. For example, the European Foods Safety Authority (EFSA) has issued a scientific opinion stating that the consumption of biophenols from olives can be advertised with claims regarding the protection of low density lipoprotein (LDL) particles from oxidative damage, as well as the maintenance of a normal concentration of blood high density lipoprotein (HDL) and cholesterol [9]. In addition, it has been reported that olive extracts from the leaf and fruit have been shown to possess utility as an obesity management tool [10]. Regarding citrus fruits, it has been suggested that flavonoids, the main compounds that are present in citrus, are associated with a reduced risk of cardiovascular disease and possess anti-inflammatory proprieties [11]. In addition, it has been observed that flavonoids possess health-promoting benefits, improving cholesterol levels in rats that were fed high-fat diets [12]. Although studies on flavonoids and biophenols have demonstrated their bioactivity in the management of lipid metabolism, no studies have been carried out to test their chemical combination on TAG, lipoprotein, or lipid metabolism management. This research may lend further support for the employment of the extract as an obesity prevention tool.

Taking into account the health benefits of the compounds that are present in the extract, and the lack of information about the effects of Citrolive™ on lipid metabolism, in the present study, two independent studies were carried out: a short-term experiment and a long-term experiment. The former included a pancreatic lipase inhibition activity test in vitro, and a postprandial TAG level test in vivo. The latter evaluated the chronic administration effects of Citrolive™ over 60 days. Wistar rats were employed in both experiments as a biomodel. In the long-term study, they were fed with a high-fat diet in order to induce obesity. Finally, several clinical parameters related to obesity status were evaluated.

In order to elucidate these findings, the aim of the present study was to ascertain the effect of a combination of citrus fruit and olive leaf extract consumption on lipid metabolism in rats with a high-fat diet with induced obesity in order to identify the effect of Citrolive™ intake on the management of lipid metabolism.

2. Material and Methods

2.1. Reagents and Chemicals

Virgin olive oil was obtained from a local market and was employed without further purification (Hacendado, Spain); egg yolk lecithin was obtained from Sigma-Aldrich (St. Louis, MO, USA), capillary tubes from Sarsted (CB 300-microvette, Germany), and isofluorane (Baxter, Germany). Two commercial presentations of orlistats were used in the experiment: a pharmaceutical orlistat presentation from Alli® (Spain) was used in oral tests in rats, while the compound that was obtained from Sigma (O4139, St. Louis, MO, USA) was used in the in vitro pancreatic lipase activity test. High-performance liquid chromatography (HPLC) standards (i.e., naringin, neohesperidin, oleuropein, and hydroxytyrosol) were obtained from Extrasynthèse (Genay, France); all of the reagents that were employed in the analyses were of HPLC grade and were supplied by Sigma (Madrid, Spain).

2.2. Citrus Fruit and Olive Leaf Extract (Citrolive™)

Citrolive™, the natural extract that is used in this study, was obtained from Nutrafur S.A-Frutarom Group (Murcia, Spain) and was used without further processing. All of the other chemicals that were used in the study were of the highest commercial grade available.

2.3. High-Performance Liquid Chromatography (HPLC) Conditions

The HPLC equipment that was used was a Hewlett-Packard Series HP 1100 that was equipped with a diode array detector. The stationary phase was a C_{18}LiChrospher 100 analytical column (250 × 4 mm i.d.) with a particle size of 5 nm (Merck, Darmstadt, Germany) thermostated at 25 °C.

For the elucidation and quantification of bioactive compounds in the Citrolive™, the extract was dissolved in dimethylsulfoxide (DMSO) at a ratio of 5 mg/mL and this solution was filtered through a 0.45-nm nylon membrane. The flow rate was 1 mL/min and the absorbance changes were monitored simultaneously at 280 and 340 nm. The mobile phases for chromatographic analysis were as follows: (A) acetic acid/water (2.5:97.5) and (B) acetonitrile. A linear gradient was run from 95% (A) and 5% (B) to 75% (A) and 25% (B) during the first 20 min; which changed to 50% each of (A) and (B) after another 20 min (40 min in total); which, after 10 more minutes, changed to 20% (A) and 80% (B) (50 min in total), and finally equilibrated over the last 10 min (60 min in total) to the initial composition.

2.4. Short-Term Study In Vitro (PL Inhibition) and In Vivo (Fat Oral Test)

2.4.1. Short-Term Study: Citrolive™ Inhibition of Pancreatic Lipase Activity (PL) In Vitro Assay

The amount of inhibition of lipase activity was determined using a Bioteck plate reader by measuring the amount of 4-metyhl-umbelliferone-oleate (4-MU-oleate) product that was released by porcine PL. The method that was employed was adapted from Nakai et al. [13] with small modifications. Briefly, 25 μL of Citrolive™ or orlistat were dissolved in DMSO, and 25 μL of 0.1 mM 4-MU-oleate was dissolved in 100 μL 13 mM tris-HCl, 150 mM NaCl, and 1.3 mM CaCl2 at a pH of 7.8. The reaction was started by adding 50 μL of fresh porcine pancreatic lipase (50 U/mL). The 96-well microplate was read at 25 °C for 10 min at 20-s intervals at an excitation of 355 nm and an emission of 460 nm. The 50% inhibitory concentration (IC50) of each test sample was obtained from the least-squares regression line plots of the logarithm of the sample concentrations ($X = \log[X]$) versus the normalized pancreatic lipase activity (%). The experiment was carried out in triplicate.

2.4.2. Short-Term Study: Citrolive™ Postprandial TAG Levels (Fat Oral Test) In Vivo Assay

Animals

Eighteen 8-week-old male Wistar rats, weighing 250–300 g, were provided by the Animal Research Centre of Murcia University. The rats were maintained under controlled temperature (22 °C), air humidity (60 ± 5%), and light–dark cycle (12-h) conditions for 2 weeks (adaptation period) with free access to the standard laboratory diet (standard laboratory diet, Rodent Maintenance Diet, RMD, Teklad Global 14% Protein Rodent Maintenance diet, Harland Laboratories) and water ad libitum before starting the experiment. Animals were randomly divided into three groups (Control, Orlistat, and Citrolive™ diet) and placed individually in metabolic cages under the same environmental conditions described above (Figure 1).

Experimental Design

The postprandial triglyceride level's procedure was carried out according Merola et al. [14]. Animals were divided into three groups (six in each group) and fasted overnight (Figure 1). Afterwards, each group was orally administered an emulsion with different dietary compositions. The group 'Control' was administered a "vehicle control diet" that consisted of 50% olive oil (w/v) and 50% physiological saline water (w/v) with 3% lecithin (w/v). The second group, 'Orlistat' [15], was treated with a "vehicle control diet" plus orlistat, which was solubilised prior the emulsification step at 60 mg/rat. The third group, 'Citrolive', was administered a "vehicle control diet" plus Citrolive™ extract, which was solubilised prior to the emulsification step at a concentration of 250 mg/kg body weight. Every diet was emulsified with a polytron at 18,000 rpm and then orally administered to the rats via an intragastric tube at a concentration of 5 mL/kg body weight, always within 30 min from preparation. Blood samples were collected via a tail vein incision before the oral administration, and at 30, 60, 120, and 180 min thereafter. The procedure was carried out under isofluorane anaesthesia (3% *w/v*), which was used only during the blood collection procedure. Finally, the blood was centrifuged so that plasma and TAG levels could be measured with a commercial colorimetric assay (Olympus triglyceride OSR6133) in an Olympus AU600 autoanalyser (Olympus, Tokyo, Japan). The technique is based on a series of combined enzymatic reactions with the formation of a product with a maximum absorbance at 500 nm. Finally, the TAG in the samples was measured proportionally with the rise of the absorbance until 520/600 nm. Plasma TAG was expressed incrementally from the baseline. Incremental areas under the response curves (AUC) during the whole time under study were calculated using the trapezoidal rule, with fasting levels as the baseline. The animal study was carried out under appropriate guidelines and was approved by the Bioethics Committee of Murcia University (authorization number: CEEA-572).

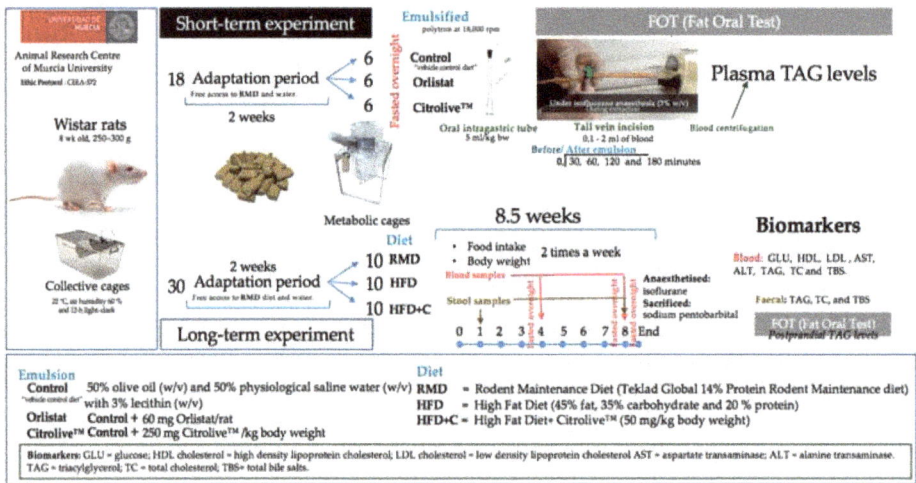

Figure 1. Study design of short- and long-term experiments to evaluate the effect of the consumption of citrus fruit and olive leaf extract on lipid metabolism.

2.5. Long-Term Study: Citrolive™ Chronic Administration during 60 Days in Wistar Rats

2.5.1. Animals

Thirty 8-week-old male Wistar rats, weighing 250–300 g, were provided by the Animal Research Centre of Murcia University. The rats were maintained under controlled temperature (22 °C), air humidity (60% ± 5%), and light–dark cycle conditions (12-h each) for 2 weeks before starting the experiment (Figure 1). Water and RMD were provided ad libitum. In order to mimic overweight-obesity, animals were fed with hypercaloric diet (high-fat diet, HFD, RMD with 45% fat, 35% carbohydrate, and 20% protein, D12451 research diet; New Brunswick, USA). Animals were divided into three groups (10 rats each): the first group was fed with RMD, the second with a HFD, and the third with the HFD plus Citrolive™ extract (60 mg/kg body weight, the same dose as that of orlistat) (HFD + C). The animal study was carried out under appropriate guidelines and was approved by the Bioethics Committee of Murcia University as mentioned above.

2.5.2. Experimental Design

The experimental period was 8.5 weeks, and during this time, the animals were given free access to feed and drink. Food intake and body weight were measured twice a week (Figure 1). Blood samples were collected before the experiment and at 30 and 60 days after overnight fasting. Stool samples were collected at 7 and 60 days. At the end of the experiment, all rats were deprived of food overnight, anaesthetised with isoflurane, and sacrificed using an intraperitoneal injection of sodium pentobarbital. Livers were collected from the 30 animals as biological samples and weighed with an analytical balance. Livers were immediately cut into small pieces and then frozen with liquid nitrogen. Liver samples were stored at $-80\,^\circ$C until the analytical procedures were carried out.

2.6. Anthropometrical Analysis

Body length (nose-to-anus length) was determined in all of the groups at the beginning and at the end of the experiments. The measurements were made on anaesthetized rats (3% w/v isofluorane). The body weight and body length were used in order to determine the body mass index according to the following formula: Body mass index (BMI) = Body weight (g)/Length^2 (cm^2).

2.7. Blood Sampling and Analytical Biomarkers

Blood samples were transferred into heparin-containing tubes. Plasma was obtained by centrifugation (3000 rpm, 10 min, 4 $^\circ$C). Glucose (GLU), total triglycerides, total cholesterol, HDL cholesterol and LDL cholesterol, and the activity of aspartate transaminase (AST) and alanine transaminase (ALT) enzymes were analysed in plasma samples using an automatic analyser (AU 600 Olympus Life, Hamburg, Germany). All analyses were performed in triplicate. The liver-to-body-weight ratio was calculated according to the following formula: Liver weight (g)/Body weight (g).

2.8. Faecal Extraction and Faeces Measurements and Analyses

Faecal fat extraction was adapted from Argmann et al. [16]. Briefly, the rats were placed in clean cages containing a metal floor grid instead of bedding. Faeces were collected over a 24-h period in parallel with a food intake measurement in order to determine the fat balance (lipid intake and output). The collected faeces were dried for one hour in a thermostatic oven at 70 $^\circ$C and weighed using an analytical balance. For the extraction, a 2:1 chloroform/methanol solution was added to 500 mg of dry stools and homogenised for ~30 s at a high speed in a polytron-type homogeniser. Then, methanol was added and the tubes were centrifuged for 15 min at 750× g, room temperature. After removing the supernatant chloroform, 0.73% NaCl solution was added and the samples were centrifuged again for another 3 min at 750× g, room temperature. The top phase was discarded and the phase interface was washed three times adding a chloroform/methanol/NaCl solution. Finally, the samples were evaporated to dryness under a steady stream of nitrogen and were then re-suspended in deionised water. Faecal TAG, TC, and TBS values were taken after 5 and 60 days from the beginning of the experiment and were analysed with the same reagents that were used for plasma analysis.

2.9. Postprandial TAG Levels (Oral Fat Test) In Vivo Assay

Postprandial TAG levels were assessed according to the same protocol that was used in the preliminary assay with small modifications. In order to evaluate the postprandial TAG systemic level curve at 30 and 60 days, the first group, HDF ($n = 6$), was administered the vehicle control diet as described in the preliminary study assay; the other group, HDF + C ($n = 6$), was administered the vehicle control diet plus Citrolive™ extract and the animals that did not receive the vehicle control or the treatment were administered with physiological saline. Afterwards, the vehicle control, vehicle control plus extract, and physiological saline were orally administered via an intragastric tube at 5 mL/kg body weight. Blood samples were obtained by tail incision at 120 and 180 min after the administration of the emulsion.

2.10. Statistical Analysis

The statistical analyses in the classical biomarkers were carried out using GraphPad statistical software, and the results are expressed as the mean ± SD (standard deviation). A two-way analysis of variance (ANOVA) followed by Dunnett's post-hoc tests were applied to determine the differences for all variables among the groups (RMD mean as control). A paired Student's t test was also carried out to ascertain the significant differences of means in clinical analyses of blood and faeces between groups (HFD and HFD + C) at each point in time. The threshold p value chosen for statistical significance was $p < 0.05$.

3. Results

3.1. Polyphenolic Distribution: HPLC Analysis

The phenolic compounds in Citrolive™ were screened and their contents were analysed by HPLC. The abundance (absolute content, on an as is basis, w/w) of the main compounds in Citrolive™ has been fully described in Table 1. Five groups of compounds are principally present in Citrolive™: from olive leaves origin: secoiridoids (oleuropein), phenolics (hydroxytyrosol, tyrosol, vanillic acid, and caffeic acid), and polyphenols (verbascoside), and from citrus origin: flavanones (naringin, neohesperidin, neoeriocitrin, and hesperidin) and flavones (luteolin-7-glucoside, apigenin-7-glucoside, diosmetin-7-glucoside, luteolin, and diosmetin).

Table 1. Absolute content (% w/w, according to the corresponding standards) of the main bioactive compounds classified by main chemical family and plant origin. Other bioactives are trace, less than 0.1%.

Olive Leave					Citrus		
Secoiridoids		Phenolics		Polyphenols	Flavonoid Flavanones		Flavonoid Flavones
Oleuropein 15.74		Hydroxytyrosol	0.84	Verbascoside 0.62	Naringin	3.89	Apigenin 7-O-glucoside 0.82
		Vanillic acid	0.35		Neohesperidin	1.93	Luteolin 7-O-glucoside 0.52
		Tyrosol	0.24		Hesperidin	0.61	Diosmetin 7-O-glucoside 0.31
		Caffeic acid	0.21		Neoeriocitrin	0.41	Luteolin 0.15

3.2. Short-Term Study: Citrolive™ Inhibition of Pancreatic Lipase Activity (PL) In Vitro Assay

The IC_{50} of Citrolive™ on PL inhibition is expressed in µg/mL and the related confidence interval (CI) is the following: while orlistat had an IC_{50} of 0.04 µg/mL (CI of 0.04 to 0.05), Citrolive™ had an IC_{50} of 70.92 (CI of 51.41 to 97.82). The solvent (DMSO) alone had no effect on PL activity.

3.3. Short-Term Study: Citrolive™ Postprandial TAG Levels (FOT—Fat Oral Test) In Vivo Assay

The postprandial TAG response is shown in Figure 2a. The plasma TAG level in the control group exhibited a constant increase until reaching a postprandial peak at 120 min; then it started decreasing at 180 min. Orlistat started to increase the TAG level after 60 min, and the latter remained constant until 120 min had passed; it reached significant values at 120 min, reducing the TAG level by almost 50% ($p < 0.05$). Citrolive™ extract significantly reduced TAG increment levels ($p < 0.05$) at all of the time points that were taken into account (except at 30 min). The AUC (area under curve) for TAG response is reported in Figure 2b. Citrolive™ had a major effect towards reducing plasma TAG levels ($p < 0.01$). This effect was even stronger than the one that was produced by orlistat ($p < 0.05$).

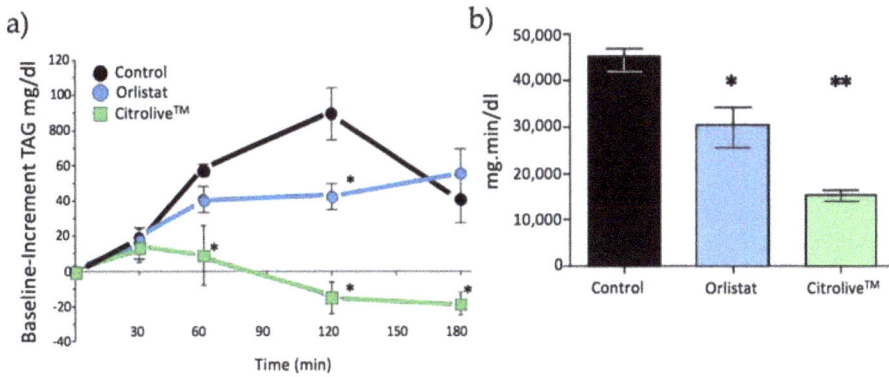

Figure 2. (**a**) Postprandial plasma TAG levels of the different oil-based emulsion vehicles during 180 min after their administration. The values were expressed as increment from basal TAG values (TAG mg/dL increment mean ± SD, $n = 6$). Stars indicate that the values were significantly different ($p < 0.05$ with *); (**b**) Area under the curve (AUC) of the test compounds. Stars indicate that the values were significantly different ($p < 0.05$ with *; $p < 0.01$ with **).

3.4. Long-Term Study: Citrolive™ Chronic Administration during 60 Days in Wistar Rats

During the experiment, no abnormal clinical signs in the rats were observed. Feeding rats with a HFD resulted in a significant body weight increment compared with the RMD by the end of the experiment (Table 2). Although no difference was found between the HFD and HFD + C groups, the treatment resulted in a 4% body weight reduction. BMI was not different between the HFD and treatment groups, but both showed differences when compared with the RMD group ($p < 0.05$). The mean food intake per week per rat was significantly different ($p < 0.01$) between the RMD and HFD diet groups (255 ± 33 kJ in the RMD group and 356 ± 52 kJ in the HFD group). No difference was found between the HFD and HFD + C diet groups (356 ± 52 kJ in the HFD group and 376 ± 47 kJ in the HFD + C group).

Table 2. Morphometric parameters and blood biomarkers of the experimental groups (mean ± SD, $n = 10$).

	Diet */Time (Day of Sampling)					
	RMD		HFD		HFD + C	
Morphometric parameters	0	60	0	60	0	60
Weight (g)	352.70 ± 29.32 [b]	453.45 ± 28.00 [a]	350.25 ± 30.66 [b]	493.00 ± 42.90 [a]	351.33 ± 26.82 [b]	487.33 ± 30.96 [a]
BMI	0.61 ± 0.03 [a]	0.62 ± 0.02 [a]	0.62 ± 0.04 [a]	0.67 ± 0.04 [a]	0.76 ± 0.06 [a]	0.75 ± 0.02 [a]
Blood biomarkers						
GLU (mg/dL)	85.01 ± 8.12 [a]	85.75 ± 6.71 [a]	84.53 ± 9.26 [b]	151.73 ± 8.85 [a]	86.81 ± 13.65 [b]	148.96 ± 16.45 [a]
TAG (mg/dL)	86.00 ± 12.00 [a]	81.20 ± 13.30 [a]	91.10 ± 15.30 [a]	102.8 ± 25.09 [a]	87.30 ± 19.90 [a]	94.08 ± 18.93 [a]
TC (mg/dL)	61.10 ± 13.52 [a]	59.15 ± 10.21 [a]	58.04 ± 13.53 [a]	56.18 ± 10.21 [a]	57.26 ± 14.12 [a]	54.44 ± 3.57 [a]
TBS (μmol/L)	5.98 ± 1.74 [a]	6.59 ± 1.35 [a]	5.85 ± 1.74 [a]	1.96 ± 0.41 [b]	5.80 ± 1.97 [a]	2.53 ± 0.81 [b]
HDL (mg/dL)	44.19 ± 5.83 [a]	39.70 ± 6.53 [a]	43.37 ± 6.30 [a]	37.24 ± 4.66 [a]	39.70 ± 6.53 [a]	35.33 ± 9.27 [a]
LDL (mg/dL)	13.04 ± 3.34 [a]	17.00 ± 4.03 [a]	13.48 ± 2.94 [b]	19.51 ± 1.97 [a]	14.21 ± 3.31 [a]	17.66 ± 2.01 [a]

* Diet: RMD = Rodent Maintenance Diet; HFD = High Fat Diet; HFD + C = High Fat Diet + Citrolive™; Morphometric parameters: BMI = body mass index Blood biomarkers: GLU = glucose; HDL = high density lipoprotein cholesterol; LDL = low density lipoprotein cholesterol; AST = aspartate transaminase; ALT = alanine transaminase; TAG = triacylglycerol; TC = total cholesterol; TBS = total bile salts; [a,b] Values not shearing the same superscript within the same diet and parameter or biomarker are significantly different for $p < 0.05$.

3.5. Clinical Measurements

A liver-weight-to-body-weight comparison between HFD and HFD + C groups is shown in Figure 3a. The HFD + C group's liver-to-body-weight ratio was significantly lower compared with the HFD group's ($p < 0.001$). Biochemical analysis showed that AST and ALT transaminase had lower values in the HFD + C group compared with the HFD group at 30 days of treatment (AST $p < 0.01$). At 60 days, both ALT and AST were significantly lower (AST $p < 0.05$ and ALT $p < 0.001$) (Figure 3b,c). The clinical analysis that was performed on the blood did not suggest a significant difference between the HFD and HFD + C groups for GLU, TC, and HDL (Table 2). However, plasma TAG was reduced by 8% in the HFD + C group at 30 and 60 days, while LDL was reduced by ~20% ($p < 0.052$) at 60 days.

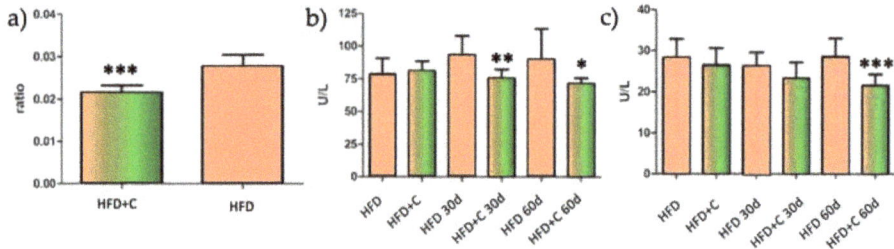

Figure 3. (**a**) Liver-to-body-weight ratio comparison between HFD and HFD + C groups; (**b**) HFD and HFD + C groups comparison of AST and (**c**) ALT transaminase at the beginning, and 30 and 60 days after the experiment. Stars indicate that the values were significantly different ($p < 0.05$ with *, $p < 0.01$ with **, and $p < 0.001$ with ***, mean $\pm$ SD $n = 10$).

3.6. Dry Weight, Total Fat, TAG, Total Bile Salts (TBS), and Cholesterol Analysis in Faeces

No diarrhoea was observed throughout the experiment. Faecal fat excretion and dry faeces weight were significantly higher in the HDF + C group compared with the HFD group ($p < 0.05$) (Table 3). Faecal fat analysis showed that TC values were significantly higher in the HDF+C group compared with the HFD group by the end of the treatment (Figure 4a). TBS values were also altered but only at 7 days after treatment (Figure 4b). On the other hand, faecal TAG values were not affected by the treatment (Table 3).

Table 3. Faecal analysis of the experimental groups (mean $\pm$ SD, $n = 10$).

	Diet */Time (Day of Sampling)					
	RMD		HFD		HFD + C	
Faecal analysis	5	60	5	60	5	60
• Dry weight (g)	1.91 ± 0.18 [a]	1.52 ± 0.41 [a]	1.76 ± 0.63 [a]	1.36 ± 0.34 [a]	2.13 ± 0.39 [a]	1.65 ± 0.25 [a]
• Total fat (mg/g)	6.11 ± 13.52 [a]	5.91 ± 10.21 [a]	8.33 ± 3.14 [a]	6.00 ± 2.10 [a]	7.33 ± 2.07 [a]	8.55 ± 1.48 [a]
Biomarkers in faeces						
• TAG (mg/dL)	289.00 ± 29.70 [a]	391.00 ± 117.10 [a]	572.00 ± 123.80 [b]	1411.00 ± 697.80 [a]	388.30 ± 104.90 [b]	1313.00 ± 564.50 [a]
• TBS (µmol/L)	7.97 ± 1.72 [b]	11.83 ± 2.48 [a]	19.21 ± 4.73 [b]	25.17 ± 9.26 [a]	19.83 ± 4.40 [a]	18.98 ± 7.80 [b]

* Diet: RMD = Rodent Maintenance Diet; HFD = High Fat Diet; HFD + C = High Fat Diet + Citrolive™; Biomarkers in faeces: TAG = triacylglycerol; TBS = total bile salts; [a,b] Values not shearing the same superscript within the same diet and parameter or biomarker are significantly different for $p < 0.05$.

3.7. Long-Term Study: Postprandial TAG Levels (Oral Fat Test) In Vivo Assay

As opposed to physiological saline, all of the treatments that were administered orally in oil-based emulsion form altered the plasma TAG kinetic curve during the experimental study trial. As shown in Figure 5a, at 30 days after the beginning of the experiment the extract significantly blunted plasma TAG increase at 120 and 180 min ($p < 0.05$). However, at 60 days, the significance was attained only at 120 min ($p < 0.05$). Although not statistically significant, the AUC HFD + C value was lower compared with the AUC HFD value at 30 ($p < 0.07$) and 60 days (Figure 5b). On the other hand, the HFD group showed a decreased AUC from 30 to 60 days during the experiment. No difference in the TAG plasma kinetic curve and AUC between the groups was found in the animals that were receiving physiological saline solution.

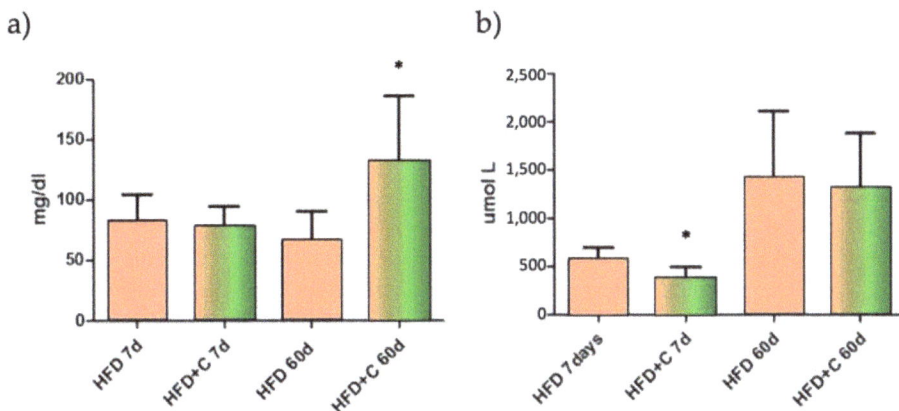

Figure 4. (**a**) Faecal TC showed that values were significantly higher in the HDF + C group compared with the HFD group by the end of the treatment; (**b**) Faecal TBS values comparison between HFD and HFD + C groups. Stars indicate that the values were significantly different ($p < 0.05$ with *, $p < 0.01$ with **, and $p < 0.001$ with ***, $n = 10$).

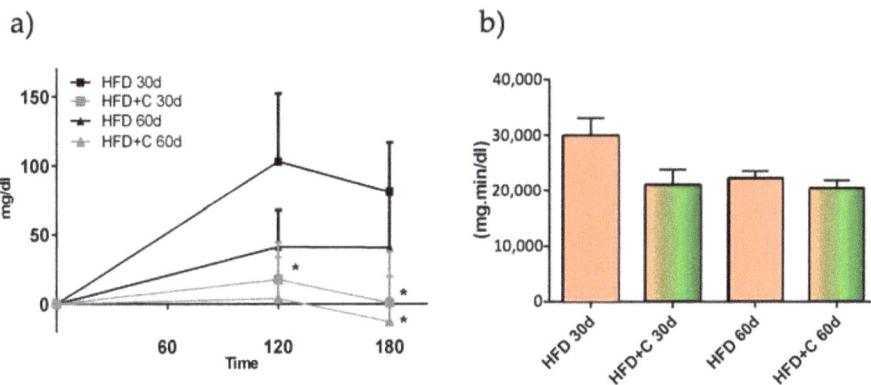

Figure 5. (**a**) Plasma TAG kinetic curve comparison between HFD (black square) and HFD + C (grey triangle) during the experimental study trial. The values were expressed as increment from basal TAG values (TAG mg/dL increment mean ± SD, $n = 6$). Stars indicate that the values were significantly different ($p < 0.05$ with *); (**b**) Area under the curve (AUC) of the experimental groups at different point of times.

4. Discussion

In this paper, we carried out two studies: a short-term (1 day) and a long-term (60 days) evaluation of Citrolive™ on lipid metabolism. The activity of Citrolive™ was tested short-term towards pancreatic lipase activity in vitro and on TAG plasma levels in vivo. In both tests, orlistat was employed as a positive control, considering that several studies positively associate orlistat treatment with reduced total and low density lipoprotein (LDL) cholesterol levels and postprandial TAG [17]. Therefore, we tested the in vivo postprandial TAG plasma levels, which is a common test for evaluating gastrointestinal lipid malabsorption [18] and is increasingly being employed to corroborate the potential activity of natural extracts as pancreatic lipase (PL) enzyme inhibitors [19].

FOT (short-term study) showed that Citrolive™ reduced postprandial TAG to lower levels than orlistat did (Figure 2). To ascertain the mechanism of action, we tested whether this result was due to an orlistat-like effect of the extract on PL, or to the reduction of TAG systemic production. The key to evaluating the orlistat-like process is to study the activity of the PL, which splits triglycerides into two products (more hydrophilic than their precursors): two molecules of fatty acid and one of 2-monoglyceride. These lipolytic products are still sparingly soluble in the aqueous environment of the intestine, and require bile salts and phospholipids for their incorporation into micelles, which are polymolecular aggregates which act as shuttles, delivering fatty acids and monoglycerides to the intestinal microvilli, where they dissociate from micelles and diffuse inside the enterocyte. Subsequently, intestinal cells re-synthesise triglyceride molecules and incorporate them into chylomicrons which are secreted to the intestinal lymph. However, the IC50 value of Citrolive™ for PL inhibition was negligible compared with the PL inhibitor, thus invalidating the hypothesis that the postprandial TAG plasma reduction effect was due to orlistat-like PL inhibition.

This result, given the need to understand it, led us to carry out a long-term trial in order to evaluate the metabolic consequences of this physiological effect during the chronic administration of the extract in Wistar rats. Interestingly, the main result that was derived from this study is a reduced hepatic injury level from decreasing liver inflammatory parameters in rats that were fed a high fat diet plus Citrolive™ for 60 days. As shown in Figure 3a, the liver-to-body-weight ratio of the Citrolive™ group was 21% lower compared with the high-fat diet control group. In addition, Citrolive™ decreased the liver intracellular enzymes ALT and AST (Figure 3b,c). The former had already displayed significantly lower values at 30 days after the beginning of the experiment.

On the other hand, in the Citrolive™ group, the LDL concentration was reduced in plasma by almost 20% with respect to the control group. This led us to consider that Citrolive™ may reduce intestinal cholesterol uptake, influencing LDL synthesis and clearance, rather than inhibiting synthetic hepatic enzymes. According to cholesterol homeostasis, cholesterol metabolism regulates LDL receptor activities and contributes accordingly to the regulation of serum cholesterol levels [20]. Therefore, the extract that was studied may have influenced intestinal cholesterol uptake, reducing the rate of LDL formation in plasma. A decreased chylomicron cholesterol uptake may have caused hepatocytes to respond by upregulating the LDL receptors on their plasma membranes. This, in turn, decreases LDL concentrations by facilitating the clearance of LDL particles from the plasma [21]. Consequently, these results suggest that one of the mechanisms whereby Citrolive™ extracts beneficially reduce hepatic inflammation in rats that are fed it is by reducing intestinal cholesterol absorption.

Lipid faecal analysis showed that, compared with the control, faecal total fat was increased in the Citrolive™ treatment and that this effect was significantly ($p < 0.05$) associated with the high content of TC (Figure 4a). However, this parameter was not affected by plasma TC, so faecal hypercholesterolemia is not dependent upon reduced hepatic synthesis or increased total faecal biliary salts (Figure 4b).

As mentioned previously, postprandial TAG plasma uptake alteration is a process that may have contributed towards reducing hepatic inflammation, as evidenced by the reduced blood postprandial TAG level after the administration of an oral emulsion. In addition, a TAG decrement might have increased faecal TAG; however, in this study, TAG excretion in faeces was not significantly altered by the treatment. Another process that might have affected TAG absorption is reduced TAG production in the liver. With good pancreatic function enabling the efficient hydrolysis of dietary TAG and normal biliary function translating into a correct micellar solubilisation of lipolytic products, the healthy intestine has a great capacity to absorb a dietary fat load in such a way that up to 95% of the fatty acids and monoglycerides are absorbed and the faecal fat content is usually below 5% [22]. On the other hand, a diminished activity of the pancreatic lipase enzyme would increase the unmetabolised faecal TAG concentration, and will decrease plasmatic TAG values.

This effect was not due to diminished plasma TAG absorption since AUC values were not affected by the Citrolive™ treatment, and the plasma TAG in Citrolive™-treated rats in the long-term experiment was reduced by only 8% compared with the high-fat diet group (Figure 5). Consequently, regarding the Citrolive™ in vivo study, and considering that the extract did not significantly affect body weight and food intake throughout the experiment, the mechanism by which Citrolive™ reduced plasma TAG levels remains elusive.

Citrolive™ extract's composition is shown in Table 1. According to HPLC analysis, oleuropein accounts for more than 15% of the total extract, which was its most prevalent compound. The study of Hur et al. [23] determined in mice whether oleuropein shows a protective effect against hepatic steatosis that was induced by a high-fat diet in order to elucidate its underlying molecular mechanisms, along with several key transcription factors. It was also studied whether their target genes were involved in adipogenesis, and whether they were downregulated by oleuropein. The conclusion of these authors is that oleuropein decreased the number and size of lipid droplets in free fatty acid treated cells, and reduced intracellular triglyceride accumulation. It is interesting that, in accordance with our results, the protective effects of oleuropein were against Free fatty acids (FFA)-induced hepatocellular steatosis, and the effects that are shown in both studies may be explained by the mechanism of oleuropein [23,24].

Another mechanism by which oleuropein may have reduced inflammation is through interference with some digestive enzymes other than pancreatic lipase. In this sense, in a study by Polzonetti et al. [25], oleuropein was shown to modulate some digestive enzymes like pepsin and trypsin. Future findings are in progress to evaluate and quantify the possible synergistic contribution of citrus flavonoids to each one of these mechanisms.

As above mentioned, hepatic injury tends to be associated with the development of inflammation, progressive metabolic dysregulation, and steatosis (fatty infiltration of the liver leading to cirrhosis and hepatocellular carcinoma). This reflects a difference between the rate at which fatty acids reach hepatocytes and the rate at which they are metabolised, stored, or assembled [26].

Out of all these findings we hypothesise that Citrolive™ treatment reduced inflammation and the accumulation of TAG in the steatotic liver through different physiological processes, namely increasing faecal total cholesterol values and altering postprandial TAG plasma uptake.

However, as whole, the results that were obtained from this study did not offer a clear answer regarding the alteration of plasma postprandial TAG uptake. Since intestinal TAG digestion is a complicated process, including phases such as emulsification, the hydrolysis of fatty acid ester bonds by specific esterases, the aqueous dispersion of lipolytic products in bile acid micelles, and absorption, mainly in the proximal jejunum but also in more distal parts of the small intestine, in order to explain our study results we have to look more carefully into these digestion processes.

5. Conclusions

In conclusion, Citrolive™ treatment ameliorates liver inflammation symptoms worsened by the negative effects of high fat diet. This was demonstrated by lower liver to body weight, decreased ALT and AST transaminases, increased faecal TC, and altered postprandial TAG plasma uptake. From the study of the extract composition, oleuropein is reputed to be one of the main candidates to exhibit those effects, by influencing systemic anti-inflammatory capacity and digestive enzyme activity. Citrus flavonoids may exert a form of background influence, and this should be evaluated in future studies

Acknowledgments: The corresponding author would like to thank Theo Niewold for his guidance during the manuscript development.

Author Contributions: N.M. designed the experiment, conducted the research, and wrote the manuscript; J.C. wrote and checked the manuscript, G.N. conducted the research and checked the manuscript, O.B.G. checked the manuscript and G.R. analysed the data. All authors read and approved the final manuscript.

Conflicts of Interest: On behalf of all authors, the corresponding author states that there is no conflict of interest.

References

1. Mitchell, N.S.; Catenacci, V.A.; Wyatt, H.R.; Hill, J.O. Obesity: Overview of an Epidemic. *Psychiatr. Clin. N. Am.* **2011**, *34*, 717–732. [CrossRef] [PubMed]
2. Sun, B.; Karin, M. Obesity, inflammation, and liver cancer. *J. Hepatol.* **2012**, *56*, 704–713. [CrossRef] [PubMed]
3. Ferramosca, A.; Conte, A.; Damiano, F.; Siculella, L.; Zara, V. Differential effects of high-carbohydrate and high-fat diets on hepatic lipogenesis in rats. *Eur. J. Nutr.* **2014**, *53*, 1103–1114. [CrossRef] [PubMed]
4. Borén, J.; Matikainen, N.; Adiels, M.; Taskinen, M.R. Postprandial hypertriglyceridemia as a coronary risk factor. *Clin. Chim. Acta* **2014**, *431*, 131–142. [CrossRef] [PubMed]
5. Cannon, C.P.; Kumar, A. Treatment of overweight and obesity: Lifestyle, pharmacologic, and surgical options. *Clin. Cornerstone* **2009**, *9*, 55–71. [CrossRef]
6. Wyatt, H.R. Update on treatment strategies for obesity. *J. Clin. Endocrinol. Metab.* **2013**, *98*, 1299–1306. [CrossRef] [PubMed]
7. Vermerris, W.; Nicholson, R. *Phenolic Compound Biochemistry*; Springer: Dordrecht, The Netherlands, 2007; pp. 202–227.
8. Yun, J.W. Possible anti-obesity therapeutics from nature-a review. *Phytochemistry* **2010**, *71*, 1625–1641. [CrossRef] [PubMed]
9. Agostoni, C.; Bresson, J.-L.; Fairweather-Tait, S.; Flynn, A.; Golly, I.; Korhonen, H.; Lagiou, P.; Løvik, M.; Marchelli, R.; Martin, A.; et al. Olive, polyphenols, LDL, HDL, cholesterol, oxidative damage, blood pressure, inflammation, upper respiratory tract, gastrointestinal tract, body defences, external agents, health claims. *EFSA J.* **2011**, *9*, 2033–2058.
10. Therios, I. Oleuropein, olive leaf extract, olive oil and the benefits of the Mediterranean diet to human health. In *Olives, Chapter 26, Olive Mill Products and Environmental Impact of Olive Oil Production*; Therios, I., Ed.; CABI: Wallingford, UK, 2008; pp. 303–317.
11. Benavente-García, O.; Castillo, J. Update on uses and properties of citrus flavonoids: New findings in anticancer, cardiovascular, and anti-inflammatory activity. *J. Agric. Food Chem.* **2008**, *56*, 6185–6205. [CrossRef] [PubMed]
12. Raasmaja, A.; Lecklin, A.; Li, X.M.; Zou, J.; Zhu, G.G.; Laakso, I.; Hiltunen, R. A water-alcohol extract of *Citrus grandis* whole fruits has beneficial metabolic effects in the obese Zucker rats fed with high fat/high cholesterol diet. *Food Chem.* **2013**, *138*, 1392–1399. [CrossRef] [PubMed]
13. Nakai, M.; Fukui, Y.; Asami, S.; Toyoda-Ono, Y.; Iwashita, T.; Shibata, H.; Kiso, Y. Inhibitory effects of oolong tea biophenols on pancreatic lipase in vitro. *J. Agric. Food Chem.* **2005**, *53*, 4593–4598. [CrossRef] [PubMed]
14. Merola, N.; Medrano Chávez, A.; Ros, G. Oral lipid loading test in Wistar rats as a reliable method for the study of the triacylglycerol malabsorption. *An. Vet. Murcia* **2011**, *27*, 65–72.
15. Al-Suwailem, A.K.; Al-Tamimi, A.S.; Al-Omar, M.A.; Al-Suhibani, M.S. Safety and Mechanism of Action of Orlistat (Tetrahydrolipstatin) as the First Local Antiobesity Drug. *J. Appl. Sci. Res.* **2006**, *2*, 205–208.
16. Argmann, C.A.; Houten, S.M.; Champy, M.-F.; Auwerx, J. Lipid and bile acid analysis. In *Current Protocols in Molecular Biology*; Ausubel, F.M., Brent, R., Kingston, R.E., Moore, D.D., Seidman, J.G., Struhl, K., Eds.; John Wiley & Sons: Hoboken, NJ, USA, 2006; Chapter 29.
17. Kiortsis, D.N.; Filippatos, T.D.; Elisaf, M.S. The effects of Orlistat on metabolic parameters and other cardiovascular risk factors. *Diabetes Metab.* **2005**, *31*, 15–22. [CrossRef]
18. Lairon, D.; Lopez-Miranda, J.; Williams, C. Methodology for studying postprandial lipid metabolism. *Eur. J. Clin. Nutr.* **2007**, *61*, 1145–1161. [CrossRef] [PubMed]
19. Birari, R.B.; Bhutani, K.K. Pancreatic lipase inhibitors from natural sources: Unexplored potential. *Drug Discov. Today* **2007**, *12*, 879–889. [CrossRef] [PubMed]
20. Rajaratnam, R.A.; Gylling, H.; Miettinen, T.A. Cholesterol absorption, synthesis, and faecal output in postmenopausal women with and without coronary artery disease. *Arterioscler. Thromb. Vasc. Biol.* **2001**, *21*, 1650–1655. [CrossRef] [PubMed]
21. Cohen, D.E. Balancing cholesterol synthesis and absorption in the gastrointestinal tract. *J. Clin. Lipidol.* **2008**, *2*, S1–S3. [CrossRef] [PubMed]
22. Ros, E. Intestinal absorption of triglyceride and cholesterol. Dietary and pharmacological inhibition to reduce cardiovascular risk. *Atherosclerosis* **2000**, *151*, 357–379. [CrossRef]

23. Hur, W.; Kim, S.W.; Lee, Y.K.; Choi, J.E.; Hong, S.W.; Song, M.J.; Bae, S.H.; Park, T.; Um, S.J.; Yoon, S.K. Oleuropein reduces free fatty acid-induced lipogenesis via lowered extracellular signal-regulated kinase activation in hepatocytes. *Nutr. Res.* **2012**, *32*, 778–786. [CrossRef] [PubMed]

24. Oi-Kano, Y.; Kawada, T.; Watanabe, T.; Koyama, F.; Watanabe, K.; Senbongi, R.; Iwai, K. Extra virgin olive oil increases uncoupling protein 1 content in brown adipose tissue and enhances noradrenaline and adrenaline secretions in rats. *J. Nutr. Biochem.* **2007**, *18*, 685–692. [CrossRef] [PubMed]

25. Polzonetti, V.; Natalini, P.; Vincenzetti, S.; Vita, A.; Pucciarelli, S. Olives and Olive Oil in Health and Disease Prevention. In *Modulatory Effect of Oleuropeinon Digestive Enzymes*; Preedy, V.R., Watson, R.R., Eds.; Academic Press: San Diego, CA, USA, 2010; pp. 1327–1333, Chapter 14.

26. Postic, C.; Girard, J. The role of the lipogenic pathway in the development of hepatic steatosis. *Diabetes Metab.* **2008**, *34*, 643–648. [CrossRef]

nutrients

Article

Association of Vitamin E Levels with Metabolic Syndrome, and MRI-Derived Body Fat Volumes and Liver Fat Content

Sabina Waniek [1], Romina di Giuseppe [1], Sandra Plachta-Danielzik [1], Ilka Ratjen [1], Gunnar Jacobs [1,2], Manja Koch [1,3], Jan Borggrefe [4], Marcus Both [5], Hans-Peter Müller [6], Jan Kassubek [6], Ute Nöthlings [7], Tuba Esatbeyoglu [8], Sabrina Schlesinger [9], Gerald Rimbach [8] and Wolfgang Lieb [1,2,*]

[1] Institute of Epidemiology, Christian-Albrechts University of Kiel, 24105 Kiel, Germany;
    sabina.waniek@epi.uni-kiel.de (S.W.); romina.digiuseppe@epi.uni-kiel.de (R.d.G.);
    sandra.plachta-danielzik@epi.uni-kiel.de (S.P.-D.); ilka.ratjen@epi.uni-kiel.de (I.R.); jacobs@popgen.de (G.J.);
    mkoch@hsph.harvard.edu (M.K.)
[2] Biobank PopGen, University Hospital Schleswig-Holstein, Campus Kiel, 24105 Kiel, Germany
[3] Department of Nutrition, Harvard T.H. Chan School of Public Health, Boston, MA 02115, USA
[4] Institute of Diagnostic and Interventional Radiology, University Hospital Cologne, 50937 Cologne, Germany;
    jan.borggrefe@uk-koeln.de
[5] Department of Radiology and Neuroradiology, University Hospital Schleswig-Holstein, Campus Kiel,
    24105 Kiel, Germany; Marcus.both@uksh.de
[6] Department of Neurology, University of Ulm, 89081 Ulm, Germany;
    hans-peter.mueller@uni-ulm.de (H.-P.M.); jan.kassubek@uni-ulm.de (J.K.)
[7] Department of Nutrition and Food Science, University of Bonn, 53113 Bonn, Germany; noethlings@uni-bonn.de
[8] Institute of Human Nutrition and Food Science, Christian-Albrechts University of Kiel, 24118 Kiel, Germany;
    tuba.esatbeyoglu@mri.bund.de (T.E.); rimbach@foodsci.uni-kiel.de (G.R.)
[9] Institute for Biometrics and Epidemiology, German Diabetes Center (DDZ) at Heinrich Heine University
    Düsseldorf, 40225 Düsseldorf, Germany; sabrina.schlesinger@DDZ.uni-duesseldorf.de
* Correspondence: wolfgang.lieb@epi.uni-kiel.de; Tel.: +49-431-500-30200; Fax: +49-431-500-30204

Received: 21 July 2017; Accepted: 11 October 2017; Published: 18 October 2017

Abstract: We aimed to relate circulating α- and γ-tocopherol levels to a broad spectrum of adiposity-related traits in a cross-sectional Northern German study. Anthropometric measures were obtained, and adipose tissue volumes and liver fat were quantified by magnetic resonance imaging in 641 individuals (mean age 61 years; 40.6% women). Concentrations of α- and γ-tocopherol were measured using high performance liquid chromatography. Multivariable-adjusted linear and logistic regression were used to assess associations of circulating α- and γ-tocopherol/cholesterol ratio levels with visceral (VAT) and subcutaneous adipose tissue (SAT), liver signal intensity (LSI), fatty liver disease (FLD), metabolic syndrome (MetS), and its individual components. The α-tocopherol/cholesterol ratio was positively associated with VAT (β scaled by interquartile range (IQR): 0.036; 95%Confidence Interval (CI): 0.0003; 0.071) and MetS (Odds Ratio (OR): 1.83; 95% CI: 1.21–2.76 for 3rd vs. 1st tertile), and the γ-tocopherol/cholesterol ratio was positively associated with VAT (β scaled by IQR: 0.066; 95% CI: 0.027; 0.104), SAT (β scaled by IQR: 0.048; 95% CI: 0.010; 0.087) and MetS (OR: 1.87; 95% CI: 1.23–2.84 for 3rd vs. 1st tertile). α- and γ-tocopherol levels were positively associated with high triglycerides and low high density lipoprotein cholesterol levels (all $P_{trend} < 0.05$). No association of α- and γ-tocopherol/cholesterol ratio with LSI/FLD was observed. Circulating vitamin E levels displayed strong associations with VAT and MetS. These observations lay the ground for further investigation in longitudinal studies.

Keywords: vitamin E; α- and γ-tocopherol; metabolic syndrome; body fat volumes; liver fat content

1. Introduction

Metabolic conditions like metabolic syndrome (MetS), fatty liver disease (FLD), and obesity have been linked to increased inflammation and oxidative stress [1–3]. Vitamin E is a lipid-soluble vitamin, encompassing different tocopherols (α-, β-, γ-, and δ-tocopherol) with important anti-oxidative and potentially anti-inflammatory functions [4,5]. In a recent randomized trial [6], vitamin E administration over 96 weeks performed better than pioglitazone and better than placebo in patients with non-alcoholic steatohepatitis (NASH). It is, therefore, conceivable that vitamin E levels are altered in patients with MetS or FLD and that vitamin E levels are correlated with other adiposity-related traits.

Previous clinical and epidemiologic studies on the association of circulating vitamin E levels with different anthropometric adiposity measures (e.g., waist circumference, body mass index (BMI)) [7–11], the MetS [12–16], and NASH [17–19] produced partially conflicting results. Among the different fat depots, subcutaneous (SAT) and, particularly, visceral adipose tissue (VAT) are considered relevant for metabolic conditions, such as MetS. Whether vitamin E levels are associated with MetS, FLD and other adiposity measures, including SAT and VAT, as determined by magnetic resonance imaging (MRI), is unknown.

Therefore, we aimed to relate circulating levels of α- and γ-tocopherol to a broad spectrum of adiposity-related traits in a community-based sample from Northern Germany. Specifically, we assessed the associations of plasma α- and γ-tocopherol levels with MetS and its individual components, MRI-determined VAT, SAT, and liver signal intensity (LSI), as well as with the presence or absence of FLD. We hypothesize that vitamin E levels are altered in individuals with MetS and that vitamin E levels are associated with VAT, SAT, and liver fat, as determined by MRI.

2. Materials and Methods

2.1. Study Sample

Between 2005 and 2007, a total of 1316 individuals from Northern Germany were recruited by the PopGen biobank [20]. Specifically, the sample consisted of 747 individuals who were identified through official population registries and from 569 blood donors. The first follow-up examination, conducted between 2010 and 2012, was attended by 952 individuals, who received a physical examination conducted by trained personnel and provided blood and urine samples. Furthermore, all participants filled-in a standardized questionnaire on demographic and health-related characteristics (including dietary intake, education, smoking status, and physical activity) and medical history [20,21]. A subsample of participants (n = 641) agreed to undergo whole-body MRI. From these participants concentrations of circulating α- and γ-tocopherol levels were measured. Thus, the association between plasma vitamin E concentrations and MetS was investigated in 641 individuals. A total of 91 individuals had to be excluded from the analyses related to MRI phenotypes because of insufficient imaging quality (n = 35), non-adherence to the MRI breathing protocol (n = 40), and missing information on quality of MRI assessment (n = 16). Further, individuals with self-reported liver disease (hepatitis A, B, C, or D virus infection, hemochromatosis, autoimmune liver disease, or liver cirrhosis (n = 29)) were excluded. Thus, the association between circulating vitamin E biomarkers and VAT and SAT was assessed in 591 individuals and the association with liver fat and FLD was evaluated in 571 individuals.

The study has been approved by the Ethics Committee of the Medical Faculty of the Christian-Albrechts University Kiel. Written informed consent was obtained from all study participants.

2.2. Physical Examination and Standardized Questionnaires

Weight and height were measured with the participant wearing light clothing and no shoes, and 2.0 kg were subtracted to correct for the remaining clothes. BMI was calculated as body weight (kg)/height (m^2). Waist circumference was measured at the midpoint between the lower ribs and iliac crest on the anterior axillary line in a resting expiratory position. After the participants had rested 5 min in a sitting position, blood pressure was measured twice (2 min interval) using a sphygmomanometer [22]. Dietary intake, including information on alcohol consumption during the last 12 months, was assessed by a self-administered semi-quantitative 112-item food-frequency questionnaire (FFQ) designed and validated especially for the German population [23]. The German Food Code and Nutrient Data Base (version II.3) was used to determine energy intake and was provided by the Department of Epidemiology of the German Institute of Human Nutrition Potsdam Rehbrücke [24]. Participants were asked to report their use of vitamin E supplements in the FFQ.

2.3. Assessment of SAT, VAT, and Liver Fat Using MRI

Liver fat and adipose tissue (AT) volumes (defined as VAT and SAT) were measured by MRI using a 1.5-T scanner (Magnetom Avanto; Siemens Medical solution, Erlangen, Germany), as described in detail elsewhere [25–27]. VAT was determined as the sum of VAT voxels from the top of the liver to the femoral heads inside the abdominal muscular wall as anatomical border and SAT was determined as the sum of AT voxels underneath the skin layer surrounding the abdomen from the top of the liver to the femur heads. To obtain the volumes (in dm^3) of VAT and SAT the voxel size ($3.9 \times 2 \times 8 \ mm^3$) was multiplied by the number of voxels [26].

Liver fat was quantified as relative LSI difference of the liver on out-of-phase compared with in-phase images in arbitrary units. Both in- and out- of phase images were acquired during a breath hold by using axial T1-weighted gradient echo sequences. Signal intensities were obtained by measuring the average of three circular regions of interest in the liver parenchyma [27].

2.4. Definitions

Hypertension was defined as systolic blood pressure $\geq$140 mmHg or diastolic blood pressure $\geq$90 mmHg, or self-reported hypertension history or use of antihypertensive medication.

MetS was defined according to the harmonized criteria [28] and was considered present when at least three of the following five criteria were met: (1) elevated triglyceride concentration ($\geq$150 mg/dL); (2) reduced high density lipoprotein (HDL)-concentration (<40 mg/dL in men or <50 mg/dL in women); (3) elevated blood pressure (systolic blood pressure $\geq$130 mmHg and/or diastolic blood pressure $\geq$85 mmHg or anti-hypertensive treatment); (4) dysglycaemia, defined as elevated plasma fasting glucose ($\geq$100 mg/dL) or anti-diabetic treatment; and (5) abdominal obesity (waist circumference $\geq$94 cm for men and $\geq$80 cm for women). In the present definition, information about triglyceride-lowering and HDL-increasing medications were not included because this information was not available in detail in our sample. Type 2 diabetes was defined as glycated hemoglobin (HbA1c) $\geq$6.5% (48 mmol/mol) or fasting glucose $\geq$126 mg/dL, or use of anti-diabetic medication or self-report physician diagnosis. FLD was defined as log liver signal intensity $\geq$3.0 according to a cut-off, which corresponds to the maximum Youden Index and was derived using spectroscopically determined FLD (liver fat $\geq$ 5.56%) as the reference method [27,29].

Total physical activity was defined as the reported frequency (hour/week) of different activities (leisure and working-time) [30], multiplied by the corresponding metabolic equivalent (MET)-value, and summed up for all activities [22,31]. Participants were classified into 3 categories to determine smoking status: no-smokers if they had never smoked; former smokers if they had smoked in the past and quit smoking more than 1 year ago; and current smokers if they were currently smoking 1 or more cigarettes per day. Participants were categorized according to the level of education into three categories: low ($\leq$9 years), middle (10 years), or high ($\geq$11 years).

2.5. Laboratory Analyses

Fasting blood (EDTA whole-blood and lithium heparin) samples were obtained from participants in a sitting position. All blood samples were centrifuged, aliquoted, and stored at $-80\ °C$. In fresh blood samples, concentrations of C-reactive protein (CRP), triglycerides, HDL-cholesterol, and total cholesterol were analyzed by enzymatic colorimetry (Roche Diagnostic, Mannheim, Germany); the concentration of glucose was determined by using enzymatic ultraviolet tests (Roche Diagnostic,

Mannheim, Germany), and HbA1c concentrations were determined by using high performance liquid chromatography (HPLC) and photometric detection (Bio-Rad Laboratories, Munich, Germany) in EDTA plasma.

Laboratory blood analyses were performed in the laboratory for clinical chemistry of the University Hospital Schleswig-Holstein, Campus Kiel in Germany.

The Institute of Human Nutrition and Food Science at the Christian-Albrechts-University of Kiel in Germany measured plasma vitamin E (α- and γ-tocopherol) levels using a HPLC with fluorescence detection. An external standard curve was used to quantify vitamin E concentrations using a Jasco HPLC system (Jasco GmbH Deutschland, Gross-Umstadt, Germany; equipped with an autosampler (Jasco AS-2057), pump (PU-2080), ternary gradient unit (LG-2080-02), 3 line degasser (DG-2080-53), and fluorescence detector (FP2020 Plus)) with a Waters Spherisorb ODS-2,3 μm column (100 × 4.6 mm) using methanol:water (98:2, *v/v*) as mobile phase. The fluorescence detector operated an excitation wavelength of 290 nm and emission wavelength of 325 nm. The flow rate of the mobile phase was set at 1.2 mL/min. Duplicate measurements were performed and the injection volume was set at 40 μL. Plasma (50 μL) was homogenized in 2 mL 1% ascorbic acid (in ethanol), 700 μL deionised water, 50 μL 0.1% butylated hydroxytoluol (in ethanol), and 2 mL n-hexane were prepared for analysing the samples. The samples were centrifuged (1000× *g* for 5 min at 4 °C). After separating the phases, 1000 μL of the upper phase was dried under vacuum in a RC-1010 centrifugal evaporator (Jouan, Saint-Herblain, France). The samples were re-suspended in 200 μL mobile phase (methanol:water, 98:2, *v/v*) [32]. The coefficients of variation for α- and γ-tocopherol were 1.05% and 1.29%, respectively.

2.6. Statistical Analyses

Some few missing values of covariates were replaced by a simple imputation, as follows: When values of categorical variables were missing, they were replaced by the most commonly observed category of that respective variable (*n* = 10). Normally distributed continuous missing variables were imputed by the respective mean and skewed variables by the sex-specific median (*n* = 2). Detailed information of missing covariates are provided in Supplementary Materials Table S1.

CRP values below 0.9 mg/dL (detection limit) were assigned a value equal to the half of the detection limit (*n* = 247). Values of γ-tocopherol (*n* = 14, respectively) were imputed by the lowest γ-tocopherol concentration measured in our sample. Detailed information of missing covariates are provided in Supplementary Materials Table S1.

Because vitamin E is bound to lipoproteins in the blood stream [33], cholesterol-adjusted α- and γ-tocopherol levels (μmol/mmol) were calculated by dividing α- and γ-tocopherol concentrations (μmol/L) by total cholesterol (mmol/L) [34].

We performed the following analyses: For descriptive purposes, anthropometric, lifestyle, and clinical factors were compared across tertiles of the α- and γ-tocopherol/cholesterol ratios. Differences in median of continuous variables were tested by using Wilcoxoń s rank-sum test, and differences in categorical variables were assessed by using a chi-square test.

Restricted cubic splines analyses displayed linear associations between vitamin E biomarkers and continuous (VAT, SAT, LSI) and binary (MetS, and FLD) outcomes. Third, linear and logistic regression models were used to relate circulating vitamin E (α- and γ-tocopherol/cholesterol ratio, each biomarker considered separately) levels to continuous outcomes (VAT, SAT, LSI) and binary outcomes (MetS, individual components of MetS, FLD), respectively. In linear regression models, both α- and γ-tocopherol levels were scaled to their interquartile range (IQR) and β coefficients interpreted as comparing VAT, SAT, and LSI values of a person with a typical "high" α- or γ-tocopherol value to a person with a typical "low" value.

Adjusted means of VAT, SAT, and LSI were calculated by general linear models, respectively. We ran age- and sex-adjusted, as well as multivariable-adjusted, models which included, based on literature research [12,13], age (continuous in years) and sex, education (low, medium, high), physical activity (continuous in MET-hour/week), smoking status (never, current, former), vitamin E supplementation (yes, no), alcohol intake (continuous in g/day), and total energy intake (continuous in kJ/day) as potential confounders. Furthermore, the models with continuous VAT and SAT as outcome variables were additionally adjusted for BMI (continuous in kg/m^2) to assess whether VAT and SAT were associated with circulating vitamin E levels independent of BMI. Individual components of MetS were adjusted for each of the other four criteria for the MetS.

Potential interactions of age, sex, and vitamin E supplementation with each metabolic outcome (VAT, SAT, LSI, MetS, individual components of MetS, FLD) were tested by including multiplicative interaction terms into the regression models. In a sensitivity analysis, we excluded vitamin E supplement users and we related α- and γ-tocopherol/cholesterol ratio levels (each biomarker considered separately) to each selected metabolic outcome (Supplementary Materials Tables S4–S6). Furthermore, we excluded individuals who reported a consumption of alcohol more than 20 g/day ($n = 134$) when examining the association of α- and γ-tocopherol/cholesterol ratio with FLD.

Categorical variables with more than two categories were included as indicator variables. P_{trend} was calculated across tertiles using median values of α- and γ-tocopherol/cholesterol ratio within each tertile and we used these values as continuous variables.

All statistical tests were two-sided and considered to be significant when *p* values < 0.05. All analyses were performed with SAS 9.4 (SAS Institute, Cary, NC, USA).

3. Results

3.1. General Characteristics

General characteristics of the study sample according to tertiles of the α- and γ-tocopherol/cholesterol ratio are depicted in Tables 1 and 2, respectively. Triglycerides levels were higher in the 3rd tertile compared to the 1st tertile of the α-tocopherol/cholesterol ratio. Furthermore, the proportion of individuals with MetS was higher in the upper tertiles of the α-tocopherol/cholesterol ratio. Similarly, triglycerides levels were higher in the 3rd tertile compared to the 1st tertile of the γ-tocopherol/cholesterol ratio. Furthermore, BMI and waist circumference, CRP levels, as well as VAT and SAT were higher in the 3rd tertile compared to the 1st tertile of the γ-tocopherol/cholesterol ratio. In addition, the prevalence of the MetS and of diabetes rose with tertiles of the γ-tocopherol/cholesterol ratio. The proportion of vitamin E supplement users was highest in the bottom tertile of γ-tocopherol/cholesterol ratio.

Table 1. General characteristics of the PopGen control study population ($n = 641$) according to tertiles (T) of α-tocopherol/cholesterol ratio.

Characteristics	Tertiles α-Tocopherol/Cholesterol Ratio						*p*
	T1 (*n* = 213)		T2 (*n* = 214)		T3 (*n* = 214)		
Median α-tocopherol/cholesterol ratio (IQR), μmol/mmol	4.63	(4.25–4.88)	5.53	(5.36–5.72)	6.74	(6.33–7.59)	
Men, %	55.9		61.7		60.8		0.422
Age, years	63.0	(56.0–70.0)	61.5	(54.0–71.0)	62.0	(51.0–71.0)	0.411
Body mass index, kg/m^2	26.6	(23.3–29.8)	26.7	(24.8–29.4)	26.7	(24.6–29.2)	0.633
Waist circumference, cm							
Men	100.0	(92.8–107.4)	100.2	(92.7–105.9)	99.4	(93.5–106.8)	0.956
Women	87.1	(78.5–96.4)	88.5	(83.2–97.4)	92.4	(80.2–99.6)	0.199

Table 1. *Cont.*

Characteristics	Tertiles α-Tocopherol/Cholesterol Ratio						*p*
	T1 (*n* = 213)		T2 (*n* = 214)		T3 (*n* = 214)		
Systolic blood pressure, mm/Hg	139.0	(127.5–150.0)	140.0	(125.0–150.0)	138.3	(125.0–150.0)	0.856
Diastolic blood pressure, mm/Hg	85.0	(80.0–90.0)	85.0	(80.0–90.0)	82.3	(80.0–90.0)	0.341
Prevalent hypertension, %	68.1		71.0		67.8		0.723
Current smokers, %	10.1		9.8		12.2		0.640
Physical activity, MET-hour/week	98.3	(61.5–141.6)	84.2	(54.8–120.1))	90.0	(59.3–131.7)	0.074
High education (≥11 years), %	29.1		40.7		37.9		0.143
Alcohol consumption, g/day	8.67	(2.76–17.0)	8.58	(4.09–17.95)	10.96	(4.15–20.05)	0.114
Vitamin E supplementation, %	5.6		6.5		10.3		0.154
Prevalent diabetes, %	8.9		8.9		14.5		0.099
Metabolic syndrome, %	36.6		36.0		48.1		0.016
C-reactive protein, mg/dL	1.10	(0.45–2.60)	1.20	(0.45–2.40)	1.40	(0.45–2.20)	0.531
HDL-cholesterol, mg/dL	67.0	(56.0–82.0)	63.5	(54.0–76.0)	57.5	(49.0–72.0)	<0.0001
Triglycerides, mg/dL	96.0	(71.0–123.0)	104.0	(78.0–132.0)	123.0	(84.0–169.0)	<0.0001
Diabetes medication, % *	3.6		7.1		14.6		0.015
Lipid-lowering medication, % *	13.6		29.3		45.8		<0.0001
Fatty liver disease, % [†]	38.9		38.5		40.3		0.928
Liver signal intensity [†]	18.6	(14.9–23.4)	18.2	(15.0–22.1)	18.0	(14.5–24.7)	0.925
Visceral adipose tissue, dm³ [‡]	3.70	(2.18–5.02)	3.90	(2.41–5.25)	3.94	(2.54–5.37)	0.478
Subcutaneous adipose tissue, dm³ [‡]	5.91	(4.45–8.23)	6.45	(4.75–8.53)	6.10	(4.88–8.24)	0.546

Data are reported as percentages (%) or median and interquartile range (IQR). * *n* = 305, [†] *n* = 571, [‡] *n* = 591; MET: Metabolic equivalent; HDL: High density lipoprotein

Table 2. General characteristics of the PopGen control study population (*n* = 641) according to tertiles of γ-tocopherol/cholesterol ratio.

Characteristics	Tertiles (T) γ-Tocopherol/Cholesterol Ratio						*p*
	T1 (*n* = 213)		T2 (*n* = 214)		T3 (*n* = 214)		
Median γ-tocopherol/cholesterol ratio (IQR), μmol/mmol	0.16	(0.13–0.18)	0.24	(0.22–0.26)	0.35	(0.31–0.42)	
Men, %	60.06		57.9		59.8		0.851
Age, years	63.0	(55.0–71.0)	61.5	(55.0–71.0)	62.0	(54.0–69.0)	0.709
Body mass index, kg/m2	26.1	(23.4–28.9)	27.3	(24.8–29.6)	26.8	(24.9–30.7)	0.005
Waist circumference, cm							
Men	98.9	(91.5–105.3)	100.8	(93.5–108.3)	100.7	(94.6–106.9)	0.271
Women	85.3	(77.4–93.6)	89.0	(82.4–98.0)	91.8	(80.2–103.5)	0.002
Systolic blood pressure, mm/Hg	139.0	(126.5–150.0)	140.0	(125.0–150.0)	139.0	(127.5–150.0)	0.858
Diastolic blood pressure, mm/Hg	85.0	(80.0–90.0)	85.0	(80.0–90.0)	85.0	(80.0–90.0)	0.853
Prevalent hypertension, %	67.1		71.5		68.2		0.598
Current smokers, %	8.0		14.5		10.3		0.278
Physical activity, MET-hour/week	86.3	(58.8–130.0)	89.5	(59.8–138.2)	90.8	(56.8–125.4)	0.932
High education (≥11 years), %	40.9		31.3		35.5		0.168
Alcohol consumption, g/d	8.87	(3.20–16.79)	10.18	(3.82–20.3)	9.74	(4.0–20.13)	0.504
Vitamin E supplementation, %	14.6		2.8		5.1		<0.0001
Prevalent diabetes, %	8.0		8.9		15.4		0.026
Metabolic syndrome, %	32.9		41.6		46.3		0.017
C-reactive protein, mg/dL	1.0	(0.45–1.90)	1.30	(0.45–2.80)	1.40	(0.45–2.60)	0.009
HDL-cholesterol, mg/dL	66.0	(54.0–79.0)	62.0	(53.0–79.0)	60.0	(51.0–74.0)	0.023
Triglycerides, mg/dL	100.0	(76.0–131.0)	103.0	(72.0–133.0)	115.5	(80.0–158.0)	0.004
Diabetes medication, % *	4.0		3.9		17.0		0.0005
Lipid-lowering medication, % *	22.8		26.9		37.0		0.073
Fatty liver disease, % [†]	37.4		36.7		43.7		0.303
Liver signal intensity [†]	18.5	(14.7–22.4)	17.9	(14.5–24.1)	18.8	(14.8–24.2)	0.599
Visceral adipose tissue, dm³ [‡]	3.55	(2.26–4.95)	3.82	(2.46–5.16)	4.15	(2.71–5.77)	0.013
Subcutaneous adipose tissue, dm³ [‡]	5.85	(4.33–7.70)	6.33	(4.81–8.46)	6.30	(4.89–9.09)	0.018

Data are reported as percentages (%) or median and interquartile range (IQR). * *n* = 305, [†] *n* = 571, [‡] *n* = 591; MET: Metabolic equivalent; HDL: High density lipoprotein.

3.2. Association of α-Tocopherol/Cholesterol Ratio with Metabolic Traits

In multivariable-adjusted linear regression models, plasma α-tocopherol/cholesterol ratio displayed a statistically significant association with VAT (β scaled by IQR: 0.036; 95% Confidence Interval (CI): 0.0003; 0.071), including a model additionally adjusted for BMI (β scaled by IQR: 0.026; 95% CI: 0.002; 0.050) (Table 3). Furthermore, consistent associations of plasma α-tocopherol/cholesterol ratio with the MetS were observed (Odds Ratio (OR): 1.83; 95% CI: 1.21–2.76 for 3rd vs. 1st tertile; P_{trend} = 0.003) (Table 5), driven by positive associations with high triglycerides (OR: 3.02; 95% CI: 1.80–5.06 for 3rd vs. 1st tertile; P_{trend} < 0.0001) and low HDL-cholesterol levels (OR: 2.52; 95% CI: 0.97–6.56 for 3rd vs. 1st tertile; P_{trend} = 0.033) (Supplementary Materials Table S2). The α-tocopherol/cholesterol ratio was neither associated with SAT, nor with LSI, modeled as a continuous or binary trait (FLD) (Table 3 and Table 5).

Table 3. Multivariable-adjusted means and 95% CI of VAT, SAT, and LSI according to tertiles of α-tocopherol/cholesterol ratio, and scaled by IQR.

Outcome	Tertiles (T) α-Tocopherol/Cholesterol Ratio			P_{trend}	β Scaled by IQR and 95% CI
	T1	T2	T3		
N	196	199	196		
Median α-tocopherol/cholesterol ratio (IQR), μmol/mmol	4.49 (4.41–4.57)	5.53 (5.44–5.63)	7.18 (7.05–7.30)		
VAT, dm³ (*n* = 591)					
Model 1	2.99 (2.74–3.26)	3.20 (2.94–3.48)	3.29 (3.04–3.57)	0.056	0.035 (−0.002; 0.071)
Model 2	2.92 (2.63–3.26)	3.13 (2.82–3.47)	3.23 (2.93–3.57)	0.043	0.036 (0.0003; 0.071)
Model 3	3.09 (2.87–3.32)	3.31 (3.09–3.32)	3.34 (3.13–3.57)	0.016	0.026 (0.002; 0.050)
SAT, dm³ (*n* = 591)					
Model 1	6.07 (5.78–6.61)	6.38 (5.88–6.93)	6.32 (5.83–6.84)	0.437	0.025 (−0.011; 0.062)
Model 2	5.98 (5.38–6.66)	6.23 (5.62–6.91)	6.22 (5.64–6.87)	0.433	0.026 (−0.009; 0.062)
Model 3	6.36 (6.00–6.74)	6.64 (6.28–7.02)	6.46 (6.12–6.81)	0.572	0.015 (−0.004; 0.034)
N	190	191	190		
Median α-tocopherol/cholesterol ratio (IQR), μmol/mmol	4.50 (4.42–4.58)	5.54 (5.44–5.64)	7.19 (7.05–7.32)		
LSI (*n* = 571)					
Model 1	16.86 (15.57–18.24)	16.67 (15.43–18.01)	17.70 (16.41–19.10)	0.491	0.014 (−0.019; 0.047)
Model 2	17.10 (15.47–18.90)	16.91 (15.43–18.01)	17.70 (16.41–19.10)	0.486	0.011 (−0.023; 0.045)

VAT: Visceral adipose tissue; SAT: Subcutaneous adipose tissue; LSI: Liver signal intensity; BMI: Body mass index; IQR: Interquartile range; CI: Confidence Interval. Model 1: Adjusted for age and sex. Model 2 is model 1 but additionally adjusted for education, physical activity, smoking status, vitamin E supplementation, alcohol intake, and total energy intake. Model 3 is model 2 but additionally adjusted for BMI.

3.3. Association of γ-Tocopherol/Cholesterol Ratio with Metabolic Traits

In multivariable-adjusted linear and logistic regression models, plasma γ-tocopherol/cholesterol ratio showed statistically significant associations with VAT (β scaled by IQR: 0.066; 95% CI: 0.027; 0.104), SAT (β scaled by IQR: 0.048; 95% CI: 0.010; 0.087), and the MetS (OR: 1.87; 95% CI: 1.23–2.84 for 3rd vs. 1st tertile; P_{trend} = 0.004) (Tables 4 and 5). The association with VAT persisted upon additional adjustment for BMI (β scaled by IQR: 0.037; 95% CI: 0.011; 0.063), whereas adding BMI to the model rendered the association of γ-tocopherol/cholesterol ratio with SAT statistically non-significant (β scaled by IQR: 0.015; 95% CI: −0.006; 0.037) (Table 4). Regarding the individual components of the MetS, the γ-tocopherol/cholesterol ratio was positively related to hypertriglyceridemia (OR: 1.81; 95% CI: 1.08–3.06 for 3rd vs. 1st tertile; P_{trend} = 0.014) and low HDL-cholesterol levels (OR: 4.67; 95% CI: 1.42–15.41 for 3rd vs. 1st tertile; P_{trend} = 0.018) in multivariable-adjusted models (Supplementary Materials Table S3).

Table 4. Multivariable-adjusted means and 95% CI of VAT, SAT, and LSI according to tertiles of γ-tocopherol/cholesterol ratio, and scaled by IQR.

Outcome	Tertiles (T) γ-Tocopherol/Cholesterol Ratio			P_trend	β Scaled by IQR and 95% CI
	T1	T2	T3		
N	196	199	196		
Median γ-tocopherol/cholesterol ratio (IQR), µmol/mmol	0.14 (0.13–0.14)	0.24 (0.23–0.25)	0.37 (0.36–0.39)		
VAT, dm³ (n = 591)					
Model 1	2.90 (2.68–3.15)	3.17 (2.92–3.44)	3.48 (3.21–3.78)	0.0002	0.073 (0.034; 0.111)
Model 2	2.92 (2.65–3.21)	3.21 (2.88–3.57)	3.45 (3.11–3.83)	0.0006	0.066 (0.027; 0.104)
Model 3	3.16 (2.97–3.37)	3.25 (3.02–3.49)	3.48 (3.24–3.73)	0.0034	0.037 (0.011; 0.063)
SAT, dm³ (n = 591)					
Model 1	5.80 (5.35–6.29)	6.40 (5.90–6.95)	6.65 (6.13–7.21)	0.006	0.059 (0.020; 0.099)
Model 2	5.81 (5.28–6.39)	6.51 (5.84–7.25)	6.58 (5.92–7.30)	0.011	0.048 (0.010; 0.087)
Model 3	6.36 (6.04–6.70)	6.61 (6.23–7.01)	6.64 (6.27–7.03)	0.103	0.015 (−0.006; 0.037)
N	190	191	190		
Median γ-tocopherol/cholesterol ratio (IQR), µmol/mmol	0.14 (0.14–0.15)	0.24 (0.23–0.25)	0.38 (0.36–0.39)		
LSI (n = 571)					
Model 1	16.32 (15.00–17.61)	17.41 (16.11–18.82)	17.60 (16.30–19.00)	0.193	0.017 (−0.020; 0.055)
Model 2	16.80 (15.35–18.40)	18.05 (16.30–20.00)	17.96 (16.28–19.82)	0.304	0.012 (−0.026; 0.051)

VAT: Visceral adipose tissue; SAT: Subcutaneous adipose tissue; LSI: Liver signal intensity; BMI: Body mass index; IQR: Interquartile range; CI: Confidence interval. Model 1: Adjusted for age and sex. Model 2 is model 1 but additionally adjusted for education, physical activity, smoking status, vitamin E supplementation, alcohol intake, and total energy intake. Model 3 is model 2 but additionally adjusted for BMI.

Table 5. Odds Ratio and 95% Confidence Interval for the association of α- and γ-tocopherol/cholesterol ratio with metabolic syndrome (MetS) and fatty liver disease (FLD).

Outcome	Tertiles (T) of α-Tocopherol/Cholesterol Ratio			P_trend
	T1	T2	T3	
Median α-tocopherol/cholesterol ratio (IQR), µmol/mmol	4.63 (4.25–4.88)	5.53 (5.36–5.72)	6.74 (6.33–7.59)	
MetS (yes/no) (258/383)	(78/135)	(77/137)	(103/111)	
Model 1	Reference	1.01 (0.67–1.51)	1.72 (1.15–2.58)	0.006
Model 2	Reference	1.09 (0.72–1.65)	1.83 (1.21–2.76)	0.003
Median α-tocopherol/cholesterol ratio (IQR), µmol/mmol	4.61 (4.25–4.87)	5.52 (5.35–5.73)	6.75 (6.29–7.57)	
FLD (yes/no) (224/347)	(72/113)	(75/120)	(77/114)	
Model 1	Reference	1.03 (0.67–1.58)	1.11 (0.72–1.70)	0.631
Model 2	Reference	1.01 (0.65–1.55)	1.09 (0.70–1.68)	0.694
	Tertiles (T) of γ-Tocopherol/Cholesterol Ratio			P_trend
	T1	T2	T3	
Median γ-tocopherol/cholesterol ratio (IQR), µmol/mmol	0.16 (0.13–0.18)	0.24 (0.22–0.26)	0.35 (0.31–0.41)	
MetS (yes/no) (258/383)	(70/143)	(89/125)	(99/115)	
Model 1	Reference	1.58 (1.05–2.39)	1.92 (1.28–2.89)	0.002
Model 2	Reference	1.50 (0.98–2.29)	1.87 (1.23–2.84)	0.004
Median γ-tocopherol/cholesterol ratio (IQR), µmol/mmol	0.16 (0.13–0.18)	0.24 (0.22–0.26)	0.34 (0.31–0.42)	
FLD (yes/no) (224/347)	(72/113)	(75/120)	(77/114)	
Model 1	Reference	0.99 (0.65–1.52)	1.38 (0.90–2.10)	0.124
Model 2	Reference	0.97 (0.62–1.51)	1.31 (0.85–2.02)	0.204

IQR: Interquartile range; MetS: Metabolic syndrome; FLD: Fatty liver disease model 1: adjusted for age and sex. Model 2 is model 1 but additionally adjusted for education, physical activity, smoking status, vitamin E supplementation, alcohol intake, and total energy intake.

No association of γ-tocopherol/cholesterol ratio with LSI (continuous trait) or FLD (binary trait) was observed (Tables 4 and 5).

3.4. Assessment of Interactions and Sensitivity Analyses

No significant interactions between circulating α- and γ-tocopherol/cholesterol concentrations and age, sex, or vitamin E supplementation in relation to each selected outcome were observed ($p > 0.05$ for all). In a sensitivity analysis, excluding vitamin E supplement users, the magnitude and the direction of the associations were essentially unchanged (Supplementary Materials Tables S4–S6).

With respect of the association of vitamin E levels with FLD, the results were essentially unchanged after excluding individuals with an alcohol consumption of more than 20 g/day. In multivariable-adjusted models, the α- and γ-tocopherol/cholesterol ratio were not statistically significantly related to the probability of having FLD (OR: 1.09; 95% CI: 0.66–1.80 for 3rd vs. 1st tertile; P_{trend} = 0.268 and OR: 1.51; 95% CI: 0.94–2.55 for 3rd vs. 1st tertile; P_{trend} = 0.239, respectively).

4. Discussion

4.1. Principal Observations

In a community-based sample from Northern Germany, the α-tocopherol/cholesterol ratio and the γ-tocopherol/cholesterol ratio were positively associated with VAT, SAT, MetS, and its components, high triglycerides and low HDL-cholesterol levels. No significant associations were observed when α- and γ-tocopherol/cholesterol ratios were studied in relation to LSI or FLD.

4.2. In the Context of the Published Literature

4.2.1. Vitamin E Levels and Measures of Adiposity and Adipose Tissue Volumes

We observed a consistent association of α- and γ tocopherol/cholesterol ratio with VAT; the γ-tocopherol/cholesterol ratio was also associated with SAT. This is in line with several prior studies that reported positive associations of circulating vitamin E levels with adiposity measures (e.g., BMI, waist circumference, waist-to-hip ratio, and waist-to-height ratio) [7–9]. For example, in a subsample of participants in the Women's Health Initiative (n = 2672), circulating γ-tocopherol levels were positively and strongly associated with BMI, waist circumference, and waist-to-height ratio, while α-tocopherol levels were only positively associated with waist-to-hip ratio [7]. Likewise, Chai et al. [8] reported in 180 premenopausal women that γ-tocopherol levels were significantly higher in obese individuals, whereas α-tocopherol levels did not differ among BMI subgroups.

With respect to α-tocopherol, Wallström et al. [9] reported that serum levels of α-tocopherol were positively related to central adiposity (defined as waist circumference and waist-to-hip ratio), but BMI was only associated with α-tocopherol in men. Body fat percentage (determined by bioelectrical impedance analysis), however, was not significantly associated with vitamin E [9]. Interestingly, in a subsample of healthy postmenopausal women (n = 48), α-tocopherol was identified as predictor of MRI-determined total fat [35]. By contrast, in a sample of 580 women, no association of vitamin E levels with measures of adiposity (BMI, waist circumference, waist-to-height ratio, visceral adiposity, and total body fat) determined by dual-energy x-ray absorptiometry was observed [10].

4.2.2. Vitamin E Levels and the Metabolic Syndrome

We observed consistent positive associations of α- and γ-tocopherol/cholesterol ratio levels with MetS in different multivariable-adjusted statistical models. These associations were driven by a positive association with low HDL-cholesterol levels and high triglycerides levels. The association of vitamin E levels with lipid traits is biologically plausible because the lipid-soluble vitamin E is transported in the blood by lipoproteins [33].

In contrast to our observations, in one study, lower levels of plasma α-tocopherol levels were reported in individuals with MetS (n = 182) compared to healthy adults (n = 91) [14]. In a much larger sample from the 2001–2006 National Health Examination Survey (NHANES; n = 3008), no association of vitamin E levels with MetS was reported. However, in further analyses, the authors observed that vitamin E concentrations were significantly positively related to the number of MetS criteria [12], which lends some support to our results.

Regarding individual components of MetS, vitamin E controlled for lipids showed a positive association with hypertriglyceridemia but not with low HDL-cholesterol levels in NHANES samples [12,13]. A possible explanation for the discrepancies might be that we considered each biomarker separately (α- and γ-tocopherol/cholesterol ratio, respectively), whereas in the other studies vitamin E was defined as the sum of α- and γ-tocopherol. Additionally, we only had a low number of individuals (*n* = 37) with low HDL-cholesterol levels, whereas both of the other studies [12,13] included more participants (*n* = 4322 and *n* = 8465, respectively).

4.2.3. Vitamin E Levels, Fatty Liver Disease, and Liver Fat Content

The association of vitamin E levels with NASH has been assessed in some prior clinical settings with rather small samples sizes [17–19]. One study [17] reported, on average, higher serum vitamin E levels in 43 patients with histologically proven NASH as compared to 33 healthy controls. In two other studies [18,19], however, vitamin E levels were lower in biopsy-proven NASH patients (*n* = 50 and *n* = 29, respectively) than in controls (*n* = 40 and *n* = 10, respectively).

Regarding liver fat content in postmenopausal healthy women, α-tocopherol was identified as a predictor of MRI-determined liver fat, along with other biomarkers [35]. However, this study was based on a rather small sample (*n* = 48) of postmenopausal women, with lack of generalizability to other women and to men [35]. We expand these analyses by assessing in a much larger sample (*n* = 571) from the general population, including men and women, the associations of circulating α- and γ-tocopherol/cholesterol ratio levels with MRI-determined LSI, a proxy for liver fat, modeled on a continuous scale and as a binary trait (FLD).

Yet, albeit FLD is commonly subdivided into non-alcoholic FLD and alcoholic FLD [36], others questioned such a distinction, e.g., because of, in part, similar pathological findings in alcoholic and non-alcoholic FLD, in part overlapping pathophysiological features, sharing of alcohol and other risk factors for FLD in a substantial fraction of the population, and a lack of a consensus regarding harmless alcohol consumption [37]. Furthermore, both non-alcoholic FLD and alcoholic FLD have been associated with premature atherosclerosis, and these findings support the paradigm that steatosis might be a precursor of an increased cardiovascular risk [38]. Therefore, considering FLD as a complex, multifactorial condition [37], we did not distinguish between alcoholic and non-alcoholic FLD but focused on MRI-derived LSI as a proxy for liver fat content.

Interestingly, vitamin E therapy (800 UI per day) for 96 weeks performed better than pioglitazone and placebo in a randomized trial in patients with NASH [6]. The primary endpoint of the study was a histological improvement of NASH features [6].

However, in contrast to the studies mentioned above, we observed no association between vitamin E levels and LSI or FLD in our sample. One potential explanation is that vitamin E levels are altered preferentially in patients with advanced liver disease [39], but not in relatively healthy men and women from the general population with rather modest alterations in liver fat, a premise that merits further investigations.

4.3. Potential Mechanisms for the Observed Associations

Our data suggest that circulating vitamin E (α- and γ-tocopherol/cholesterol ratio) levels are positively associated with MetS and MRI-determined body fat volumes (particularly VAT).

Circulating vitamin E levels are affected by several factors: Dietary vitamin E is absorbed in the gastro-intestinal system (the efficiency of vitamin E absorption is widely variable, ranging from 20–80%) and transported via chylomicrons to the liver [40]. Taken up by the liver, α-tocopherol has several possible metabolic pathways. The hepatic α-tocopherol transfer protein (α-TTP) is the major regulator for maintaining normal plasma α-tocopherol concentrations [41]. γ-tocopherol has much less affinity (α-tocopherol = 100%, γ-tocopherol = 9% [42]) for α-TTP and is largely metabolized in the liver and secreted in the bile [43]. Experimental evidence indicates that α-TTP activity is modulated by oxidative stress potentially influencing vitamin E status [44–47].

α-TTP can facilitate α-tocopherol transfer to very low density lipoproteins (VLDL), whereas the underlying mechanism is still not understood, and facilitate its return to the liver [40]. It is suggested that VLDL is enriched with α-tocopherol to a lesser extent in MetS compared to healthy adults and therefore transported in VLDL to a lesser degree to extrahepatic tissues in individuals with MetS because of a slower α-tocopherol catabolism in MetS compared to healthy adults [16,48]. Therefore, it might be that the disappearance of α-tocopherol from plasma is slower in individuals with MetS [16,48], which would explain the positive association of plasma α-tocopherol levels with MetS as observed in our analyses.

Looking at the excretion, α-tocopherol can be secreted in bile for fecal excretion, but it is not known if this pathway is altered in individuals with MetS [48]. Interestingly, data from a recent study [48] indicate that MetS may inhibit the hepatic metabolism of α-tocopherol to the α-tocopherol metabolite α-carboxyethyl hydroxychromanol (CEHC) (secreted in bile for elimination via feces or excreted via urine [43]). In a recent clinical trial, Traber et al. [48] observed that individuals with MetS ($n = 10$) excrete less vitamin E (lower amounts of α- and γ-CEHC were detected in the urine) as compared to healthy adults ($n = 10$). The authors speculated that individuals with MetS might need more vitamin E because of increased oxidative and inflammatory stressors, thereby suggesting they had increased requirements for α-tocopherol and therefore retained more vitamin E compared to healthy adults [48]. We observed that α- and γ-tocopherol/cholesterol ratios were more strongly and positively associated with VAT than with SAT. Indeed, it is known that adipose tissue, as an endocrine organ, contains a large number of pro-inflammatory cytokines including tumor necrosis factor-α, interleukin (IL)-1β, and IL-6-promoting inflammatory response and oxidative stress [49,50]. Of note, VAT has been shown to release more inflammatory markers (e.g., two to three times more IL-6) than SAT [50]; a rise in concentration of inflammatory markers could be responsible for increased oxidative stress leading to higher vitamin E levels as a compensatory mechanism. Furthermore, in our study, adjustment for BMI rendered the association of γ-tocopherol/cholesterol with SAT statistically non-significant, but not the association with VAT. This might be explained by the fact that BMI is more strongly correlated with SAT than with VAT [51].

4.4. Strengths and Limitation

Strengths of the present study include the assessment of VAT, SAT and liver fat by MRI in a moderate-sized sample from the general population, the measurement of vitamin E in plasma, and the detailed assessment of covariates. The following limitation merits consideration. The cross-sectional study design precludes causal inferences, because exposure and outcome were assessed at the same time. Furthermore, we cannot completely rule out reverse causality. Besides, the cross-sectional study design and the small regression coefficients observed for the associations of VAT and SAT, with both α- and γ-tocopherol values, limit our ability to quantify and translate the observed associations into clinical meaningful findings. Moreover, we had no information about why individuals were taking vitamin E supplements and about the use of statins. However, we did have self-reported information regarding the use of lipid-lowering medications for a subsample ($n =305$).

In summary, we observed significant associations of circulating vitamin E concentrations with MetS and MRI-determined body fat volumes (particularly VAT). Further investigations of longitudinal relationships between α- and γ-tocopherol levels and metabolic conditions and liver fat are warranted.

Supplementary Materials: The following are available online at www.mdpi.com/2072-6643/9/10/1143/s1, Table S1: Missing covariates information; Table S2: Odds ratio and 95% Confidence Interval for the association of α-tocopherol/cholesterol ratio with individual components of MetS; Table S3: Odds ratio and 95% Confidence Interval for the association of γ-tocopherol/cholesterol ratio with individual components of MetS; Table S4: Sensitivity analysis: multivariable-adjusted means and 95% CI of VAT, SAT, and LSI according to tertiles of α-tocopherol/cholesterol ratio, and scaled by IQR after excluding vitamin E supplement users; Table S5: Sensitivity analysis: multivariable-adjusted means and 95% CI of VAT, SAT, and LSI according to tertiles of γ-tocopherol/cholesterol ratio, and scaled by IQR after excluding vitamin E supplement users; Table S6: Sensitivity analysis: odds Ratio and 95% Confidence Interval for the association of α- and γ-tocopherol/cholesterol ratio with MetS and FLD after excluding vitamin E supplement users.

Acknowledgments: Manja Koch is recipient of a Postdoctoral Research Fellowship from the German Research Foundation (Deutsche Forschungsgemeinschaft, DFG, KO 5187/1-1). Romina di Giuseppe is supported by the Deutsche Forschungsgemeinschaft Excellence Cluster "Inflammation at Interfaces" (grants EXC306 and EXC306/2). The PopGen 2.0 Network is supported by the German Federal Ministry of Education and Research (grant 01GR0468 and 01EY1103). The founding sponsors had no role in the design of the study; in the collection, analyses, or interpretation of data; in the writing of the manuscript, and in the decision to publish the results.

Author Contributions: W.L. and G.R. formulated the research question; W.L., S.S. and U.N. designed the study; T.E. performed the vitamin E measurements, S.W. and R.d.G. performed the statistical analyses, S.W., W.L., R.D.G., S.P.D., M.K., I.R. and S.S. contributed to the interpretation of the data; S.W. and W.L. drafted the manuscript. All authors (S.W., R.d.G., S.P.D., T.E., M.K., S.S., I.R., M.B., J.B., J.K., H.-P. M., G.J., U.N., G.R. and W.L.) critically reviewed and approved the final version of the manuscript.

Conflicts of Interest: The authors declare no conflict of interest.

References

1. Roberts, C.K.; Sindhu, K.K. Oxidative stress and metabolic syndrome. *Life Sci.* **2009**, *84*, 705–712. [CrossRef] [PubMed]
2. Bonomini, F.; Rodella, L.F.; Rezzani, R. Metabolic syndrome, aging and involvement of oxidative stress. *Aging Dis.* **2015**, *6*, 109–120. [CrossRef] [PubMed]
3. Polimeni, L.; Del Ben, M.; Baratta, F.; Perri, L.; Albanese, F.; Pastori, D.; Violi, F.; Angelico, F. Oxidative stress: New insights on the association of non-alcoholic fatty liver disease and atherosclerosis. *World J. Hepatol.* **2015**, *7*, 1325–1336. [CrossRef] [PubMed]
4. Jiang, Q. Natural forms of vitamin E: Metabolism, antioxidant, and anti-inflammatory activities and their role in disease prevention and therapy. *Free Radic. Biol. Med.* **2014**, *72*, 76–90. [CrossRef] [PubMed]
5. Borel, P.; Preveraud, D.; Desmarchelier, C. Bioavailability of vitamin E in humans: An update. *Nutr. Rev.* **2013**, *71*, 319–331. [CrossRef] [PubMed]
6. Sanyal, A.J.; Chalasani, N.; Kowdley, K.V.; McCullough, A.; Diehl, A.M.; Bass, N.M.; Neuschwander-Tetri, B.A.; Lavine, J.E.; Tonascia, J.; Unalp, A.; et al. Pioglitazone, vitamin E, or placebo for nonalcoholic steatohepatitis. *N. Engl. J. Med.* **2010**, *362*, 1675–1685. [CrossRef] [PubMed]
7. Kabat, G.C.; Heo, M.; Ochs-Balcom, H.M.; LeBoff, M.S.; Mossavar-Rahmani, Y.; Adams-Campbell, L.L.; Nassir, R.; Ard, J.; Zaslavsky, O.; Rohan, T.E. Longitudinal association of measures of adiposity with serum antioxidant concentrations in postmenopausal women. *Eur. J. Clin. Nutr.* **2016**, *70*, 47–53. [CrossRef] [PubMed]
8. Chai, W.; Conroy, S.M.; Maskarinec, G.; Franke, A.A.; Pagano, I.S.; Cooney, R.V. Associations between obesity and serum lipid-soluble micronutrients among premenopausal women. *Nutr. Res.* **2010**, *30*, 227–232. [CrossRef] [PubMed]
9. Wallstrom, P.; Wirfalt, E.; Lahmann, P.H.; Gullberg, B.; Janzon, L.; Berglund, G. Serum concentrations of beta-carotene and alpha-tocopherol are associated with diet, smoking, and general and central adiposity. *Am. J. Clin. Nutr.* **2001**, *73*, 777–785. [PubMed]
10. Garcia, O.P.; Ronquillo, D.; Caamano Mdel, C.; Camacho, M.; Long, K.Z.; Rosado, J.L. Zinc, vitamin A, and vitamin C status are associated with leptin concentrations and obesity in Mexican women: Results from a cross-sectional study. *Nutr. Metab.* **2012**, *9*, 59. [CrossRef] [PubMed]
11. Galan, P.; Viteri, F.E.; Bertrais, S.; Czernichow, S.; Faure, H.; Arnaud, J.; Ruffieux, D.; Chenal, S.; Arnault, N.; Favier, A.; et al. Serum concentrations of beta-carotene, vitamins C and E, zinc and selenium are influenced by sex, age, diet, smoking status, alcohol consumption and corpulence in a general French adult population. *Eur. J. Clin. Nutr.* **2005**, *59*, 1181–1190. [CrossRef] [PubMed]

12. Beydoun, M.A.; Shroff, M.R.; Chen, X.; Beydoun, H.A.; Wang, Y.; Zonderman, A.B. Serum antioxidant status is associated with metabolic syndrome among U.S. Adults in recent national surveys. *J. Nutr.* **2011**, *141*, 903–913. [CrossRef] [PubMed]

13. Ford, E.S.; Mokdad, A.H.; Giles, W.H.; Brown, D.W. The metabolic syndrome and antioxidant concentrations: Findings from the third national health and nutrition examination survey. *Diabetes* **2003**, *52*, 2346–2352. [CrossRef] [PubMed]

14. Godala, M.M.; Materek-Kusmierkiewicz, I.; Moczulski, D.; Rutkowski, M.; Szatko, F.; Gaszynska, E.; Tokarski, S.; Kowalski, J. Lower plasma levels of antioxidant vitamins in patients with metabolic syndrome: A case control study. *Adv. Clin. Exp. Med.* **2016**, *25*, 689–700. [CrossRef] [PubMed]

15. Li, Y.; Guo, H.; Wu, M.; Liu, M. Serum and dietary antioxidant status is associated with lower prevalence of the metabolic syndrome in a study in Shanghai, China. *Asia Pac. J. Clin. Nutr.* **2013**, *22*, 60–68. [PubMed]

16. Mah, E.; Sapper, T.N.; Chitchumroonchokchai, C.; Failla, M.L.; Schill, K.E.; Clinton, S.K.; Bobe, G.; Traber, M.G.; Bruno, R.S. Alpha-tocopherol bioavailability is lower in adults with metabolic syndrome regardless of dairy fat co-ingestion: A randomized, double-blind, crossover trial. *Am. J. Clin. Nutr.* **2015**, *102*, 1070–1080. [CrossRef] [PubMed]

17. Machado, M.V.; Ravasco, P.; Jesus, L.; Marques-Vidal, P.; Oliveira, C.R.; Proenca, T.; Baldeiras, I.; Camilo, M.E.; Cortez-Pinto, H. Blood oxidative stress markers in non-alcoholic steatohepatitis and how it correlates with diet. *Scand. J. Gastroenterol.* **2008**, *43*, 95–102. [CrossRef] [PubMed]

18. Erhardt, A.; Stahl, W.; Sies, H.; Lirussi, F.; Donner, A.; Haussinger, D. Plasma levels of vitamin E and carotenoids are decreased in patients with nonalcoholic steatohepatitis (NASH). *Eur. J. Med. Res.* **2011**, *16*, 76–78. [CrossRef] [PubMed]

19. Bahcecioglu, I.H.; Yalniz, M.; Ilhan, N.; Ataseven, H.; Ozercan, I.H. Levels of serum vitamin A, alpha-tocopherol and malondialdehyde in patients with non-alcoholic steatohepatitis: Relationship with histopathologic severity. *Int. J. Clin. Pract.* **2005**, *59*, 318–323. [CrossRef] [PubMed]

20. Nothlings, U.; Krawczak, M. PopGen. A population-based biobank with prospective follow-up of a control group. *Bundesgesundhbl. Gesundheitsforsch. Gesundheitsschutz* **2012**, *55*, 831–835.

21. Krawczak, M.; Nikolaus, S.; Von Eberstein, H.; Croucher, P.J.; El Mokhtari, N.E.; Schreiber, S. Popgen: Population-based recruitment of patients and controls for the analysis of complex genotype-phenotype relationships. *Community Genet.* **2006**, *9*, 55–61. [CrossRef] [PubMed]

22. Barbaresko, J.; Siegert, S.; Koch, M.; Aits, I.; Lieb, W.; Nikolaus, S.; Laudes, M.; Jacobs, G.; Nothlings, U. Comparison of two exploratory dietary patterns in association with the metabolic syndrome in a northern German population. *Br. J. Nutr.* **2014**, *112*, 1364–1372. [CrossRef] [PubMed]

23. Nothlings, U.; Hoffmann, K.; Bergmann, M.M.; Boeing, H. Fitting portion sizes in a self-administered food frequency questionnaire. *J. Nutr.* **2007**, *137*, 2781–2786. [PubMed]

24. Dehne, L.I.; Klemm, C.; Henseler, G.; Hermann-Kunz, E. The German food code and nutrient data base (BLS II.2). *Eur. J. Epidemiol.* **1999**, *15*, 355–359. [CrossRef] [PubMed]

25. Muller, H.P.; Raudies, F.; Unrath, A.; Neumann, H.; Ludolph, A.C.; Kassubek, J. Quantification of human body fat tissue percentage by MRI. *NMR Biomed.* **2011**, *24*, 17–24. [CrossRef] [PubMed]

26. Fischer, K.; Moewes, D.; Koch, M.; Muller, H.P.; Jacobs, G.; Kassubek, J.; Lieb, W.; Nothlings, U. Mri-determined total volumes of visceral and subcutaneous abdominal and trunk adipose tissue are differentially and sex-dependently associated with patterns of estimated usual nutrient intake in a northern German population. *Am. J. Clin. Nutr.* **2015**, *101*, 794–807. [CrossRef] [PubMed]

27. Koch, M.; Borggrefe, J.; Barbaresko, J.; Groth, G.; Jacobs, G.; Siegert, S.; Lieb, W.; Muller, M.J.; Bosy-Westphal, A.; Heller, M.; et al. Dietary patterns associated with magnetic resonance imaging-determined liver fat content in a general population study. *Am. J. Clin. Nutr.* **2014**, *99*, 369–377. [CrossRef] [PubMed]

28. Alberti, K.G.; Eckel, R.H.; Grundy, S.M.; Zimmet, P.Z.; Cleeman, J.I.; Donato, K.A.; Fruchart, J.C.; James, W.P.; Loria, C.M.; Smith, S.C., Jr.; et al. Harmonizing the metabolic syndrome: A joint interim statement of the international diabetes federation task force on epidemiology and prevention; national heart, lung, and blood institute; American heart association; World heart federation; International atherosclerosis society; and International association for the study of obesity. *Circulation* **2009**, *120*, 1640–1645. [PubMed]

29. Szczepaniak, L.S.; Nurenberg, P.; Leonard, D.; Browning, J.D.; Reingold, J.S.; Grundy, S.; Hobbs, H.H.; Dobbins, R.L. Magnetic resonance spectroscopy to measure hepatic triglyceride content: Prevalence of hepatic steatosis in the general population. *Am. J. Physiol. Endocrinol. Metab.* **2005**, *288*, E462–E468. [CrossRef] [PubMed]

30. Haftenberger, M.; Schuit, A.J.; Tormo, M.J.; Boeing, H.; Wareham, N.; Bueno-de-Mesquita, H.B.; Kumle, M.; Hjartaker, A.; Chirlaque, M.D.; Ardanaz, E.; et al. Physical activity of subjects aged 50–64 years involved in the European prospective investigation into cancer and nutrition (EPIC). *Public Health Nutr.* **2002**, *5*, 1163–1176. [CrossRef] [PubMed]

31. Ainsworth, B.E.; Haskell, W.L.; Herrmann, S.D.; Meckes, N.; Bassett, D.R., Jr.; Tudor-Locke, C.; Greer, J.L.; Vezina, J.; Whitt-Glover, M.C.; Leon, A.S. 2011 compendium of physical activities: A second update of codes and MET values. *Med. Sci. Sports Exerc.* **2011**, *43*, 1575–1581. [CrossRef] [PubMed]

32. Augustin, K.; Blank, R.; Boesch-Saadatmandi, C.; Frank, J.; Wolffram, S.; Rimbach, G. Dietary green tea polyphenols do not affect vitamin E status, antioxidant capacity and meat quality of growing pigs. *J. Anim. Physiol. Anim. Nutr.* **2008**, *92*, 705–711. [CrossRef] [PubMed]

33. Kayden, H.J.; Traber, M.G. Absorption, lipoprotein transport, and regulation of plasma concentrations of vitamin E in humans. *J. Lipid Res.* **1993**, *34*, 343–358. [PubMed]

34. Thurnham, D.I.; Davies, J.A.; Crump, B.J.; Situnayake, R.D.; Davis, M. The use of different lipids to express serum tocopherol: Lipid ratios for the measurement of vitamin E status. *Ann. Clin. Biochem.* **1986**, *23*, 514–520. [CrossRef] [PubMed]

35. Lim, U.; Turner, S.D.; Franke, A.A.; Cooney, R.V.; Wilkens, L.R.; Ernst, T.; Albright, C.L.; Novotny, R.; Chang, L.; Kolonel, L.N.; et al. Predicting total, abdominal, visceral and hepatic adiposity with circulating biomarkers in Caucasian and Japanese American women. *PLoS ONE* **2012**, *7*, e43502. [CrossRef] [PubMed]

36. Italian Association for the Study of the Liver. AISF position paper on nonalcoholic fatty liver disease (NAFLD): Updates and future directions. *Dig. Liver Dis.* **2017**, *49*, 471–483.

37. Volzke, H. Multicausality in fatty liver disease: Is there a rationale to distinguish between alcoholic and non-alcoholic origin? *World J. Gastroenterol.* **2012**, *18*, 3492–3501. [CrossRef] [PubMed]

38. Loria, P.; Marchesini, G.; Nascimbeni, F.; Ballestri, S.; Maurantonio, M.; Carubbi, F.; Ratziu, V.; Lonardo, A. Cardiovascular risk, lipidemic phenotype and steatosis. A comparative analysis of cirrhotic and non-cirrhotic liver disease due to varying etiology. *Atherosclerosis* **2014**, *232*, 99–109. [CrossRef] [PubMed]

39. Pacana, T.; Sanyal, A.J. Vitamin E and nonalcoholic fatty liver disease. *Curr. Opin. Clin. Nutr. Metab. Care* **2012**, *15*, 641–648. [CrossRef] [PubMed]

40. Schmolz, L.; Birringer, M.; Lorkowski, S.; Wallert, M. Complexity of vitamin E metabolism. *World J. Biol. Chem.* **2016**, *7*, 14–43. [CrossRef] [PubMed]

41. Traber, M.G. Vitamin E regulatory mechanisms. *Annu. Rev. Nutr.* **2007**, *27*, 347–362. [CrossRef] [PubMed]

42. Hosomi, A.; Arita, M.; Sato, Y.; Kiyose, C.; Ueda, T.; Igarashi, O.; Arai, H.; Inoue, K. Affinity for alpha-tocopherol transfer protein as a determinant of the biological activities of vitamin E analogs. *FEBS Lett.* **1997**, *409*, 105–108. [CrossRef]

43. Traber, M.G. Mechanisms for the prevention of vitamin E excess. *J. Lipid Res.* **2013**, *54*, 2295–2306. [CrossRef] [PubMed]

44. Etzl, R.P.; Vrekoussis, T.; Kuhn, C.; Schulze, S.; Poschl, J.M.; Makrigiannakis, A.; Jeschke, U.; Rotzoll, D.E. Oxidative stress stimulates alpha-tocopherol transfer protein in human trophoblast tumor cells bewo. *J. Perinat. Med.* **2012**, *40*, 373–378. [CrossRef] [PubMed]

45. Ulatowski, L.; Dreussi, C.; Noy, N.; Barnholtz-Sloan, J.; Klein, E.; Manor, D. Expression of the alpha-tocopherol transfer protein gene is regulated by oxidative stress and common single-nucleotide polymorphisms. *Free Radic. Biol. Med.* **2012**, *53*, 2318–2326. [CrossRef] [PubMed]

46. Usenko, C.Y.; Harper, S.L.; Tanguay, R.L. Fullerene C60 exposure elicits an oxidative stress response in embryonic zebrafish. *Toxicol. Appl. Pharmacol.* **2008**, *229*, 44–55. [CrossRef] [PubMed]

47. Otulakowski, G.; Engelberts, D.; Arima, H.; Hirate, H.; Bayir, H.; Post, M.; Kavanagh, B.P. Alpha-tocopherol transfer protein mediates protective hypercapnia in murine ventilator-induced lung injury. *Thorax* **2017**, *72*, 538–549. [CrossRef] [PubMed]

48. Traber, M.G.; Mah, E.; Leonard, S.W.; Bobe, G.; Bruno, R.S. Metabolic syndrome increases dietary alpha-tocopherol requirements as assessed using urinary and plasma vitamin E catabolites: A double-blind, crossover clinical trial. *Am. J. Clin. Nutr.* **2017**, *105*, 571–579. [CrossRef] [PubMed]

49. Marseglia, L.; Manti, S.; D'Angelo, G.; Nicotera, A.; Parisi, E.; Di Rosa, G.; Gitto, E.; Arrigo, T. Oxidative stress in obesity: A critical component in human diseases. *Int. J. Mol. Sci.* **2014**, *16*, 378–400. [CrossRef] [PubMed]
50. Fonseca-Alaniz, M.H.; Takada, J.; Alonso-Vale, M.I.; Lima, F.B. Adipose tissue as an endocrine organ: From theory to practice. *J. Pediatr.* **2007**, *83*, S192–S203. [CrossRef]
51. Camhi, S.M.; Bray, G.A.; Bouchard, C.; Greenway, F.L.; Johnson, W.D.; Newton, R.L.; Ravussin, E.; Ryan, D.H.; Smith, S.R.; Katzmarzyk, P.T. The relationship of waist circumference and BMI to visceral, subcutaneous, and total body fat: Sex and race differences. *Obesity* **2011**, *19*, 402–408. [CrossRef] [PubMed]

nutrients

MDPI

Review

Natural Dietary Pigments: Potential Mediators against Hepatic Damage Induced by Over-The-Counter Non-Steroidal Anti-Inflammatory and Analgesic Drugs

Herson Antonio González-Ponce [1], Ana Rosa Rincón-Sánchez [2], Fernando Jaramillo-Juárez [3] and Han Moshage [1,4,*]

[1] Department of Gastroenterology and Hepatology, University Medical Center Groningen,
   University of Groningen, 9713GZ Groningen, The Netherlands; herson_qfbd@hotmail.com
[2] Department of Molecular Biology and Genomics, University Center of Health Sciences,
   Universidad de Guadalajara, Guadalajara 44340, Mexico; anarosarincon@yahoo.com.mx
[3] Department of Physiology and Pharmacology, Basic Science Center, Universidad Autónoma de Aguascalientes,
   Aguascalientes 20131, Mexico; jara@att.net.mx
[4] Department of Laboratory Medicine, University Medical Center Groningen, University of Groningen,
   9713GZ Groningen, The Netherlands
* Correspondence: a.j.moshage@umcg.nl; Tel.: +31-(0)50-361-2364

Received: 7 December 2017; Accepted: 14 December 2017; Published: 24 January 2018

Abstract: Over-the-counter (OTC) analgesics are among the most widely prescribed and purchased drugs around the world. Most analgesics, including non-steroidal anti-inflammatory drugs (NSAIDs) and acetaminophen, are metabolized in the liver. The hepatocytes are responsible for drug metabolism and detoxification. Cytochrome P450 enzymes are phase I enzymes expressed mainly in hepatocytes and they account for ≈75% of the metabolism of clinically used drugs and other xenobiotics. These metabolic reactions eliminate potentially toxic compounds but, paradoxically, also result in the generation of toxic or carcinogenic metabolites. Cumulative or overdoses of OTC analgesic drugs can induce acute liver failure (ALF) either directly or indirectly after their biotransformation. ALF is the result of massive death of hepatocytes induced by oxidative stress. There is an increased interest in the use of natural dietary products as nutritional supplements and/or medications to prevent or cure many diseases. The therapeutic activity of natural products may be associated with their antioxidant capacity, although additional mechanisms may also play a role (e.g., anti-inflammatory actions). Dietary antioxidants such as flavonoids, betalains and carotenoids play a preventive role against OTC analgesics-induced ALF. In this review, we will summarize the pathobiology of OTC analgesic-induced ALF and the use of natural pigments in its prevention and therapy.

Keywords: analgesics; liver; acute liver failure; oxidative stress; antioxidant capacity

1. Introduction

Humans have always relied on nature for their basic needs. For thousands of years, plants and their derivatives have formed the basis of sophisticated traditional medicine and have been an invaluable source of bioactive compounds with therapeutic potential. They play an important role all over the world in the treatment and prevention of human diseases [1,2]. The first records of the use of plants in medicine are from Mesopotamia and date from about 2600 BC. Most of the plant derivatives reported and used by the Mesopotamians are still in use as antibiotics and anti-inflammatory treatments [3]. Plant organs such as roots, leaves, fruits, seeds, and other sub-products contain a vast array of biological activities related to the presence of many chemically-diverse components.

Therefore, plants represent an enormous resource for many kinds of bioactive molecules with highly specific biological activities for different diseases [4]. Many of the pharmacological activities of natural products are related to the presence of bioactive compounds with excellent capacity to reduce oxidative stress [5–8]. Although bioactive components are not always essential for the normal development and/or reproduction of a plant, they may play an important role as protective agents against environmental factors and predators, thus enhancing their survival [9,10]. The potential beneficial health effects of these dietary constituents are highly dependent upon their uptake from natural sources, their metabolism and their disposition in target tissues and cells [11,12]. Thus, dietary phytochemicals are important in human nutrition, medicinal chemistry and drug development [13]. It has been estimated that 25–50% of marketed drugs are derived from natural products and almost 50% of novel FDA-approved drugs between 1981–2006 have a natural product origin [2,14]. Since bioactive components from natural sources have evolved through natural selection, they are often perceived as showing more "drug-likeness and biological friendliness than totally synthetic molecules" [15]. Because these compounds have been selected for optimal interactions with cellular macromolecules in plants, they are likely to induce highly specific biological actions in mammals making them good candidates for further drug development [16]. Therefore, these compounds have proven to be a rich source of novel compounds for biological studies and an essential source for drug discovery [13].

In recent years, the self-consumption of over-the-counter (OTC) drugs such as non-steroidal anti-inflammatory and analgesic drugs has rapidly increased. It now represents a serious health problem around the world due to the high rates of morbidity and mortality both from conscious and unconscious overdoses [17–20]. Most of these intoxications cause either acute liver damage or chronic gastrointestinal, cardiovascular and renal diseases, with severe oxidative stress as a cause or consequence [21,22]. In this review, we will discuss the importance of dietary plant pigments in human health and their use as a preventive or therapeutic modality in the treatment of OTC drugs-induced acute liver failure.

2. Over-The-Counter (OTC) Non-Steroidal Anti-Inflammatory and Analgesic Drugs

Pain in daily life is the most common complaint among patients seeking care in an emergency department [23]. It is also a common experience for adolescents who suffer frequently from headache, abdominal and musculoskeletal types of pain [24]. There are two main classes of drugs recognized by the FDA: prescription and non-prescription (OTC) drugs. OTC drugs can be purchased and self-administered without a prescription or guidance of a general practitioner [25]. OTC drugs are sold worldwide, although the regulatory systems differ between countries [26]. OTC medications are the most commonly purchased and used drugs in the United States for the control of pain in patients with arthritis, minor surgery, headache, dysmenorrhea, backache, strains and sprains [27]. Due to the limited health care access of patients in developing countries, these drugs are often used inappropriately, increasing the risk of adverse effects, acute intoxications, and deaths [28]. OTC medications can be classified according to the World Health Organization Anatomical Therapeutic Chemical (ATC) classification into ten categories: analgesics, laxatives, antithrombotic agents, antacids, cough and cold preparations, antihistamines, dermatologicals, throat preparations, nasal preparations and antidiarrheals [29]. These medications are normally safe when used properly, but when used for extended periods or at high doses the incidence of adverse effects increases [30]. The OTC non-steroidal anti-inflammatory drugs (NSAIDs) ibuprofen, naproxen, and aspirin, as well as the analgesic acetaminophen are the most frequently used medications. They are used by approximately 23% of the population around the world. OTC analgesics and NSAIDs are mainly used by elderly patients to relief pain and inflammation [31]. However, there is a lack of information about the potential toxicity or adverse drug interactions associated with the long-term use and misuse of OTC analgesics. Not all consumers realize that prolonged daily use and intake of high doses of single OTC analgesics or combinations dramatically increase the risk of toxicity or adverse drug events, particularly for the hepatic, gastrointestinal, cardiovascular and renal systems. In addition, patients may not be aware that common cough, cold, or flu medications may contain OTC analgesics again increasing the risk of toxicity [18,32]. It is well known that use of analgesics is critically relevant for public health, but there are no representative population-based data on their actual use [33]. The use and prescription of OTC analgesics in patients known to be at high risk to develop adverse effects has been regulated to reduce the incidence of intoxications and mortality [34,35].

Epidemiology of OTC Non-Steroidal Anti-Inflammatory and Analgesic Drug (Over) Use

Pain in the United States is one of the most important causes of lost labor productivity. In 1985, it was estimated that the total loss of work days due to pain equaled $55 billion [36]. Likewise, a significant percentage of the adult population in Canada lost work days because of the high prevalence of headaches, including migraine [37]. Because of the high incidence of pain, the use and sales of OTC analgesic drugs has steeply increased. A survey in UK demonstrated that 60% of OTC NSAIDs prescribed were for elderly patients. Thirty-eight percent were taking drugs that can interact with NSAIDs, 46% had one or more conditions that may be aggravated by NSAIDs, and 18% had side effects [38]. Most of these analgesics and NSAIDs are sold over-the-counter [39]. In the UK, acetaminophen (APAP) has been reported as the most commonly drug used for self-poisoning and overdoses with an increase of ≈28% from 1976 to 1990, and in 1993, 48% of all overdoses reported in the UK involved acetaminophen or acetaminophen-containing drugs (reviewed in [40]). A survey of medication use in the United States under participants over 18 years old demonstrated that 81% of them used at least one OTC or prescription drug in the preceding week, and the highest prevalence of medication use was observed in women over 65 years old. In addition, six of the ten most frequently used drugs are OTC drugs. The most frequently used drugs were acetaminophen, ibuprofen and aspirin [31]. An Emergency Department survey at the USA showed that 56.2% of 546 patients interviewed took OTC analgesics, including acetaminophen (53%), ibuprofen (34%), aspirin (17%) and naproxen (7.8%); and 6.2% of these 546 patients exceeded the manufacturer's maximum recommended daily dose for at least one medication for at least one day during the three days preceding the evaluation [28]. A study in the Czech Republic from 2007 to 2011 to determine the toxicological characteristics of suicide attempts by deliberate self-poisoning reported that acetaminophen, diclofenac and ibuprofen were related to ≈16% of the cases from 2393 calls concerning children and adolescents and 30.3% of these cases were related to drug combinations including acetaminophen [41]. A Dutch survey performed in 2014 revealed that almost one-third of the general population used NSAIDs prior to the survey and 31% of those used two or more NSAIDs. In addition, 23% of NSAID users consumed these drugs for more than seven days and 9% of this population exceeded the daily maximum dose. These results suggest that at least 333,000 Dutch adults exceeded the maximum dose of OTC NSAIDs [35]. In Germany, analgesic use increased from 19.2% of the population in 1998 to 21.4% in the period 2008–2011. This increase is due exclusively to the increase in OTC analgesic use from 10.0% to 12.2%. Ibuprofen was the most commonly used analgesic. In the period 2008–2011, the use of analgesics was significantly higher in women than men (25.1% vs. 17.6%) with ibuprofen and acetaminophen being the most commonly consumed analgesics. From all the analgesic users, 4.9% (74/1490) used a combination of acetaminophen and ibuprofen. NSAIDs were used mostly in combination by 6.0% of the participants during the seven-day period [33]. A recent Swedish survey concluded that it is important to inform the population about the therapeutic use and risks of the consumption of OTC analgesics since there is a significant influence of parents and peers on the young population [26]. Due to the risk of unintentional overdoses with OTC medications, the prevalence of the problem and the frequent lack of an expert to guide and inform consumers on the proper use, a report in the USA concluded that OTC (over)consumption is a serious public health threat. This health issue requires urgent attention because many consumers are not able to identify or differentiate the active component(s) in OTC analgesic medications and their biological activities, nor do they adhere to the recommended intake instructions [42]. In addition, there is a high prevalence of drug–drug interactions resulting from the co-administration of NSAIDs and other commonly used medications by patients with osteoarthrosis (OA) and rheumatoid arthritis (RA). Therefore, it is necessary to maintain medical supervision of those patients with OA and RA receiving OTC NSAIDs and other medications, as well as to inform them about the risk of toxicity and how to identify toxicity [43].

In summary, when OTC analgesics are taken as recommended by the general practitioner, they are a safe, effective, and economical treatment to relief pain, inflammation, and fever. Nevertheless, because of their wide availability and perceived safety, OTC analgesics are frequently overconsumed resulting in their hepatic, gastrointestinal and cardiovascular side effects [32,44].

3. Liver Histology and Structures

The liver is the main site for the biotransformation of exogenous chemical compounds (xenobiotics) consumed by humans such as drugs. Therefore, understanding drug toxicity is only possible with a thorough understanding of liver structure and function. The liver is a large, solid and highly vascularized organ with a pivotal function in metabolic homeostasis, detoxification, and immunity in the human body [45]. The liver has endocrine and exocrine properties. The main endocrine functions are the secretion of different hormones such as insulin-like growth factors, angiotensinogen, and thrombopoietin; and bile secretion as the major exocrine function [46]. The liver is capable to synthesize and degrade a wide variety of molecules in a regulated way such as carbohydrates, lipids, amino acids, bile acids and xenobiotics [47]. The hepatic parenchyma is organized in lobules, composed of functional units consisting of epithelial cells (hepatocytes) and non-parenchymal cells including sinusoidal endothelial cells, Kupffer cells, stellate cells, cholangiocytes (biliary epithelial cells) and immune cells (Figure 1) [48]. The anatomical features that characterize the liver architecture are: the portal triad (bile duct, hepatic artery, and portal vein), the central vein, and the hepatic sinusoids [47]. Hepatocytes form plates and are in contact with the blood since they are juxtaposed with hepatic sinusoids [49]. The hepatic sinusoid is a microvascular unit formed by endothelial cells distinguished by fenestrations and separated from the hepatocytes by the subendothelial space of Disse, where stellate cells reside. This mass of cells is pervaded by a system of secretory channels (bile ducts) which empty into the intestine (Figure 1) [50].

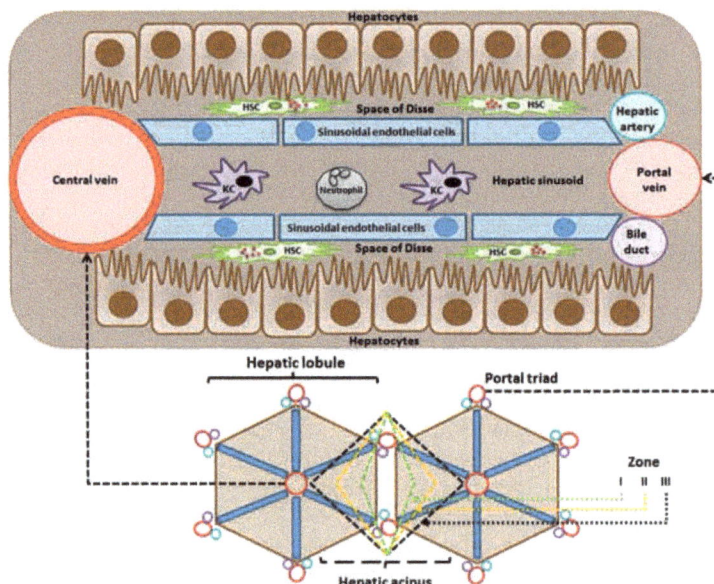

Figure 1. Structure and cell types of a normal liver. HSC, hepatic stellate cells; KC, Kupffer cells.

The hepatocytes produce many circulating plasma proteins such as albumin, coagulation factors, and acute phase proteins. Hepatocytes metabolize and store gut-derived nutrients and glycogen and generate glucose under conditions of starvation. Hepatocytes have a central role in the regulation of lipid metabolism and synthetize lipoproteins. Bile acids are de novo synthetized by the hepatocytes using cholesterol as precursor [51]. Hepatocytes can adapt to the metabolic needs in the body through regulation of protein synthesis and/or zonation. This is controlled by both hormonal and metabolic signals [52].

3.1. Drug Biotransformation

When a foreign compound (xenobiotic) enters the body, it is metabolized by members of a group of hepatic enzymes known as xenobiotic-metabolizing enzymes, which include phase I oxidative enzymes and phase II conjugating enzymes [53]. Hepatocytes carry out most of the metabolic functions of the liver. Both endobiotics and xenobiotics are metabolized across the liver cell plate and secreted into bile [47]. The portal vein brings blood to the liver from the splenic, superior mesenteric, inferior mesenteric, gastric, and cystic veins. Portal blood flow comprises 75–85% of hepatic blood supply, and the remaining 15–25% is delivered by the hepatic artery [54]. Therefore, all absorbed xenobiotics eventually reach the liver to be metabolized and excreted. Due to this first pass metabolism, the concentration of xenobiotics in the systemic circulation is low compared to the portal circulation [51,55]. Xenobiotics must be converted into polar (hydrophilic) metabolites to facilitate their excretion. These water-soluble conjugates can be excreted from the body through the kidneys [56]. The cytochrome P450 (CYP450) system is a phase I microsomal enzyme-family that participates in the metabolism of xenobiotics via oxidation, reduction or hydrolysis, yielding more polar metabolites whereas phase II metabolism via conjugation reactions like glucuronidation or sulfation, facilitates their excretion together with the phase III drug transporters.

The phase I enzymes belong mainly to the flavin-containing monooxygenase (FMO) superfamily and the CYP superfamily. These enzymes are important in the metabolism of most clinically used drugs, but they also participate in the metabolic activation of chemical carcinogens and toxins. The phase II enzymes conjugate the primary metabolites into more polar species, facilitating their excretion from the body. Therefore, the accumulation of (intermediary) metabolites is dependent on the relative expression of phase I and phase II enzymes. This, in turn, is dependent on the extent of induction of these enzymes and gene polymorphisms [57,58]. Although these reactions are meant to detoxify xenobiotics into less toxic metabolites, they can sometimes generate reactive intermediates (electrophilic metabolites) inducing cell toxicity [59]. The phase III drug/metabolite transporters such as P-glycoprotein (P-gp), multidrug resistant-associated proteins (MRPs) and organic anion transporting polypeptide 2 (OATP2) play an important role in the determination of the systemic bioavailability of many drugs since they are capable to reduce drug absorption, prevent their access to the systemic circulation and enhance their excretion to the gut lumen. The induction of these transporters is regulated by the activation of several nuclear transcription factors like the orphan nuclear receptors such as pregnane X receptor (PXR), farnesoid X receptor (FXR) and constitutive androstane receptor (CAR). Thus, the activation or induction of phase I and II enzymes and phase III transporters provide an important way to protect the body from xenobiotics and other cellular stressors (reviewed in [60]).

It has been demonstrated that the expression of CYP450 enzymes is higher in the pericentral area than in the periportal area, and that pericentral hepatocytes have a larger area of smooth endoplasmic reticulum and a higher surface density of CYP450 compared to periportal hepatocytes [61]. Thus, the pericentral area is more susceptible to toxic metabolites generated by CYP enzymes.

Since most of the CYPs are inducible, hepatic drug metabolism is regulated at the level of gene expression. However, posttranslational modifications as well as, e.g., alterations in blood flow, also contribute to the regulation of drug-metabolizing activity [56].

In addition, the expression of enzymes and xenobiotic transporters may be regulated through the activation of specific receptors by xenobiotics [53].

3.2. Free Radicals and Reactive Oxygen Species

Production of energy, ATP, in cells requires oxygen consumption. Free radicals are produced as a result of aerobic ATP production via the mitochondrial electron transport chain [62–64].

Harman proposed the "free-radical theory" of ageing in the mid-1950s, suggesting that endogenous reactive oxygen species, produced by the metabolism of mammalian cells, induce cumulative damage [65]. This concept was initially controversial until superoxide dismutase (SOD) was identified [66]. SOD is an enzyme that inactivates superoxide anions produced by the aerobic metabolism of cells, providing a mechanistic link to support Harman's hypothesis. Ageing and the development of age-related diseases appears to be a consequence of increased levels of intracellular oxidants that induce significant effects such as the activation of signaling pathways and the damage of cellular components [67].

Reactive oxygen species and reactive nitrogen species (ROS and RNS, respectively) are fundamental in modulating mitochondrial functions via the regulation of electron transfer chain enzymes and mitochondrial membrane potential [68]. ROS are crucial for various cellular processes, including cell proliferation [69], apoptosis [70,71], cytotoxicity against bacteria and other pathogens [72], cell adhesion and immune responses [73]. ROS and RNS also act as second messengers in redox signaling [74].

Mitochondrial metabolism, although essential for cellular homeostasis, is also considered the main source of intracellular ROS: superoxide radicals are mainly generated by complex I (NADH:ubiquinone oxidoreductase) and complex III (ubiquinol-cytochrome *c* reductase) of the electron transport chain [75]. However, mitochondrial metabolism is not the only source of oxidants. Under physiological conditions, cytosolic enzyme systems including NAPDH oxidases (NOX), microsomal monooxygenases (cytochromes P450), xanthine oxidase (XO), nitric oxide synthases (NOS), lipoxygenases (LOX), cyclooxygenases (COX) and myeloperoxidases can also produce ROS and RNS [67,76–78]. ROS and RNS are generated in excess in some pathological conditions such as neurodegenerative disorders, cancer, diabetes and cardiovascular and liver diseases, and cause cell damage due to their high reactivity with cellular biomolecules [79].

ROS and RNS comprise a group of different molecules, including free radicals such as superoxide anion ($O_2^{\bullet-}$), hydroxyl radical ($^{\bullet}OH$) and nitric oxide ($NO^{\bullet}$), and non-radicals, such as hydrogen peroxide (H_2O_2), singlet oxygen ($^1O_2$) and peroxynitrite ($ONOO^-$). Many free radicals are extremely unstable, whereas others are freely diffusible and relatively long-lived [64,79].

Additional endogenous non-mitochondrial sources of free radicals include Fenton's reaction, peroxisomal beta-oxidation, and the respiratory burst of phagocytic cells [22,80]. In addition, the production of pro-inflammatory cytokines by activated macrophages and neutrophils and viral proteins stimulate the generation of ROS [81]. The auto-oxidation of many biologically important molecules and the electron delocalization that takes place in reactions of heme-containing proteins, also results in the production of oxidants [64]. The most relevant exogenous sources of free radical production are pollutants/toxins such as cigarette smoke, alcohol, ionizing and UV radiation, pesticides, and ozone [22]. Moreover, several OTC anti-inflammatory and analgesic drugs induce excess generation of ROS and RNS when used at high or prolonged doses due to their metabolism in the liver, e.g., acetaminophen [82,83], diclofenac [84,85], aspirin [86], and ibuprofen [87].

3.3. Cellular Oxidative Stress

The excessive generation of ROS and RNS cause damage to cellular macromolecules such as nuclear and mitochondrial DNA, RNA, lipids and proteins by nitration, oxidation, and halogenation reactions, leading to impaired cellular functions and increased mutagenesis [88,89]. Oxidative damage to essential cellular components (macromolecules, organelles) is generally considered as an important mechanism in the pathophysiology of inflammatory diseases [90]. Oxidative stress is the result of either increased generation of ROS and RNS by endogenous and/or exogenous factors, or the result of a decline of the cellular antioxidant capacity (Figure 2) [91].

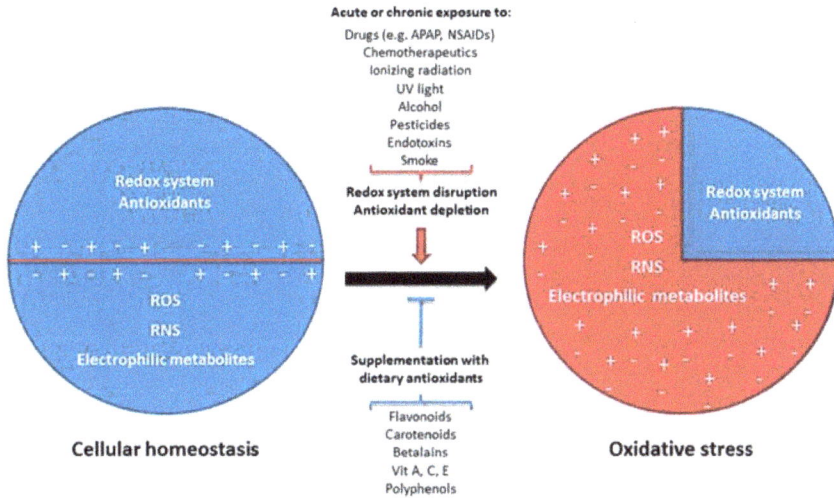

Figure 2. Disruption of the redox system leads to oxidative stress and cellular injury. ROS, reactive oxygen species; RNS, reactive nitrogen species; APAP, acetaminophen; NSAIDs, non-steroidal anti-inflammatory drugs.

Cells are protected against ROS and RNS by both enzymatic and non-enzymatic antioxidant defense systems [92,93]. Excessive generation of oxidants might saturate the antioxidant pathways leading to cellular injury. Thus, free radicals react with membrane phospholipids or lipids from dietary intake, inducing lipid peroxidation (LPO) and the generation of highly toxic products such as trans-4-hydroxy-2-nonenal (4-HNE), 4-hydroperoxy-2-nonenal (HPNE) and malondialdehyde (MDA). 4-HNE, HPNE and MDA can subsequently react with DNA bases such as deoxyadenosine, deoxycytidine and deoxyguanine to form various mutagenic exocyclic adducts implicated in, e.g., hepatocarcinogenesis [94,95]. ROS, RNS and LPO products can also induce expression of genes implicated in the inflammatory response and the pathogenesis of several diseases such as the transcription factor Nuclear Factor NF-κβ, iNOS, and cyclooxygenase-2 (COX-2) [79].

Protein carbonylation is not only associated with an age-related diminished capacity of the antioxidant defense systems, but also with increased generation of ROS and RNS. Protein carbonylation results in altered protein structure and function [96].

Oxidative stress also induces a cellular stress response through the activation of the main stress signaling pathways such as the extracellular signal-regulated kinase (ERK), c-Jun N-terminal protein kinase (JNK) and p38 mitogen-activated protein kinase (MAPK) signaling cascades, the phosphoinositide 3-kinase (PI(3)K)/Akt pathway, the nuclear factor NF-κB signaling system, p53 pathway, and the heat shock response [67]. Some of these pathways can also be activated through other mechanisms like DNA damage [97,98] and stimulation of growth-factor receptors [99,100].

In addition, the complex formed by the redox regulatory protein thioredoxin (Trx) and the apoptosis signal-regulating kinase (ASK1) can be dissociated by oxidative stress and induce the subsequent activation of the JNK and p38 kinases [101]. Glutathione S-transferase binds to JNK to keep it inactivated under normal conditions, but under oxidative stress conditions this interaction can be disrupted [102]. These results suggest that there is a link between alterations in the intracellular redox system and the activity of stress-activated pathways [67].

4. Drug-Induced Liver Injury

Drug-induced oxidative stress is a frequent cause of hepatotoxicity, liver injury and failure. In this regard, drug-induced oxidative stress is considered an important event that can lead to the initiation or progression of liver injury [103]. It is frequently accompanied by clinical signs of acute hepatitis and/or cholestasis [104]. Drug-induced liver injury (DILI) accounts for almost 50% of the cases of acute liver failure (ALF) in the United States [105,106], and for more than 50% in UK. Several factors have been identified that predict drug-induced liver injury (DILI), e.g., dose, alcohol consumption, use of concomitant drugs, nature of the drug, time of exposure, age, preconditioning diseases, and congenital anomalies [104,107]. DILI may be classified as non-idiosyncratic or idiosyncratic. Idiosyncratic drug reactions are unpredictable and independent, and can occur from intermediate to long periods of exposure by an activation of the immune response, inflammation, and cell death (mostly apoptosis). Drug-induced predictable liver injury, such as from acetaminophen, can occur within few hours or days and are mediated by the production of free radicals or electrophilic metabolites from drug biotransformation inducing organelle stress and cell death (both necrosis and apoptosis) [104,108,109]. Necrosis involves the depletion or inactivation of endogenous antioxidants and the induction of cellular stress including mitochondrial stress and decreased ATP synthesis which leads to cellular dysfunction and ATP-independent death. In contrast, apoptosis is an ATP-dependent mechanism involving activation of nucleases [108,110]. Therefore, the activation of death-signaling pathways such as JNK is an important event in DILI [111].

The liver plays a critical role in the disposition of orally administered therapeutic agents. It is the port-of-entry of most orally taken drugs and it represents the major site of drug biotransformation making the liver susceptible to drug-induced toxicity. Products of drug biotransformation (electrophilic compounds and free radicals) have been implicated as causative agents of liver toxicity through direct injury to the hepatocytes by interfering with critical cellular functions (e.g., ATP production), modifying important biomolecules (e.g., proteins, lipids, or nucleic acids), depleting cellular antioxidants, inducing cellular oxidative stress [112,113].

Cellular functions can be affected by both direct effects on organelles (e.g., the mitochondria, the endoplasmic reticulum, the cytoskeleton, microtubules, or the nucleus), and indirectly modulating signaling kinases, transcription factors, and gene expression. These cellular effects can activate the immune response via the release of pro-inflammatory cytokines and/or cell debris into the blood-stream, resulting in the recruitment of immune cells (neutrophils), cellular stress and hepatocyte death that ultimately induce liver injury and failure [56,114–116].

Mitochondria play an important role in the development of DILI since they are an important regulator of cellular homeostasis and their dysfunction can trigger liver cell toxicity resulting in mild to fulminant hepatic failure [117]. Very often, cell death is associated with depletion of mitochondrial glutathione and not with loss of cytoplasmic glutathione [118]. Therefore, it is important to elucidate whether drug metabolites have direct effects on mitochondrial function (e.g., via inhibition of electron transport chain or increasing lipid peroxidation and membrane permeability) resulting in hepatocellular death or indirect effects via activation of the mitochondrial pathways of programmed cell death [119].

4.1. OTC Non-Steroidal Anti-Inflammatory and Analgesic Drugs-Induced Acute Liver Injury

NSAIDs are the most widely used OTC drugs as well as the most prescribed class of drugs for a variety of conditions [120,121]. The group of NSAIDs is composed of a large class of chemical compounds with the same biological activity: blocking the production of prostaglandins (PGs) through the inhibition of the enzyme cyclooxygenase (COX). The COX enzyme is present as two isoforms, each with distinct functions. COX-1 is an isoenzyme constitutively expressed in the stomach, kidney, intestinal mucosa, and other tissues, and is involved in the biosynthesis of PGs serving homeostatic functions. It plays an important role in vasoconstriction and platelet aggregation.

The inducible isoenzyme COX-2 is induced during inflammation, where it causes vasodilation, and other pathologic conditions [122,123]. Acetaminophen and NSAIDs misuse or overdoses have potential significant adverse effects that include gastrointestinal ulcers with consequential bleeding, renal dysfunction, and hepatotoxicity, as well as the risk of death (Table 1) [120,124]. In fact, after antibiotics and anticonvulsants, NSAIDs are considered the most common medications associated with drug-induced liver injury mainly through an idiosyncratic form of hepatotoxicity [125,126]. NSAIDs have also been associated with increased ROS production and oxidative stress. Excessive ROS generation and disturbed cellular redox balance are considered to be important factors in the dysfunction of various biological signaling pathways (Figure 3) [121,127,128].

Current treatments for (drug-induced) acute liver failure are limited since patients usually present at a late stage. Early diagnosis improves prognosis. Treatment with N-acetylcysteine (NAC) is the only clinically used antidote against acetaminophen intoxication. NAC is a precursor of glutathione (GSH) and reduces oxidative stress but is not always effective and liver transplantation is often required.

Activated charcoal is another potential treatment to reduce the absorption of NSAIDs and liver damage but is only effective within the first few hours after intoxication (Table 1). Therefore, alternatives for the treatment of DILI are needed. Natural pigments with antioxidant and therapeutic activity from plants, fruits and their derivatives might be used as an alternative strategy to reduce the incidence and effects of DILI. The use of antioxidants from natural and dietary sources represents a rational defense to prevent or cure liver diseases related to cellular oxidative stress. Promising results of natural antioxidant compounds against different types of liver toxicity or diseases have been obtained in cell culture models and animal studies, but their efficacy in clinical studies remains uncertain (Figure 3) [103,129].

Table 1. Range of therapeutic dosage per day of acetaminophen and non-steroidal anti-inflammatory drugs (NSAIDs) in humans and current treatments against intoxications.

Drug	Therapeutic Dosage in Adults (Orally) per Day	Mechanism of Toxicity	Treatment or Antidote
Acetaminophen (Paracetamol, APAP, Tylenol®, Johnson & Johnson, New Brunswick, NJ, USA)	325–4000 mg/day	N-acetyl-p-benzoquinone imine (NAPQI)-induced mitochondrial dysfunction and oxidative stress	N-acetylcysteine (NAC) 70–140 mg/kg, and activated charcoal to reduce the absorption of the drug
Acetylsalicylic acid (ASA, Aspirin®, Bayer AG, Leverkusen, Germany)	500–4000 mg/day	Mitochondrial dysfunction and oxidative stress induced by salicylic acid and its oxidated metabolite gentisic acid	Gastric lavage and sodium bicarbonate perfusion to reduce acidity and increase excretion of salicylic acid
Diclofenac (Cataflam®, Novartis AG, East Hanover, NJ, USA)	50–200 mg/day	Thiol-reactive quinone imines-induced mitochondrial dysfunction and oxidative stress	Diuresis and dialysis to enhance the excretion of the drug
Naproxen (Aleve®, Bayer AG, Leverkusen, Germany)	220–660 mg/day	Metabolite-induced oxidative stress and liver damage	Gastric lavage and activated charcoal to reduce the absorption of the drug
Ibuprofen (Advil®, Pfizer Inc., New York, NY, USA)	200–1200 mg/day	Hypersensitivity response related to an immuno-allergic reaction	Gastric lavage and activated charcoal to reduce the absorption of the drug

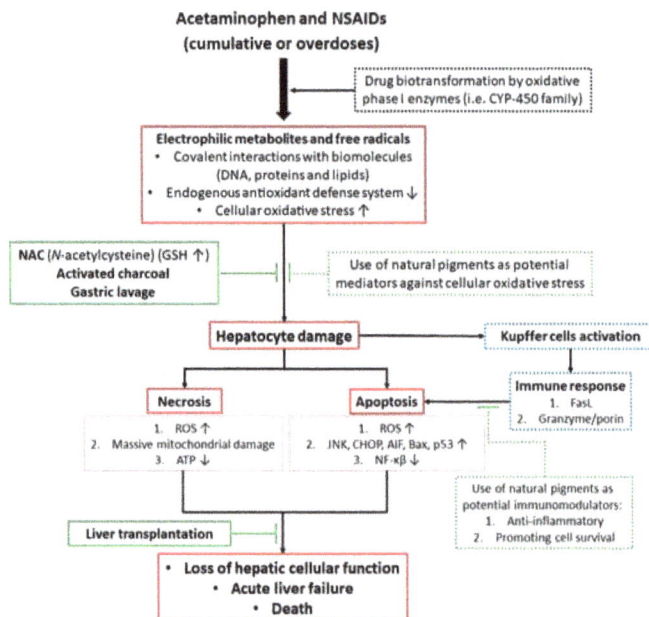

Figure 3. Biotransformation of drugs by oxidase enzymes triggers an intracellular chain reaction mediated by the overproduction of reactive metabolites and free radicals which leads to cell death. Current treatments for drug-induced liver injury are limited. Natural pigments represent a potential alternative treatment to prevent acute liver failure. CYP450, cytochrome P450; GSH, glutathione; ROS, reactive oxygen species; JNK, c-Jun N-terminal protein kinase; CHOP, C/EBP homologous protein; AIF, apoptosis-inducing factor; Bax, bcl-2-associated X protein.

Crofford [123] described that approximately 15% of patients taking NSAIDs display increased markers of liver injury such as alanine aminotransferase (ALT) and aspartate aminotransferase (AST). Dose reduction or discontinuation of the drug can normalize these markers. NSAIDs are generally grouped according to their chemical structures, plasma half-life, and COX-1 versus COX-2 selectivity. Structurally, NSAIDs include several groups such as salicylic acids, acetic acids, propionic acids, fenamic acids, pyrazolones, oxicams, sulfonamide, sulfonylurea, as well as non-acidic drugs. In this review, we will focus on liver injury induced by misuse or overdoses of acetaminophen, acetylsalicylic acid, diclofenac, naproxen, and ibuprofen.

4.1.1. Acetaminophen (APAP, Paracetamol)

Acetaminophen (APAP) is the main cause of drug intoxication and acute liver failure (ALF) worldwide. This is due primarily because of its perceived safety. Acute hepatotoxicity may be induced by a single overdose or unexpected side-effects (idiosyncratic) [130,131]. Toxicity by APAP accumulation is also common in frequent users. In fact, some patients frequently ingest different acetaminophen-containing products at the same time, generating overdoses and toxicity [132]. Thus, acetaminophen is a classical dose-dependent hepatotoxin that is responsible for almost 50% of all ALF cases in many Western countries [133,134].

At therapeutic doses, APAP is mainly metabolized by microsomal enzymes in the liver and eliminated for approximately 85–90% via glucuronidation and sulfation reactions [135]. However, at high doses (or patients with risk factors as chronic alcohol ingestion and malnutrition), the conjugation pathways are saturated, and part of the drug is converted by the CYP450 drug metabolizing system (mainly CYP1A2, CYP2E1 and CYP3A4) to the highly reactive metabolite *N*-acetyl-*p*-benzoquinone imine (NAPQI) that reacts with protein sulfhydryl groups of cysteine. Once generated, NAPQI is immediately inactivated by endogenous reduced glutathione (GSH) to form NAPQI-GSH conjugates which are excreted through the urine. However, when the hepatic reservoir of GSH is depleted, cellular organelles (e.g., mitochondria and endoplasmic reticulum) are exposed to the highly reactive metabolite NAPQI. NAPQI reacts with (membrane) biomolecules forming adducts and resulting in the disruption of cellular homeostasis [7,81,136,137].

Activation of the JNK pathway plays an important role in APAP-induced liver injury and hepatocyte death. Models of genetic JNK knock-out and hepatoprotective compounds with JNK suppressing activity may prevent APAP-induced oxidative stress, cell death and liver failure. Once activated, JNK translocates to the mitochondria inducing mitochondrial permeability transition (MPT), mitochondrial dysfunction and cell death [118]. Protein kinase C (PKC) may also play a critical role in APAP-induced hepatotoxicity via the JNK signaling pathway since treatment with PKC inhibitors (Ro-31-8245, Go6983) protected primary mouse hepatocytes. Ro-31-8245 treatment increased p-AMPK levels (phosphorylated AMP-activated kinase), and promoted autophagy. Treatment with the PKC inhibitor Go6976 inhibits JNK activation and translocation, protecting hepatocytes against APAP cytotoxicity [138].

In mouse models and in human hepatocytes, APAP-induced liver injury involves mitochondrial damage, oxidative stress, JNK activation, and nuclear DNA fragmentation. Thus, protein adducts in mitochondria damage the electron transport chain, increase oxidative stress and disturb the innate immune system of the liver [139,140]. Of note, liver injury is aggravated by subsequent oxidant stress via ROS. The enhanced superoxide formation leads to generation of the potent oxidant peroxynitrite in mitochondria [141].

In the mouse model, APAP toxicity produces very early activation (phosphorylation) of JNK in the cytoplasm. Activated JNK translocates to the mitochondria and increases oxidant stress. This causes the formation of the mitochondrial permeability transition pore and collapse of the mitochondrial membrane potential, as well as a drop in ATP production [140,142,143]. In addition to depleting intracellular GSH, APAP treatment also increases lipid peroxidation and causes hepatic DNA fragmentation. The combination of massive mitochondrial dysfunction and nuclear disintegration leads to cellular necrosis [144–146]. Alterations in hepatic innate immunity and inflammation also play a significant role in the progression of hepatic failure after acetaminophen overdose [135,147].

The primary mechanism of cell death in acetaminophen-induced liver failure is thought to be necrosis, however, some reports have shown that apoptosis may also play a significant role [140]. In the human hepatoblastoma cell line (HuH7), activation of caspases was observed and the manifestation of apoptosis was preceded by a translocation of cytochrome c from mitochondria to the cytosol [148]. It remains to be elucidated whether hepatoma cell lines accurately reflect the in vivo metabolism of APAP. Recent studies also demonstrated increased serum markers of apoptosis, such as caspase-cleaved cytokeratin-18 (M30), in the early phase of acetaminophen-induced ALF in humans [149]. The exact mechanism(s) and optimal management of APAP-induced acute liver failure still need to be clearly elucidated to reduce the high morbidity and mortality of APAP-induced ALF [150].

4.1.2. Acetylsalicylic Acid (ASA, Aspirin)

Acetylsalicylic acid (ASA) is a widely used NSAID due to its pharmacological properties (including analgesic, antipyretic, anti-inflammatory and anti-platelets effects), as well as its easy availability. ASA is an irreversible inhibitor of both cyclooxygenase isoenzymes, COX-1 and COX-2. After oral administration, ASA is absorbed from the stomach and small intestine, primarily by passive diffusion across the gastrointestinal tract [151]. ASA is rapidly deacetylated to salicylic acid by esterases in the gastrointestinal mucosa, in the blood and in the liver. The oxidation of salicylic acid in human liver microsomes produces two metabolites, 2,5-dihydroxybenzoic acid (gentisic acid) and

2,3-dihydroxybenzoic acid. The major human cytochrome P450 involved in both biotransformation reactions is CYP2E1 [152].

Liver toxicity induced by ASA is considered to be dose-dependent, although predisposing conditions may exist that increase the individual risk of liver damage [153,154]. Oxidative stress is one of the mechanisms associated with the adverse effects of ASA and salicylic acid may induce cytochrome P450-mediated lipid peroxidation in liver microsomes [155]. On the other hand, antioxidant properties of this drug have also been reported [156]. Doi and Horie [86] also analyzed mitochondrial dysfunction and oxidative stress in salicylic acid-induced liver injury. In rat hepatocytes, salicylic acid significantly increased the leakage of lactate dehydrogenase and increased thiobarbituric acid reactive substances (TBARS) formation, a marker of lipid peroxidation, whereas antioxidants (promethazine and DPPD (N,N'-diphenyl-p-phenylenediamine)) suppressed both harmful effects. TBARS formation in rat liver microsomes was also suppressed by diethyldithiocarbamate (a specific inhibitor of CYP2E1) and diclofenac (a specific inhibitor of CYP2C11). Salicylic acid also significantly decreased ATP content in isolated rat hepatocytes and mitochondrial respiration. The authors suggest that salicylic acid impairs mitochondrial function leading to lethal liver cell injury by lipid peroxidation.

Raza et al. [157] analyzed the oxidative effects of ASA in cultured human hepatoma cells (HepG2) and reported a cascade of adverse events starting with the overproduction of cellular ROS through the uncoupling of the complex I (NADH:ubiquinone oxidoreductase) and IV (cytochrome c oxidase) of the mitochondrial electron transport chain and ultimately resulting in reduced levels of GSH. The altered MPT disrupted the mitochondrial ATP synthesis, decreased the expression of the anti-apoptotic protein Bcl-2 and induced the activation and release of pro-apoptotic proteins to induce cell death. ASA-induced cytotoxicity was augmented by inhibition of GSH synthesis and attenuated by increasing the GSH pool. The authors conclude that ASA-induced toxicity in human HepG2 cells is mediated by increased metabolic and oxidative stress, accompanied by mitochondrial dysfunction, resulting in apoptosis [158].

Tassone et al. [159] reported a pilot study in 22 newly diagnosed diabetic patients treated with ASA (100 mg/daily for four weeks). The authors suggest that ASA treatment for primary prevention in diabetic patients causes oxidative stress and impairs vascular function.

In addition to mitochondrial dysfunction, ASA treatment can also lead to accumulation of free fatty acids in the liver leading to massive hepatic steatosis [154]. The mechanism of ASA-induced hepatotoxicity is different from that of other NSAIDs. As described above, ASA is first hydrolyzed by non-specific esterases into salicylic acid. In mitochondria, salicylic acid may form salicylyl-coenzyme A (CoA) conjugates, thus sequestering extramitochondrial CoA. This conjugate indirectly inhibits β-oxidation of long-chain fatty acids since CoA is necessary to transport free fatty acids into the mitochondria [160,161]. The inhibition of mitochondrial β-oxidation of long-chain fatty acids by ASA may lead to microvesicular steatosis known as Reye's syndrome [113,162]. Salicylic acid can also inhibit Krebs cycle enzymes such as α-ketoglutarate dehydrogenase and succinate dehydrogenase leading to mitochondrial dysfunction [162].

Finally, Jain et al. [151] published that treatment of female rats with ASA (100 mg/kg b.w. (body weight)) caused significant histopathological alterations in the liver, including degenerative and pyknotic changes in the nuclei, vacuolization and clear dilatations in the sinusoids and hypertrophy of hepatocytes.

4.1.3. Diclofenac

Diclofenac is a commonly prescribed NSAID of the phenyl-acetic acid class. Diclofenac has been used in a variety of inflammatory conditions and has strong anti-inflammatory activity, although analgesic and antipyretic properties have also been reported [163]. In contrast to traditional NSAIDs, diclofenac appears to have a higher selectivity for COX-2 than COX-1. Diclofenac has been associated with serious dose-dependent gastrointestinal, cardiovascular, and renal toxicity [164]. Diclofenac-induced liver injury has been used as a model of drug-related toxicity. The hepatotoxicity induced by diclofenac is mainly due to its metabolites, but genetic factors can also increase the susceptibility to produce and accumulate the reactive acylglucuronide metabolite which triggers an immune response and liver injury [84,154,165–167]. The bioactivation of diclofenac by CYP2C9 or CYP3A4 yields thiol-reactive quinone-imines which in turn are conjugated by UDP (Uridine 5′-diphospho)-glucuronosyltransferase (UGT2B7) into protein-reactive acyl-glucuronides. Therefore, both disruption of mitochondrial function and alterations in the redox state due to oxidative or nitrosative stress appear to be the main mechanisms of diclofenac-induced cell death and liver injury.

Laine et al. [168] conducted a long-term prospective clinical trial to analyze the frequency of diclofenac-induced adverse hepatic effects. A total of 17,289 arthritis patients received diclofenac for a mean duration of 18 months. Increased serum transaminase occurred primarily within the first 4–6 months of therapy and was observed in 3.1% of arthritic patients. Of note, ALT/AST ratios of >10× the upper limit of normal (ULN) were only observed in 0.5% of cases. The clinical liver symptoms requiring hospitalization were relatively rare (23/100,000 patients or 0.023% of cases). No liver failure or death was observed. These and other results indicate that diclofenac rarely causes severe liver injury in humans [169].

The United States (U.S.) Drug Induced Liver Injury Network (DILIN) is a prospective registry of severe idiosyncratic drug hepatotoxicity. Schmeltzer et al. [126] reported a study on liver injury caused by NSAIDs in the U.S. The authors conclude that hepatocellular injury is the most common manifestation seen with NSAID toxicity and diclofenac is the most frequently implicated NSAID agent (16/30 cases). Bort et al. [170] analyzed acute diclofenac cytotoxicity on human and rat hepatocytes and hepatic cell lines (HepG2, FaO). Diclofenac impaired ATP synthesis by mitochondria and the authors suggested that toxicity might be related to drug metabolism because diclofenac was more cytotoxic to drug metabolizing cells (rat and human primary hepatocytes) than to non-metabolizing cell lines (HepG2, FaO). The toxic effect was reduced by the addition of cytochrome P450 inhibitors and in vitro cytotoxicity of diclofenac correlated well with the generation of two metabolites: 5-hydroxydiclofenac and N,5-dihydroxydiclofenac.

Gómez-Lechón et al. [85] also analyzed the generation of ROS and the apoptotic effect of diclofenac after exposure of human and rat hepatocytes to diclofenac. Antioxidants were able to prevent caspase-8 and -9 activation by diclofenac and maintain mitochondrial integrity. The authors concluded that the mitochondrial pathway of apoptosis is the only (or major) pathway involved in diclofenac-induced apoptosis and that the strongest apoptotic effect was produced by the metabolite 5-hydroxydiclofenac.

Other studies reported similar findings, such as diclofenac (metabolite)-induced ROS generation, lipid peroxidation, mitochondrial injury, ATP depletion, GSH depletion, lysosomal fragmentation and DNA fragmentation [171–173]. Many of these signs of toxicity were reversed by antioxidants, MPT pore sealing agents, lysosomotropic agents and inhibitors of cytochrome P450 isoenzymes. The final common pathway of all these events is the leakage of cytochrome c from the mitochondrial intermembrane space into the cytosol, resulting in the activation of caspases-9 and 3, mitochondrial/lysosomal cross-talk and apoptosis.

Yano et al. [169] investigated the immune response in diclofenac-induced idiosyncratic hepatotoxicity in mice. Gene expression of the main interleukins (ILs) and chemokines involved in the inflammatory response and the expression of helper T (Th) 17 cell-derived factors in the liver were significantly increased, as well as the levels of IL-17 in plasma. The results suggest at least a partial involvement of IL-17 in the development of diclofenac-induced liver injury since antagonizing IL-17 reduced toxicity. In addition, both gene expression and plasma levels of IL-1β were rapidly increased after diclofenac administration suggesting its involvement in the pathogenesis of diclofenac-induced hepatotoxicity.

Interaction between drug-induced toxicity pathways and the pro-inflammatory cytokine tumor necrosis factor alpha (TNFα) was investigated in HepG2 cells by Fredriksson et al. [174]. Transcriptomics of the stress response pathways initiated by diclofenac and carbamazepine, revealed the endoplasmic reticulum (ER) stress/translational initiation signaling and nuclear factor-erythroid

2 (NF-E2)-related factor 2 (Nrf2) antioxidant signaling as two important affected pathways. Inhibition of the Nrf2-dependent adaptive oxidative stress response enhanced drug/TNFα-induced cytotoxicity but did not affect C/EBP homologous protein (CHOP) expression. Both hepatotoxic drugs enhanced expression of the translational initiation factor EIF4A1, which was essential for CHOP expression and drug/TNFα-mediated cell killing. The authors conclude from their data that diclofenac initiates PERK-mediated CHOP signaling in an EIF4A1 dependent manner, thereby sensitizing the hepatocyte towards caspase-8-dependent TNFα-induced apoptosis.

4.1.4. Naproxen

Naproxen is a propionic acid derivative NSAID and has been available as OTC medication since 1994. Naproxen is an analgesic, antipyretic and anti-inflammatory drug. It is a non-selective inhibitor of the enzymes COX-1 and COX-2 and decreases synthesis of prostaglandins, important mediators in inflammatory and pain pathways. Currently, more than 10 million prescriptions for naproxen are filed each year but these numbers do not include the large-scale OTC sales. Side effects of naproxen include dizziness, dyspepsia, nausea, and abdominal discomfort, but rarely liver injury. Naproxen is metabolized by the cytochrome P450 system and idiosyncratic liver injury may be due to a toxic metabolite although the mechanism has not been completely elucidated yet [175,176].

The absorption of naproxen is rapid and complete when given orally and the biotransformation of this drug includes demethylation and glucuronidation, as well as sulfate conjugation reactions. The metabolites are excreted in urine, with only a small proportion of the drug being eliminated unchanged [177]. In human liver, CYP2C9 and CYP1A2 are involved in naproxen metabolism [178,179] and subsequent glucuronidation takes place via UDP-glucuronosyltransferase (UGT2B7) [180]. There is some evidence that naproxen metabolism is related to hepatotoxicity: Yokoyama et al. [181] reported that naproxen induces lipid peroxidation in isolated rat hepatocytes, resulting in cell death and formation of high molecular weight protein aggregates in the hepatocytes. Oxidative stress was demonstrated by the formation of TBARS, a marker of lipid peroxidation. The increase of TBARS strongly correlated with the decrease of intracellular GSH. The authors concluded that ROS production and lipid peroxidation are induced by the metabolism of naproxen in rat primary hepatocytes.

A follow-up study confirmed the pivotal role of the GSH/GSSG (glutathione/glutathione disulfide) ratio in naproxen-induced toxicity. Increased GSSG levels preceded lipid peroxidation and LDH release [182]. Ji et al. [183] reported that ferrous iron release contributes to naproxen-induced microsomal lipid peroxidation and that naproxen and salicylic acid are not uncouplers of cytochrome P450. Naproxen toxicity and disposition were also investigated in the isolated perfused liver. Lo et al. [184] investigated the disposition of naproxen, its reactive acyl glucuronide metabolite (NAG) and a mixture of NAG rearrangement isomers (isoNAG) and concluded that covalent protein-adducts were formed in the liver, with isoNAG being the more important substrate for adduct formation. Using the same experimental model, Yokoyama et al. [185] demonstrated that naproxen increased liver damage (AST, ALT) and peroxidation (TBARS). GSSG and TBARS content were significantly increased in naproxen-perfused liver. In addition, the biliary excretion (clearance) of indocyanine green, a compound used for testing liver function, was decreased. The authors concluded that the biliary excretion system was disrupted due to naproxen-induced hepatic oxidative stress. Naproxen-induced hepatotoxicity and liver injury is very rare (≈1–3 per 100,000 users), although cases of acute hepatitis have been reported within six weeks after the start of naproxen intake. Once naproxen intake is terminated, biochemical markers of liver injury such as AST and ALT usually return to normal levels [186].

Andrejak et al. [175] described a patient using naproxen at 500 mg/day with nausea, abdominal pain, malaise, jaundice and increased AST, ALT and alkaline phosphatase levels. Histology showed moderate hepatocellular necrosis. After termination of naproxen, the patient recovered rapidly. Ali et al. [176] described a patient who developed jaundice and intractable pruritus shortly after taking naproxen. Histological analysis showed inflammatory infiltration and a progressive loss of the small interlobular bile ducts (ductopenia). The authors suggest that normalization of liver function and histology after termination of naproxen consumption may take up to 10 years.

4.1.5. Ibuprofen

Ibuprofen was the first member of propionic acid derivatives to be introduced in 1969 as a better alternative to ASA. Ibuprofen is the most frequently prescribed NSAID and it is also an OTC drug. Ibuprofen has excellent analgesic and anti-inflammatory properties as well as antipyretic activity because it inhibits both cyclooxygenases (COX-1 and COX-2). Ibuprofen is frequently used to relief pain related to dysmenorrhea, headache, and osteoarthritis or rheumatoid arthritis. Adverse reactions to ibuprofen appear to be dose and duration dependent and the major adverse effects are related to the gastrointestinal tract, the kidneys and blood coagulation. In addition, ibuprofen may produce dizziness, dyspepsia, bronchospasm, and hypersensitivity reactions, but rarely causes clinically apparent and serious acute liver injury [43,187,188].

In fact, ibuprofen at OTC doses does not represent a risk for developing liver injury, because it has a short plasma half-life and it does not give rise to toxic metabolites (e.g., covalent modification of liver proteins by the quinine-imine metabolites of paracetamol or irreversible acetylation of biomolecules by ASA) [189]. High doses of ibuprofen (2400 to 3200 mg daily) may produce increased ALT plasma levels (<100 U/L), although clinically apparent liver injury due to ibuprofen is very rare. Only a few cases of ibuprofen-induced acute liver toxicity have been reported. The mechanism of toxicity has not been completely elucidated and may be multifactorial. Most of the ibuprofen-induced cases of hepatotoxicity have a rapid onset suggesting the production of a reactive metabolite as well as the involvement of a hypersensitivity response related to an immuno-allergic reaction [188]. Underlying liver diseases like hepatitis C may increase the risk of ibuprofen-induced acute liver injury [154,190]. Finally, Basturk et al. [191] published a case report of a seven-year-old patient who developed toxic epidermal necrolysis and vanishing bile duct syndrome (VBDS) after oral ibuprofen intake. Acute VBDS is a rare disease with unknown etiology. The patient was treated with supportive care (a steroid and ursodeoxycholic acid), with complete recovery after eight months.

5. Endogenous Antioxidant Defense Systems

Reactive Oxygen Species (ROS) are produced during normal intracellular metabolism and from exogenous substances. They play an important role in a range of biological processes, e.g., in the defense against microorganisms, as second messengers in several signaling pathways, and in modulating gene expression. Moreover, when generated in excess and when redox balance is disturbed, these free radicals can damage cellular organelles and induce inflammation, ischemia, apoptosis, and necrosis. Since living organisms are continuously exposed to free radicals, cells have developed antioxidant defense mechanisms. These antioxidant defense mechanisms include molecules and enzymes formed endogenously and bioactive molecules obtained from food (reviewed in [192]).

Any substance or compound that scavenges free radicals or non-radical ROS or RNS, or inhibits cellular oxidation reactions can be considered an antioxidant [193]. Endogenous antioxidants are capable to counteract the deleterious effects of free radicals and maintain cellular homeostasis. These endogenous antioxidants include both enzymatic antioxidants such as catalase, superoxide dismutases, glutathione peroxidases, peroxiredoxins, and thioredoxins [194] as well as non-enzymatic antioxidants (glutathione, urate, bilirubin, melatonin). The coordinated action of antioxidant enzymes ensures efficient ROS removal and promotes repair [195].

Free radicals, ROS and RNS are very reactive and since they are generated in different cell organelles, enzymatic defense systems have evolved in different cellular compartments. One of the most effective intracellular enzymatic antioxidants is superoxide dismutase (SOD) which catalyzes the dismutation of $O_2^{\bullet-}$ (superoxide anions) to O_2 and to the less-reactive species H_2O_2 with remarkably high reaction rates. This is accomplished by successive oxidation and reduction of the transition metal ions incorporated in the SOD enzymes (reviewed in [196]). The enzyme SOD is widespread in nature and present in all oxygen-metabolizing cells [197]. In humans, there are three superoxide dismutases: mitochondrial Mn-SOD, cytosolic Cu/Zn-SOD, and extracellular ecSOD. The main function of all SODs is to protect the cell from harmful effects of superoxide anions. In addition, extracellular SOD (ecSOD) is known to affect endothelial cells by preventing NO from reacting with superoxide anions (reviewed by [198]).

Glutathione peroxidase (GPx), catalase (CAT) and peroxiredoxins (Prx) control the ultimate fate of hydrogen peroxide produced from the superoxide anions by SODs.

Peroxisomes are the major storage site of catalase, an enzyme that catalyzes the biotransformation of H_2O_2 into water and O_2. In addition, catalase has a peroxidative activity promoting the reaction between H_2O_2 and hydrogen donors to generate water and oxidize the reduced donor [199]. Although catalase might not be essential for all cell types, under normal conditions its deficiency in conditions of oxidative stress can increase cell damage and death [200].

GPx enzymes protect the cells against oxidative stress through the reduction of hydroperoxides using GSH as electron donor, yielding water and oxidized GSSG (glutathione disulfide). This antioxidant property is important for cellular homeostasis since hydroperoxides can be substrates for the Fenton reaction which induces oxidative stress through the production of the highly reactive hydroxyl radical. Five different isoenzymes of GPx have been identified but their expression depends on the tissue and species (reviewed in [200]).

Peroxiredoxins (Prx) are a group of 25-kDa proteins present in organisms of all kingdoms including humans. They contain a conserved cysteine (Cys) residue giving them the capacity to donate electrons and inactivate hydroperoxides and peroxynitrite. Six different isoforms of Prx (PrxI to PrxVI) have been identified in mammals, varying in number and position of the Cys residues. These enzymes may protect cellular components against hydroperoxides produced by normal metabolism and prevent cellular oxidative damage (reviewed in [201]).

The thioredoxin system is composed of thioredoxin (Trx) and thioredoxin reductases (TrxR), a group of enzymes that belong to the pyridine nucleotide-disulfide oxidoreductases family. The thioredoxin system plays an important role in DNA synthesis, defense against oxidative stress, redox signaling and apoptosis. Trx expression is very high in the intestine and has an important role in the gut immune response [202].

The tripeptide GSH (γ-glutamylcysteinylglycine) is synthetized in the cytosol and is an essential non-enzymatic regulator of intracellular redox homeostasis. The cysteine residue present in this compound allows GSH to inactivate oxidants through the reversible oxidation of its active thiol forming the oxidized form GSSG. GSH is a ubiquitous antioxidant present in many organelles and has two important functions: (1) to scavenge or inactivate ROS, RNS and electrophilic compounds generated in cellular metabolism; and (2) to function as a substrate for GPx to inactivate hydroperoxides and prevent the generation of hydroxyl radicals. Mitochondria contain large stores of GSH. Mitochondria are important organelles that produce a substantial amount of ROS. It has been reported that mitochondrial GSH (mGSH) has a vital role in maintaining the integrity and function of the mitochondria and its depletion may trigger undesirable events that promote mitochondrial dysfunction and cell death (reviewed in [203,204]).

Other important molecules in the cell that have antioxidant capacity include metal-binding proteins. The function of these proteins is to sequester metals such as iron and copper, preventing transition-metal catalyzed generation of radicals, e.g., transferrin and lactoferrin bind iron while albumin binds copper. Other molecules, such as bilirubin, melatonin, lipoic acid, coenzyme Q and uric acid, have also been proposed to act as antioxidants [192].

Protection of cellular systems against oxidative stress is also achieved via transcriptional regulation of antioxidant enzymes. Various transcription factors are involved in the regulation of the expression of antioxidant enzymes such as SOD, catalase, GPx, Prx and Trx. These transcription factors include Nrf1/2, PGC1-α and Foxo3a, often acting together, e.g., as co-activators [195,205,206].

The nuclear transcription factor E2-related factor 2 (Nrf2) belongs to the Cap'n'Collar/basic leucine zipper (CNC-bZIP) family of proteins [207]. Nrf2 protein is normally inactive and located in the cytoplasm bound to its inhibitor Kelch-like ECH-associated protein 1 (Keap1) [208]. Nrf2 is an oxidative stress sensor which, once activated by oxidants, dissociates from Keap1 and translocates to the nucleus to induce transcription of target genes involved in the protection against oxidative stress (reviewed in [209]). Nrf2 has also been demonstrated to play a role in the protection against drug-related toxicity and xenobiotics-induced carcinogenesis via the induction of phase II enzymes such as UDP-glucuronosyltransferase (UGT), sulfotransferases (SULT), NAD(P)H:quinone oxidoreductase 1 (NQO1) and glutathione S-transferase (GST). These enzymes are involved in the metabolic inactivation and detoxification of drugs and carcinogens [210]. Nrf2 is also involved in the transcriptional induction of the antioxidant enzymes γ-glutamylcysteine synthetase (γ-GCS) and heme oxygenase-1 (HO-1) [211]. Nrf2 induces transcription via binding to the antioxidant response element (ARE) within the $5'$-flanking region of its target genes [212].

The concerted actions of the enzymatic and non-enzymatic antioxidant defense systems are essential for the effective detoxification of ROS, RNS, electrophilic compounds and carcinogens and the maintenance of cellular redox homeostasis and cell survival (Figure 4).

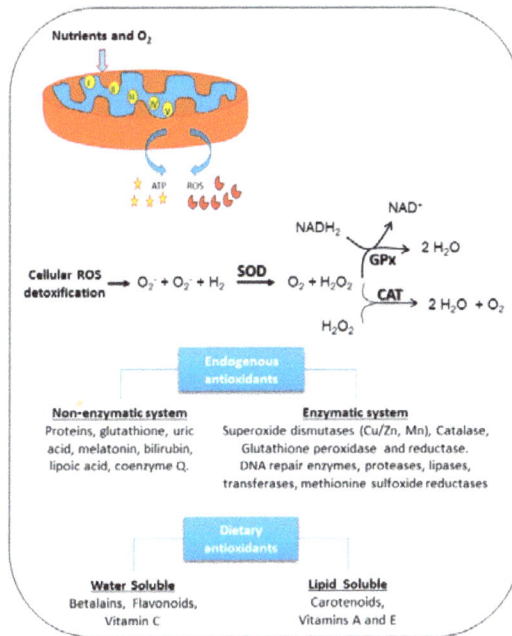

Figure 4. Cellular antioxidant defense systems. ROS, reactive oxygen species; SOD, superoxide dismutase.

This also implies that when endogenous antioxidant systems are compromised or exhausted, cells will no longer be able to cope with oxidative stress. This will ultimately result in cellular dysfunction and death. Therefore, it is important to obtain antioxidants from dietary intake to provide an extra line of defense against cellular oxidative stress.

6. Dietary Natural Pigments with Biological Activity

Dietary natural products with therapeutic activity have been used for centuries for the treatment of human diseases. Vegetables and fruits are very important in human nutrition as sources of nutrients and phytochemicals that reduce the risk of several diseases [213] such as cancer (reviewed in [214]), cardiovascular disease [215], neurodegenerative diseases (reviewed in [216]), type 2 diabetes [217] and hypertension [218].

It is estimated that the frequent intake of vegetables and fruits might prevent up to one third of cancer-related deaths in the United States [219]. Several in vitro and animal studies suggest that the biological activities of vegetables, fruits and derivatives are frequently related to their antioxidant capacity (reviewed in [40,220]). Therefore, the health benefits of diets rich in vegetables, fruits and derived products such as beverages are due not only to fibers, vitamins and minerals, but also to a diversity of plant pigments and secondary metabolites with potential biological activity in humans (reviewed in [12]). Wang et al. reviewed different studies of natural extracts such as *Dunaliella salina*, *Hydnora africana*, *Calotropis procera*, *Zingiber officinale* Rosc, *Hibiscus sabdariffa* L., *Aloe barbadensis* Miller, *Lycium barbarum* and food-derived compounds such as resveratrol, apigenin, silymarin, lupeol, sesamol, ellagic acid, gallic acid, picroliv, curcumin, β-carotene, sulforaphane, and α-lipoic acid, against acetaminophen-induced hepatotoxicity. All of them showed a protective effect against the cellular oxidative stress induced by the highly reactive acetaminophen metabolite NAPQI but they appear to have different mechanisms of detoxification depending on the method of extraction, reagents and dosage used (reviewed in [40]). In addition, an in vitro study with 20 fruits commonly consumed in the American diet (apple, avocado, banana, blueberry, cantaloupe, cherry, cranberry, red and white grape, grapefruit, lemon, melon, nectarine, orange, peach, pear, pineapple, plum, strawberry, and watermelon) suggest that there is synergism between these compounds that improves the antioxidant capacity compared with single antioxidant compounds [221].

Vitamins, pigments and other phenolic compounds are widely distributed in the plant kingdom and can be found in vegetables, fruits, roots, seeds, leaves, and other natural products such as honey, propolis and royal jelly. Natural products show a vast diversity of phytochemical compounds, including carotenoids, betalains, flavonoids, and other phenolic compounds [222]. They include structurally simple molecules as well as complex oligo/polymeric assemblies with very specific bio-physicochemical properties making them important candidates for new drugs [223].

Colored compounds or pigments, are generated from natural sources such as plants, animals, fungi and microorganisms and serve vital functions in cellular processes such as photosynthesis, protection of cells against UV light and oxidative stress and transport of oxygen in blood (heme). Dietary natural pigments can be classified by their structural characteristics as benzopyran derivatives (e.g., flavonoids), *N*-heterocyclic compounds (e.g., betalains) and isoprenoid derivatives (e.g., carotenoids) (reviewed in [224]).

The presence or absence of many bioactive compounds in natural products depends on geographical and environmental factors, such as humidity, temperature, season, pollution and altitude, as well as conditions of growth and storage, and the presence of genetically different varieties [225]. Therefore, it is difficult to standardize the chemical composition of natural products, but it is possible to link their biological activity to the presence of specific compounds with demonstrated therapeutic activity [226].

Plant-derived antioxidants, in particular pigments, have gained considerable importance due to their health benefits. Recently, plant foods and food-derived antioxidants have received growing attention, because they are known to function as chemopreventive agents and to protect against oxidative damage and genotoxicity [227]. The National Research Council (NRC) has recommended eating five or more servings of fruits and vegetables to increase the health benefits [228].

In addition, there is still a lack of knowledge about the potential therapeutic effect of plant-derived compounds against OTC analgesic and NSAIDs-induced acute liver failure. Therefore, additional animal studies and clinical trials are needed to discover new antidotes and therapeutic doses from natural sources to prevent or reduce drug-induced liver toxicity and acute failure.

6.1. Flavonoids

Flavonoids belong to the class of plant phenolic pigments with low molecular weight with a characteristic flavan nucleus as their main structure. These plant components have a protective function against UV radiation, pathogens and herbivores. They are distributed throughout the entire plant and more than 10,000 different flavonoid compounds have been identified [229,230].

All green plant cells are capable of synthesizing flavonoids. They may participate in the light-dependent phase of photosynthesis to catalyze electron transport, regulating the ion channels involved in photophosphoregulation [231]. Biosynthesis invariably starts with the ubiquitous amino acid phenylalanine through the phenylpropanoid pathway and takes different pathways depending on the kind of flavonoid that is synthesized (reviewed in [232–234]). The chemical nature of flavonoids depends on their structural class, degree of hydroxylation, additional substitutions and conjugations, and degree of polymerization [235]. They frequently occur attached to sugars (glycosides) increasing their water solubility [236]. Humans and animals cannot produce these plant-based antioxidants and they are normally obtained through the dietary consumption of vegetables, fruits, and teas [237].

Flavonoids are all derived from from a flavan nucleus composed of two phenolic rings (A and B) and one oxane (C) with 15 carbon atoms (C6-C3-C6) [238]. Flavonoids can be divided in different subclasses, based on the connecting position of the B and C rings as well as the degree of saturation, oxidation and hydroxylation of the C ring. These sub-classes include flavonols, isoflavonols, flavanones, flavan-3-ols (or catechins), flavones, isoflavones, anthocyanidins, aurones, flavandiols and flavanonols (Figure 5) (reviewed in [232,233,239]).

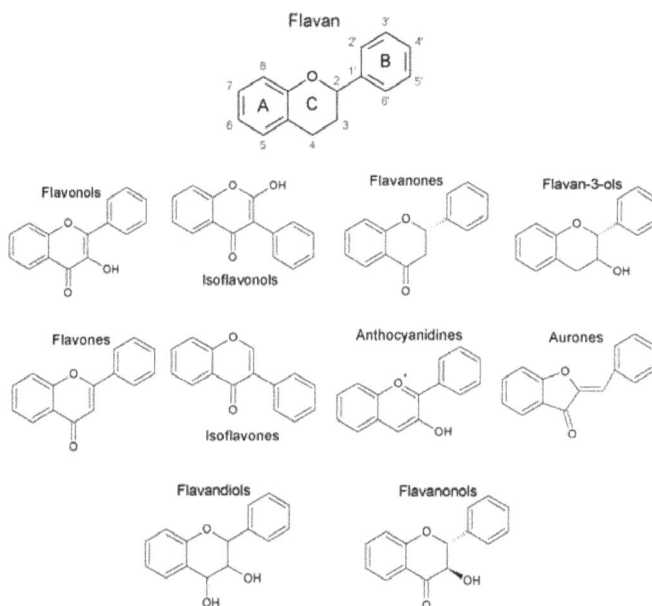

Figure 5. Basic flavonoid structure (flavan) and main classification of flavonoids.

The total dietary intake of flavonoids such as flavanones, flavonols, flavones, anthocyanins, catechins, and biflavans was initially estimated to be up to 1 g/day in the United States [240]. Subsequent studies suggested that those amounts were somewhat overestimated and the daily human intake of flavonoids was re-estimated to be approximately 23 mg/day [241,242]. Human consumption of flavonoids is not limited to only plant foods [243] but also via intake of plant extracts such as tea, coffee, and red wine [244]. Bioavailability of flavonoids depends on the food source, dosage and their physicochemical properties such as molecular size, lipophilicity, solubility and pKa. In nature, flavonoids are usually bound to sugars such as β-glycosides except for flavan-3-ols and proanthocyanidins. Glycoside conjugates need to be hydrolyzed in the small intestine to release the aglycone flavonoid and be absorbed by passive diffusion through the epithelium. After absorption, flavonoids can react with oxidants to inactivate them or they are metabolized in the liver, e.g., via conjugation reactions. Therefore, aglycone flavonoids cannot be detected in urine or plasma (reviewed in [239]). Recent evidence suggests that the colon plays an important role in the bioavailability of flavonoids since analysis of urinary flavonoid metabolites indicate the presence of colonic metabolites. These results suggest that some of these compounds and their metabolites may play an important role in the protective properties of a vegetable and fruit-rich diet (reviewed in [12]). Flavonoids have a short half-life (≈2–3 h) and they do not accumulate in the body explaining their low toxicity even when ingested for prolonged periods. Despite this, flavonoids might interfere in the biotransformation of many commonly used drugs such as losartan, digoxin, cyclosporine, vinblastine and fexofenadine by the inhibition of CYP450 enzymes, thus increasing their pharmacological potency (reviewed in [232,245]).

Flavonoids have played a major role as medical treatment in ancient times and their use has continued to present day [232]. Flavonoids can inhibit several important enzymes including ATPase, aldose reductase, hexokinase and tyrosine kinase [246]. Flavonoids can also induce several enzymes, e.g., aryl hydroxylase and epoxide hydroxylase. Thus, flavonoids seem to possess several pharmacological properties that make them excellent agents to serve as natural biological response modifiers [231]. In addition, modulation of the activity of pro-inflammatory enzymes is one of the most important mechanisms of action for flavonoids. Pro-inflammatory enzymes, such as cytosolic phospholipase A2 (cPLA2), cyclooxygenases (COX), lipoxygenases (LOX), and inducible NO synthase (iNOS), produce very potent inflammatory mediators and therefore their inhibition contributes to the overall anti-inflammatory potential of flavonoids (reviewed in [247]).

The flavonoid-related antioxidant activity is mainly due to their free radical scavenging potential. The rate and efficiency of this biological activity is determined by the presence of structural features such as the number and position of hydroxyl groups (A, B and C rings) and 2,3-double bond conjugated with a 4-oxo function (C ring). These structural features facilitate electron delocalization and radical absorption [248]. A quantum chemical study of flavonoids explains that the 3-hydroxyl group in ring C plays an important role in the activity of flavonols due to its capacity to interact with the positions $2'$ and $6'$ of ring B and the 4-keto group of ring C, making them the most potent flavonoid antioxidants [249]. In addition, in vivo studies have demonstrated that flavonoids can also indirectly induce antioxidant enzymes and increase the concentration of uric acid in plasma (reviewed in [250]). Based on these structural characteristics and biological activities, flavonoids have been evaluated in various diseases and toxicity models, including liver injury and toxicity to determine their protective effect (reviewed in [239]). Quercetin, one of the most abundant flavonoids, has been tested in different models of liver toxicity including carbon tetrachloride, ethanol, clivorine, thioacetamide and paracetamol. In these studies, orally or intraperitoneally administered quercetin, at doses of 40–90 mg/kg for mice and 20–50 mg/kg for rat, showed a strong inhibition of oxidative stress-induced injury, enhancing the redox status and reducing inflammation [251–255]. A mixture of flavonoids from *German chamomile* (160 mg/kg p.o. (per os) in vivo, and 250–750 μg/mL in vitro) and apigenin-7-glucoside (AP7Glu) (10–60 μM in vitro) protected against ethanol- and CCl$_4$-induced hepatocellular injury in rats and primary rat hepatocytes by normalizing ceramide content [256]. Another study showed that a mixture of nine flavonoids identified from *Laggera alata* extract had hepatoprotective effects at doses (orally) of 50–200 mg/kg in vivo and 1–100 μg/mL in vitro against CCl$_4$-induced liver injury in rats and primary rat hepatocytes due to its potent antioxidative and anti-inflammatory activity. These results also suggest that flavonoids can scavenge reactive oxygen species by non-enzymatic mechanisms and enhance the activity of hepatic antioxidant enzymes [257]. High glucose-induced oxidative stress contributes diabetes-related liver pathology. Anthocyanins have been reported to reduce intracellular reactive oxygen species (ROS) levels and cyanidin-3-O-β-glucoside (C3G) had the highest antioxidant activity. This flavonoid contributed to the prevention of hyperglycemia-induced hepatic oxidative damage, both in HepG2 cells (1–100 μM) and in mice (100 mg/kg p.o.). This compound induced the synthesis of glutathione (GSH) via increasing the expression of the glutamate-cysteine ligase catalytic subunit (Gclc) gene through the activation of protein kinase A (PKA) and the phosphorylation of cAMP-response element binding protein (CREB) as the target transcription factor (independently of Nrf1/2 transcription factors) [258]. In addition, in a study of anti-retroviral drugs-induced oxidative stress and liver toxicity in rats, silibinin (100 mg/kg p.o.), one of the most well-known and potent hepatoprotective flavonolignanes isolated from *Silybum marianum*, improved all biochemical and pathological parameters, confirming its hepatoprotective and antioxidant potential [259]. These studies suggest that dietary intake of flavonoids contributes to the prevention or treatment of hepatotoxicity induced by the exposure to xenobiotics and other environmental factors.

6.2. Betalains

Betalains are water-soluble plant pigments that contain nitrogen. They provide protection against UV radiation and pathogens and can act as optical attractants to pollinators. They are synthesized from the amino acid tyrosine by the condensation of betalamic acid, which is a common chromophore of all betalains. Betalains can be classified into betacyanins (red-violet) or betaxanthins (yellow-orange) depending on the nature of the groups conjugated to the betalamic acid (Figure 6). Betacyanins (e.g., betanin and betanidin) and betaxanthins (e.g., vulgaxanthin I and II) are divided into several subclasses, based on the chemical characteristics of the betalamic acid conjugate [260].

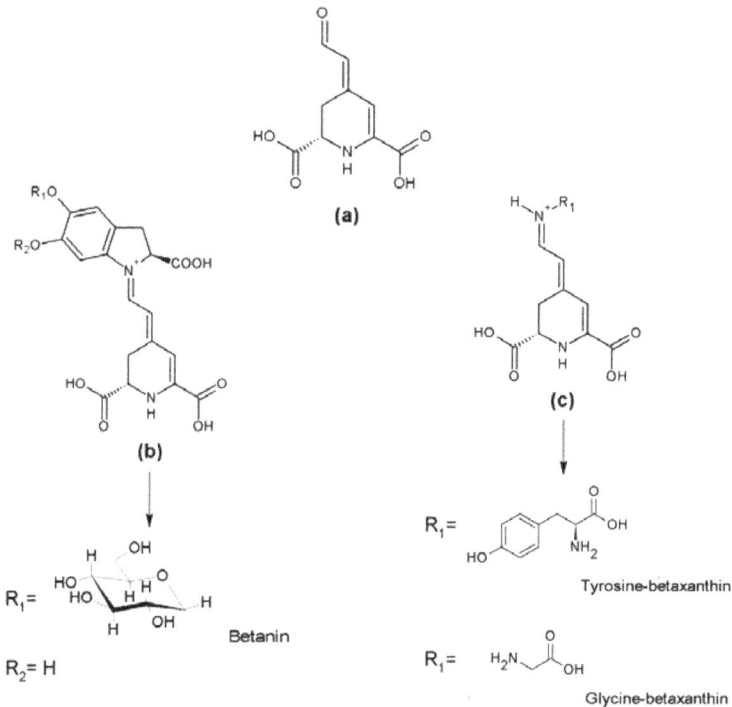

Figure 6. Betalamic acid (**a**), precursor of betalains. Betacyanins (**b**) and betaxanthins (**c**), as the main classes with some derivatives.

Betalamic acid can be conjugated with cyclo-3,4-dihydroxyphenylalanine (*cyclo*-Dopa) to form betacyanins and with amino acids or amines to form betaxanthins. The groups conjugated to the betalamic acid determine the absorption wavelength of around 480 nm for betaxanthins and around 540 nm for betacyanins. Structural modifications and the presence of side chains (e.g., sugars and amino acids) cause hypso- or bathochromic shifts (shorter or longer) in the absorption wavelengths [260–262].

Betalain pigments are particularly abundant in the Caryophyllales order and can be found in roots, flowers, fruits and some vegetative tissues of plants [263]. However, betalains are not produced in the Caryophyllaceae and Molluginaceae families, since these families accumulate anthocyanins, flavonoid-derived pigments. Betalains and anthocyanins have never been reported together in the same plant, suggesting they are mutually exclusive [264]. This appears to be an evolutionary mechanism of adaptation to protect against environmental factors and predators [265].

Betalains are cationic compounds with a high affinity for negatively charged membranes, thus improving their efficacy as antioxidants. The active cyclic amine group of betalains functions as hydrogen donor and confers reducing properties to these compounds. In addition, the betacyanins such as betanin and betanidin have an enhanced antioxidant capacity in vitro compared to betaxanthins, catechin and α-tocopherol. This is due to the presence of a phenolic ring which increases their electron transfer capability, making them superior antioxidants [266]. It has been suggested that the free radical scavenging activity of betanin is pH-dependent. It is also important to consider the contribution of anionic forms of cyclo-DOPA-5-O-β-D-glucoside to the antioxidant activity of betanin at basic pH. These data suggest that betanin is a better hydrogen and electron donor at higher pH, contributing to its enhanced free radical-scavenging activity [267].

Several studies have demonstrated potent radical scavenging activity of betalains and their derivatives in vitro [268–272]. The antioxidant protection against reactive radical species may be mediated either by direct free radical scavenging or by the induction of endogenous antioxidant defense mechanisms under the control of redox-regulated transcription factors such as Nrf2 [266]. Nrf2 orchestrates the expression of genes encoding antioxidant and phase II enzymes such as heme-oxygenase 1 (HO-1), NAD(P)H:quinone oxidoreductase 1 (NQO1) and glutathione S-transferases (GST). The chemo-preventive effect of betanin is due to the induction of endogenous redox-system enzymes and glutathione synthesis mediated by Nrf2 dependent signal transduction pathways [272]. Chemopreventive compounds may enhance the transcriptional activity of Nrf2 independently from Keap1 [273].

The dietary intake of betalains in humans is limited, since they can be obtained only from red beet, swiss chard, amaranthus, cactus pear, pitaya and some tubers and their derived products. They are also used as food colorants providing another dietary source of these phytochemicals. The intake of red beet juice or cactus pear revealed the bioavailability of betalains. They can be absorbed by the small intestine into the systemic circulation in their intact forms, indicating that hydrolysis reactions are not always necessary for absorption as reported for some glycosylated flavonoids, although the bioavailability of betalains has been reported to be only $\approx$1% of the consumed amount. The maximum concentration of betalains is detected in plasma 3 h after intake. Moreover, some unknown metabolites have been detected in urine after betalain intake. The biological actions of these metabolites are unknown. Betalains are capable to bind to biological membranes reducing their bioavailability and capacity to react with other macromolecules. The major commercially exploited source of betalain is red beetroot (*Beta vulgaris*), which contains two major soluble pigments: betanin (red) and vulgaxanthine I (yellow). Previous studies reported that the betacyanin and betaxanthin content of red beet roots varies between 0.04–0.21% and 0.02–0.14%, respectively, depending on the cultivar, although some new varieties have higher betalain content (reviewed in [274]). Red beet roots contain a large amount of betanin, 300–600 mg/kg, and lower concentrations of isobetanin, betanidin and betaxanthins [266]. The prickly pear (*Opuntia ficus indica*) contains about 50 mg/kg of betanin and 26 mg/kg of indicaxanthin [269]. *Opuntia robusta* and *Opuntia streptacantha* fruit juices also contain large amounts of betacyanins (333 and 87 mg/L, respectively) and betaxanthin (134 and 36 mg/L, respectively) [7]. The concentration of betacyanins in red pitaya (*Hylocereus* cacti), expressed as betanin equivalents ranges from 100–400 mg/kg depending on the species [275,276].

Thanks to the increasing interest in the antioxidant properties of betalains, some studies have focused on their benefits against oxidative stress-related organ damage, although there are only a few studies on the hepatoprotective effect of betalains. Red beet juice and other red beet products, frequently used in dietary products, contain many bioactive compounds, the most important being betalains and may provide protection against certain oxidative stress-related disorders in humans [266]. Betalains from the berries of *Rivina humilis* showed dose-dependent cytotoxicity (0–40 μg/mL) in HepG2 hepatoma cells [277]. Frequent consumption of purple *Opuntia* cactus fruit juices rich in betalains protects hepatocytes against oxidative stress and improves the redox balance in acetaminophen-induced acute liver toxicity in vivo (800 mg/kg of lyophilized juice p.o.) and in vitro (8% v/v) [7]. In vitro studies have subsequently demonstrated that the free radical scavenging capacity of betanin (1–35 μM) via activation of the Nrf2 pathway and subsequent induction of Nrf2 target genes, is responsible for its hepatoprotective and anticarcinogenic effects [272,273]. Betanin (1–4% in fodder) also attenuated CCl_4-induced liver damage in common carp (*Cyprinus carpio* L.) by the inhibition of CYP2E1 activity and reduction of oxidative stress [278]. The role of betalains as hepatoprotective and chemopreventive compounds can be summarized by their ability to stabilize cellular membranes, scavenge free radicals or electrophilic metabolites, and improve the cellular redox-balance. Nevertheless, the optimal daily intake of betalains to achieve hepatoprotection in humans has not been elucidated yet due to their apparent poor stability and bioavailability. In addition, there is still a lack of information regarding the safety of consumption of processed betalains and their possible pharmacological interactions, necessitating more toxicological and clinical studies.

6.3. Carotenoids

Carotenoids are colored (red, orange, and yellow) pigments widely distributed in nature with over 700 structurally different compounds [279]. These lipophilic molecules are synthesized in photosynthetic organelles as well as in fruits, flowers, seeds, and storage roots. Carotenoids have important functions in photosynthesis in plant and algae, protection against UV light and in photomorphogenesis. Some microorganisms such as yeast and bacteria can also produce carotenoids that function in alleviating photo-oxidative damage [280,281]. Structurally, carotenoids are composed of eight isoprenoid units and can be considered lycopene ($C_{40}H_{56}$) derivatives. The synthesis of this group of pigments involves hydrogenation, dehydrogenation, cyclization, oxidation, double bond migration, methyl migration, chain elongation or chain shortening [282].

The natural functions and actions of carotenoids are determined by their physical and chemical properties, which are defined by the molecular structure. Carotenoids must have a specific molecular geometry to ensure their biological activity. The presence and number of double bonds in their structures define to a large extent the chemical properties, e.g., lipophilicity that facilitates their ability to inactivate free radicals and protect cellular membranes [283]. The carotenoid classification is based on their chemical composition: carotenoids composed of carbon and hydrogen and carotenoids composed of carbon, hydrogen and oxygen, also called oxycarotenoids or xanthophylls. Alternatively, carotenoids may also be classified as primary carotenoids that are required for plant photosynthesis (β-carotene, violaxanthin, and neoxanthin) and secondary carotenoids that serve as UV protectors and attractants present in flowers and fruits (α-carotene, β-cryptoxanthin, zeaxanthin, lutein, astaxanthin, and capsanthin) [284] (Figure 7).

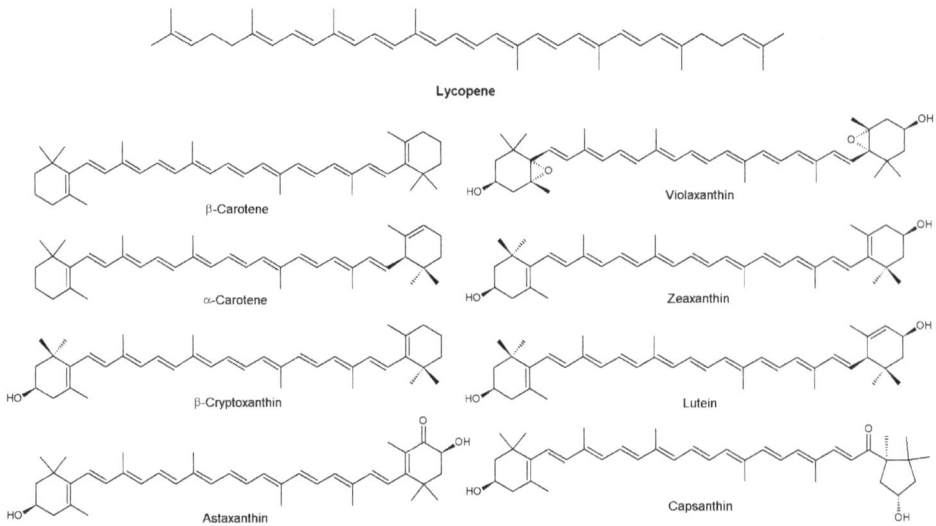

Figure 7. Chemical structure of lycopene with the main primary and secondary carotenoids.

Carotenoid biosynthesis in plants is highly regulated, although all processes involved are not completely elucidated yet. In recent decades, most carotenogenic genes from various plants, yeasts and algae have been identified and their functions elucidated [281]. Carotenoids are synthesized via the isoprenoid pathway [285] in a highly regulated process using isopentenyl pyrophosphate (IPP) as a common precursor of many isoprenoid compounds. The synthesis and amount of these pigments in plants are based on the requirements to maintain vital functions and to protect them against environmental factors [286]. Humans and animals depend on dietary intake for carotenoid supply, since they cannot synthesize carotenoids de novo [287].

Humans only have access to 40–50 of all structurally identified carotenoids via the diet, although only ≈20 have been detected in human blood plasma including α-, β- and γ-carotene, lycopene, α- and β-cryptoxanthin, zeaxanthin, lutein, neurosporene, phytofluene and phytoene (reviewed in [288]). These carotenoids are abundant in tomatoes, carrots, oranges, beans, beet, broccoli, Brussel sprouts, coleslaw, celery, zucchini, pepper, spinach, cucumber, mango, and watermelon [289]. Recommended dietary intake is based on the contribution of carotenoids to (pro)vitamin A content. Vitamin A (retinol) can be obtained preformed from the diet through meat and dairy products, or from vegetables and fruits as carotenoids that are subsequently converted into provitamin A. It has been estimated that 14 µg of β-carotene is necessary to yield 1 µg of retinol (i.e., 1 retinol activity equivalent). Based on FAO and WHO recommendations, an intake of 700–900 µg of retinol per day is recommended, equal to approximately 11 mg/day of β-carotene [290]. It has been suggested that only ≈5% of the total intake of carotenoids is absorbed in the small intestine. Once absorbed, carotenoids can be detected in plasma lipoproteins, mainly LDL. Carotenoid absorption, bioavailability, breakdown, transport and storage is dependent on a number of factors, including the type of carotenoid and dietary source, as well as genetic factors, nutritional status, age, gender and diseases. Moreover, it has been reported that individual carotenoids may inhibit the absorption of each other. In addition, carotenoid availability may be decreased by interactions with ASA and sulfonamides (reviewed in [280,288]). Carotenoids are also found in nutritional supplements and other dietary sources since they are also used as food colorants [291], since they have been increasingly exploited by food, nutraceutical and pharmaceutical companies due to the recent interest in their biological potential as antioxidants and treatment against chronic or age-related diseases in humans [280].

The direct antioxidant activity of carotenoids, both in vitro and in vivo, depends on the presence and number of functional groups such as carbonyl and hydroxyl groups. Thus, capsanthin and astaxanthin display better antioxidant activity than β-carotene or zeaxanthin. Besides, carotenoids appear to modulate the expression of antioxidant enzymes and it has been suggested that carotenoids might have a synergistic effect with other dietary antioxidants (reviewed in [292]). The antioxidant action of carotenoids is due to their ability to inactivate singlet oxygen and other free radicals. Moreover, carotenoids preferentially quench peroxyl radicals even in the presence of other oxidizable substrates. They also act as chain-breaking antioxidants, thus preventing lipid peroxidation and protecting cell membranes, with an almost equally potency as α-tocopherol [293].

Carotenoids accumulate mainly in the liver and are incorporated into lipoproteins for release into the circulation. Carotenoids contribute to the antioxidant defense system in the liver as scavengers of free radicals. In patients with chronic liver diseases, micronutrient antioxidants are severely depleted in serum and liver tissue and liver injury is associated with decreased antioxidant levels, particularly carotenoids. Carotenoids may have beneficial effect in non-alcoholic fatty liver disease (NAFLD) probably via multiple mechanisms, including antioxidant and anti-inflammatory effects and regulation of M1/M2 macrophage polarization (reviewed in [294]).

Carotenoids also demonstrate benefits in cancer and stroke since in these diseases free radicals play an important role. It has been suggested that carotenoids influence the strength and fluidity of membranes, thus affecting its permeability to oxygen and other molecules. In vivo and in vitro studies have shown that the photo-protective role of carotenoids is related to its antioxidant capacity (reviewed in [283,295]).

Recent studies have demonstrated the hepatoprotective action of carotenoids in different models of liver toxicity and hepatitis. Bixin, the main carotenoid from *Bixa Orellana* L. (annatto) seeds, used prophylactically for seven days at 5 mg/kg p.o. protected against CCl_4-induced liver toxicity by reducing lipid peroxidation [296]. In the same model of liver toxicity, similar results were obtained with a prophylactic treatment for 14 days with astaxanthin (ASX) and astaxanthin esters (ASXEs) at 100 and 250 μg/kg p.o. from the green microalga *Haematococcus pluvialis* [297]. Likewise, lycopene-enriched tomato paste (at lycopene equivalence doses of 0.5 and 2.5 mg/kg p.o. for 28 days prior intoxication) increased the activity of cellular antioxidant enzymes such as superoxide dismutase, catalase and GSH-peroxidase and reduced microsomal lipid peroxidation induced by *N*-nitrosodiethylamine in rat liver [298]. In a rat high fat diet model of NAFLD, lycopene at 5, 10 and 20 mg/kg p.o. for six weeks, reduced liver damage, increased both enzymatic and non-enzymatic antioxidant defenses and reduced levels of the pro-inflammatory cytokine TNFα [299]. Prophylactic treatment of lycopene at 10 mg/kg i.p. (intraperitoneally) for six days was also protective against D-galactosamine/Lipopolysaccharide-induced hepatitis in rats [300]. These studies demonstrate that carotenoids are potent antioxidants and protect cellular membranes, suggesting that they can also be protective against oxidative liver damage induced by acute and chronic exposure to analgesic and non-steroidal anti-inflammatory OTC drugs.

7. Concluding Remarks

The overconsumption and misuse of OTC analgesic drugs is a clinical problem of epidemic proportion. It is the cause of a significant rise in acute and chronic liver diseases, especially in developing countries, due to a lack of information about the risk of self-medication and a lack of affordable access to general practitioners. Plant and natural products as traditional medicine have been used for many centuries throughout the world. Currently, there is a trend towards increased consumption of these products to treat or prevent many diseases. It has been demonstrated that many of these products and their secondary metabolites have beneficial effects in many different pathologies, especially those associated with inflammation and oxidative stress. An important issue that still needs attention is the bioavailability of natural pigments. For example, there is still considerable controversy about the bioavailability of hydrophilic flavonoids and betalains, in contrast to, e.g., carotenoids. Therefore, additional information about the pharmacokinetic and pharmacodynamic properties of these compounds are needed. Natural pigments appear to be biologically active at very low concentrations and their effects go beyond that of simple anti-oxidants. Natural pigments, or their metabolites, often induce highly specific intracellular responses modulating specific signaling pathways to protect cells. These specific qualities make natural dietary pigments excellent candidates for the treatment and/or prevention of OTC-drug-induced liver diseases. Several studies have evaluated the use of crude extracts or isolated compounds of natural products and observed beneficial effects in cell culture or animal models. However, these studies were mostly performed using the natural pigments as prophylactic. The challenges for the future application of these products as therapeutic or preventive interventions are: (1) identifying the most effective compounds in crude extracts; (2) to reconstitute the most effective mixtures of purified compounds that preserve synergy between individual components; (3) the reproducible cultivation and preparation of (extracts of) natural products with adequate bioanalytical characterization and quality control; and (4) pre-clinical studies in humans to extend the experimental findings to clinical application.

Acknowledgments: Financial support was obtained from the Graduate School of Medical Sciences of the University of Groningen, The Netherlands.

Author Contributions: All authors contributed to the preparation of this review and approved the text.

Conflicts of Interest: The authors declare no conflict of interest.

References

1. Chin, Y.-W.; Balunas, M.J.; Chai, H.B.; Kinghorn, A.D. Drug discovery from natural sources. *AAPS J.* **2006**, *8*, E239–E253. [CrossRef] [PubMed]
2. Newman, D.J.; Cragg, G.M. Natural Products as Sources of New Drugs over the Last 25 Years. *J. Nat. Prod.* **2007**, *70*, 461–477. [CrossRef] [PubMed]
3. Newman, D.J.; Cragg, G.M.; Snader, K.M. The influence of natural products upon drug discovery (Antiquity to late 1999). *Nat. Prod. Rep.* **2000**, *17*, 215–234. [CrossRef] [PubMed]
4. Gu, J.; Gui, Y.; Chen, L.; Yuan, G.; Lu, H.-Z.; Xu, X. Use of Natural Products as Chemical Library for Drug Discovery and Network Pharmacology. *PLoS ONE* **2013**, *8*, e62839. [CrossRef] [PubMed]
5. Sreekanth, D.; Arunasree, M.K.; Roy, K.R.; Chandramohan Reddy, T.; Reddy, G.V.; Reddanna, P. Betanin a betacyanin pigment purified from fruits of Opuntia ficus-indica induces apoptosis in human chronic myeloid leukemia Cell line-K562. *Phytomedicine* **2007**, *14*, 739–746. [CrossRef] [PubMed]
6. Lapshina, E.A.; Zamaraeva, M.; Cheshchevik, V.T.; Olchowik-Grabarek, E.; Sekowski, S.; Zukowska, I.; Golovach, N.G.; Burd, V.N.; Zavodnik, I.B. Cranberry flavonoids prevent toxic rat liver mitochondrial damage in vivo and scavenge free radicals in vitro: Cranberry flavonoids prevent mitochondrial damage and scavenge free radicals. *Cell Biochem. Funct.* **2015**, *33*, 202–210. [CrossRef] [PubMed]
7. González-Ponce, H.; Martínez-Saldaña, M.; Rincón-Sánchez, A.; Sumaya-Martínez, M.; Buist-Homan, M.; Faber, K.N.; Moshage, H.; Jaramillo-Juárez, F. Hepatoprotective Effect of *Opuntia robusta* and *Opuntia streptacantha* Fruits against Acetaminophen-Induced Acute Liver Damage. *Nutrients* **2016**, *8*, 607. [CrossRef] [PubMed]
8. Pang, C.; Zheng, Z.; Shi, L.; Sheng, Y.; Wei, H.; Wang, Z.; Ji, L. Caffeic acid prevents acetaminophen-induced liver injury by activating the Keap1-Nrf2 antioxidative defense system. *Free Radic. Biol. Med.* **2016**, *91*, 236–246. [CrossRef] [PubMed]
9. Dewick, P.M. *Medicinal Natural Products: A Biosynthetic Approach*, 3rd ed.; John Wiley and Sons, Ltd.: Chichester, UK, 2009; 539p.
10. Dias, D.A.; Urban, S.; Roessner, U. A Historical Overview of Natural Products in Drug Discovery. *Metabolites* **2012**, *2*, 303–336. [CrossRef] [PubMed]
11. Donovan, J.L.; Manach, C.; Faulks, R.M.; Kroon, P.A. Absorption and metabolism of dietary plant secondary metabolites. In *Plant Secondary Metabolites Occurrence, Structure and Role in the Human Diet*; Blackwell Publishing Ltd.: Hoboken, NJ, USA, 2006; pp. 303–351.

12. Crozier, A.; Del Rio, D.; Clifford, M.N. Bioavailability of dietary flavonoids and phenolic compounds. *Mol. Asp. Med.* **2010**, *31*, 446–467. [CrossRef] [PubMed]
13. Hong, J. Role of natural product diversity in chemical biology. *Curr. Opin. Chem. Biol.* **2011**, *15*, 350–354. [CrossRef] [PubMed]
14. Kingston, D.G.I. Modern Natural Products Drug Discovery and Its Relevance to Biodiversity Conservation. *J. Nat. Prod.* **2011**, *74*, 496–511. [CrossRef] [PubMed]
15. Koehn, F.E.; Carter, G.T. The evolving role of natural products in drug discovery. *Nat. Rev. Drug Discov.* **2005**, *4*, 206–220. [CrossRef] [PubMed]
16. Feher, M.; Schmidt, J.M. Property Distributions: Differences between Drugs, Natural Products, and Molecules from Combinatorial Chemistry. *J. Chem. Inf. Comput. Sci.* **2003**, *43*, 218–227. [CrossRef] [PubMed]
17. Vonkeman, H.E.; van de Laar, M.A.F.J. Nonsteroidal Anti-Inflammatory Drugs: Adverse Effects and Their Prevention. *Semin. Arthritis Rheum.* **2010**, *39*, 294–312. [CrossRef] [PubMed]
18. Fendrick, A.M.; Pan, D.E.; Johnson, G.E. OTC analgesics and drug interactions: Clinical implications. *Osteopath. Med. Prim. Care* **2008**, *2*. [CrossRef] [PubMed]
19. Bronstein, A.C.; Spyker, D.A.; Cantilena, L.R.; Rumack, B.H.; Dart, R.C. 2011 Annual Report of the American Association of Poison Control Centers' National Poison Data System (NPDS): 29th Annual Report. *Clin. Toxicol.* **2012**, *50*, 911–1164. [CrossRef] [PubMed]
20. Nourjah, P.; Ahmad, S.R.; Karwoski, C.; Willy, M. Estimates of acetaminophen (paracetamol)-associated overdoses in the United States. *Pharmacoepidemiol. Drug Saf.* **2006**, *15*, 398–405. [CrossRef] [PubMed]
21. Fontana, R.J. Acute Liver Failure including Acetaminophen Overdose. *Med. Clin. N. Am.* **2008**, *92*, 761–794. [CrossRef] [PubMed]
22. Poljsak, B.; Šuput, D.; Milisav, I. Achieving the Balance between ROS and Antioxidants: When to Use the Synthetic Antioxidants. *Oxid. Med. Cell. Longev.* **2013**, *2013*. [CrossRef] [PubMed]
23. Cordell, W.H.; Keene, K.K.; Giles, B.K.; Jones, J.B.; Jones, J.H.; Brizendine, E.J. The high prevalence of pain in emergency medical care. *Am. J. Emerg. Med.* **2002**, *20*, 165–169. [CrossRef] [PubMed]
24. Lagerløv, P.; Rosvold, E.O.; Holager, T.; Helseth, S. How adolescents experience and cope with pain in daily life: A qualitative study on ways to cope and the use of over-the-counter analgesics. *BMJ Open* **2016**, *6*, e010184. [CrossRef] [PubMed]
25. Hersh, E.V.; Moore, P.A.; Ross, G.L. Over-the-counter analgesics and antipyretics: A critical assessment. *Clin. Ther.* **2000**, *22*, 500–548. [CrossRef]
26. Holmström, I.K.; Bastholm-Rahmner, P.; Bernsten, C.; Röing, M.; Björkman, I. Swedish teenagers and over-the-counter analgesics—Responsible, casual or careless use. *Res. Soc. Adm. Pharm.* **2014**, *10*, 408–418. [CrossRef] [PubMed]
27. Hersh, E.V.; Pinto, A.; Moore, P.A. Adverse drug interactions involving common prescription and over-the-counter analgesic agents. *Clin. Ther.* **2007**, *29*, 2477–2497. [CrossRef] [PubMed]
28. Heard, K.; Sloss, D.; Weber, S.; Dart, R.C. Overuse of Over-the-Counter Analgesics by Emergency Department Patients. *Ann. Emerg. Med.* **2006**, *48*, 315–318. [CrossRef] [PubMed]
29. Goh, L.Y.; Vitry, A.I.; Semple, S.J.; Esterman, A.; Luszcz, M.A. Self-medication with over-the-counter drugs and complementary medications in South Australia's elderly population. *BMC Complement. Altern. Med.* **2009**, *9*, 42. [CrossRef] [PubMed]
30. Daly, F.F.; O'malley, G.F.; Heard, K.; Bogdan, G.M.; Dart, R.C. Prospective evaluation of repeated supratherapeutic acetaminophen (paracetamol) ingestion. *Ann. Emerg. Med.* **2004**, *44*, 393–398. [CrossRef] [PubMed]
31. Kaufman, D.W.; Kelly, J.P.; Rosenberg, L.; Anderson, T.E.; Mitchell, A.A. Recent patterns of medication use in the ambulatory adult population of the United States: The slone survey. *JAMA* **2002**, *287*, 337–344. [CrossRef] [PubMed]
32. Bjarnason, I. Gastrointestinal safety of NSAIDs and over-the-counter analgesics: Gastrointestinal safety of NSAIDs and over-the-counter analgesics. *Int. J. Clin. Pract.* **2013**, *67*, 37–42. [CrossRef] [PubMed]
33. Sarganas, G.; Buttery, A.K.; Zhuang, W.; Wolf, I.-K.; Grams, D.; Rosario, A.S.; Scheidt-Nave, C.; Knopf, H. Prevalence, trends, patterns and associations of analgesic use in Germany. *BMC Pharmacol. Toxicol.* **2015**, *16*, 28. [CrossRef] [PubMed]
34. Warlé-van Herwaarden, M.F.; Kramers, C.; Sturkenboom, M.C.; van den Bemt, P.M.; De Smet, P.A. Targeting Outpatient Drug Safety. *Drug Saf.* **2012**, *35*, 245–259. [CrossRef] [PubMed]

35. Koffeman, A.R.; Valkhoff, V.E.; Celik, S.; Jong, G.W.; Sturkenboom, M.C.; Bindels, P.J.; van der Lei, J.; Luijsterburg, P.A.; Bierma-Zeinstra, S.M. High-risk use of over-the-counter non-steroidal anti-inflammatory drugs: A population-based cross-sectional study. *Br. J. Gen. Pract.* **2014**, *64*, e191–e198. [CrossRef] [PubMed]

36. Sternbach, R.A. Survey of Pain in the United States: The Nuprin Pain Report. *Clin. J. Pain* **1986**, *2*, 49–53. [CrossRef]

37. Edmeads, J.; Findlay, H.; Tugwell, P.; Pryse-Phillips, W.; Nelson, R.F.; Murray, T.J. Impact of Migraine and Tension-Type Headache on Life-Style, Consulting Behaviour, and Medication Use: A Canadian Population Survey. *Can. J. Neurol. Sci.* **1993**, *20*, 131–137. [CrossRef] [PubMed]

38. Steele, K.; Mills, K.A.; Gilliland, A.E.; Irwin, W.G.; Taggart, A. Repeat prescribing of non-steroidal anti-inflammatory drugs excluding aspirin: How careful are we? *Br. Med. J. Clin. Res. Ed.* **1987**, *295*, 962–964. [CrossRef] [PubMed]

39. Abbott, F.V.; Fraser, M.I. Use and abuse of over-the-counter analgesic agents. *J. Psychiatry Neurosci.* **1998**, *23*, 13. [PubMed]

40. Wang, X.; Wu, Q.; Liu, A.; Anadón, A.; Rodríguez, J.-L.; Martínez-Larrañaga, M.-R.; Yuan, Z.; Martínez, M.-A. Paracetamol: Overdose-induced oxidative stress toxicity, metabolism and protective effects of various compounds in vivo and in vitro. *Drug Metabol. Rev.* **2017**, *49*, 395–437. [CrossRef] [PubMed]

41. Zakharov, S.; Navratil, T.; Pelclova, D. Suicide attempts by deliberate self-poisoning in children and adolescents. *Psychiatry Res.* **2013**, *210*, 302–307. [CrossRef] [PubMed]

42. Wolf, M.S.; King, J.; Jacobson, K.; Di Francesco, L.; Bailey, S.C.; Mullen, R.; McCarthy, D.; Serper, M.; Davis, T.C.; Parker, R.M. Risk of Unintentional Overdose with Non-Prescription Acetaminophen Products. *J. Gen. Intern. Med.* **2012**, *27*, 1587–1593. [CrossRef] [PubMed]

43. Moore, N.; Pollack, C.; Butkerait, P. Adverse drug reactions and drug-drug interactions with over-the-counter NSAIDs. *Ther. Clin. Risk Manag.* **2015**, *11*, 1061–1075. [PubMed]

44. Wazaify, M.; Kennedy, S.; Hughes, C.M.; McElnay, J.C. Prevalence of over-the-counter drug-related overdoses at Accident and Emergency departments in Northern Ireland—A retrospective evaluation. *J. Clin. Pharm. Ther.* **2005**, *30*, 39–44. [CrossRef] [PubMed]

45. Duarte, S.; Baber, J.; Fujii, T.; Coito, A.J. Matrix metalloproteinases in liver injury, repair and fibrosis. *Matrix Biol.* **2015**, *44–46*, 147–156. [CrossRef] [PubMed]

46. Si-Tayeb, K.; Lemaigre, F.P.; Duncan, S.A. Organogenesis and Development of the Liver. *Dev. Cell* **2010**, *18*, 175–189. [CrossRef] [PubMed]

47. Correa, P.R.A.; Nathanson, M.H. Functions of the liver. In *Textbook of Hepatology: From Basic Science to Clinical Practice*, 3rd ed.; Blackwell Publishing Ltd.: Hoboken, NJ, USA, 2007; Chapter 2.2, pp. 89–128.

48. Kmieć, Z. Introduction—Morphology of the Liver Lobule. In *Cooperation of Liver Cells in Health and Disease*; Kmieć, Z., Ed.; Springer: Berlin/Heidelberg, Germany, 2001; pp. 1–6. Available online: http://dx.doi.org/10.1007/978-3-642-56553-3_1 (accessed on 4 May 2017).

49. DeLeve, L. Hepatic Microvasculature in Liver Injury. *Semin. Liver Dis.* **2007**, *27*, 390–400. [CrossRef] [PubMed]

50. Hernandez-Gea, V.; Friedman, S.L. Pathogenesis of Liver Fibrosis. *Annu. Rev. Pathol. Mech. Dis.* **2011**, *6*, 425–456. [CrossRef] [PubMed]

51. Guicciardi, M.E.; Malhi, H.; Mott, J.L.; Gores, G.J. Apoptosis and Necrosis in the Liver. In *Comprehensive Physiology*; Terjung, R., Ed.; John Wiley & Sons, Inc.: Hoboken, NJ, USA, 2013. Available online: http://doi.wiley.com/10.1002/cphy.c120020 (accessed on 4 May 2017).

52. Gebhardt, R. Metabolic zonation of the liver: Regulation and implications for liver function. *Pharmacol. Ther.* **1992**, *53*, 275–354. [CrossRef]

53. Gonzalez, F.J.; Yu, A.-M. Cytochrome P450 and xenobiotic receptor humanized mice. *Annu. Rev. Pharmacol. Toxicol.* **2006**, *46*, 41–64. [CrossRef] [PubMed]

54. Vollmar, B.; Menger, M.D. The Hepatic Microcirculation: Mechanistic Contributions and Therapeutic Targets in Liver Injury and Repair. *Physiol. Rev.* **2009**, *89*, 1269–1339. [CrossRef] [PubMed]

55. Protzer, U.; Maini, M.K.; Knolle, P.A. Living in the liver: Hepatic infections. *Nat. Rev. Immunol.* **2012**, *12*, 201–213. [CrossRef] [PubMed]

56. Njoku, D. Drug-Induced Hepatotoxicity: Metabolic, Genetic and Immunological Basis. *Int. J. Mol. Sci.* **2014**, *15*, 6990–7003. [CrossRef] [PubMed]

57. Nebert, D.W.; Russell, D.W. Clinical importance of the cytochromes P450. *Lancet* **2002**, *360*, 1155–1162. [CrossRef]

58. Guengerich, F.P. Cytochromes P450, drugs, and diseases. *Mol. Interv.* **2003**, *3*, 194–204. [CrossRef] [PubMed]

59. Shimada, T. Xenobiotic-Metabolizing Enzymes Involved in Activation and Detoxification of Carcinogenic Polycyclic Aromatic Hydrocarbons. *Drug Metab. Pharmacokinet.* **2006**, *21*, 257–276. [CrossRef] [PubMed]

60. Xu, C.; Li, C.Y.-T.; Kong, A.-N.T. Induction of phase I, II and III drug metabolism/transport by xenobiotics. *Arch. Pharm. Res.* **2005**, *28*, 249. [CrossRef] [PubMed]

61. Kanai, K.; Kanamura, S.; Watanabe, J. Peri- and postnatal development of heterogeneity in the amounts of endoplasmic reticulum in mouse hepatocytes. *Am. J. Anat.* **1986**, *175*, 471–480. [CrossRef] [PubMed]

62. Forman, H.J.; Kennedy, J.A. Role of superoxide radical in mitochondrial dehydrogenase reactions. *Biochem. Biophys. Res. Commun.* **1974**, *60*, 1044–1050. [CrossRef]

63. Loschen, G.; Azzi, A.; Richter, C.; Flohé, L. Superoxide radicals as precursors of mitochondrial hydrogen peroxide. *FEBS Lett.* **1974**, *42*, 68–72. [CrossRef]

64. Halliwell, B. Reactive Species and Antioxidants. Redox Biology Is a Fundamental Theme of Aerobic Life. *Plant Physiol.* **2006**, *141*, 312–322. [CrossRef] [PubMed]

65. Harman, D. Aging: A Theory Based on Free Radical and Radiation Chemistry. *J. Gerontol.* **1956**, *11*, 298–300. [CrossRef] [PubMed]

66. McCord, J.M.; Fridovich, I. Superoxide dismutase an enzymic function for erythrocuprein (hemocuprein). *J. Biol. Chem.* **1969**, *244*, 6049–6055. [PubMed]

67. Finkel, T.; Holbrook, N.J. Oxidants, oxidative stress and the biology of ageing. *Nature* **2000**, *408*, 239–247. [CrossRef] [PubMed]

68. Zorov, D.B.; Filburn, C.R.; Klotz, L.-O.; Zweier, J.L.; Sollott, S.J. Reactive oxygen species (ROS-induced) ROS release. *J. Exp. Med.* **2000**, *192*, 1001–1014. [CrossRef] [PubMed]

69. Boonstra, J.; Post, J.A. Molecular events associated with reactive oxygen species and cell cycle progression in mammalian cells. *Gene* **2004**, *337*, 1–13. [CrossRef] [PubMed]

70. Wong, G.H. Protective roles of cytokines against radiation: Induction of mitochondrial MnSOD. *Biochim. Biophys. Acta (BBA) Mol. Basis Dis.* **1995**, *1271*, 205–209. [CrossRef]

71. Mokim Ahmed, K.; Li, J.J. NF-κB-mediated adaptive resistance to ionizing radiation. *Free Radic. Biol. Med.* **2008**, *44*, 1–13. [CrossRef] [PubMed]

72. Muriel, P. Role of free radicals in liver diseases. *Hepatol. Int.* **2009**, *3*, 526–536. [CrossRef] [PubMed]

73. Dröge, W. Free Radicals in the Physiological Control of Cell Function. *Physiol. Rev.* **2002**, *82*, 47–95. [CrossRef] [PubMed]

74. Forman, H.J. Redox signaling: Thiol chemistry defines which reactive oxygen and nitrogen species can act as second messengers. *Am. J. Physiol.* **2004**, *287*, C246–C256. [CrossRef] [PubMed]

75. Turrens, J.F. Superoxide production by the mitochondrial respiratory chain. *Biosci. Rep.* **1997**, *17*, 3–8. [CrossRef] [PubMed]

76. Halliwell, B. Reactive Oxygen Species and the Central Nervous System. *J. Neurochem.* **1992**, *59*, 1609–1623. [CrossRef] [PubMed]

77. Geiszt, M.; Kopp, J.B.; Várnai, P.; Leto, T.L. Identification of Renox, an NAD(P)H oxidase in kidney. *Proc. Natl. Acad. Sci. USA* **2000**, *97*, 8010–8014. [CrossRef] [PubMed]

78. Suh, Y.-A.; Arnold, R.S.; Lassegue, B.; Shi, J.; Xu, X.; Sorescu, D.; Chung, A.; Griendling, K.; Lambeth, J. Cell transformation by the superoxide-generating oxidase Mox1. *Nature* **1999**, *401*, 79–82. [CrossRef] [PubMed]

79. Bartsch, H.; Nair, J. Chronic inflammation and oxidative stress in the genesis and perpetuation of cancer: Role of lipid peroxidation, DNA damage, and repair. *Langenbeck's Arch. Surg.* **2006**, *391*, 499–510. [CrossRef] [PubMed]

80. Boveris, A.; Oshino, N.; Chance, B. The cellular production of hydrogen peroxide. *Biochem. J.* **1972**, *128*, 617–630. [CrossRef] [PubMed]

81. Jaeschke, H. Reactive oxygen and mechanisms of inflammatory liver injury: Present concepts: Reactive oxygen and liver inflammation. *J. Gastroenterol. Hepatol.* **2011**, *26*, 173–179. [CrossRef] [PubMed]

82. Adamson, G.M.; Harman, A.W. Oxidative stress in cultured hepatocytes exposed to acetaminophen. *Biochem. Pharmacol.* **1993**, *45*, 2289–2294. [CrossRef]

83. Knockaert, L.; Descatoire, V.; Vadrot, N.; Fromenty, B.; Robin, M.-A. Mitochondrial CYP2E1 is sufficient to mediate oxidative stress and cytotoxicity induced by ethanol and acetaminophen. *Toxicol. In Vitro* **2011**, *25*, 475–484. [CrossRef] [PubMed]

84. Boelsterli, U. Diclofenac-induced liver injury: A paradigm of idiosyncratic drug toxicity. *Toxicol. Appl. Pharmacol.* **2003**, *192*, 307–322. [CrossRef]

85. Gómez-Lechón, M.J.; Ponsoda, X.; O'Connor, E.; Donato, T.; Castell, J.V.; Jover, R. Diclofenac induces apoptosis in hepatocytes by alteration of mitochondrial function and generation of ROS. *Biochem. Pharmacol.* **2003**, *66*, 2155–2167. [CrossRef] [PubMed]

86. Doi, H.; Horie, T. Salicylic acid-induced hepatotoxicity triggered by oxidative stress. *Chem. Biol. Interact.* **2010**, *183*, 363–368. [CrossRef] [PubMed]

87. Al-Nasser, I.A. Ibuprofen-induced liver mitochondrial permeability transition. *Toxicol. Lett.* **2000**, *111*, 213–218. [CrossRef]

88. Vázquez-Medina, J.P.; Zenteno-Savín, T.; Elsner, R.; Ortiz, R.M. Coping with physiological oxidative stress: A review of antioxidant strategies in seals. *J. Comp. Physiol. B* **2012**, *182*, 741–750. [CrossRef] [PubMed]

89. Jones, D.P. Redefining oxidative stress. *Antioxid. Redox Signal.* **2006**, *8*, 1865–1879. [CrossRef] [PubMed]

90. Seven, A.; Güzel, S.; Aslan, M.; Hamuryudan, V. Lipid, protein, DNA oxidation and antioxidant status in rheumatoid arthritis. *Clin. Biochem.* **2008**, *41*, 538–543. [CrossRef] [PubMed]

91. Toyokuni, S.; Okamoto, K.; Yodoi, J.; Hiai, H. Persistent oxidative stress in cancer. *FEBS Lett.* **1995**, *358*, 1–3. [CrossRef]

92. Conde de la Rosa, L.; Schoemaker, M.H.; Vrenken, T.E.; Buist-Homan, M.; Havinga, R.; Jansen, P.L.M.; Moshage, H. Superoxide anions and hydrogen peroxide induce hepatocyte death by different mechanisms: Involvement of JNK and ERK MAP kinases. *J. Hepatol.* **2006**, *44*, 918–929. [CrossRef] [PubMed]

93. Halliwell, B. Free radicals and antioxidants—*Quo vadis*? *Trends Pharmacol. Sci.* **2011**, *32*, 125–130. [CrossRef] [PubMed]

94. Marnett, L.J. Oxy radicals, lipid peroxidation and DNA damage. *Toxicology* **2002**, *181–182*, 219–222. [CrossRef]

95. Zhou, L.; Yang, Y.; Tian, D.; Wang, Y. Oxidative stress-induced 1, N6-ethenodeoxyadenosine adduct formation contributes to hepatocarcinogenesis. *Oncol. Rep.* **2013**, *29*, 875–884. [CrossRef] [PubMed]

96. Dukan, S.; Farewell, A.; Ballesteros, M.; Taddei, F.; Radman, M.; Nyström, T. Protein oxidation in response to increased transcriptional or translational errors. *Proc. Natl. Acad. Sci. USA* **2000**, *97*, 5746–5749. [CrossRef] [PubMed]

97. Johnson, T.M.; Yu, Z.-X.; Ferrans, V.J.; Lowenstein, R.A.; Finkel, T. Reactive oxygen species are downstream mediators of p53-dependent apoptosis. *Proc. Natl. Acad. Sci. USA* **1996**, *93*, 11848–11852. [CrossRef] [PubMed]

98. Polyak, K.; Xia, Y.; Zweier, J.L.; Kinzler, K.W.; Vogelstein, B. A model for p53-induced apoptosis. *Nature* **1997**, *389*, 300–305. [CrossRef] [PubMed]

99. Knebel, A.; Rahmsdorf, H.J.; Ullrich, A.; Herrlich, P. Dephosphorylation of receptor tyrosine kinases as target of regulation by radiation, oxidants or alkylating agents. *EMBO J.* **1996**, *15*, 5314–5325. [PubMed]

100. Wang, X.; McCullough, K.D.; Franke, T.F.; Holbrook, N.J. Epidermal growth factor receptor-dependent Akt activation by oxidative stress enhances cell survival. *J. Biol. Chem.* **2000**, *275*, 14624–14631. [CrossRef] [PubMed]

101. Saitoh, M.; Nishitoh, H.; Fujii, M.; Takeda, K.; Tobiume, K.; Sawada, Y.; Kawabata, M.; Miyazono, K.; Ichijo, H. Mammalian thioredoxin is a direct inhibitor of apoptosis signal-regulating kinase (ASK) 1. *EMBO J.* **1998**, *17*, 2596–2606. [CrossRef] [PubMed]

102. Adler, V.; Yin, Z.; Fuchs, S.Y.; Benezra, M.; Rosario, L.; Tew, K.D.; Pincus, M.; Sardana, M.; Henderson, C.; Wolf, C.R.; et al. Regulation of JNK signaling by GSTp. *EMBO J.* **1999**, *18*, 1321–1334. [CrossRef] [PubMed]

103. Li, S.; Tan, H.-Y.; Wang, N.; Zhang, Z.-J.; Lao, L.; Wong, C.-W.; Feng, Y. The Role of Oxidative Stress and Antioxidants in Liver Diseases. *Int. J. Mol. Sci.* **2015**, *16*, 26087–26124. [CrossRef] [PubMed]

104. Kaplowitz, N. Drug-induced liver injury. *Clin. Infect. Dis.* **2004**, *38* (Suppl. 2), S44–S48. [CrossRef] [PubMed]

105. Reuben, A.; Koch, D.G.; Lee, W.M. Drug-induced acute liver failure: Results of a U.S. Multicenter, prospective study. *Hepatology* **2010**, *52*, 2065–2076. [CrossRef] [PubMed]

106. Larson, A.M.; Polson, J.; Fontana, R.J.; Davern, T.J.; Lalani, E.; Hynan, L.S.; Reisch, J.S.; Schiødt, F.V.; Ostapowicz, G.; Shakil, A.O.; et al. Acetaminophen-induced acute liver failure: Results of a United States multicenter, prospective study. *Hepatology* **2005**, *42*, 1364–1372. [CrossRef] [PubMed]

107. Bernal, W.; Wendon, J. Acute Liver Failure. *N. Engl. J. Med.* **2013**, *369*, 2525–2534. [CrossRef] [PubMed]

108. Grattagliano, I. Biochemical mechanisms in drug-induced liver injury: Certainties and doubts. *World J. Gastroenterol.* **2009**, *15*, 4865. [CrossRef] [PubMed]

109. Iorga, A.; Dara, L.; Kaplowitz, N. Drug-Induced Liver Injury: Cascade of Events Leading to Cell Death, Apoptosis or Necrosis. *Int. J. Mol. Sci.* **2017**, *18*, 1018. [CrossRef] [PubMed]

110. Roth, A.D.; Lee, M.-Y. Idiosyncratic Drug-Induced Liver Injury (IDILI): Potential Mechanisms and Predictive Assays. *BioMed Res. Int.* **2017**, *2017*, 9176937. [CrossRef] [PubMed]

111. Han, D.; Dara, L.; Win, S.; Than, T.A.; Yuan, L.; Abbasi, S.Q.; Liu, Z.-X.; Kaplowitz, N. Regulation of drug-induced liver injury by signal transduction pathways: Critical role of mitochondria. *Trends Pharmacol. Sci.* **2013**, *34*, 243–253. [CrossRef] [PubMed]

112. Cohen, S.D.; Pumford, N.R.; Khairallah, E.A.; Boekelheide, K.; Pohl, L.R.; Amouzadeh, H.R.; Hinson, J.A. Selective Protein Covalent Binding and Target Organ Toxicity. *Toxicol. Appl. Pharmacol.* **1997**, *143*, 1–12. [CrossRef] [PubMed]

113. Baillie, T.A.; Rettie, A.E. Role of biotransformation in drug-induced toxicity: Influence of intra-and inter-species differences in drug metabolism. *Drug Metab. Pharmacokinet.* **2011**, *26*, 15–29. [CrossRef] [PubMed]

114. Kaplowitz, N. Mechanisms of liver cell injury. *J. Hepatol.* **2000**, *32*, 39–47. [CrossRef]

115. Singal, A.K.; Jampana, S.C.; Weinman, S.A. Antioxidants as therapeutic agents for liver disease. *Liver Int.* **2011**, *31*, 1432–1448. [CrossRef] [PubMed]

116. Han, D.; Shinohara, M.; Ybanez, M.D.; Saberi, B.; Kaplowitz, N. Signal Transduction Pathways Involved in Drug-Induced Liver Injury. In *Adverse Drug Reactions*; Uetrecht, J., Ed.; Springer: Berlin/Heidelberg, Germany, 2010; pp. 267–310. Available online: http://dx.doi.org/10.1007/978-3-642-00663-0_10 (accessed on 27 June 2017).

117. Begriche, K.; Massart, J.; Robin, M.-A.; Borgne-Sanchez, A.; Fromenty, B. Drug-induced toxicity on mitochondria and lipid metabolism: Mechanistic diversity and deleterious consequences for the liver. *J. Hepatol.* **2011**, *54*, 773–794. [CrossRef] [PubMed]

118. Meredith, M.J.; Reed, D.J. Status of the mitochondrial pool of glutathione in the isolated hepatocyte. *J. Biol. Chem.* **1982**, *257*, 3747–3753. [PubMed]

119. Jones, D.P.; Lemasters, J.J.; Han, D.; Boelsterli, U.A.; Kaplowitz, N. Mechanisms of pathogenesis in drug hepatotoxicity putting the stress on mitochondria. *Mol. Interv.* **2010**, *10*, 98–111. [CrossRef] [PubMed]

120. Simon, L.S. Nonsteroidal anti-inflammatory drugs and their risk: A story still in development. *BMC Arthritis Res. Ther.* **2013**, *15*. [CrossRef] [PubMed]

121. Ghosh, R.; Alajbegovic, A.; Gomes, A.V. NSAIDs and Cardiovascular Diseases: Role of Reactive Oxygen Species. *Oxid. Med. Cell. Longev.* **2015**, *2015*, 536962. [CrossRef] [PubMed]

122. Dubois, R.N.; Abramson, S.B.; Crofford, L.; Gupta, R.A.; Simon, L.S.A.; Van De Putte, L.B.; Lipsky, P.E. Cyclooxygenase in biology and disease. *FASEB J.* **1998**, *12*, 1063–1073. [CrossRef] [PubMed]

123. Crofford, L.J. Use of NSAIDs in treating patients with arthritis. *BMC Arthritis Res. Ther.* **2013**, *15*. [CrossRef]

124. Schmeltzer, P.A.; Kosinski, A.S.; Kleiner, D.E.; Hoofnagle, J.H.; Stolz, A.; Fontana, R.J.; Russo, M.W. Liver injury from nonsteroidal anti-inflammatory drugs in the United States. *Liver Int.* **2016**, *36*, 603–609. [CrossRef] [PubMed]

125. Leise, M.D.; Poterucha, J.J.; Talwalkar, J.A. Drug-Induced Liver Injury. *Mayo Clin. Proc.* **2014**, *89*, 95–106. [CrossRef] [PubMed]

126. Unzueta, A.; Vargas, H.E. Nonsteroidal Anti-Inflammatory Drug–Induced Hepatoxicity. *Drug Hepatotoxic.* **2013**, *17*, 643–656. [CrossRef] [PubMed]

127. Liou, J.-Y.; Wu, C.-C.; Chen, B.-R.; Yen, L.B.; Wu, K.K. Nonsteroidal Anti-Inflammatory Drugs Induced Endothelial Apoptosis by Perturbing Peroxisome Proliferator-Activated Receptor-δ Transcriptional Pathway. *Mol. Pharmacol.* **2008**, *74*, 1399–1406. [CrossRef] [PubMed]

128. Watson, A.J.M.; Askew, J.N.; Benson, R.S.P. Poly(adenosine diphosphate ribose) polymerase inhibition prevents necrosis induced by H_2O_2 but not apoptosis. *Gastroenterology* **1995**, *109*, 472–482. [CrossRef]

129. Eugenio-Pérez, D.; Montes de Oca-Solano, H.A.; Pedraza-Chaverri, J. Role of food-derived antioxidant agents against acetaminophen-induced hepatotoxicity. *Pharm. Biol.* **2016**, *54*, 2340–2352. [CrossRef] [PubMed]

130. Castro, M.P.V. Intoxicación por acetaminofén en adultos. *Med. Leg. Costa Rica* **2016**, *33*, 103–109.

131. Bunchorntavakul, C.; Reddy, K.R. Acetaminophen-related Hepatotoxicity. *Drug Hepatotoxic.* **2013**, *17*, 587–607. [CrossRef] [PubMed]

132. Hodgman, M.J.; Garrard, A.R. A Review of Acetaminophen Poisoning. *Toxicology* **2012**, *28*, 499–516. [CrossRef] [PubMed]
133. Lee, W. Acute Liver Failure. *Semin. Respir. Crit. Care Med.* **2012**, *33*, 36–45. [CrossRef] [PubMed]
134. Jaeschke, H. Acetaminophen: Dose-Dependent Drug Hepatotoxicity and Acute Liver Failure in Patients. *Dig. Dis.* **2015**, *33*, 464–471. [CrossRef] [PubMed]
135. Lancaster, E.M.; Hiatt, J.R.; Zarrinpar, A. Acetaminophen hepatotoxicity: An updated review. *Arch. Toxicol.* **2015**, *89*, 193–199. [CrossRef] [PubMed]
136. Jaeschke, H. The role of oxidant stress and reactive nitrogen species in acetaminophen hepatotoxicity. *Toxicol. Lett.* **2003**, *144*, 279–288. [CrossRef]
137. McGill, M.R.; Jaeschke, H. Metabolism and Disposition of Acetaminophen: Recent Advances in Relation to Hepatotoxicity and Diagnosis. *Pharm. Res.* **2013**, *30*, 2174–2187. [CrossRef] [PubMed]
138. Saberi, B.; Ybanez, M.D.; Johnson, H.S.; Gaarde, W.A.; Han, D.; Kaplowitz, N. Protein kinase C (PKC) participates in acetaminophen hepatotoxicity through JNK dependent and independent signaling pathways. *Hepatology* **2014**, *59*, 1543–1554. [CrossRef] [PubMed]
139. Jaeschke, H.; McGill, M.R.; Ramachandran, A. Oxidant stress, mitochondria, and cell death mechanisms in drug-induced liver injury: Lessons learned from acetaminophen hepatotoxicity. *Drug Metab. Rev.* **2012**, *44*, 88–106. [CrossRef] [PubMed]
140. Jaeschke, H.; Xie, Y.; McGill, M.R. Acetaminophen-induced Liver Injury: From Animal Models to Humans. *J. Clin. Transl. Hepatol.* **2014**, *2*, 153–161. [PubMed]
141. Cover, C. Peroxynitrite-Induced Mitochondrial and Endonuclease-Mediated Nuclear DNA Damage in Acetaminophen Hepatotoxicity. *J. Pharmacol. Exp. Ther.* **2005**, *315*, 879–887. [CrossRef] [PubMed]
142. Masubuchi, Y.; Suda, C.; Horie, T. Involvement of mitochondrial permeability transition in acetaminophen-induced liver injury in mice. *J. Hepatol.* **2005**, *42*, 110–116. [CrossRef] [PubMed]
143. Hanawa, N.; Shinohara, M.; Saberi, B.; Gaarde, W.A.; Han, D.; Kaplowitz, N. Role of JNK Translocation to Mitochondria Leading to Inhibition of Mitochondria Bioenergetics in Acetaminophen-induced Liver Injury. *J. Biol. Chem.* **2008**, *283*, 13565–13577. [CrossRef] [PubMed]
144. Gujral, J.S.; Knight, T.R.; Farhood, A.; Bajt, M.L.; Jaeschke, H. Mode of Cell Death after Acetaminophen Overdose in Mice: Apoptosis or Oncotic Necrosis? *Toxicol. Sci.* **2002**, *67*, 322–328. [CrossRef] [PubMed]
145. Das, J.; Ghosh, J.; Manna, P.; Sil, P.C. Acetaminophen induced acute liver failure via oxidative stress and JNK activation: Protective role of taurine by the suppression of cytochrome P450 2E1. *Free Radic. Res.* **2010**, *44*, 340–355. [CrossRef] [PubMed]
146. Arai, T.; Koyama, R.; Yuasa, M.; Kitamura, D.; Mizuta, R. Acrolein, a highly toxic aldehyde generated under oxidative stress in vivo, aggravates the mouse liver damage after acetaminophen overdose. *Biomed. Res.* **2014**, *35*, 389–395. [CrossRef] [PubMed]
147. Kon, K.; Kim, J.-S.; Jaeschke, H.; Lemasters, J.J. Mitochondrial permeability transition in acetaminophen-induced necrosis and apoptosis of cultured mouse hepatocytes. *Hepatology* **2004**, *40*, 1170–1179. [CrossRef] [PubMed]
148. Kass, G.E.N.; Macanas-Pirard, P.; Lee, P.C.; Hinton, R.H. The Role of Apoptosis in Acetaminophen-Induced Injury. *Ann. N. Y. Acad. Sci.* **2003**, *1010*, 557–559. [CrossRef] [PubMed]
149. Possamai, L.A.; McPhail, M.J.; Quaglia, A.; Zingarelli, V.; Abeles, R.D.; Tidswell, R.; Puthucheary, Z.; Rawal, J.; Karvellas, C.; Leslie, E.M.; et al. Character and Temporal Evolution of Apoptosis in Acetaminophen-Induced Acute Liver Failure. *Crit. Care Med.* **2013**, *41*, 2543–2550. [CrossRef] [PubMed]
150. Yoon, E.; Babar, A.; Choudhary, M.; Kutner, M.; Pyrsopoulos, N. Acetaminophen-Induced Hepatotoxicity: A Comprehensive Update. *J. Clin. Transl. Hepatol.* **2016**, *4*, 131–142. [PubMed]
151. Jain, N.; Shrivastava, R.; Raghuwanshi, A.; Shrivastava, V. Aspirin induced changes in serum ACP, ALP, GOT, GPT, bilirubin and creatinine in corelation with histopathological changes in liver and kidney of female albino rat. *Int. J. Appl. Pharm.* **2012**, *4*, 9–11.
152. Bojić, M.; Sedgeman, C.A.; Nagy, L.D.; Guengerich, F.P. Aromatic Hydroxylation of Salicylic Acid and Aspirin by Human Cytochromes P450. *Eur. J. Pharm. Sci.* **2015**, *73*, 49–56. [CrossRef] [PubMed]
153. Zimmerman, H.J. Aspirin-induced hepatic injury. *Ann. Intern. Med.* **1974**, *80*, 103–105. [CrossRef] [PubMed]
154. Bessone, F. Non-steroidal anti-inflammatory drugs: What is the actual risk of liver damage? *World J. Gastroenterol.* **2010**, *16*, 5651–5661. [CrossRef] [PubMed]

155. Doi, H.; Masubuchi, Y.; Narimatsu, S.; Nishigaki, R.; Horie, T. Salicylic acid-induced lipid peroxidation in rat liver microsomes. *Res. Commun. Mol. Pathol. Pharmacol.* **1998**, *100*, 265–271. [PubMed]

156. Berk, M.; Dean, O.; Drexhage, H.; McNeil, J.J.; Moylan, S.; O'Neil, A.; Davey, C.; Sanna, L.; Maes, M. Aspirin: A review of its neurobiological properties and therapeutic potential for mental illness. *BMC Med.* **2013**, *11*, 74. [CrossRef] [PubMed]

157. Raza, H.; John, A.; Benedict, S. Acetylsalicylic acid-induced oxidative stress, cell cycle arrest, apoptosis and mitochondrial dysfunction in human hepatoma HepG2 cells. *Eur. J. Pharmacol.* **2011**, *668*, 15–24. [CrossRef] [PubMed]

158. Raza, H.; John, A. Implications of Altered Glutathione Metabolism in Aspirin-Induced Oxidative Stress and Mitochondrial Dysfunction in HepG2 Cells. *PLoS ONE* **2012**, *7*, e36325. [CrossRef] [PubMed]

159. Tassone, E.J.; Perticone, M.; Sciacqua, A.; Mafrici, S.F.; Settino, C.; Malara, N.; Mollace, V.; Sesti, G.; Perticone, F. Low dose of acetylsalicylic acid and oxidative stress-mediated endothelial dysfunction in diabetes: A short-term evaluation. *Acta Diabetol.* **2015**, *52*, 249–256. [CrossRef] [PubMed]

160. Deschamps, D.; Fisch, C.; Fromenty, B.; Berson, A.; Degott, C.; Pessayre, D. Inhibition by salicylic acid of the activation and thus oxidation of long chain fatty acids. Possible role in the development of Reye's syndrome. *J. Pharmacol. Exp. Ther.* **1991**, *259*, 894. [PubMed]

161. Lauterburg, B.; Grattagliano, I.; Gmür, R.; Stalder, M.; Hildebrand, P. Noninvasive assessment of the effect of xenobiotics on mitochondrial function in human beings: Studies with acetylsalicylic acid and ethanol with the use of the carbon 13-labeled ketoisocaproate breath test. *J. Lab. Clin. Med.* **1995**, *125*, 378–383. [PubMed]

162. Fromenty, B.; Pessayre, D. Inhibition of mitochondrial beta-oxidation as a mechanism of hepatotoxicity. *Pharmacol. Ther.* **1995**, *67*, 101–154. [CrossRef]

163. Gan, T.J. Diclofenac: An update on its mechanism of action and safety profile. *Curr. Med. Res. Opin.* **2010**, *26*, 1715–1731. [CrossRef] [PubMed]

164. Altman, R.; Bosch, B.; Brune, K.; Patrignani, P.; Young, C. Advances in NSAID Development: Evolution of Diclofenac Products Using Pharmaceutical Technology. *Drugs* **2015**, *75*, 859–877. [CrossRef] [PubMed]

165. Aithal, G.P.; Day, C.P. Nonsteroidal Anti-Inflammatory Drug–Induced Hepatotoxicity. *Drug Induc. Liver Dis.* **2007**, *11*, 563–575. [CrossRef] [PubMed]

166. Watkins, P.B.; Seligman, P.J.; Pears, J.S.; Avigan, M.I.; Senior, J.R. Using controlled clinical trials to learn more about acute drug-induced liver injury. *Hepatology* **2008**, *48*, 1680–1689. [CrossRef] [PubMed]

167. Banks, A.T.; Zimmerman, H.J.; Ishak, K.G.; Harter, J.G. Diclofenac-associated hepatotoxicity: Analysis of 180 cases reported to the food and drug administration as adverse reactions. *Hepatology* **1995**, *22*, 820–827. [CrossRef] [PubMed]

168. Laine, L.; Goldkind, L.; Curtis, S.P.; Connors, L.G.; Yanqiong, Z.; Cannon, C.P. How Common Is Diclofenac-Associated Liver Injury? Analysis of 17,289 Arthritis Patients in a Long-Term Prospective Clinical Trial. *Am. J. Gastroenterol.* **2009**, *104*, 356–362. [CrossRef] [PubMed]

169. Yano, A.; Higuchi, S.; Tsuneyama, K.; Fukami, T.; Nakajima, M.; Yokoi, T. Involvement of immune-related factors in diclofenac-induced acute liver injury in mice. *Toxicology* **2012**, *293*, 107–114. [CrossRef] [PubMed]

170. Bort, R.; Ponsoda, X.; Jover, R.; Gómez-Lechón, M.J.; Castell, J.V. Diclofenac Toxity to Hepatocytes: A Role for Drug Metabolism in Cell Toxicity. *J. Pharmacol. Exp. Ther.* **1999**, *288*, 65–72. [PubMed]

171. Amin, A.; Hamza, A.A. Oxidative stress mediates drug-induced hepatotoxicity in rats: A possible role of DNA fragmentation. *Toxicology* **2005**, *208*, 367–375. [CrossRef] [PubMed]

172. Vickers, A.E.M. Tissue slices for the evaluation of metabolism-based toxicity with the example of diclofenac. *Eval. Metab. Based Drug Toxic Drug Dev.* **2009**, *179*, 9–16. [CrossRef] [PubMed]

173. Pourahmad, J.; Mortada, Y.; Eskandari, M.R.; Shahraki, J. Involvement of Lysosomal Labilisation and Lysosomal/mitochondrial Cross-Talk in Diclofenac Induced Hepatotoxicity. *Iran. J. Pharm. Res.* **2011**, *10*, 877–887. [PubMed]

174. Fredriksson, L.; Wink, S.; Herpers, B.; Benedetti, G.; Hadi, M.; de Bont, H.; Groothuis, G.; Luijten, M.; Danen, E.; de Graauw, M.; et al. Drug-Induced Endoplasmic Reticulum and Oxidative Stress Responses Independently Sensitize toward TNFα-Mediated Hepatotoxicity. *Toxicol. Sci.* **2014**, *140*, 144–159. [CrossRef] [PubMed]

175. Andrejak, M.; Davion, T.; Gineston, J.L.; Capron, J.P. Cross hepatotoxicity between non-steroidal anti-inflammatory drugs. *Br. Med. J. Clin. Res. Ed.* **1987**, *295*, 180–181. [CrossRef] [PubMed]

176. Ali, S.; Pimentel, J.D.; Ma, C. Naproxen-induced liver injury. *Hepatobiliary Pancreat. Dis. Int.* **2011**, *10*, 552–556. [CrossRef]

177. Davies, N.M.; Anderson, K.E. Clinical Pharmacokinetics of Naproxen. *Clin. Pharmacokinet.* **1997**, *32*, 268–293. [CrossRef] [PubMed]

178. Miners, J.O.; Coulter, S.; Tukey, R.H.; Veronese, M.E.; Birkett, D.J. Cytochromes P450, 1A2, and 2C9 are responsible for the human hepatic O-demethylation of R- and S-naproxen. *Biochem. Pharmacol.* **1996**, *51*, 1003–1008. [CrossRef]

179. Tracy, T.S.; Marra, C.; Wrighton, S.A.; Gonzalez, F.J.; Korzekwa, K.R. Involvement of multiple cytochrome P450 isoforms in naproxen O-demethylation. *Eur. J. Clin. Pharmacol.* **1997**, *52*, 293–298. [CrossRef] [PubMed]

180. Bowalgaha, K.; Elliot, D.J.; Mackenzie, P.I.; Knights, K.M.; Swedmark, S.; Miners, J.O. S-Naproxen and desmethylnaproxen glucuronidation by human liver microsomes and recombinant human UDP-glucuronosyltransferases (UGT): Role of UGT2B7 in the elimination of naproxen. *Br. J. Clin. Pharmacol.* **2005**, *60*, 423–433. [CrossRef] [PubMed]

181. Yokoyama, H.; Horie, T.; Awazu, S. Oxidative stress in isolated rat hepatocytes during naproxen metabolism. *Biochem. Pharmacol.* **1995**, *49*, 991–996. [CrossRef]

182. Yokoyama, H.; Horie, T.; Awazu, S. Glutathione disulfide formation during naproxen metabolism in the isolated rat hepatocytes. *Res. Commun. Mol. Pathol. Pharmacol.* **1998**, *99*, 143–154. [PubMed]

183. Ji, B.; Masubuchi, Y.; Horie, T. A Possible Mechanism of Naproxen-Induced Lipid Peroxidation in Rat Liver Microsomes. *Pharmacol. Toxicol.* **2001**, *89*, 43–48. [CrossRef] [PubMed]

184. Lo, A.; Addison, R.S.; Hooper, W.D.; Dickinson, R.G. Disposition of naproxen, naproxen acyl glucuronide and its rearrangement isomers in the isolated perfused rat liver. *Xenobiotica* **2001**, *31*, 309–319. [CrossRef] [PubMed]

185. Yokoyama, H.; Horie, T.; Awazu, S. Naproxen-induced oxidative stress in the isolated perfused rat liver. *Chem. Biol. Interact.* **2006**, *160*, 150–158. [CrossRef] [PubMed]

186. Naproxen. LiverTox, 2017. Available online: https://livertox.nlm.nih.gov//Naproxen.htm (accessed on 27 June 2017).

187. Bushra, R.; Aslam, N. An Overview of Clinical Pharmacology of Ibuprofen. *Oman Med. J.* **2010**, *25*, 155–1661. [PubMed]

188. Ibuprofen. LiverTox. 2017. Available online: https://livertox.nlm.nih.gov//Ibuprofen.htm (accessed on 27 June 2017).

189. Rainsford, K.D. Ibuprofen: Pharmacology, efficacy and safety. *Inflammopharmacology* **2009**, *17*, 275–342. [CrossRef] [PubMed]

190. Riley, T.R.; Smith, J.P. Ibuprofen-induced hepatotoxicity in patients with chronic hepatitis c: A case series. *Am. J. Gastroenterol.* **1998**, *93*, 1563–1565. [CrossRef] [PubMed]

191. Basturk, A.; Artan, R.; Yılmaz, A.; Gelen, M.T.; Duman, O. Acute vanishing bile duct syndrome after the use of ibuprofen. *Arab. J. Gastroenterol.* **2016**, *17*, 137–139. [CrossRef] [PubMed]

192. Sánchez-Valle, V.; Carlos Chavez-Tapia, N.; Uribe, M.; Méndez-Sánchez, N. Role of oxidative stress and molecular changes in liver fibrosis: A review. *Curr. Med. Chem.* **2012**, *19*, 4850–4860. [CrossRef] [PubMed]

193. Halliwell, B. Biochemistry of oxidative stress. *Biochem. Soc. Trans.* **2007**, *35*, 1147–1150. [CrossRef] [PubMed]

194. Matés, J.M. Effects of antioxidant enzymes in the molecular control of reactive oxygen species toxicology. *Toxicology* **2000**, *153*, 83–104. [CrossRef]

195. Marí, M.; Colell, A.; Morales, A.; von Montfort, C.; Garcia-Ruiz, C.; Fernández-Checa, J.C. Redox control of liver function in health and disease. *Antioxid. Redox Signal.* **2010**, *12*, 1295–1331. [CrossRef] [PubMed]

196. Flora, S.J. Structural, chemical and biological aspects of antioxidants for strategies against metal and metalloid exposure. *Oxid. Med. Cell. Longev.* **2009**, *2*, 191–206. [CrossRef] [PubMed]

197. Gregory, E.M.; Goscin, S.A.; Fridovich, I. Superoxide dismutase and oxygen toxicity in a eukaryote. *J. Bacteriol.* **1974**, *117*, 456–460. [PubMed]

198. Gao, F.; Kinnula, V.L.; Myllärniemi, M.; Oury, T.D. Extracellular Superoxide Dismutase in Pulmonary Fibrosis. *Antioxid. Redox Signal.* **2008**, *10*, 343–354. [CrossRef] [PubMed]

199. Ha, H.-L. Oxidative stress and antioxidants in hepatic pathogenesis. *World J. Gastroenterol.* **2010**, *16*, 6035. [CrossRef] [PubMed]

200. Matés, J.M.; Pérez-Gómez, C.; De Castro, I.N. Antioxidant enzymes and human diseases. *Clin. Biochem.* **1999**, *32*, 595–603. [CrossRef]

201. Rhee, S.G.; Chae, H.Z.; Kim, K. Peroxiredoxins: A historical overview and speculative preview of novel mechanisms and emerging concepts in cell signaling. *Free Radic. Biol. Med.* **2005**, *38*, 1543–1552. [CrossRef] [PubMed]

202. Gasdaska, J.R.; Gasdaska, P.Y.; Gallegos, A.; Powis, G. Human thioredoxin reductase gene localization to chromosomal position 12q23–q24.1 and mRNA distribution in human tissue. *Genomics* **1996**, *37*, 257–259. [CrossRef] [PubMed]

203. Fernandez-Checa, J.C.; Kaplowitz, N. Hepatic mitochondrial glutathione: Transport and role in disease and toxicity. *Toxicol. Appl. Pharmacol.* **2005**, *204*, 263–273. [CrossRef] [PubMed]

204. Masella, R.; Di Benedetto, R.; Varì, R.; Filesi, C.; Giovannini, C. Novel mechanisms of natural antioxidant compounds in biological systems: Involvement of glutathione and glutathione-related enzymes. *J. Nutr. Biochem.* **2005**, *16*, 577–586. [CrossRef] [PubMed]

205. St Pierre, J.; Drori, S.; Uldry, M.; Silvaggi, J.M.; Rhee, J.; Jäger, S.; Handschin, C.; Zheng, K.; Lin, J.; Yang, W.; et al. Suppression of Reactive Oxygen Species and Neurodegeneration by the PGC-1 Transcriptional Coactivators. *Cell* **2006**, *127*, 397–408. [CrossRef] [PubMed]

206. Olmos, Y.; Valle, I.; Borniquel, S.; Tierrez, A.; Soria, E.; Lamas, S.; Monsalve, M. Mutual Dependence of Foxo3a and PGC-1α in the Induction of Oxidative Stress Genes. *J. Biol. Chem.* **2009**, *284*, 14476–14484. [CrossRef] [PubMed]

207. Moi, P.; Chan, K.; Asunis, I.; Cao, A.; Kan, Y.W. Isolation of NF-E2-related factor 2 (Nrf2), a NF-E2-like basic leucine zipper transcriptional activator that binds to the tandem NF-E2/AP1 repeat of the beta-globin locus control region. *Proc. Natl. Acad. Sci. USA* **1994**, *91*, 9926–9930. [CrossRef] [PubMed]

208. Kaspar, J.W.; Niture, S.K.; Jaiswal, A.K. Nrf2:INrf2 (Keap1) signaling in oxidative stress. *Free Radic. Biol. Med.* **2009**, *47*, 1304–1309. [CrossRef] [PubMed]

209. Zhang, H.; Davies, K.J.A.; Forman, H.J. Oxidative stress response and Nrf2 signaling in aging. *Free Radic. Biol. Med.* **2015**, *88*, 314–336. [CrossRef] [PubMed]

210. Shen, G.; Kong, A.-N. Nrf2 plays an important role in coordinated regulation of Phase II drug metabolism enzymes and Phase III drug transporters. *Biopharm. Drug Dispos.* **2009**, *30*, 345–355. [CrossRef] [PubMed]

211. Nguyen, T.; Sherratt, P.J.; Pickett, C.B. Regulatory mechanisms controlling gene expression mediated by the antioxidant response element. *Annu. Rev. Pharmacol. Toxicol.* **2003**, *43*, 233–260. [CrossRef] [PubMed]

212. Rushmore, T.H.; Morton, M.R.; Pickett, C.B. The antioxidant responsive element. Activation by oxidative stress and identification of the DNA consensus sequence required for functional activity. *J. Biol. Chem.* **1991**, *266*, 11632–11639. [PubMed]

213. Boeing, H.; Bechthold, A.; Bub, A.; Ellinger, S.; Haller, D.; Kroke, A.; Leschik-Bonnet, E.; Müller, M.J.; Oberritter, H.; Schulze, M.; et al. Critical review: Vegetables and fruit in the prevention of chronic diseases. *Eur. J. Nutr.* **2012**, *51*, 637–663. [CrossRef] [PubMed]

214. Roleira, F.M.F.; Tavares-da-Silva, E.J.; Varela, C.L.; Costa, S.C.; Silva, T.; Garrido, J.; Borges, F. Plant derived and dietary phenolic antioxidants: Anticancer properties. *Food Chem.* **2015**, *183*, 235–258. [CrossRef] [PubMed]

215. Rimm, E.B.; Ascherio, A.; Giovannucci, E.; Spiegelman, D.; Stampfer, M.J.; Willett, W.C. Vegetable, fruit, and cereal fiber intake and risk of coronary heart disease among men. *JAMA* **1996**, *275*, 447–451. [CrossRef] [PubMed]

216. Kelsey, N.A.; Wilkins, H.M.; Linseman, D.A. Nutraceutical Antioxidants as Novel Neuroprotective Agents. *Molecules* **2010**, *15*, 7792–7814. [CrossRef] [PubMed]

217. Liu, Y.-J.; Zhan, J.; Liu, X.-L.; Wang, Y.; Ji, J.; He, Q.-Q. Dietary flavonoids intake and risk of type 2 diabetes: A meta-analysis of prospective cohort studies. *Clin. Nutr.* **2014**, *33*, 59–63. [CrossRef] [PubMed]

218. Ascherio, A.; Stampfer, M.J.; Colditz, G.A.; Willett, W.C.; McKinlay, J. Nutrient intakes and blood pressure in normotensive males. *Int. J. Epidemiol.* **1991**, *20*, 886–891. [CrossRef] [PubMed]

219. Doll, R.; Peto, R. The causes of cancer: Quantitative estimates of avoidable risks of cancer in the United States today. *J. Natl. Cancer Inst.* **1981**, *66*, 1192–1308. [CrossRef]

220. Ames, B.N.; Shigenaga, M.K.; Hagen, T.M. Oxidants, antioxidants, and the degenerative diseases of aging. *Proc. Natl. Acad. Sci. USA* **1993**, *90*, 7915–7922. [CrossRef] [PubMed]

221. Vinson, J.A.; Su, X.; Zubik, L.; Bose, P. Phenol Antioxidant Quantity and Quality in Foods: Fruits. *J. Agric. Food Chem.* **2001**, *49*, 5315–5321. [CrossRef] [PubMed]

222. Heber, D. Vegetables, fruits and phytoestrogens in the prevention of diseases. *J. Postgrad. Med.* **2004**, *50*, 145–149. [PubMed]

223. Quideau, S.; Deffieux, D.; Douat-Casassus, C.; Pouységu, L. Plant Polyphenols: Chemical Properties, Biological Activities, and Synthesis. *Angew. Chem. Int. Ed.* **2011**, *50*, 586–621. [CrossRef] [PubMed]

224. Delgado-Vargas, F.; Jiménez, A.R.; Paredes-López, O. Natural pigments: Carotenoids, anthocyanins, and betalains—Characteristics, biosynthesis, processing, and stability. *Crit. Rev. Food Sci. Nutr.* **2000**, *40*, 173–289. [CrossRef] [PubMed]

225. Hartmann, A.; Patz, C.-D.; Andlauer, W.; Dietrich, H.; Ludwig, M. Influence of Processing on Quality Parameters of Strawberries. *J. Agric. Food Chem.* **2008**, *56*, 9484–9489. [CrossRef] [PubMed]

226. Kujumgiev, A.; Tsvetkova, I.; Serkedjieva, Y.; Bankova, V.; Christov, R.; Popov, S. Antibacterial, antifungal and antiviral activity of propolis of different geographic origin. *J. Ethnopharmacol.* **1999**, *64*, 235–240. [CrossRef]

227. Brahmi, D.; Bouaziz, C.; Ayed, Y.; Mansour, H.B.; Zourgui, L.; Bacha, H. Chemopreventive effect of cactus Opuntia ficus indica on oxidative stress and genotoxicity of aflatoxin B1. *Nutr. Metabol.* **2011**, *8*, 73. [CrossRef] [PubMed]

228. Dewanto, V.; Wu, X.; Adom, K.K.; Liu, R.H. Thermal Processing Enhances the Nutritional Value of Tomatoes by Increasing Total Antioxidant Activity. *J. Agric. Food Chem.* **2002**, *50*, 3010–3014. [CrossRef] [PubMed]

229. Harborne, J.B.; Williams, C.A. Advances in flavonoid research since 1992. *Phytochemistry* **2000**, *55*, 481–504. [CrossRef]

230. Pollastri, S.; Tattini, M. Flavonols: Old compounds for old roles. *Ann. Bot.* **2011**, *108*, 1225–1233. [CrossRef] [PubMed]

231. Das, D.K. Naturally occurring flavonoids: Structure, chemistry, and high-performance liquid chromatography methods for separation and characterization. *Oxyg. Radic. Biol. Syst. Part D* **1994**, *234*, 410–420.

232. Havsteen, B.H. The biochemistry and medical significance of the flavonoids. *Pharmacol. Ther.* **2002**, *96*, 67–202. [CrossRef]

233. Falcone Ferreyra, M.L.; Rius, S.P.; Casati, P. Flavonoids: Biosynthesis, biological functions, and biotechnological applications. *Front. Plant. Sci.* **2012**, *3*, 222. [CrossRef] [PubMed]

234. Martens, S.; Preuß, A.; Matern, U. Multifunctional flavonoid dioxygenases: Flavonol and anthocyanin biosynthesis in *Arabidopsis thaliana* L. *Phytochemistry* **2010**, *71*, 1040–1049. [CrossRef] [PubMed]

235. Heim, K.E.; Tagliaferro, A.R.; Bobilya, D.J. Flavonoid antioxidants: Chemistry, metabolism and structure-activity relationships. *J. Nutr. Biochem.* **2002**, *13*, 572–584. [CrossRef]

236. Croft, K.D. The chemistry and biological effects of flavonoids and phenolic acids. *Ann. N. Y. Acad. Sci.* **1998**, *854*, 435–442. [CrossRef] [PubMed]

237. Hässig, A.; Linag, W.X.; Schwabl, H.; Stampfli, K. Flavonoids and tannins: Plant-based antioxidants with vitamin character. *Med. Hypotheses.* **1999**, *52*, 479–481. [CrossRef] [PubMed]

238. Stalikas, C.D. Extraction, separation, and detection methods for phenolic acids and flavonoids. *J. Sep. Sci.* **2007**, *30*, 3268–3295. [CrossRef] [PubMed]

239. Kumar, S.; Pandey, A.K. Chemistry and Biological Activities of Flavonoids: An Overview. *Sci. World J.* **2013**, *2013*, 162750. [CrossRef] [PubMed]

240. Kühnau, J. The flavonoids. A class of semi-essential food components: Their role in human nutrition. *Word Rev. Nutr. Diet.* **1976**, *24*, 117–191.

241. Hertog, M.G.L.; Hollman, P.C.H.; Katan, M.B.; Kromhout, D. Intake of potentially anticarcinogenic flavonoids and their determinants in adults in the Netherlands. *Nutr. Cancer* **1993**, *20*, 21–29. [CrossRef] [PubMed]

242. Hertog, M.L.; Kromhout, D.; Aravanis, C.; Blackburn, H.; Buzina, R.; Fidanza, F.; Giampaoli, S.; Jansen, A.; Menotti, A.; Nedeljkovic, S.; et al. Flavonoid intake and long-term risk of coronary heart disease and cancer in the seven countries study. *Arch. Intern. Med.* **1995**, *155*, 381–386. [CrossRef] [PubMed]

243. Crozier, A.; Lean, M.E.; McDonald, M.S.; Black, C. Quantitative analysis of the flavonoid content of commercial tomatoes, onions, lettuce, and celery. *J. Agric. Food Chem.* **1997**, *45*, 590–595. [CrossRef]

244. Hertog, M.G.; Hollman, P.C.; Van de Putte, B. Content of potentially anticarcinogenic flavonoids of tea infusions, wines, and fruit juices. *J. Agric. Food Chem.* **1993**, *41*, 1242–1246. [CrossRef]

245. Bjørklund, G.; Dadar, M.; Chirumbolo, S.; Lysiuk, R. Flavonoids as detoxifying and pro-survival agents: What's new? *Food Chem. Toxicol.* **2017**, *110*, 240–250. [CrossRef] [PubMed]

246. Ferry, D.R.; Smith, A.; Malkhandi, J.; Fyfe, D.W.; Anderson, D.; Baker, J.; Kerr, D.J. Phase I clinical trial of the flavonoid quercetin: Pharmacokinetics and evidence for in vivo tyrosine kinase inhibition. *Clin. Cancer Res.* **1996**, *2*, 659–668. [PubMed]

247. García-Lafuente, A.; Guillamón, E.; Villares, A.; Rostagno, M.A.; Martínez, J.A. Flavonoids as anti-inflammatory agents: Implications in cancer and cardiovascular disease. *Inflamm. Res.* **2009**, *58*, 537–552. [CrossRef] [PubMed]

248. Bors, W.; Heller, W.; Michel, C.; Saran, M. Flavonoids as antioxidants: Determination of radical-scavenging efficiencies. *Oxyg. Radic. Biol. Syst. Part B Oxyg. Radic. Antioxid.* **1990**, *186*, 343–355.

249. Van Acker, S.A.; de Groot, M.J.; van den Berg, D.-J.; Tromp, M.N.; Donné-Op den Kelder, G.; van der Vijgh, W.J.; Bast, A. A quantum chemical explanation of the antioxidant activity of flavonoids. *Chem. Res. Toxicol.* **1996**, *9*, 1305–1312. [CrossRef] [PubMed]

250. Georgiev, V.; Ananga, A.; Tsolova, V. Recent Advances and Uses of Grape Flavonoids as Nutraceuticals. *Nutrients* **2014**, *6*, 391–415. [CrossRef] [PubMed]

251. Domitrović, R.; Jakovac, H.; Marchesi, V.V.; Vladimir-Knežević, S.; Cvijanović, O.; Tadić, Ž.; Romić, Ž.; Rahelić, D. Differential hepatoprotective mechanisms of rutin and quercetin in CCl4-intoxicated BALB/cN mice. *Acta Pharmacol. Sin.* **2012**, *33*, 1260–1270. [CrossRef] [PubMed]

252. Jie, Q.; Tang, Y.; Deng, Y.; Li, Y.; Shi, Y.; Gao, C.; Xing, M.; Wang, D.; Liu, L.; Yao, P. Bilirubin participates in protecting of heme oxygenase-1 induction by quercetin against ethanol hepatotoxicity in cultured rat hepatocytes. *Alcohol* **2013**, *47*, 141–148. [CrossRef] [PubMed]

253. Ji, L.; Ma, Y.; Wang, Z.; Cai, Z.; Pang, C.; Wang, Z. Quercetin Prevents Pyrrolizidine Alkaloid Clivorine-Induced Liver Injury in Mice by Elevating Body Defense Capacity. *PLoS ONE* **2014**, *9*, e98970. [CrossRef] [PubMed]

254. De David, C.; Rodrigues, G.; Bona, S.; Meurer, L.; González-Gallego, J.; Tuñón, M.J.; Marroni, M. Role of Quercetin in Preventing Thioacetamide-Induced Liver Injury in Rats. *Toxicol. Pathol.* **2011**, *39*, 949–957. [CrossRef] [PubMed]

255. Yousef, M.I.; Omar, S.A.M.; El-Guendi, M.I.; Abdelmegid, L.A. Potential protective effects of quercetin and curcumin on paracetamol-induced histological changes, oxidative stress, impaired liver and kidney functions and haematotoxicity in rat. *Food Chem. Toxicol.* **2010**, *48*, 3246–3261. [CrossRef] [PubMed]

256. Babenko, N.A.; Shakhova, E.G. Effects of flavonoids on sphingolipid turnover in the toxin-damaged liver and liver cells. *Lipids Health Dis.* **2008**, *7*. [CrossRef] [PubMed]

257. Wu, Y.; Wang, F.; Zheng, Q.; Lu, L.; Yao, H.; Zhou, C.; Wu, X.; Zhao, Y. Hepatoprotective effect of total flavonoids from Laggera alata against carbon tetrachloride-induced injury in primary cultured neonatal rat hepatocytes and in rats with hepatic damage. *J. Biomed. Sci.* **2006**, *13*, 569–578. [CrossRef] [PubMed]

258. Zhu, W.; Jia, Q.; Wang, Y.; Zhang, Y.; Xia, M. The anthocyanin cyanidin-3-*O*-β-glucoside, a flavonoid, increases hepatic glutathione synthesis and protects hepatocytes against reactive oxygen species during hyperglycemia: Involvement of a cAMP-PKA-dependent signaling pathway. *Free Radic. Biol. Med.* **2012**, *52*, 314–327. [CrossRef] [PubMed]

259. Raghu, R.; Karthikeyan, S. Zidovudine and isoniazid induced liver toxicity and oxidative stress: Evaluation of mitigating properties of silibinin. *Environ. Toxicol. Pharmacol.* **2016**, *46*, 217–226. [CrossRef] [PubMed]

260. Strack, D.; Vogt, T.; Schliemann, W. Recent advances in betalain research. *Phytochemistry* **2003**, *62*, 247–269. [CrossRef]

261. Jackman, R.L.; Smith, J.L. Anthocyanins and betalains. In *Natural Food Colorants*; Hendry, G.A.F., Houghton, J.D., Eds.; Springer: Boston, MA, USA, 1996; pp. 244–309. Available online: http://dx.doi. org/10.1007/978-1-4615-2155-6_8 (accessed on 16 June 2017).

262. Stintzing, F.C.; Schieber, A.; Carle, R. Identification of Betalains from Yellow Beet (*Beta vulgaris* L.) and Cactus Pear [*Opuntia ficus-indica* (L.) Mill.] by High-Performance Liquid Chromatography—Electrospray Ionization Mass Spectrometry. *J. Agric. Food Chem.* **2002**, *50*, 2302–2307. [CrossRef] [PubMed]

263. Steglich, W.; Strack, D. Chapter 1 Betalains. *Alkaloids Chem. Pharmacol.* **1990**, *39*, 1–62.

264. Stafford, H.A. Anthocyanins and betalains: Evolution of the mutually exclusive pathways. *Plant Sci.* **1994**, *101*, 91–98. [CrossRef]

265. Clement, J.S.; Mabry, T.J. Pigment evolution in the Caryophyllales: A systematic overview. *Plant Biol.* **1996**, *109*, 360–367. [CrossRef]

266. Kanner, J.; Harel, S.; Granit, R. Betalains. A New Class of Dietary Cationized Antioxidants. *J. Agric. Food Chem.* **2001**, *49*, 5178–5185. [CrossRef] [PubMed]

267. Gliszczynska-Swiglo, A.; Szymusiak, H.; Malinowska, P. Betanin, the main pigment of red beet: Molecular origin of its exceptionally high free radical-scavenging activity. *Food Addit. Contam.* **2006**, *23*, 1079–1087. [CrossRef] [PubMed]

268. Escribano, J.; Pedreño, M.A.; García-Carmona, F.; Muñoz, R. Characterization of the antiradical activity of betalains from *Beta vulgaris* L. roots. *Phytochem. Anal.* **1998**, *9*, 124–127. [CrossRef]

269. Butera, D.; Tesoriere, L.; Di Gaudio, F.; Bongiorno, A.; Allegra, M.; Pintaudi, A.M.; Kohen, R.; Livrea, M.A. Antioxidant Activities of Sicilian Prickly Pear (*Opuntia ficus indica*) Fruit Extracts and Reducing Properties of Its Betalains: Betanin and Indicaxanthin. *J. Agric. Food Chem.* **2002**, *50*, 6895–6901. [CrossRef] [PubMed]

270. Pavlov, A.; Kovatcheva, P.; Tuneva, D.; Ilieva, M.; Bley, T. Radical Scavenging Activity and Stability of Betalains from Beta vulgaris Hairy Root Culture in Simulated Conditions of Human Gastrointestinal Tract. *Plant Foods Hum. Nutr.* **2005**, *60*, 43–47. [CrossRef] [PubMed]

271. Swarna, J.; Lokeswari, T.S.; Smita, M.; Ravindhran, R. Characterisation and determination of in vitro antioxidant potential of betalains from *Talinum triangulare* (Jacq.) Willd. *Food Chem.* **2013**, *141*, 4382–4390. [CrossRef] [PubMed]

272. Esatbeyoglu, T.; Wagner, A.E.; Motafakkerazad, R.; Nakajima, Y.; Matsugo, S.; Rimbach, G. Free radical scavenging and antioxidant activity of betanin: Electron spin resonance spectroscopy studies and studies in cultured cells. *Food Chem. Toxicol.* **2014**, *73*, 119–126. [CrossRef] [PubMed]

273. Krajka-Kuzniak, V.; Paluszczak, J.; Szaefer, H.; Baer-Dubowska, W. Betanin, a beetroot component, induces nuclear factor erythroid-2-related factor 2-mediated expression of detoxifying/antioxidant enzymes in human liver cell lines. *Br. J. Nutr.* **2013**, *110*, 2138–2149. [CrossRef] [PubMed]

274. Moreno, D.A.; García-Viguera, C.; Gil, J.I.; Gil-Izquierdo, A. Betalains in the era of global agri-food science, technology and nutritional health. *Phytochem. Rev.* **2008**, *7*, 261–280. [CrossRef]

275. Wybraniec, S.; Mizrahi, Y. Fruit Flesh Betacyanin Pigments in *Hylocereus* Cacti. *J. Agric. Food Chem.* **2002**, *50*, 6086–6089. [CrossRef] [PubMed]

276. Wu, L.; Hsu, H.-W.; Chen, Y.-C.; Chiu, C.-C.; Lin, Y.-I.; Ho, J.A. Antioxidant and antiproliferative activities of red pitaya. *Food Chem.* **2006**, *95*, 319–327. [CrossRef]

277. Khan, M.I.; Sri Harsha, P.S.C.; Giridhar, P.; Ravishankar, G.A. Pigment identification, nutritional composition, bioactivity, and in vitro cancer cell cytotoxicity of *Rivina humilis* L. berries, potential source of betalains. *LWT Food Sci. Technol.* **2012**, *47*, 315–323. [CrossRef]

278. Han, J.; Gao, C.; Yang, S.; Wang, J.; Tan, D. Betanin attenuates carbon tetrachloride (CCl_4)-induced liver injury in common carp (*Cyprinus carpio* L.). *Fish Physiol. Biochem.* **2014**, *40*, 865–874. [CrossRef] [PubMed]

279. Britton, G.; Liaaen-Jensen, S.; Pfander, H. Introduction and guidelines on the use of the Handbook. In *Carotenoids: Handbook*; Britton, G., Liaaen-Jensen, S., Pfander, H., Eds.; Birkhäuser Basel: Basel, Switzerland, 2004; pp. 1–33. Available online: http://dx.doi.org/10.1007/978-3-0348-7836-4_1 (accessed on 27 June 2017).

280. Fraser, P.D.; Bramley, P.M. The biosynthesis and nutritional uses of carotenoids. *Prog. Lipid Res.* **2004**, *43*, 228–265. [CrossRef] [PubMed]

281. Stange, C. (Ed.) *Carotenoids in Nature*; Subcellular Biochemistry Book Series; Springer International Publishing: Cham, Switzerland, 2016; Available online: http://link.springer.com/10.1007/978-3-319-39126-7 (accessed on 20 June 2017).

282. Goodwin, T.W. Nature and Properties. In *The Biochemistry of the Carotenoids: Volume I Plants*; Goodwin, T.W., Ed.; Springer: Dordrecht, The Netherlands, 1980; pp. 1–32. Available online: http://dx.doi.org/10.1007/978-94-009-5860-9_1 (accessed on 21 June 2017).

283. Britton, G. Structure and properties of carotenoids in relation to function. *FASEB J.* **1995**, *9*, 1551–1558. [CrossRef] [PubMed]

284. Lichtenthaler, H.K. Chlorophylls and carotenoids: Pigments of photosynthetic biomembranes. *Plant Cell Membr.* **1987**, *148*, 350–382.

285. Gray, J.C. Control of Isoprenoid Biosynthesis in Higher Plants. *Adv. Bot. Res.* **1988**, *14*, 25–91.

286. McGarvey, D.J.; Croteau, R. Terpenoid metabolism. *Plant Cell* **1995**, *7*, 1015–1026. [CrossRef] [PubMed]

287. Goodwin, T.W. Mammals. In *The Biochemistry of the Carotenoids: Volume II Animals Goodwin*; Goodwin, T.W., Ed.; Springer: Dordrecht, The Netherlands, 1984; pp. 173–195. Available online: http://dx.doi.org/10.1007/978-94-009-5542-4_11 (accessed on 21 June 2017).

288. Fiedor, J.; Burda, K. Potential Role of Carotenoids as Antioxidants in Human Health and Disease. *Nutrients* **2014**, *6*, 466–488. [CrossRef] [PubMed]

289. O'Neill, M.E.; Carroll, Y.; Corridan, B.; Olmedilla, B.; Granado, F.; Blanco, I.; Van den Berg, H.; Hininger, I.; Rousell, A.-M.; Chopra, M.; et al. A European carotenoid database to assess carotenoid intakes and its use in a five-country comparative study. *Br. J. Nutr.* **2001**, *85*, 499–507.

290. Xavier, A.A.O.; Pérez-Gálvez, A. Carotenoids as a Source of Antioxidants in the Diet. In *Carotenoids in Nature: Biosynthesis, Regulation and Function*; Stange, C., Ed.; Springer International Publishing: Cham, Switzerland, 2016; pp. 359–375. Available online: http://dx.doi.org/10.1007/978-3-319-39126-7_14 (accessed on 21 June 2017).

291. Bramley, P. The genetic enhancement of phytochemicals: The case of carotenoids. In *Phytochemical Functional Foods*; Woodhead Publishing: Cambridge, UK, 2003; pp. 253–279. Available online: http://www.sciencedirect.com/science/article/pii/B9781855736726500174 (accessed on 21 June 2017).

292. Palozza, P.; Krinsky, N.I. Antioxidant effects of carotenoids in vivo and in vitro: An overview. *Methods Enzymol.* **1992**, *213*, 403–420. [PubMed]

293. Liebler, D.C. Antioxidant Reactions of Carotenoids. *Ann. N. Y. Acad. Sci.* **1993**, *691*, 20–31. [CrossRef] [PubMed]

294. Ni, Y.; Zhuge, F.; Nagashimada, M.; Ota, T. Novel Action of Carotenoids on Non-Alcoholic Fatty Liver Disease: Macrophage Polarization and Liver Homeostasis. *Nutrients* **2016**, *8*, 391. [CrossRef] [PubMed]

295. Hong, W.K.; Sporn, M.B. Recent Advances in Chemoprevention of Cancer. *Science* **1997**, *278*, 1073–1077. [CrossRef] [PubMed]

296. Moreira, P.R.; Maioli, M.A.; Medeiros, H.C.; Guelfi, M.; Pereira, F.T.; Mingatto, F.E. Protective effect of bixin on carbon tetrachloride-induced hepatotoxicity in rats. *Biol. Res.* **2014**, *47*, 49. [CrossRef] [PubMed]

297. Rao, A.R.; Sarada, R.; Shylaja, M.D.; Ravishankar, G.A. Evaluation of hepatoprotective and antioxidant activity of astaxanthin and astaxanthin esters from microalga-Haematococcus pluvialis. *J. Food Sci. Technol.* **2015**, *52*, 6703–6710. [CrossRef] [PubMed]

298. Kujawska, M.; Ewertowska, M.; Adamska, T.; Sadowski, C.; Ignatowicz, E.; Jodynis-Liebert, J. Antioxidant effect of lycopene-enriched tomato paste on *N*-nitrosodiethylamine-induced oxidative stress in rats. *J. Physiol. Biochem.* **2014**, *70*, 981–990. [CrossRef] [PubMed]

299. Jiang, W.; Guo, M.-H.; Hai, X. Hepatoprotective and antioxidant effects of lycopene on non-alcoholic fatty liver disease in rat. *World J. Gastroenterol.* **2016**, *22*, 10180–10188. [CrossRef] [PubMed]

300. Sheriff, S.A.; Devaki, T. Lycopene stabilizes lipoprotein levels during D-galactosamine/lipopolysaccharide induced hepatitis in experimental rats. *Asian Pac. J. Trop. Biomed.* **2012**, *2*, 975–980. [CrossRef]

nutrients

MDPI

Article

In Vivo Protective Effects of Nootkatone against Particles-Induced Lung Injury Caused by Diesel Exhaust Is Mediated via the NF-κB Pathway

Abderrahim Nemmar [1,*], Suhail Al-Salam [2], Sumaya Beegam [1], Priya Yuvaraju [1], Naserddine Hamadi [3] and Badreldin H. Ali [4]

[1] Department of Physiology, College of Medicine and Health Sciences, United Arab Emirates University, P.O. Box 17666 Al Ain, UAE; sumayab@uaeu.ac.ae (S.B.); priyay@uaeu.ac.ae (P.Y.)
[2] Department of Pathology, College of Medicine and Health Sciences, United Arab Emirates University, P.O. Box 17666 Al Ain, UAE; suhaila@uaeu.ac.ae
[3] Department of Pharmacology, College of Medicine and Health Sciences, United Arab Emirates University, P.O. Box 17666 Al Ain, UAE; hamadinasro@uaeu.ac.ae
[4] Department of Pharmacology and Clinical Pharmacy, College of Medicine & Health Sciences, Sultan Qaboos University, P.O. Box 35, Muscat 123, Al-Khod, Oman; alibadreldin@hotmail.com
* Correspondence: anemmar@uaeu.ac.ae or anemmar@hotmail.com; Tel.: +971-3-713-7533; Fax: +971-3-767-1966

Received: 2 January 2018; Accepted: 23 January 2018; Published: 26 February 2018

Abstract: Numerous studies have shown that acute particulate air pollution exposure is linked with pulmonary adverse effects, including alterations of pulmonary function, inflammation, and oxidative stress. Nootkatone, a constituent of grapefruit, has antioxidant and anti-inflammatory effects. However, the effect of nootkatone on lung toxicity has not been reported so far. In this study we evaluated the possible protective effects of nootkatone on diesel exhaust particles (DEP)-induced lung toxicity, and the possible mechanisms underlying these effects. Mice were intratracheally (i.t.) instilled with either DEP (30 μg/mouse) or saline (control). Nootkatone was given to mice by gavage, 1 h before i.t. instillation, with either DEP or saline. Twenty-four hours following DEP exposure, several physiological and biochemical endpoints were assessed. Nootkatone pretreatment significantly prevented the DEP-induced increase in airway resistance in vivo, decreased neutrophil infiltration in bronchoalveolar lavage fluid, and abated macrophage and neutrophil infiltration in the lung interstitium, assessed by histopathology. Moreover, DEP caused a significant increase in lung concentrations of 8-isoprostane and tumor necrosis factor α, and decreased the reduced glutathione concentration and total nitric oxide activity. These actions were all significantly alleviated by nootkatone pretreatment. Similarly, nootkatone prevented DEP-induced DNA damage and prevented the proteolytic cleavage of caspase-3. Moreover, nootkatone inhibited nuclear factor-kappaB (NF-κB) induced by DEP. We conclude that nootkatone prevented the DEP-induced increase in airway resistance, lung inflammation, oxidative stress, and the subsequent DNA damage and apoptosis through a mechanism involving inhibition of NF-κB activation. Nootkatone could possibly be considered a beneficial protective agent against air pollution-induced respiratory adverse effects.

Keywords: diesel exhaust particles; nootkatone; airway resistance; Lung; oxidative stress; inflammation; NF-κB

---

1. Introduction

Particulate air pollution consists of a complex mixture of solid and liquid particles of organic and inorganic substances suspended in the air [1]. Fine particulate matter and particulate matter with an aerodynamic diameter of less than 2.5 μm ($PM_{2.5}$) has been associated with significant adverse health effects, including increases in morbidity and mortality, particularly in the cardiovascular and respiratory systems [1,2].

Exhaust from diesel powered vehicles is one of the main sources of diesel exhaust particles (DEP), one of the major constituents of $PM_{2.5}$ and nanoparticles (diameter ≤ 0.1 μm) in urban areas [1,3]. These nanoparticles are characterized by their small size, allowing them to penetrate deeply into the respiratory tract, and their high alveolar deposition and large surface area, permitting the carrying of significant amounts of toxic compounds, (e.g., hydrocarbons and metals) [3]. These particles were reported to cause pulmonary oxidative stress and inflammation, and alter cardiac autonomic function in humans and experimental animals [1,3,4]. Acute controlled exposure to emissions from diesel engines in healthy human volunteers has been shown to cause a decrease in lung function, and induce respiratory irritation, inflammation, oxidative stress, and adverse cardiovascular effects [5–7]. Likewise, experimental studies in mice and rats reported the occurrence of lung inflammation, oxidative stress and increase in airway resistance [8–10].

Nootkatone, a sesquiterpene, is a recognized bioactive compound, isolated from the rhizomes of *Cyperus rotundus* [11]. Nootkatone is also naturally found in grapefruit oil [11,12]. It has been reported that nootkatone has several pharmacological properties, such as antiseptic, antioxidant, and antiallergic activities [11,13,14].

Since respiratory toxicity, related to DEP, involves oxidative stress and inflammation [6–8,15], and nootkatone has palliative effects against some experimental diseases involving inflammation and oxidative stress [11,13], we considered that it was relevant to assess the possible ameliorative effects of nootkatone on DEP-induced lung injury and the mechanisms underlying these effects in mice. This is the first study on such an interaction.

2. Material and Methods

2.1. Animals and Treatments

This project was reviewed and approved by the Institutional Review Board of the United Arab Emirates University, College of Medicine and Health Sciences, and experiments were performed in accordance with protocols approved by the Institutional Animal Care and Research Advisory Committee.

2.2. Diesel Exhaust Particles (DEP) and Animal Treatments

The DEP (SRM 2975) were obtained from the National Institute of Standards and Technology (NIST, Gaithersburg, MD, USA), and were suspended in sterile saline (NaCl 0.9%), containing Tween 80 (0.01%). To minimize aggregation, particle suspensions were sonicated (Clifton Ultrasonic Bath, Clifton, NJ, USA) for 15 min and vortexed before their dilution, and prior to intratracheal (i.t.) administration. Control animals received saline containing Tween 80 (0.01%). These particles were previously analysed by transmission electron microscopy, and shown to have a substantial amount of ultrafine (nano) sized particle aggregates, and larger particle aggregates [16].

Animals and treatments: BALB/C mice (Taconic Farms Inc., Germantown, NY, USA), weighing 20–25 g, were housed in light (12-h light: 12-h dark cycle) and temperature-controlled (22 $\pm$ 1 °C) rooms. They had free access to commercial laboratory chow and were provided with tap water ad libitum.

The pulmonary exposure to DEP was achieved by i.t. administration [17]. Mice were first anesthetized with sodium pentobarbital [60 mg/kg, intraperitoneal (i.p.)] and placed supine with an extended neck on an angled board. A Becton Dickinson 24 Gauge cannula was inserted via the mouth into the trachea. Either the DEP suspension (30 μg/mouse) [17] or vehicle was instilled i.t. (100 μL) via a sterile syringe and followed by an air bolus of 100 μL. Nootkatone was administered by gavage (90 mg/kg), 1 h before exposure to either DEP or vehicle. The dose of nootkatone used here has been chosen from our pilot experiments, which showed its effectiveness in preventing cellular infiltration in the lung as compared with 10 and 30 mg/kg (Supplementary Table S1). The animals were randomly divided into four equal groups and were treated as follows:

- Group 1: Normal saline administered by gavage 1 h prior to pulmonary exposure to vehicle;
- Group 2: Normal saline administered by gavage 1 h prior to pulmonary exposure to DEP (30 µg/mouse);
- Group 3: Nootkatone (90 mg/kg) administered by gavage 1 h prior to pulmonary exposure to vehicle;
- Group 4: Nootkatone (90 mg/kg) administered by gavage 1 h prior to pulmonary exposure to DEP (30 µg/mouse).

Twenty-four hours after the pulmonary exposure to either DEP or vehicle, airway hyperresponsiveness and lung inflammation, oxidative stress, DNA damage, apoptosis and nuclear factor-kappa-B (NF-κB) activation were assessed.

2.3. Airway Reactivity to Methacholine

Airway hyperreactivity responses were measured using a forced oscillation technique (FlexiVent, SCIREQ, Montreal, QC, Canada). Airway resistance (R) was assessed after increasing exposures to methacholine. Mice were anesthetized with an intraperitoneal injection of pentobarbital (70 mg/kg). The trachea was exposed and an 18-gauge metal needle was inserted into the trachea. Mice were connected to a computer-controlled small animal ventilator and quasi-sinusoidally ventilated, with a tidal volume of 10 mL/kg at a frequency of 150 breaths/min and a positive end-expiratory pressure of 2 cm H_2O, to achieve a mean lung volume close to that observed during spontaneous breathing. After measurement of a baseline, each mouse was challenged with methacholine aerosol, generated with an in-line nebulizer and administered directly through the ventilator for 5 s, with increasing concentrations (0, 0.625, 2.5, 10 and 40 mg/mL). Airway resistance (R) was measured using a "snapshot" protocol each 20 s for 2 min. The mean of these five values was used for each methacholine concentration, unless the coefficient of determination of a measurement was smaller than 0.95. For each mouse, R was plotted against methacholine concentration (from 0 to 40 mg/mL) [18,19].

2.4. Collection and Analysis of Bronchoalveolar (BAL) Fluid

The collection and analysis of BAL was performed, in separate animals, according to a previously described method [8,18,20,21]. In brief, mice were sacrificed with an overdose of sodium pentobarbital after either DEP or saline administration, with or without nootkatone treatment. The trachea was cannulated and lungs were lavaged three times with 0.7 mL (a total volume of 2.1 mL) of sterile NaCl 0.9% solution. The recovered fluid aliquots were pooled. No difference in the volume of collected fluid was observed between the different groups. BAL fluid was centrifuged (1000× g 10 min, 4 °C). Cells were counted in a Thoma hemocytometer after resuspension of the pellets and staining with 1% gentian violet. The cell differentials were microscopically performed on cytocentrifuge preparations fixed in methanol and stained with Diff Quick (Dade, Brussels, Belgium).

2.5. Histology

In separate groups of animals, the lungs were excised, washed with ice-cold saline, blotted with filter paper and weighed. Each lung was dissected, casseted and fixed directly in 10% neutral formalin for 24 h, which was followed by dehydration with increasing concentrations of ethanol, clearing with xylene and embedding with paraffin. Three-µm sections were prepared from paraffin blocks and stained with haematoxylin and eosin. The stained sections were evaluated by the histopathologist, who participated in this project, using light microscopy [8].

2.6. Measurement of 8-Isoprostane, Reduced Glutathione (GSH), Total Nitric Oxide (NO) and Tumor Necrosis Factor (TNFα) in Lung Homogenates

Following the exposure to either DEP or saline, with or without nootkatone pretreatment, individual mice were sacrificed by an overdose of sodium pentobarbital, and their lungs were quickly collected and rinsed with ice-cold PBS (pH 7.4) before homogenization, as described before [9]. The homogenates were centrifuged for 10 min at $3000\times g$ to remove cellular debris, and the supernatants were used for further analysis [9]. Protein content was measured by Bradford's method. Concentrations of 8-isoprostane were determined using an ELISA Kit (Cayman Chemicals, Ann Arbor, MI, USA) [22]. The concentrations of TNFα were determined using ELISA Kits (Duo Set, R & D Systems, Minneapolis, MN, USA). Measurement of GSH concentrations was carried out according to the method described for the commercially available kit (Sigma-Aldrich Fine Chemicals, Schnelldorf, Germany). The determination of nitric oxide (NO) was performed with a total NO assay kit from R & D Systems (Minneapolis, MN, USA), which measures the more stable NO metabolites: NO_2^- and NO_3^- [23,24].

2.7. Western Blot Analysis

Protein expressions for NF-κB, p65 and cleaved caspase-3 were measured using Western blotting techniques. Lung tissues, harvested from the mice, were snap frozen immediately with liquid nitrogen and stored at −80 °C. Later, the tissues were weighed, rinsed with saline and homogenized with lysis buffer (pH 7.4), containing NaCl (140 mM), KCl (300 mM), trizma base (10 mM), EDTA (1 mM), Triton X-100 0.5% (v/v), sodium deoxycholate 0.5% (w/v), protease and phosphatase inhibitor. The homogenates were centrifuged for 20 min at 4 °C. The supernatants were collected and protein estimation was made using a Pierce bicinchoninic acid protein assay kit (Thermo Scientific, Waltham, MA, USA). A 35 μg sample of protein was electrophoretically separated by 10% sodium dodecyl sulfate polyacrylamide gel electrophoresis and then transferred onto polyvinylidene difluoride membranes. The immunoblots were then blocked with 5% non-fat milk and subsequently probed with either the rabbit monoclonal NF-κB p65 antibody (1:25,000 dilution, Abcam, Hong Kong, China) or rabbit monoclonal cleaved caspase-3 antibody (1:250 dilution, Cell Signalling Technology, Danvers, MA, USA), at 4 °C, overnight. The blots were then incubated with goat anti-rabbit IgG horseradish peroxidase conjugated secondary antibody (1:5000 dilution, Abcam), for 2 h, at room temperature, and developed using Pierce enhanced chemiluminescent plus Western blotting substrate Kit (Thermo Scientific). The densitometric analysis of the protein bands was performed for NF-κB p65 and caspase-3 with Typhoon FLA 9500 (GE Healthcare Bio-Sciences AB, Uppsala, Sweden). Blots were then re-probed with either mouse monoclonal GAPDH antibody (1:5000 dilution, Abcam) or mouse monoclonal β actin antibody (1:1000 dilution, Abcam) and used as a control.

2.8. DNA Damage Assessment by COMET Assay

Promptly after sacrifice, the lungs from control and DEP-exposed mice, with or without nootkatone pretreatment ($n = 5$ in each group), were removed. The COMET assay was performed, as described before [25,26], and the assessment of length of the DNA migration (i.e., diameter of the nucleus plus migrated DNA) was measured using image analysis Axiovision 3.1 software (Carl Zeiss, Toronto, ON, Canada) [27–29].

2.9. Statistics

All statistical analyses were performed with GraphPad Prism Software version 5. Comparisons between groups were performed by one way analysis of variance (ANOVA), followed by Newman–Keuls multiple range tests. All the data in figures were reported as mean ± SEM. p-Values < 0.05 are considered significant.

3. Results

3.1. Airway Hyperreactivity to Methacholine

The airway hyperreactivity to methacholine (0–40 mg/mL), measured by the forced oscillations technique, after i.t. instillation of either saline or DEP, with or without nootkatone pretreatment, is shown in Figure 1. I.t. instillation of DEP induced a dose-dependent and significant increase in airway resistance, compared with saline-instilled mice. No differences were noticed between saline and nootkatone+saline groups. Remarkably, nootkatone pretreatment induced a significant prevention of DEP-induced augmentation of airway resistance after increasing concentrations of methacholine (Figure 1A). The airway resistance to the methacholine dose–response curve was used to calculate an index of airway reactivity as the slope of the linear regression, using 0–40 mg/mL concentrations (Figure 1B). A significant augmentation of airway hyperreactivity to methacholine in mice i.t. instilled with DEP, compared with those instilled with saline ($p < 0.01$), and a complete abrogation of this effect, were observed following nootkatone pretreatment ($p < 0.01$) (Figure 1B).

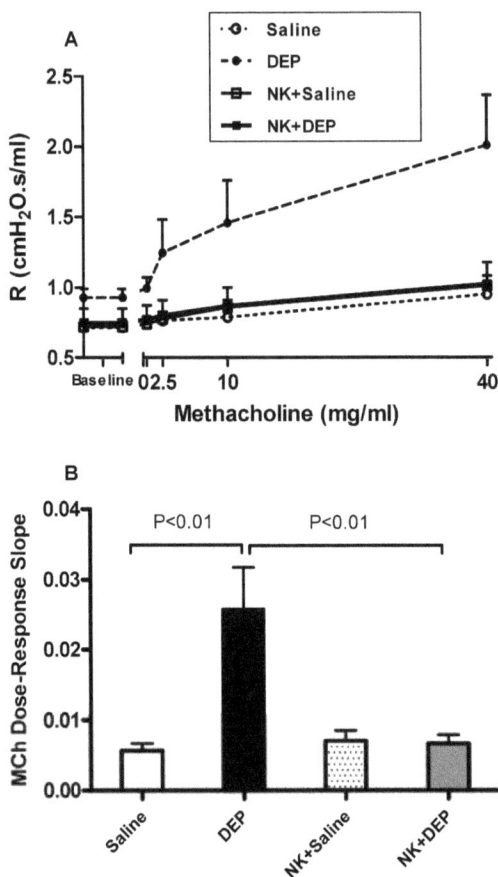

Figure 1. Airway hyper-responsiveness. The airway resistance (R), after increasing concentrations of methacholine (0–40 mg/mL), was measured via the forced oscillation technique (FlexiVent), 24 h after intratracheal instillation of either saline or diesel exhaust particles (DEP, 30 μg/animal), with or without nootkatone (NK) pretreatment. There was a dose–response relationship of total respiratory system resistance to increasing doses of MCh (**A**). From the resistance MCh dose–response curve in (**A**), an index of airway responsiveness was calculated as the slope of the linear regression, using 0–40 mg/mL concentrations (**B**). Data are mean ± SEM (*n* = 6–8).

3.2. Lung Histology

Light microscopy analysis of the lung sections obtained from saline-exposed mice displayed normal structures (Figure 2A). Likewise, lung sections from the nootkatone+saline group showed normal appearances and structures (Figure 2B). The lung sections of mice i.t. instilled with DEP showed particles inside alveolar macrophages which crossed to the alveolar interstitial space (Figure 2C,D). In this group, there was focal damage to the alveolar wall and severe expansion of the alveolar interstitial space, due to heavy neutrophil polymorphs and macrophage infiltration of the interstitium (Figure 2C,D). In the nootkatone+DEP group (Figure 2E,F), histological analysis of the lung showed the presence of DEP within the alveolar macrophages, which crossed to the alveolar interstitial space. There was a marked reduction in the inflammatory infiltrate (neutrophil polymorphs and macrophages) when compared with the DEP-exposed group. Moreover, the interstitial space was also reduced in size when compared with the DEP group.

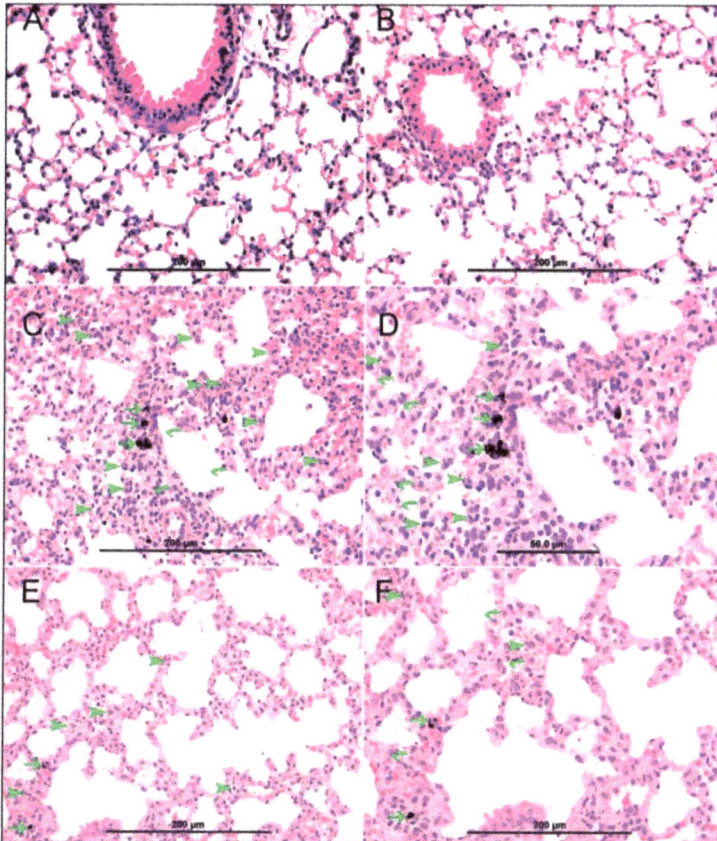

Figure 2. Representative light microscopy sections of lung tissues of mice, 24 h after administration of saline (**A**), nootkatone (NK)+saline (**B**), diesel exhaust particles (DEP; 30 µg/animal; **C,D**), and NK+DEP (**E,F**). (**A,B**) Both saline and NK+saline groups show normal lung tissue with unremarkable changes. (**C,D**) DEP-exposed lungs show particles within alveolar macrophages (thin arrows). There is severe expansion of the alveolar interstitial space with many neutrophil polymorphs (arrow head), and many macrophages (curved arrow). (**E,F**) The NK+DEP group shows DEP particles within alveolar macrophages (thin arrows). There is mild expansion of the alveolar interstitial space with a few neutrophil polymorphs (arrow head), and a few macrophages (curved arrow).

3.3. Cell Composition and Number in BAL Fluid

The total number of cells and neutrophils in BAL fluid was significantly augmented by pulmonary exposure to DEP (Figure 3). The latter effects were significantly mitigated by nootkatone pretreatment (Figure 3).

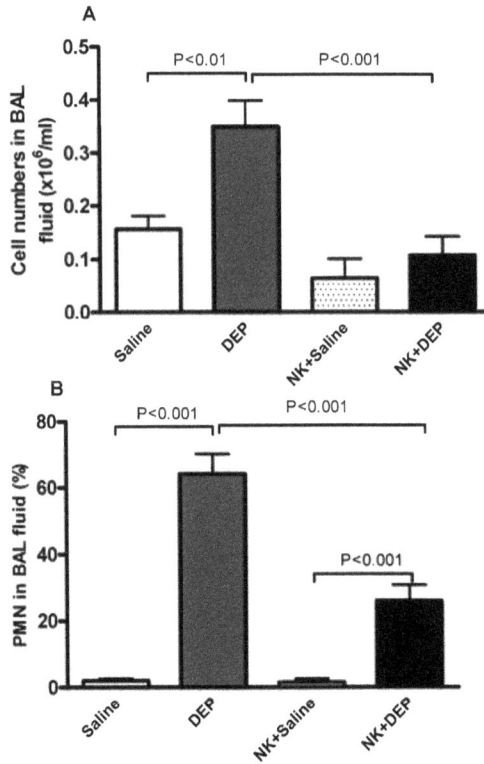

Figure 3. Number of cells (**A**) and polymorphonuclear neutrophils (PMN) (**B**) in bronchoalveolar lavage, 24 h after intratracheal instillation of either saline or diesel exhaust particles (DEP, 30 µg/animal), with or without nootkatone (NK) pretreatment. Data are mean ± SEM (n = 6–8 in each group).

3.4. 8-Isoprostane, GSH, Total NO and TNFα in Lung Homogenates

As shown in Figure 4A–C, compared with the control group, 8-isoprostane, GSH and total NO levels in lung homogenates were significantly affected following DEP exposure.

DEP induced a significant increase of 8-isoprostane in the lung. Nootkatone pretreatment abrogated the effect of DEP and reduced the concentrations of 8-isoprostane to control values in the lung (Figure 4A).

DEP caused a significant decrease in the antioxidant, GSH, and this effect was significantly prevented by pretreatment with nootkatone (Figure 4B).

Likewise, total NO activity was decreased by pulmonary exposure to DEP. This effect was significantly mitigated by nootkatone pretreatment (Figure 4C).

TNFα concentration in lung homogenates was significantly augmented after DEP exposure compared with the control group (Figure 4D). TNFα concentration was significantly reduced in the nootkatone+DEP versus the DEP group ($p < 0.01$), and in the nootkatone+saline group versus the saline group ($p < 0.001$; Figure 4D).

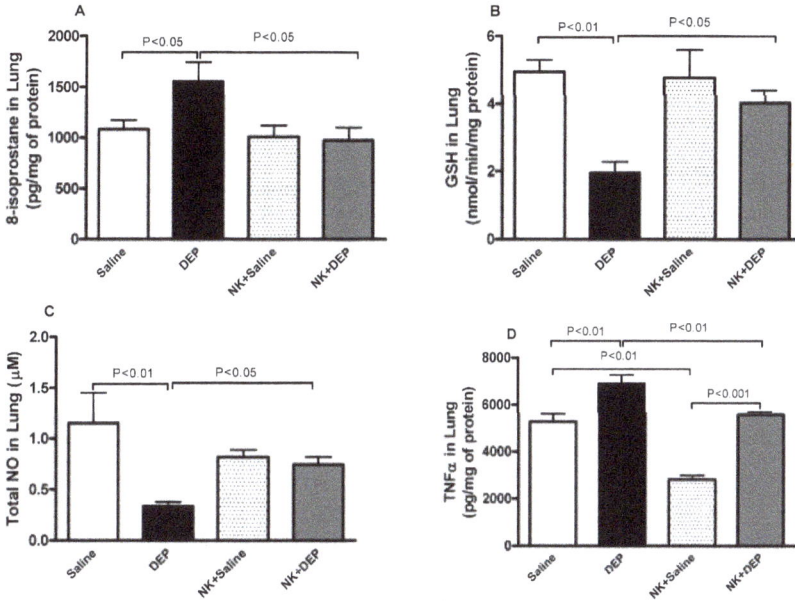

Figure 4. Lung homogenate levels of 8-isoprostane (**A**); reduced glutathione (GSH, **B**); total nitric oxide (NO, **C**) and tumor necrosis factor α (TNFα, **D**); 24 h after intratracheal instillation of either saline or diesel exhaust particles (DEP, 30 μg/animal) with or without nootkatone (NK) pretreatment. Data are mean ± SEM (*n* = 5–8 in each group).

3.5. Lung DNA Damage

Figure 5 illustrates the impact of DEP exposure on lung DNA damage and the effect of nootkatone pretreatment therein. Compared with the control group, DEP exposure caused a significant increase in DNA migration ($p < 0.001$). Pretreatment with nootkatone significantly abated this effect ($p < 0.001$).

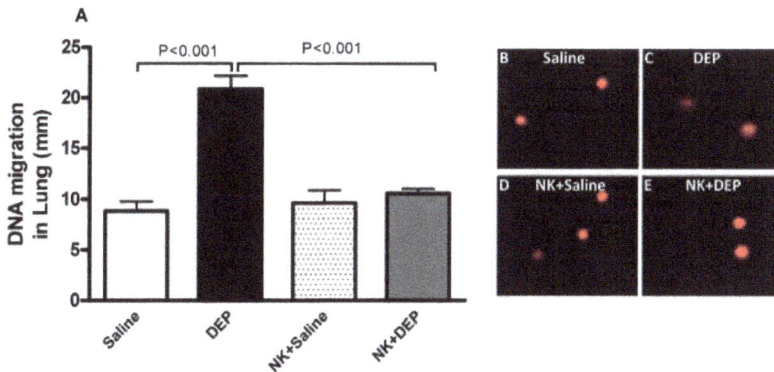

Figure 5. DNA migration (mm) in the lung tissues (**A**) evaluated by Comet assay, 24 h after intratracheal instillation of either saline or diesel exhaust particles (DEP, 30 μg/animal), with or without nootkatone (NK) pretreatment. Data are mean ± SEM (*n* = 5 in each group). Representative images, illustrating the quantification of the DNA migration by the Comet assay, under alkaline conditions, in control (**B**); DEP (**C**); NK+saline (**D**) and NK+DEP (**E**).

3.6. Western Blot Analysis for the Detection of Caspase-3 and NF-κB

The pulmonary exposure to DEP caused a significant increase in cleaved caspase-3 ($p < 0.05$. This effect was significantly prevented by nootkatone pretreatment ($p < 0.01$; Figure 6).

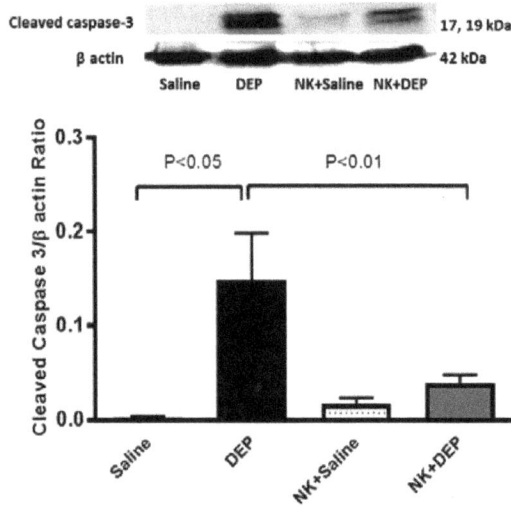

Figure 6. Cleaved caspase-3 levels in the lung tissues determined by Western blotting, 24 h after intratracheal instillation of either saline or diesel exhaust particles (DEP, 30 µg/animal), with or without nootkatone (NK) pretreatment. Data are mean ± SEM ($n = 4$ in each group).

Figure 7 shows that, compared with the control group, the i.t. instillation of DEP induced a significant increase in the expression of NF-κB ($p < 0.05$), and that the pretreatment with nootkatone significantly inhibited this effect ($p < 0.01$).

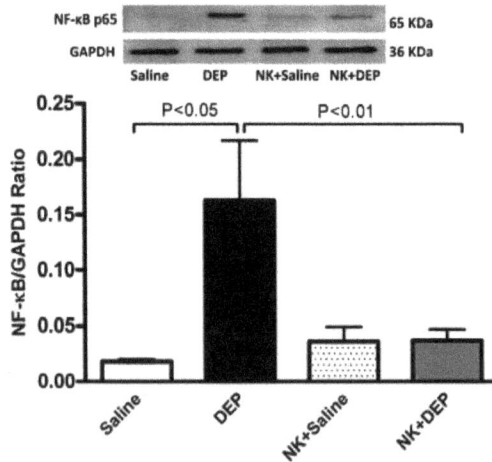

Figure 7. Nuclear factor-kappaB (NF-κB) levels in the lung tissues determined by Western blotting, 24 h after intratracheal instillation of either saline or diesel exhaust particles (DEP, 30 µg/animal), with or without nootkatone (NK) pretreatment. Data are mean ± SEM ($n = 4$ in each group).

4. Discussion

In this work, nootkatone prevented a DEP-induced increase in airway resistance and pulmonary inflammation, oxidative stress, DNA damage and apoptosis. Moreover, nootkatone abrogated NF-κB activation in DEP-induced lung toxicity, suggesting that this compound exerts its protective effects by inhibiting the NF-κB activating pathway.

Human studies demonstrated a decrease in lung function, inflammation and oxidative stress, following short-term (hours to days) exposure to ambient particulate matter [30,31]. Thus, the current acute (24 h) study is relevant to human exposure scenarios. We studied the effect of 30 μg DEP/mouse, which is similar to doses studied previously in the context of ambient particulate matter exposure [32,33]. The United States Environmental Protection Agency stated a range of maximal city particulate matter concentrations, with aerodynamic diameter of less than 10 μm (PM_{10}), as between 26 and 534 μg/m^3 [34]. Numerous big cities in the world have much greater levels of PM_{10}, with yearly averages of 200 to 600 μg/m^3 and peak concentrations regularly surpassing 1000 μg/m^3 [35]. Taking into consideration the highest value in the USA and supposing a minute ventilation of 6 l/min (~8.6 m^3 over 24 h) for a healthy adult at rest, the total dose of PM inhaled over 24 h would be 4614 μg [32]. Human exposure to a daily dose of 4614 μg of PM would represent more than 35 μg of PM exposure for a mouse (25 g body weight), with a minute ventilation of 35–50 mL/min [32]. The dose we tested presently is similar to the comparative human dose of ±35 μg/mouse, reported by Mutlu et al. [32]. The pulmonary exposure to DEP was achieved by i.t. instillation, which represents a valid method of exposure because mice breathe through the nose, which filters the majority of inhaled particles [36,37].

We, and others, have previously reported that acute exposure to DEP induces lung inflammation in mice [8,15,38]. Likewise, human studies have shown the occurrence of lung inflammation and alteration of pulmonary function following acute exposure to DEP in healthy volunteers [5,7]. Nootkatone, naturally found in grapefruit oil, is currently used in flavor and fragrance applications [11]. Recent reports have suggested that nootkatone could be potentially used in insect control [11]. While the effects of nootkatone on allergy [39], obesity [12], platelet aggregation [40] and ischemia reperfusion [41] have been reported, its effects on lung injury induced by DEP have not been reported before.

Our study shows that DEP exposure induced a significant increase in airway hyperreactivity to methacholine, and that pretreatment with nootkatone abrogated this effect. Along with the increase in airway resistance, we found that nootkatone significantly reduced the influx of neutrophils in BAL fluid and the alveolar interstitial infiltration of macrophages and neutrophils assessed by histopathology. This effect is novel and expands the list of beneficial effects of this natural product. It has been reported that omega-3 polyunsaturated fatty acids protect against fine particle induced lung inflammation [42]. It has been shown also that emodin, an anthraquinone derivative from the Chinese herb, *Radix et Rhizoma Rhei*, mitigates DEP-induced lung inflammation and decreases lung function [9].

In order to assess the mechanism by which nootkatone exerts its protective effects on lung inflammation and airway resistance, we measured markers of oxidative stress and inflammation in lung homogenate. Our data show that nootkatone significantly prevented the increase in 8-isoprostane, a stable marker of lipid peroxidation which is generated by the peroxidation of arachidonic acid, catalyzed by free radicals [43]. Moreover, nootkatone significantly prevented the decrease in the GSH and NO, suggesting their consumption in the course of the breakdown of free radicals. It is well established that NO possesses oxygen radical scavenging properties, and that low levels of NO• are required to suppress lipid peroxidation [44]. Moreover, it has been reported that oxidative stress leads to reduced bioactivity of NO [45]. In fact, a decrease in NO following pulmonary exposure to nanoparticles has been reported in rats and mice [28,46,47]. The latter effect was explained by a significant decrease in NO production which coincided with the reduction of NO synthase activity and the excessive generation of reactive oxygen species [28,46,47]. The present findings suggest that the DEP induced oxidative stress in the lung, evidenced by an increase in 8-isoprostane and a decrease of the antioxidant GSH and the inhibitor of the lipid peroxidation chain reaction NO, could be mitigated by nootkatone, which exerts a potent antioxidant effect, by abrogating the increase in 8-isoprostane and replenishing the levels of GSH and NO in the lungs. Likewise, the increase in the pro-inflammatory cytokine TNFα was significantly reduced by nootkatone, compared with the DEP group. The concentration of TNFα in the nootkatone+saline group was even lower than that observed in the saline group. These effects suggest that nootkatone possesses anti-inflammatory and antioxidant properties. Nootkatone has been reported to improve the survival rate of septic mice through the induction of heme oxygenase-1, which has important antioxidant and anti-inflammatory effects [13]. The same authors also showed that nootkatone prevents the expression of inducible nitric oxide synthase and NO generation in RAW264.7 cells, following stimulation with lipopolysaccharide [13]. More recently, it has been reported that nootkatone inhibits TNFα/interferon γ-induced production of chemokines in HaCaT cells [48].

It has been suggested that ambient particulate matter and nanoparticles induce DNA oxidation injury resulting from oxidative stress and inflammation [28,29,49]. Here, we show that DEP exposure caused DNA damage assessed by COMET assay, and that this effect has been significantly inhibited by nootkatone pretreatment. It is well established that DNA damage can induce apoptosis [50]. The latter is the process of programmed cell death, which happens within the cells as a result of their shrinking, chromatin condensation, blebbing, nuclear fragmentation and chromosomal DNA fragmentation [50]. Caspases are key mediators of apoptosis. Among them, caspase-3 is a frequently activated death protease, catalyzing the specific cleavage of many key cellular proteins [51]. It is well established that reactive oxygen species can produce/regulate apoptosis, typically through caspase-3 activation [51]. We found that DEP induced a significant increase in cleaved caspase-3. Interestingly, we showed that nootkatone pretreatment significantly prevented the DEP-induced increase in cleaved caspase-3. It has been reported that eugenol, a methoxyphenol component of clove oil, with in vitro and in vivo anti-inflammatory and antioxidant properties, attenuated caspase-3 activation induced by DEP and prevented changes in lung mechanics, pulmonary inflammation, and alveolar collapse [52]. An in vitro study reported that essential oil from grapefruit, namely the dichloromethane fraction containing aldehyde compounds and nootkatone, induced apoptosis in human leukemic HL-60 cells [53]. The finding of the latter study is in disagreement with ours. This discrepancy could be related to the fact that we used pure nootkatone versus the use of the dichloromethane fraction containing aldehyde compounds and nootkatone, and the fact that our study has been performed in vivo as compared with the use of HL-60 cells [53].

To gain more insight into the mechanisms underlying the protective effects of nootkatone, we quantified NF-κB by Western blot. The latter has been implicated in the pathogenesis of a number of inflammatory diseases, including chronic obstructive pulmonary disease and asthma, and is crucial for activating the transcription of proinflammatory cytokines that cause inflammatory events, as well as oxidative stress and immunity [54]. Moreover, the activation of NF-κB has been shown to be associated with the onset of pulmonary inflammation following exposure to various environmental pollutants, including gaseous and particulate air pollution, cigarette smoke and engineered nanoparticles [4,55]. Activation of the NF-κB signalling pathway has been identified in human lung biopsies exposed to diesel exhaust [56] and in in vitro cell models [57]. Our data show that i.t. instillation of DEP stimulates NF-κB in the lung and the pretreatment with nootkatone significantly inhibited NF-κB activation. Our in vivo study corroborates an in vitro study, which showed that nootkatone inhibits TNFα/interferon γ-induced production of chemokines in HaCaT cells by inhibiting PKCζ and p38 MAPK signaling pathways that lead to activation of NF-κB [48]. Moreover, it has been shown that eugenol efficiently reduced LPS-induced lung toxicity by modulating pulmonary inflammation and remodeling, via a mechanism involving the prevention of TNF-α release and NF-κB activation [58]. It has also been demonstrated that targeting NF-κB could be an efficient therapeutic strategy for the treatment of septic shock, since the inhibition of NF-κB activation selectively inhibited the augmentation in inducible nitric oxide synthase expression (iNOS) activity and iNOS-mediated NO release [59]. At the cellular level, exposure to nanoparticles has been shown to activate several signaling pathways (i.e., NF-κB, NADPH oxidase) to coordinate pathophysiological responses, characterized by the production of inflammatory mediators and reactive oxygen species [4]. This is followed by a cascade of events that leads to the death of cells by different mechanisms (apoptosis, necrosis, autophagy) [4]. Additional work is needed to clarify the downstream chain signaling events involved in DEP induced lung toxicity, and the impact of nootkatone therein.

In conclusion, taken together, our data show that nootkatone pretreatment prevented lung inflammation assessed by BAL fluid analysis and histopathology and inhibited the production of TNFα and oxidative stress, and the subsequent DNA damage and apoptosis through a mechanism involving inhibition of NF-κB activation. The present study shows that in vitro anti-inflammatory antioxidant effects of nootkatone reported by others before [48] translate into an in vivo model of lung inflammation and could lead to possible prevention of lung toxicity, induced by ambient air pollution, and to new therapies for lung inflammatory diseases, in general.

Supplementary Materials: The following are available online at www.mdpi.com/2072-6643/10/3/263/s1. Table S1: Pilot experiments used for the selection of nootkatone (NK) dose which was based on the assessment of cell numbers in bronchoalveolar lavage (BAL) fluid, 24 h after intratracheal instillation of either saline or diesel exhaust particles (DEP, 30 µg/animal) with or without pretreatment with various doses of NK (10, 30 or 90 mg/kg).

Acknowledgments: This work was supported by the CMHS, UPAR and Center-based interdisciplinary grants.

Author Contributions: All authors have read and approved the manuscript. A.N. designed, planned, supervised all the experiments, and wrote the manuscript. S.A.-S. performed and wrote the histopathological part of the work. S.B., P.Y. and N.H. performed the experiments. B.H.A. contributed to the design and the writing of the manuscript.

Conflicts of Interest: The authors declare no conflict of interest.

References

1. Atkinson, R.W.; Kang, S.; Anderson, H.R.; Mills, I.C.; Walton, H.A. Epidemiological time series studies of PM$_{2.5}$ and daily mortality and hospital admissions: A systematic review and meta-analysis. *Thorax* **2014**, *69*, 660–665. [CrossRef] [PubMed]

2. Cai, Y.; Zhang, B.; Ke, W.; Feng, B.; Lin, H.; Xiao, J.; Zeng, W.; Li, X.; Tao, J.; Yang, Z.; et al. Associations of Short-Term and Long-Term Exposure to Ambient Air Pollutants with Hypertension: A Systematic Review and Meta-Analysis. *Hypertension* **2016**, *68*, 62–70. [CrossRef] [PubMed]

3. Steiner, S.; Bisig, C.; Petri-Fink, A.; Rothen-Rutishauser, B. Diesel exhaust: Current knowledge of adverse effects and underlying cellular mechanisms. *Arch. Toxicol.* **2016**, *90*, 1541–1553. [CrossRef] [PubMed]

4. Nemmar, A.; Holme, J.A.; Rosas, I.; Schwarze, P.E.; Alfaro-Moreno, E. Recent advances in particulate matter and nanoparticle toxicology: A review of the in vivo and in vitro studies. *Biomed. Res. Int.* **2013**, *2013*, 279371. [CrossRef] [PubMed]

5. Salvi, S.; Blomberg, A.; Rudell, B.; Kelly, F.; Sandstrom, T.; Holgate, S.T.; Frew, A. Acute inflammatory responses in the airways and peripheral blood after short-term exposure to diesel exhaust in healthy human volunteers. *Am. J. Respir. Crit. Care Med.* **1999**, *159*, 702–709. [CrossRef] [PubMed]

6. Laumbach, R.J.; Kipen, H.M.; Ko, S.; Kelly-McNeil, K.; Cepeda, C.; Pettit, A.; Ohman-Strickland, P.; Zhang, L.; Zhang, J.; Gong, J.; et al. A controlled trial of acute effects of human exposure to traffic particles on pulmonary oxidative stress and heart rate variability. *Part. Fibre Toxicol.* **2014**, *11*, 45. [CrossRef] [PubMed]

7. Xu, Y.; Barregard, L.; Nielsen, J.; Gudmundsson, A.; Wierzbicka, A.; Axmon, A.; Jonsson, B.A.; Karedal, M.; Albin, M. Effects of diesel exposure on lung function and inflammation biomarkers from airway and peripheral blood of healthy volunteers in a chamber study. *Part. Fibre Toxicol.* **2013**, *10*, 60. [CrossRef] [PubMed]

8. Nemmar, A.; Al-Salam, S.; Zia, S.; Marzouqi, F.; Al-Dhaheri, A.; Subramaniyan, D.; Dhanasekaran, S.; Yasin, J.; Ali, B.H.; Kazzam, E.E. Contrasting actions of diesel exhaust particles on the pulmonary and cardiovascular systems and the effects of thymoquinone. *Br. J. Pharmacol.* **2011**, *164*, 1871–1882. [CrossRef] [PubMed]

9. Nemmar, A.; Al-Salam, S.; Yuvaraju, P.; Beegam, S.; Ali, B.H. Emodin mitigates diesel exhaust particles-induced increase in airway resistance, inflammation and oxidative stress in mice. *Respir. Physiol. Neurobiol.* **2015**, *215*, 51–57. [CrossRef] [PubMed]

10. Moon, K.Y.; Park, M.K.; Leikauf, G.D.; Park, C.S.; Jang, A.S. Diesel exhaust particle-induced airway responses are augmented in obese rats. *Int. J. Toxicol.* **2014**, *33*, 21–28. [CrossRef] [PubMed]

11. Leonhardt, R.H.; Berger, R.G. Nootkatone. *Adv. Biochem. Eng. Biotechnol.* **2015**, *148*, 391–404. [PubMed]

12. Murase, T.; Misawa, K.; Haramizu, S.; Minegishi, Y.; Hase, T. Nootkatone, a characteristic constituent of grapefruit, stimulates energy metabolism and prevents diet-induced obesity by activating AMPK. *Am. J. Physiol. Endocrinol. Metab.* **2010**, *299*, E266–E275. [CrossRef] [PubMed]

13. Tsoyi, K.; Jang, H.J.; Lee, Y.S.; Kim, Y.M.; Kim, H.J.; Seo, H.G.; Lee, J.H.; Kwak, J.H.; Lee, D.U.; Chang, K.C. (+)-Nootkatone and (+)-valencene from rhizomes of Cyperus rotundus increase survival rates in septic mice due to heme oxygenase-1 induction. *J. Ethnopharmacol.* **2011**, *137*, 1311–1317. [CrossRef] [PubMed]

14. Nemmar, A.; Al-Salam, S.; Beegam, S.; Yuvaraju, P.; Ali, B.H. Thrombosis, systemic and cardiac oxidative stress and DNA damage induced by pulmonary exposure to diesel exhaust particles, and the effect of nootkatone thereon. *Am. J. Physiol. Heart Circ. Physiol.* **2018**. [CrossRef] [PubMed]

15. Nemmar, A.; Al Salam, S.; Dhanasekaran, S.; Sudhadevi, M.; Ali, B.H. Pulmonary exposure to diesel exhaust particles promotes cerebral microvessel thrombosis: Protective effect of a cysteine prodrug L-2-oxothiazolidine-4-carboxylic acid. *Toxicology* **2009**, *263*, 84–92. [CrossRef] [PubMed]

16. Nemmar, A.; Al Maskari, S.; Ali, B.H.; Al Amri, I.S. Cardiovascular and lung inflammatory effects induced by systemically administered diesel exhaust particles in rats. *Am. J. Physiol. Lung Cell. Mol. Physiol.* **2007**, *292*, L664–L670. [CrossRef] [PubMed]

17. Nemmar, A.; Zia, S.; Subramaniyan, D.; Fahim, M.A.; Ali, B.H. Exacerbation of thrombotic events by diesel exhaust particle in mouse model of hypertension. *Toxicology* **2011**, *285*, 39–45. [CrossRef] [PubMed]

18. Nemmar, A.; Subramaniyan, D.; Zia, S.; Yasin, J.; Ali, B.H. Airway resistance, inflammation and oxidative stress following exposure to diesel exhaust particle in angiotensin II-induced hypertension in mice. *Toxicology* **2012**, *292*, 162–168. [CrossRef] [PubMed]

19. Nemmar, A.; Al-Salam, S.; Subramaniyan, D.; Yasin, J.; Yuvaraju, P.; Beegam, S.; Ali, B.H. Influence of experimental type 1 diabetes on the pulmonary effects of diesel exhaust particles in mice. *Toxicol. Lett.* **2013**, *217*, 170–176. [CrossRef] [PubMed]

20. Nemmar, A.; Dhanasekaran, S.; Yasin, J.; Ba-Omar, H.; Fahim, M.A.; Kazzam, E.E.; Ali, B.H. Evaluation of the direct systemic and cardiopulmonary effects of diesel particles in spontaneously hypertensive rats. *Toxicology* **2009**, *262*, 50–56. [CrossRef] [PubMed]

21. Hardy, R.D.; Coalson, J.J.; Peters, J.; Chaparro, A.; Techasaensiri, C.; Cantwell, A.M.; Kannan, T.R.; Baseman, J.B.; Dube, P.H. Analysis of pulmonary inflammation and function in the mouse and baboon after exposure to Mycoplasma pneumoniae CARDS toxin. *PLoS ONE* **2009**, *4*, e7562. [CrossRef]

22. Nemmar, A.; Al-Salam, S.; Beegam, S.; Yuvaraju, P.; Yasin, J.; Ali, B.H. Pancreatic effects of diesel exhaust particles in mice with type 1 diabetes mellitus. *Cell. Physiol. Biochem.* **2014**, *33*, 413–422. [CrossRef] [PubMed]

23. Tsikas, D. Methods of quantitative analysis of the nitric oxide metabolites nitrite and nitrate in human biological fluids. *Free Radic. Res.* **2005**, *39*, 797–815. [CrossRef] [PubMed]

24. Wennmalm, A.; Benthin, G.; Edlund, A.; Jungersten, L.; Kieler-Jensen, N.; Lundin, S.; Westfelt, U.N.; Petersson, A.S.; Waagstein, F. Metabolism and excretion of nitric oxide in humans. An experimental and clinical study. *Circ. Res.* **1993**, *73*, 1121–1127. [CrossRef] [PubMed]

25. De Souza, M.F.; Goncales, T.A.; Steinmetz, A.; Moura, D.J.; Saffi, J.; Gomez, R.; Barros, H.M. Cocaine induces DNA damage in distinct brain areas of female rats under different hormonal conditions. *Clin. Exp. Pharmacol. Physiol.* **2014**, *41*, 265–269. [CrossRef] [PubMed]

26. Olive, P.L.; Banath, J.P.; Fjell, C.D. DNA strand breakage and DNA structure influence staining with propidium iodide using the alkaline comet assay. *Cytometry* **1994**, *16*, 305–312. [CrossRef] [PubMed]

27. Hartmann, A.; Speit, G. The contribution of cytotoxicity to DNA-effects in the single cell gel test (comet assay). *Toxicol. Lett.* **1997**, *90*, 183–188. [CrossRef]

28. Nemmar, A.; Yuvaraju, P.; Beegam, S.; Fahim, M.A.; Ali, B.H. Cerium Oxide Nanoparticles in Lung Acutely Induce Oxidative Stress, Inflammation, and DNA Damage in Various Organs of Mice. *Oxid. Med. Cell. Longev.* **2017**, *2017*, 9639035. [CrossRef] [PubMed]

29. Nemmar, A.; Yuvaraju, P.; Beegam, S.; Yasin, J.; Kazzam, E.E.; Ali, B.H. Oxidative stress, inflammation, and DNA damage in multiple organs of mice acutely exposed to amorphous silica nanoparticles. *Int. J. Nanomed.* **2016**, *11*, 919–928. [CrossRef] [PubMed]

30. Hazucha, M.J.; Bromberg, P.A.; Lay, J.C.; Bennett, W.; Zeman, K.; Alexis, N.E.; Kehrl, H.; Rappold, A.G.; Cascio, W.E.; Devlin, R.B. Pulmonary responses in current smokers and ex-smokers following a two hour exposure at rest to clean air and fine ambient air particles. *Part. Fibre Toxicol.* **2013**, *10*, 58. [CrossRef] [PubMed]

31. Rice, M.B.; Ljungman, P.L.; Wilker, E.H.; Gold, D.R.; Schwartz, J.D.; Koutrakis, P.; Washko, G.R.; O'Connor, G.T.; Mittleman, M.A. Short-term exposure to air pollution and lung function in the Framingham Heart Study. *Am. J. Respir. Crit. Care Med.* **2013**, *188*, 1351–1357. [CrossRef] [PubMed]

32. Mutlu, G.M.; Green, D.; Bellmeyer, A.; Baker, C.M.; Burgess, Z.; Rajamannan, N.; Christman, J.W.; Foiles, N.; Kamp, D.W.; Ghio, A.J.; et al. Ambient particulate matter accelerates coagulation via an IL-6-dependent pathway. *J. Clin. Investig.* **2007**, *117*, 2952–2961. [CrossRef] [PubMed]

33. Nemmar, A.; Al, D.R.; Alamiri, J.; Al, H.S.; Al, S.H.; Beegam, S.; Yuvaraju, P.; Yasin, J.; Ali, B.H. Diesel Exhaust Particles Induce Impairment of Vascular and Cardiac Homeostasis in Mice: Ameliorative Effect of Emodin. *Cell. Physiol. Biochem.* **2015**, *36*, 1517–1526. [CrossRef] [PubMed]

34. Brook, R.D.; Franklin, B.; Cascio, W.; Hong, Y.L.; Howard, G.; Lipsett, M.; Luepker, R.; Mittleman, M.; Samet, J.; Smith, S.C.; et al. Air pollution and cardiovascular disease—A statement for healthcare professionals from the expert panel on population and prevention science of the American Heart Association. *Circulation* **2004**, *109*, 2655–2671. [CrossRef] [PubMed]

35. United Nations Environment Program; World Health Organization (WHO). Air Pollution in the world's megacities. A Report from the U.N. Environment Programme and WHO. *Environment* **1994**, *36*, 25–37.

36. Driscoll, K.E.; Costa, D.L.; Hatch, G.; Henderson, R.; Oberdorster, G.; Salem, H.; Schlesinger, R.B. Intratracheal instillation as an exposure technique for the evaluation of respiratory tract toxicity: Uses and limitations. *Toxicol. Sci.* **2000**, *55*, 24–35. [CrossRef] [PubMed]

37. Tamagawa, E.; Bai, N.; Morimoto, K.; Gray, C.; Mui, T.; Yatera, K.; Zhang, X.; Xing, L.; Li, Y.; Laher, I.; et al. Particulate matter exposure induces persistent lung inflammation and endothelial dysfunction. *Am. J. Physiol. Lung Cell. Mol. Physiol.* **2008**, *295*, L79–L85. [CrossRef] [PubMed]

38. Laks, D.; de Oliveira, R.C.; de Andre, P.A.; Macchione, M.; Lemos, M.; Faffe, D.; Saldiva, P.H.; Zin, W.A. Composition of diesel particles influences acute pulmonary toxicity: An experimental study in mice. *Inhal. Toxicol.* **2008**, *20*, 1037–1042. [CrossRef] [PubMed]

39. Jin, J.H.; Lee, D.U.; Kim, Y.S.; Kim, H.P. Anti-allergic activity of sesquiterpenes from the rhizomes of Cyperus rotundus. *Arch. Pharm. Res.* **2011**, *34*, 223–228. [CrossRef] [PubMed]

40. Seo, E.J.; Lee, D.U.; Kwak, J.H.; Lee, S.M.; Kim, Y.S.; Jung, Y.S. Antiplatelet effects of Cyperus rotundus and its component (+)-nootkatone. *J. Ethnopharmacol.* **2011**, *135*, 48–54. [CrossRef] [PubMed]

41. Chang, K.C.; Lee, D.U. Nootkatone from the rhizomes of *Cyperus rotundus* protects against ischemia-reperfusion mediated acute myocardial injury in the rat. *Int. J. Pharm.* **2016**, *12*, 845–850.

42. Li, X.Y.; Hao, L.; Liu, Y.H.; Chen, C.Y.; Pai, V.J.; Kang, J.X. Protection against fine particle-induced pulmonary and systemic inflammation by omega-3 polyunsaturated fatty acids. *Biochim. Biophys. Acta* **2017**, *1861*, 577–584. [CrossRef] [PubMed]

43. Yao, H.; Rahman, I. Current concepts on oxidative/carbonyl stress, inflammation and epigenetics in pathogenesis of chronic obstructive pulmonary disease. *Toxicol. Appl. Pharmacol.* **2011**, *254*, 72–85. [CrossRef] [PubMed]

44. Hummel, S.G.; Fischer, A.J.; Martin, S.M.; Schafer, F.Q.; Buettner, G.R. Nitric oxide as a cellular antioxidant: A little goes a long way. *Free Radic. Biol. Med.* **2006**, *40*, 501–506. [CrossRef] [PubMed]

45. Forstermann, U. Nitric oxide and oxidative stress in vascular disease. *Pflug. Arch.* **2010**, *459*, 923–939. [CrossRef] [PubMed]

46. Nurkiewicz, T.R.; Porter, D.W.; Hubbs, A.F.; Stone, S.; Chen, B.T.; Frazer, D.G.; Boegehold, M.A.; Castranova, V. Pulmonary nanoparticle exposure disrupts systemic microvascular nitric oxide signaling. *Toxicol. Sci.* **2009**, *110*, 191–203. [CrossRef] [PubMed]

47. Du, Z.; Zhao, D.; Jing, L.; Cui, G.; Jin, M.; Li, Y.; Liu, X.; Liu, Y.; Du, H.; Guo, C.; et al. Cardiovascular Toxicity of Different Sizes Amorphous Silica Nanoparticles in Rats after Intratracheal Instillation. *Cardiovasc. Toxicol.* **2013**, *13*, 194–207. [CrossRef] [PubMed]

48. Choi, H.J.; Lee, J.H.; Jung, Y.S. (+)-Nootkatone inhibits tumor necrosis factor alpha/interferon gamma-induced production of chemokines in HaCaT cells. *Biochem. Biophys. Res. Commun.* **2014**, *447*, 278–284. [CrossRef] [PubMed]

49. Moller, P.; Danielsen, P.H.; Karottki, D.G.; Jantzen, K.; Roursgaard, M.; Klingberg, H.; Jensen, D.M.; Christophersen, D.V.; Hemmingsen, J.G.; Cao, Y.; et al. Oxidative stress and inflammation generated DNA damage by exposure to air pollution particles. *Mutat. Res. Rev. Mutat. Res.* **2014**, *762*, 133–166. [CrossRef] [PubMed]

50. Nowsheen, S.; Yang, E.S. The intersection between DNA damage response and cell death pathways. *Exp. Oncol.* **2012**, *34*, 243–254. [PubMed]

51. Savitskaya, M.A.; Onishchenko, G.E. Mechanisms of Apoptosis. *Biochemistry* **2015**, *80*, 1393–1405. [CrossRef] [PubMed]

52. Zin, W.A.; Silva, A.G.; Magalhaes, C.B.; Carvalho, G.M.; Riva, D.R.; Lima, C.C.; Leal-Cardoso, J.H.; Takiya, C.M.; Valenca, S.S.; Saldiva, P.H.; et al. Eugenol attenuates pulmonary damage induced by diesel exhaust particles. *J. Appl. Physiol.* **2012**, *112*, 911–917. [CrossRef] [PubMed]

53. Hata, T.; Sakaguchi, I.; Mori, M.; Ikeda, N.; Kato, Y.; Minamino, M.; Watabe, K. Induction of apoptosis by Citrus paradisi essential oil in human leukemic (HL-60) cells. *In Vivo* **2003**, *17*, 553–559. [PubMed]

54. Schuliga, M. NF-kappaB Signaling in Chronic Inflammatory Airway Disease. *Biomolecules* **2015**, *5*, 1266–1283. [CrossRef] [PubMed]

55. Traboulsi, H.; Guerrina, N.; Iu, M.; Maysinger, D.; Ariya, P.; Baglole, C.J. Inhaled Pollutants: The Molecular Scene behind Respiratory and Systemic Diseases Associated with Ultrafine Particulate Matter. *Int. J. Mol. Sci.* **2017**, *18*, 243. [CrossRef] [PubMed]

56. Pourazar, J.; Mudway, I.S.; Samet, J.M.; Helleday, R.; Blomberg, A.; Wilson, S.J.; Frew, A.J.; Kelly, F.J.; Sandstrom, T. Diesel exhaust activates redox-sensitive transcription factors and kinases in human airways. *Am. J. Physiol. Lung Cell. Mol. Physiol.* **2005**, *289*, L724–L730. [CrossRef] [PubMed]

57. Totlandsdal, A.I.; Cassee, F.R.; Schwarze, P.; Refsnes, M.; Lag, M. Diesel exhaust particles induce CYP1A1 and pro-inflammatory responses via differential pathways in human bronchial epithelial cells. *Part. Fibre Toxicol.* **2010**, *7*, 41. [CrossRef] [PubMed]

58. Magalhaes, C.B.; Riva, D.R.; DePaula, L.J.; Brando-Lima, A.; Koatz, V.L.; Leal-Cardoso, J.H.; Zin, W.A.; Faffe, D.S. In vivo anti-inflammatory action of eugenol on lipopolysaccharide-induced lung injury. *J. Appl. Physiol.* **2010**, *108*, 845–851. [CrossRef] [PubMed]

59. Liu, S.F.; Ye, X.; Malik, A.B. In vivo inhibition of nuclear factor-kappa B activation prevents inducible nitric oxide synthase expression and systemic hypotension in a rat model of septic shock. *J. Immunol.* **1997**, *159*, 3976–3983. [PubMed]

MDPI

St. Alban-Anlage 66

4052 Basel, Switzerland

Tel. +41 61 683 77 34

Fax +41 61 302 89 18

http://www.mdpi.com

Nutrients Editorial Office

E-mail: nutrients@mdpi.com

http://www.mdpi.com/journal/nutrients

www.ingramcontent.com/pod-product-compliance
Lightning Source LLC
Chambersburg PA
CBHW051723210326

41597CB00032B/5580